Contemporary Problems in Carbonium Ion Chemistry I/II

Contemporary Problems in Carbonium Ion Chemistry I/II

Nonclassical Carbocations
by V. A. Barkhash

Rearrangements of Carbocations by 1,2-Shifts
by V. G. Shubin

With 13 Figures and 21 Tables

AKADEMIE-VERLAG · BERLIN
1984

Die Originalausgabe erscheint im Springer-Verlag
Berlin · Heidelberg · New York · Tokyo
in der Schriftenreihe "Topics in Current Chemistry", Volume 116/117

Vertrieb ausschließlich für die DDR und die sozialistischen Länder

Erschienen im Akademie-Verlag, DDR-1086 Berlin, Leipziger Straße 3—4
Lizenznummer: 202 · 100/491/84
Printed in the German Democratic Republic
Gesamtherstellung: VEB Druckerei „Thomas Müntzer", 5820 Bad Langensalza
Umschlaggestaltung: Karl Salzbrunn
LSV 1275
Bestellnummer: 763 317 1 (6808)

11800

Table of Contents

Nonclassical Carbocations

Vladimir Alexandrovich Barkhash

Institute of Organic Chemistry, Siberian Division of the USSR Academy of Sciences, 630090 Novosibirsk-90, USSR

List of Abbreviations

AO	Atomic Orbital
Bs	Brosyl- = 4-Bromotolylsulfonyl
CNDO	Complete neglect of differential overlap
DE	Delocalization Energy
DNB	Dinitrobenzoate
HFIP	Hexafluoro-iso-propanol
MO	Molecular Orbital
Ms	Mesyl- = Methylsulfonyl
Ns	Nosyl- = p-Nitrobenzenesulfonyl
p-An	p-Anisyl = 4-Methoxyphenyl
PNB	p-Nitrobenzoyl
SCFMO	Self-consistent field
TCNE	Tetracyanoethylene
TFA	Trifluoroacetyl
TFE	Tetrafluoroethylene
Ts	Tosyl- = p-tolylsulfonyl
WM	Wagner-Meerwein Rearrangement

1 Introduction

In 1965 a book by P. Bartlett appeared under the title "The Nonclassical Ions" [1]. The book is a collection of papers reprinted from various journals. The many reviews that have appeared since [2-22] are either antiquated (the book published in 1972 [12] covers the literature mainly before 1968) or relatively biased (e.g., [3,4,10]) on brief [2,7,11].

This review attempts to discuss the various points of view on the "nonclassical" carbocations. The main point is to establish the relative role of "nonclassical" and "classical" ions in various chemical processes. The author has followed P. Bartlett's advice [1] that when setting forth the achievements of the human mind one should see how we came to the modern understanding of a given problem ("... how we know what we know").

The theory of "nonclassical" ions offers an explanation of many unique chemical, stereochemical and kinetic peculiarities of bicyclic compounds. It has expanded our knowledge on chemical bonds in carbocations by introducing electron-deficient bonds (as in boron hydrides). It has accounted for many rearrangements of stable cations. As a "side" result our knowledge has been extended about ionization processes in a solution, as well as about stereochemical methods.

2 Main Terms of Nonclassical Carbocations

In 1939 Hevell, Salas and Wilson [23] assumed an intermediate, "bridge" ion *2* to be formed when camphene hydrochloride *1* is rearranged into isobornyl chloride *3*. This happened 17 years after Meerwein first postulated the intermediate formation of "carbonium" ions in chemical reactions.

1 *2* *3*

For 10 years this work attracted no attention until in 1949–1952 Winstein and Trifan, in their works on the solvolysis of 2-norbornyl brosylates [24,25], experimentally founded the subsequent development of "nonclassical" carbocations. Then followed a great number of publications suggesting (or rejecting) various nonclassical ions as intermediates in chemical reactions.

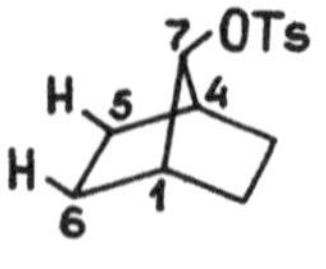

4

Many studies on nonclassical ions have been made using norbornane derivatives because of two reasons:

1) easy availability and

2) extreme chemical inertness of saturated norborane derivatives so that any "assistance" of adjacent groups leads to great readily detectable effects. Indeed, 7-norbornyl tosylate *4* is subject to acetolysis 10^4 times as slowly as the primary n-butyl tosylate, and 10^7 times as slowly as the secondary cyclopentyl tosylate.

Tanida [26] argues for the exceptional inertness of compound *4*:

a) steric hindrance to solvation due to interference by the $C^{5(6)}$—H bonds;

b) increase of steric strain when hybridizing C^7 changes from sp^3 to sp^2, firstly, due to an increase in the angle strain ($<C^1—C^7—C^4$ amounts to 93–96°) and, secondly, because the favourable gauche conformation of the fragment $C^1—C^7—C^4$ in the ground state must be replaced by the eclipsed conformation in the intermediate state;

c) prohibition of carbocation stabilization through hyperconjugation by the Bredt rule.

The advantages of the norbornane system are accompanied by disadvantages; the skeleton asymmetry explains the difference in the behaviour of e.g. 2-exo- and 2-endo-sulphonates by the different influence of the steric factors in the ground and transition states on the ionization of these epimers, especially as the steric effects are more essential for rigid bicyclic systems than for flexible aliphatic or alicyclic substrates [21].

In 1952 Winstein suggested for the 2-norbornyl ion a formula *5* [25] corresponding to three canonical forms *6*–*8* in terms of the resonance theory. Formula *5*

5 6 7 8

reflects to a certain degree the views of Dewar [27] and those of Walsh [28] on the nature of the bond in such an ion without making any choice between these views. Dewar considered a dative bond between a pair of π-electrons and a bridging atom in the ion under discussion; he believed the σ-electrons to be incapable of such a bond. In his opinion 2-norbornyl chloride *9* is solvolysed with intermediate formation of a π-complex *8* [29]. Dewar believed the formation of these complexes to be determined by:

1) the strength of the $C^1—C^6$ bond;

2) the electronegativity of the $\geq C_6-$ group; etc.

9 8

Since in a π-complex the bond between the apical atom and the two basal carbon atoms is comparatively weak, such a structure can be, without essential increase of energy, deformed by a transfer of the atom in the apex. Hence in strained systems the π-complex can be selectively stabilized relative to the classical carbocation because the I-strain in the former can be weakened by deforming the bond with the apical group.

Walsh, as distinct from Dewar believed the bonds in nonclassical ions to be best described in terms of the dative bond between a pair of σ-electrons and the third atom:

Similar ideas have been developed by Ingold [30] who used the term "synartetic ion" implying a strong interaction between σ-electrons of the C—C bond and the cation centre ("... split single bond fastening together the locations of a split charge").

Although the ideas of Walsh and Dewar differ as to the state of hybridization of the three electron-deficient atoms they agree in that each of the three atoms is directly bonded with the other two via the single electron pair. Thus there is a direct overlap not only between the atomic orbital (AO) at C^6 and those at C^1 and C^2, but also between C^1 and C^2. The dashed line in formula *5* signifies the overlapping of the two AO contained in the multicentre molecular orbital (MO) carrying two electrons. The solid line denotes the overlapping of two AO contained in the two-centre orbital carrying two electrons. The character of overlapping is confirmed by chemical data and by the MO theory [31, 32].

As far back as 1949 Longuet-Higgins [33] pointed to the possibility of forming quite a new type of bond by sharing one pair of electrons among three nuclei. The simplest example of such a bond is the stable ion H_3^+ in which two electrons occupy the three-centre MO formed by the overlapping of Is-AO of the three hydrogen atoms *10*. The arising three-centre MO is topologically equivalent to the π-MO of the cyclopropenyl cation *11*. In either case the electron pair occupies the appropriate MO with low energy.

The structure of

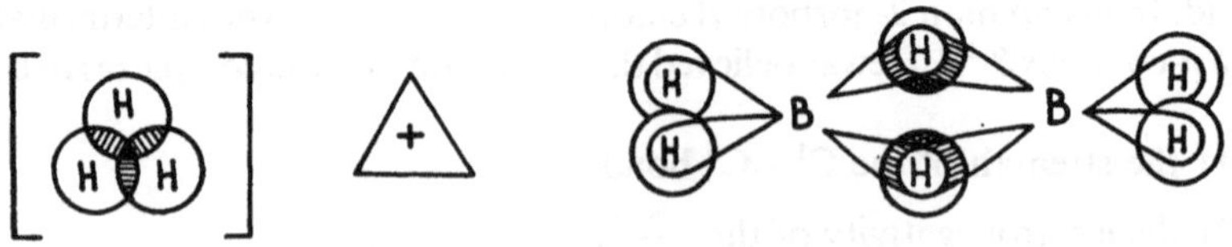

10 11 12

diborane B_2H_6 *12* is explained with the central hydrogen atoms bonded to both boron atoms by three-centre bonds. The carbocation is isoelectronic with the neutral boron atom; one can therefore expect these types of bonding to be significant in carbonium ions as well.

As Brown has rightly pointed out [13], by admitting the nonclassical structure of the 2-norbornyl and similar cations we admit a new type of bonding which was not earlier accepted in organic chemistry. At the borderline between inorganic and organic chemistry — in organometallic compounds — hundreds of examples have long been known of "electron-deficient" neutral substances (metal alkyls, carboranes, clusters etc.) not carrying a sufficient number of electrons to provide a pair of electrons for each chemical bond, e.g.

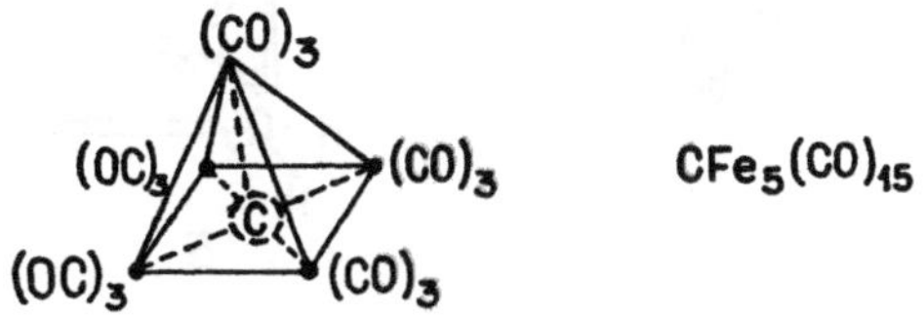

To describe the structure of this compounds it has long been accepted to use two-electron three- and multicentre bonds, while the coordination number of carbon may vary from 5 to 8.

One of the properties of nonclassical ions is the intramolecular delocalization of the positive charge. This definition, however, is not a sufficient criterion because it is also satisfied by such typically classical ions as allylic or benzylic ones. In these ions the molecular orbitals are formed by π-overlapping of atomic orbitals *13*. This kind of overlapping is the most typical of classical ions (see, e.g., ions *14–16*).

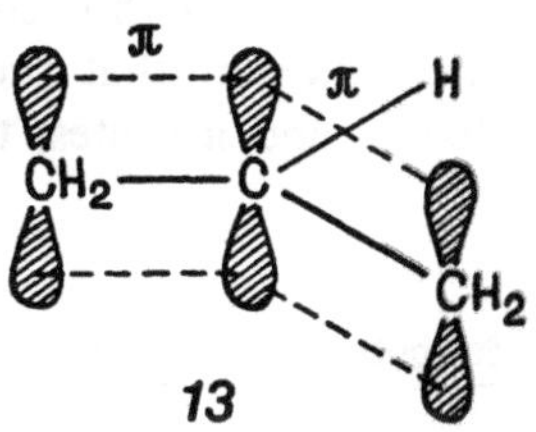

13

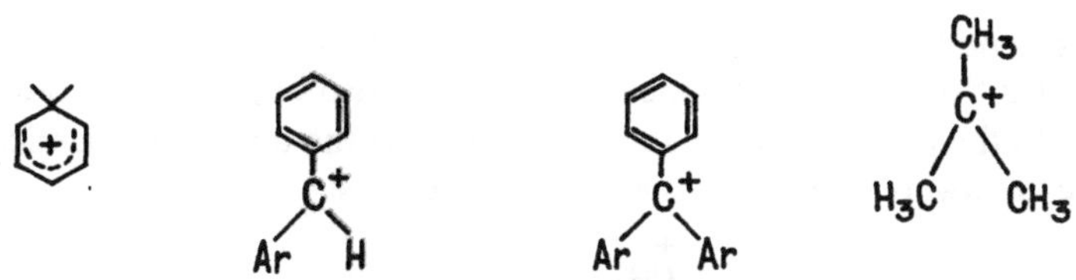

14 *15* *16* *17*

The cation *17* is stabilized by hyperconjugation and +I-effect of CH_3 groups. In nonclassical carbocations the MO is formed, at least partially, by the σ-overlapping between atomic orbitals. Such an overlapping may be relatively small, as in the unsymmetrical homoallylic ion *18* or much larger, as in the 7-norbornenyl ion *19* where the C^2—C^3 interaction is a pure σ-π-overlap, while C^7—C^2 and C^7—C^3 are practically β,σ-overlaps. In the 2-norbornyl cation *5* (according to Streitwieser) there is a σ-bonding of the sp^3-orbital of C^6 and of the two p-orbitals of C^1 and C^2 which in their turn overlap to form a π-bond between C^1 and C^2.

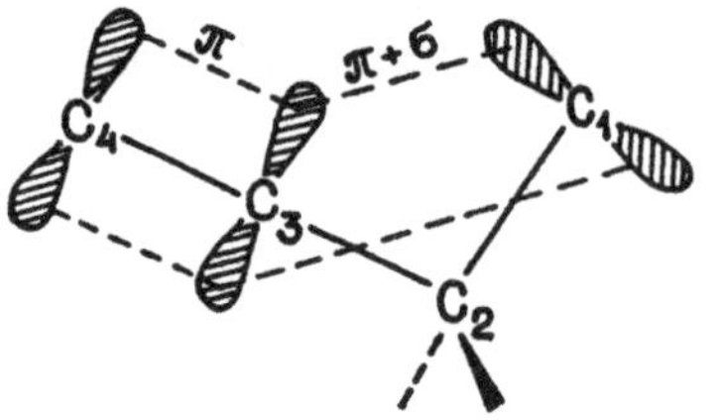

18

19

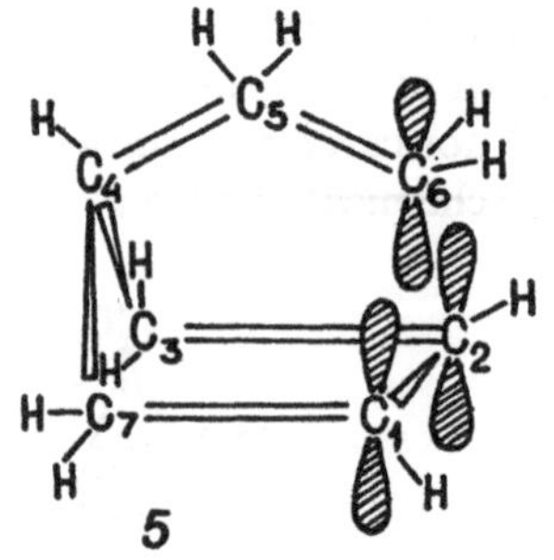

5

Nonclassical ions distinguishing themselves from classical ones delocalize the positive charge via a MO formed by overlapping the orbitals of the atoms of which at least a part are not bonded to others by one σ-bond.

Thus, the nonclassical ion is a positively charged particle in which the charge is delocalized by a multicentre molecular orbital formed, at least partly, by the σ-overlap of the orbitals of atoms at least part of which are not bonded with others by one σ-bond.

In his review G. Sargent [2] attempts to single out a separate group of "bridged" ions giving them the following definition: bridged ions are nonclassical ions having a σ-overlapping between the AO of a carbon and the appropriately oriented atomic orbitals on each of the other two C atoms which are linked with each other by a two-centre localized σ-bond. Taking into account, however, that the "bridged" 2-norbornyl ion *5* can be obtained by the so-called σ- and π-routes, the practical inexpedience of this definition becomes evident.

Hardly acceptable is also the suggestion of Bethel and Gold [11] that all nonclassical ions be called "bridged" ions, since this name is also applicable to other ions, e.g., *20* and *21*. In the present survey we will further use the term "nonclassical" ion even though we are aware of its "transient" character (see [11]).

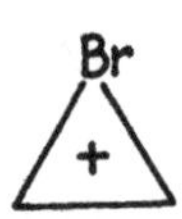

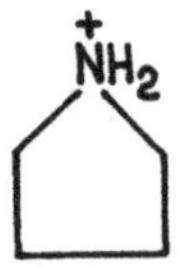

20 21

In 1972 Olah [34] proposed a classification for carbocations dividing them into trivalent ("classical") carbenium ions and penta- and tetracoordinate ("nonclassical") carbonium ions.

Carbenium ions generally have a planar, or close to planar, sp^2-hybridized electron-deficient centre. The carbocation centres in *carbonium ions* are essentially less electron-deficient carbon atoms bound with other atoms by three single bonds and by a fourth two-electron three-centre bond. The latter may be directed either to the two other cation centres (left, pentacoordinate ions) or to the carbon atom to which they are also bound by a single bond (right, tetracoordinate ions).

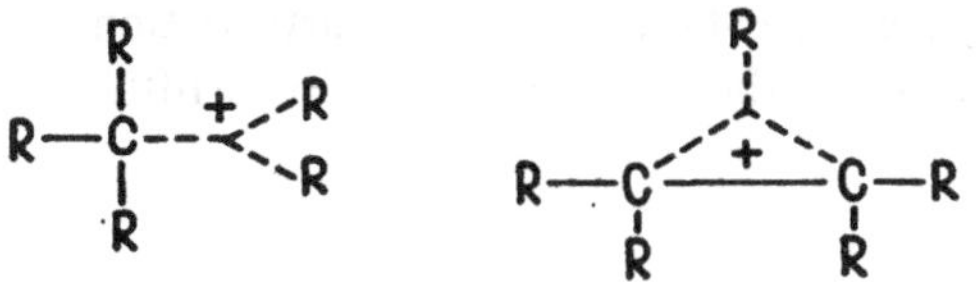

Olah emphasizes that the division of cations into classical and nonclassical is frequently arbitrary, since in many cations there is an intermediate range of delocalization ("partial carbonium-ion character") as in the 2-methylnorbornyl ion. The author does not want to name classical ions "carbonium" because it is restricted to highest valency state carbocations; this requirement is met by penta- and tetracoordinate carbocations but not trivalent ones. On the other hand, while in the formation of other "onium" ions the atom of the donor (nitrogen, oxygen etc.) increases its covalence by one unit upon addition of the acceptor (electrophile), in the formation of a classical ion the covalence of the carbon atom decreases from 4 to 3. As for the name "carbenium" ion, in the author's opinion it reflects the logical relationships between the carbene and the carbenium ion, between the alkene and the carbenium ion:

$$\uparrow\downarrow\ddot{C}H_2 + H^+ \rightleftarrows C^+H_3 \qquad RCH = CHR + H^+ \rightleftarrows RC^+HCH_2R$$

If we admit the intramolecular interaction of the developing carbocation centre with remote σ- and π-bonds often forming the same nonclassical ion (see above) it is but a step to recognize a similarity between the mechanisms of the intermolecular interactions of both π- and σ-bonds with the "external" electrophile forming intermediate structures with 2-electron 3-centre bonding arrangement (2π3C).

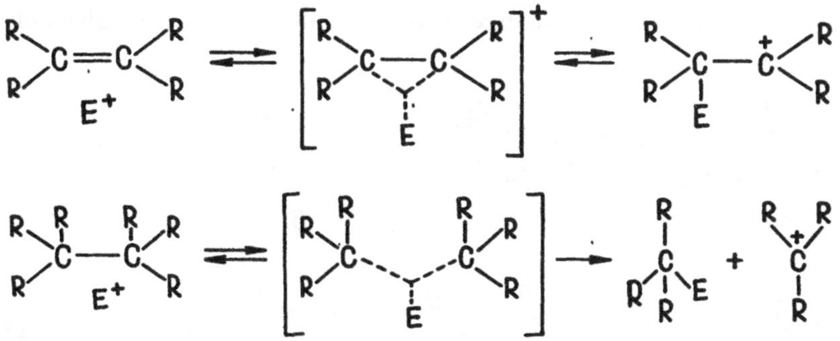

The difference is that after addition to the σ-bond, no covalent bond remains between the two atoms of the latter; the splitting of the three-centre bond results in the break up of the parent σ-bond. But if the electron donor is the π-bond, then the splitting of three-centre bond results in classical carbocations and further in addition or substitution products. In recent years Olah has experimentally proved the σ-bonds C—C and C—H to display considerable reactivity in such electrophilic reactions as protolytic processes (isomerization, hydrogen exchange, protolysis), alkylation, nitration and halogenation [35].

In the third section of this survey we will mainly discuss the interaction of the cation centre with the appropriately located σ-bond *A*, in the fourth section — that with the π-electrons of the double bond *B* or of the aromatic ring *B′*, in the fifth — with the cyclopropane ring *C*.

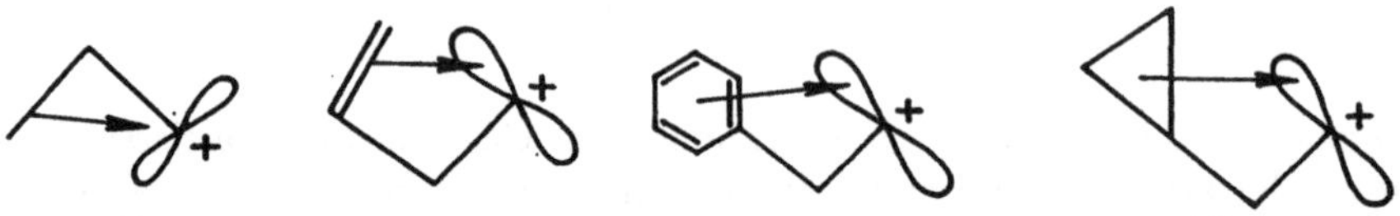

(A) (B) (B′) (C)

The survey does not touch "phenonium" ions and α-cyclopropylcarbinyl ions [12].

3 2-Norbornyl Cation

Nonclassical ions with a norbornane skeleton, as has been already noted, have been studied extensively, though this system permits a far greater number of ambiguous interpretations than the bicyclo[2,2,2]octane system [36].

Studies on the 2-norbornyl ion were started by the solvolysis of 2-norbornyl arenesulphonates [24,25,37]; as these works have served as models for all subsequent ones they will be dwelt upon in more detail.

3.1 Solvolysis of 2-Norbornyl Arenesulphonates

Acetolysis of the exo-brosylate *22* proceeds 350 times as fast as that of the endo-epimer *23*; in the first case the configuration is completely retained but the rate of the complete racemization exceeds that of solvolysis. The solvolysis of endo isomer

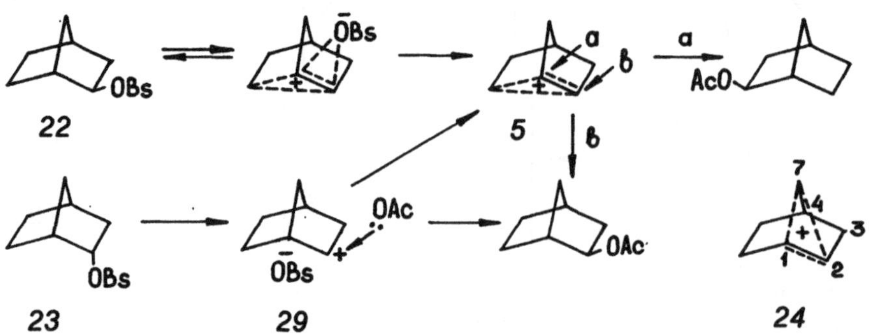

23 also results in the formation of exo acetate alone, but in this case 7–8% of optical activity is retained and no significant difference is observed between the rate of racemation and that of formation of 4-bromobenzenesulphonic acid.

Reinvestigation of the reaction products by vapour-phase chromatography [38] has confirmed the exclusive exo-configuration of the substitution products: also, the reaction mixture was found to contain hydrocarbons. Thus, exo-brosylate *22* on acetolysis (at 25 °C) yields 95.6% of acetate and 4.4% of hydrocarbons; endo-brosylate *23* at 75 °C forms a mixture of 87.3% of acetate and 12.7% of hydrocarbons. More than 99.7% of acetates have an exo configuration; hydrocarbons are a mixture of nortricyclene and norbornene (39:1).

The enhanced rate of solvolysis of the exo-isomer (350-fold) as compared to endo isomer is, according to Winstein, due to an intramolecular attack by σ-electrons of the $C^6—C^1$ bond on C^2 in the transition state of the $C^2—O$ bond heterolysis. In the endo epimer, since the stereoelectronic considerations require a backside attack, the participation of the $C^6—C^1$ bond is impossible. The σ-participation of the $C^1—C^7$ bond is stereoelectronically feasible and would yield ion *24*. This ion could only be attacked by a nucleophile from the endo-side but no such products have been isolated. Seemingly, the participation of the $C^1—C^7$ bond would bring about so high an increase in the strain of the $C^1—C^2—C^3—C^4—C^7$ cycle that it would exceed the decrease in the free energy caused by the σ-delocalization [6, 24, 25]. Besides an increase in the angular strain, the absence of σ-participation of the $C^1—C^7$ bond can be also accounted for by a less favourable relative arrangement of the developing p-orbital of C^7 and of the σ-bond $C^1—C^7$.

If the forming ion of the *24* type is essentially stabilized by the substituent at C^1, then these unfavourable factors may be overcome.

In 1970 the solvolysis of *1-methoxy-2-endo-norbornylbrosylate 25* was shown [39] to yield a product with a bicyclo[3,1,1]heptyl skeleton *26*; this seems to be the first example of this kind of contraction of the cycle in the solvolysis of norbornyl derivatives. The formation of ketones *27* and *26* from brosylate *25* points to the migration of C^6 and C^7.

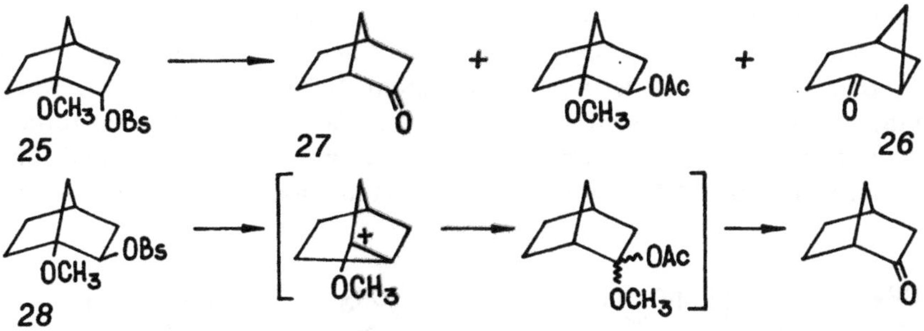

The absence of the migration of C^7 in the brosylate *28*, as well as upon solvolysis of unsubstituted 2-endo-norbornyl brosylate *23* confirms that the $C^1—C^7$ bond participates before the endo leaving group is completely separated from the molecule.

Winstein explained the complete *racemization* upon acetolysis of exo epimer *22* by the formation of an intermediate symmetric nonclassical ion *5*. At first glance the racemization could be accounted for by formation of an intermediate, optically inactive norbornene but it had never been isolated in these reactions at any stage; besides, the reaction proceeds with a complete racemization even under the conditions when the elimination is undoubtedly hindered in comparison with nucleophilic substitution. On the other hand, the racemization can be explained not only by the formation of an intermediate ion *5* but also as a consequence of the *Wagner-Meerwein rearrangement* (WM), the 6,2- or 3,2-hydride shifts; in all these cases the classical 2-norbornyl ion is transformed into its enantiomer. None of these hypotheses can account for:

1) the difference in the rates of solvolysis for exo and endo isomers;

2) the absolute stereospecificity of solvolysis for exo-brosylate. The classical 2-norbornyl cation must form a certain amount of endo isomer (see p. 59) since the endo

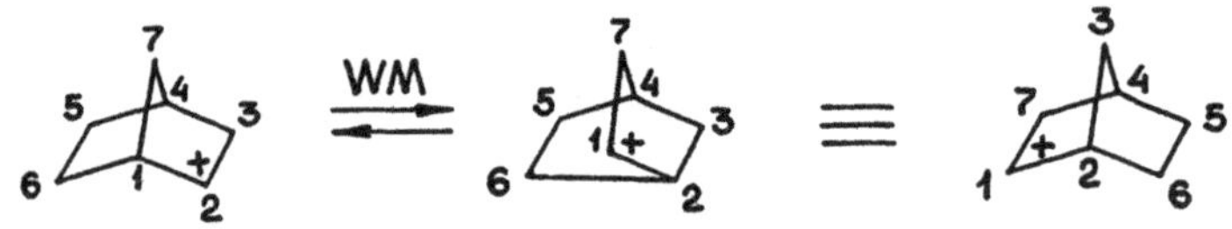

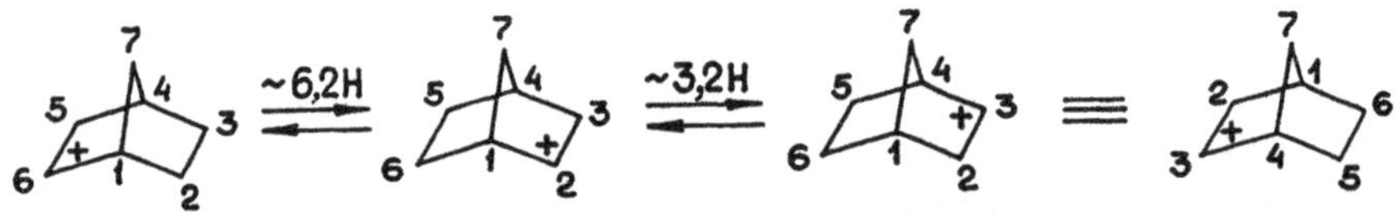

attack is favoured by the backside solvent participation in the ionization and by the leaving group shielding the exo side. The nonclassical ion *5* can be attacked by the nucleophile only from the exo side on one of its atomic orbitals involved in multicentre bonding. Such an attack on C^1 or C^2 can only result in exo-acetate.

It is interesting that though C^6 must formally bear a positive charge, no products from the attack on this atom have yet been isolated. It may be due to an increase in the steric strain brought about by introduction of a double bond into the product; also, the migrating atom C^6 is usually less substituted than C^1 and C^2. Such an attack could probably occur if C^6 were a tertiary carbon atom. Schleyer [40], however, when studying the solvolysis of 6,6-dimethyl-2-norbornyl tosylates, showed that the formation of the tertiary carbocation is not yet sufficient to break the $C^1—C^6$ bond. The results of Jones and Jones [41] show that only a combination of two factors — the

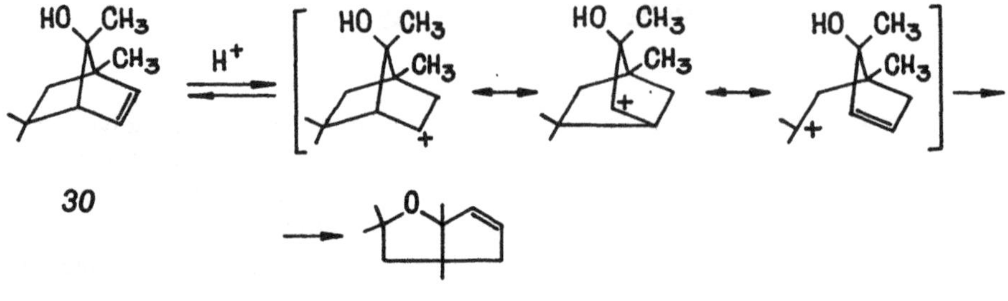

30

electron-releasing alkyl groups at C^6 and electron-withdrawing groups at C^7 — is sufficient to increase the contribution of structure *30* and thereby to fragment the C^1—C^6 bond.

Quite different is the *protonation of an isomeric olefine* with the formation of an intermediate stable tertiary classical cation and with fragmentation of the 1–7 bond.

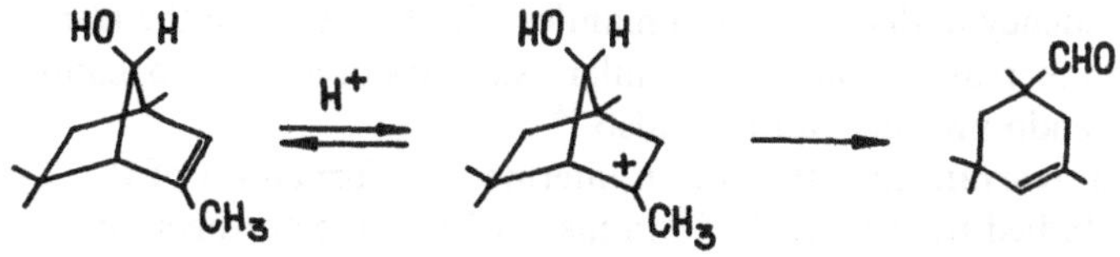

Solvolysis of endo isomer *23* in accordance with the above scheme, proceeds without anchimeric assistance with formation of a classical 2-norbornyl ion coordinated with the solvent from the side opposite to the leaving group. This ion can give exo-acetate with the retained optical activity or rearrange (to a larger degree) into the nonclassical ion *5*; the latter explains the abnormally high degree of racemization and the absolute stereospecificity upon acetolysis of endo-brosylate *23*.

Howe and Winstein [42] have shown: if the skeleton of the endo-norbornyl brosylate molecule is so much deformed that the σ-bonds C^1—C^7 and C^2—OBs become nearly transcollinear, then the σ-participation of the C^1—C^7 bond in the solvolysis of endo-ester becomes possible. Thus, the acetolysis of endo-brosylate *31* yielded 22.5% of endo-acetate *33*; this is explained by the formation of an intermediate nonclassical ion (*32*). Solvolysis without anchimeric assistance would have yielded exo-acetate as the main product since the exo side in this ion is far less sterically hindered than the endo side; it is, however, only a side product of 6.5%.

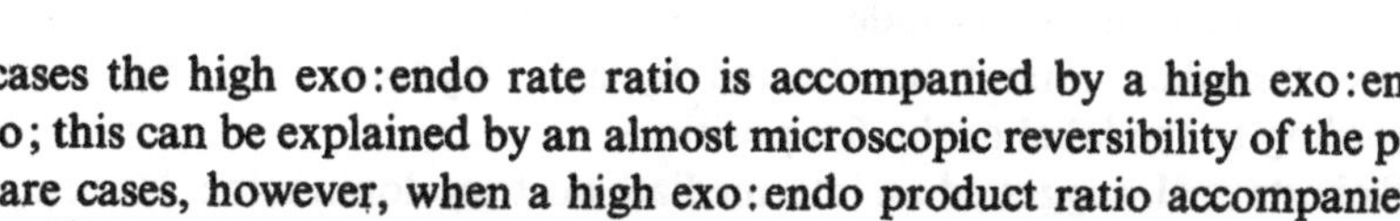

31 *32* *33*

In most cases the high exo:endo rate ratio is accompanied by a high exo:endo product ratio; this can be explained by an almost microscopic reversibility of the process. There are cases, however, when a high exo:endo product ratio accompanies a low rate ratio [43]. Apparently the difference in the energies of the transition state of

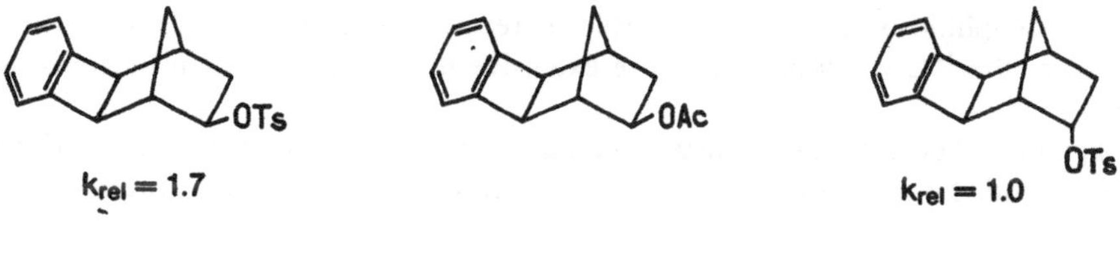

$k_{rel} = 1.7$ $k_{rel} = 1.0$

ionization is essentially smaller than that for the reaction between carbocation and solvent. According to [43] the σ-participation of the C—C bond may be of three degrees:

1) a low degree changing the "inversion/conservation of the configuration" ratio;

2) a higher one leading to rearrangements;

3) the tendency to delocalization manifests itself already in the transition state and results in an increase of rate. For similar examples of a low exo:endo rate ratio and a high exo:endo product ratio see also [44,45].

Structural rearrangements with an intermediate formation of the 2-norbornyl ion have been studied by Roberts [46,47] in his works on the acetolysis of 2-exo-norbornyl 2,3-^{14}C brosylate. The results obtained are presented in Table 1.

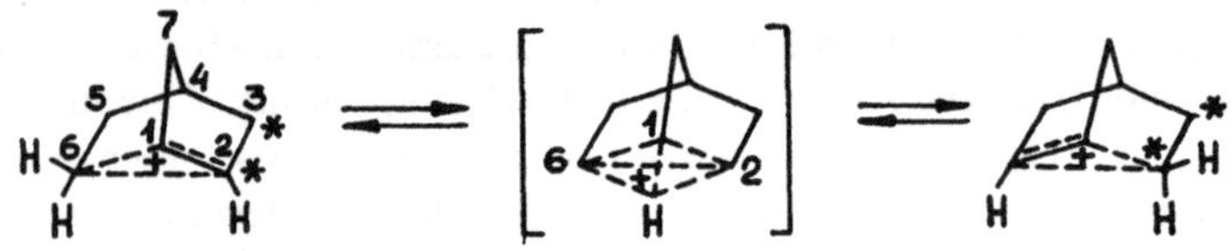

5 31 5a

Table 1. Calculated and Experimental Isotopic Distribution for the 2-Norbornyl Acetate Formed on Acetolysis of 2-exo-Norbornyl [2,3-^{14}C]brosylate [46,47]

Calculated distribution for a reaction proceeding	^{14}C activity % at			
	$C^2 + C^3$	$C^1 + C^4$	C^7	$C^5 + C^6$
Via ion (*5*)	50	25	26	0
Via ion (*31*)	33	17	17	33
By 6,2-hydride shift	50	0	0	50
By 3,2-hydride shift	100	·0	0	0
Via ion (*5*) (55%) and ion (*31*) (45%)	42.5	21.25	21.25	15
Experimental distribution	40	23	22	15

The above authors conclude from these data that the racemization in the solvolysis of exo-brosylate cannot be effected by a 3,2-hydride shift, but by the 6,2-shift, according to the label at the 5,6-positions.

Roberts et al. suggested an intermediate nortricyclonium ion *34* even more symmetrical than ion *5* and in which atoms 1,2,6 are totally equivalent. An attack at any of them results in racemic exo-acetate. The initially formed ion *5* is partially rearranged into ion *34* (45% in acetic acid); this rearrangement proceeds in the higher degree, the less nucleophilic is the solvent capable of reacting with ion *5*. As the authors have shown, the ^{14}C scrambling is more extensive in formic acid than in aqueous acetone.

To explain Robert's results Winstein assumed, instead of ion *34*, the equilibrium of ions *5* and *5a* differing only in the position of ^{14}C. As an intermediate compound (or a transition state) of such a rearrangement Winstein and Trifan proposed the edge-protonated cyclopropane *35* rather than the face-protonated (nortricyclonium) ion *34* suggested by Roberts.

34 *35*

With Robert's data one should consider the criticism of his results by Deno [48]. According to the latter the equalization of carbon and hydrogen atoms that precedes the reaction of the carbonium ion with the solvent results from intramolecular rearrangements; so it should be regarded neither as a proof of the equivalence of carbon atoms in carbonium ion structures nor as a confirmation of resonance structures. Hence, Robert's evidence that on conversion of 2-norbornyl tosylate into norbornyl acetate all the atoms except C^4 become nearly equivalent does not favour the structure of the nonclassical ion *5* because the experimental data on isotopic equalization indicate only relative rates of carbonium ion rearrangements.

3.2 The π-Route to the 2-Norbornyl Ion

The structure of the 2-norbornyl cation *5* is indirectly proved by the solvolytic cyclization of 2-(3-cyclopentenyl)ethanol. Thus the acetolysis rate of nosylate *36* exceeds that of a saturated analogue by a factor of 95 [49, 50]. This fact has been interpreted in favour of direct participation of a double bond in the substitution of an arylsulphonate anion. As shown by Sargent and Bartlett [51] the substitution of

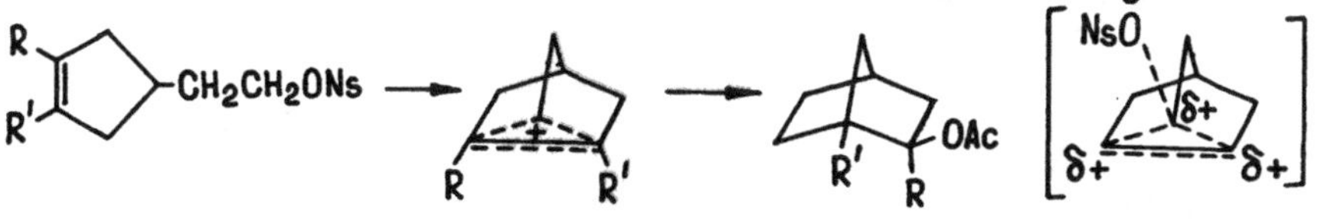

36 *5* (A)

one olefin hydrogen by a CH_3 group in *36* increases the rate of acetolysis by seven, that of two hydrogens by 38.5, i.e., the second CH_3 group produces practically the same accelerating effect 5.5 as the first one. These data point to the formation of a symmetrical transition state with an equal portion of positive charge in each of the C atoms bonded with the CH_3 groups *A*. According to the above authors, the reaction starting from the symmetrical ground state *36*, passing through the symmetrical state *A* must lead to the symmetrical initial intermediate *5*.

If 2-(3-cyclopentenyl)ethyl thiocyanate *37* is heated in various solvents it is transformed into a mixture of compounds *39–41* [52]. 2-Cyclopentylethyl thiocyanate is not rearranged under these conditions; this proves the transformation of *37* into *39*

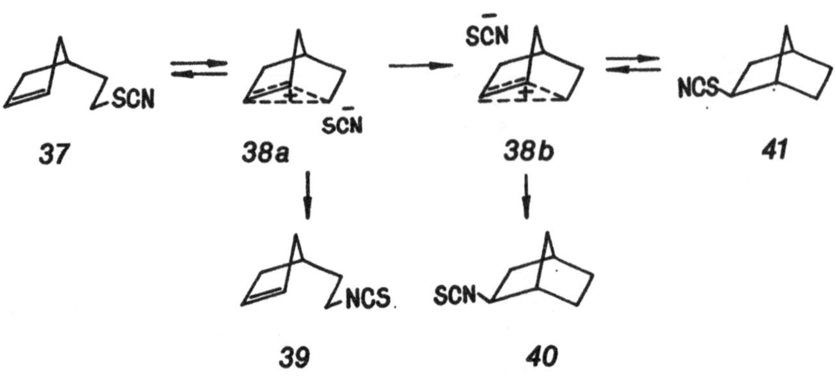

to involve the participation of a double bond. Under the reaction conditions compound *41* is slowly rearranged into *40* but it does not form cyclopentenylethyl derivatives. The authors believe these reactions to include the formation of two "isomeric" ion pairs *38a* and *38b*. If the ion pair *38a* yields the products *37* and *39* and is rearranged into *38b*, the latter is not transformed into *38a*. The absence of endo-products and products corresponding to 3,2- or 6,2(6,1)-hydride shifts corroborates the mechanism of monomolecular ionization with formation of an intermediate nonclassical norbornyl cation [53].

Bartlett [49, 51, 54, 55] has assumed that 2-norbornyl ions formed upon solvolysis of 2-exo-norbornyl brosylate ("σ-route") and 2-(cyclopentene-3-yl)ethyl nosylate ("π-route") differ essentially in the position of the leaving group:

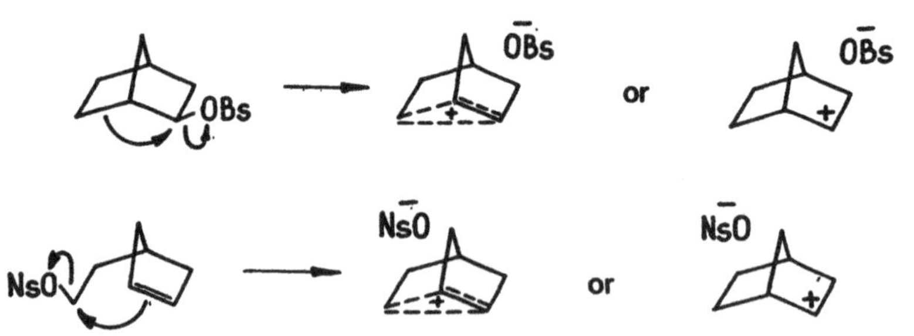

In 1969 Lee and Hahn studied the ^{14}C scrambling upon solvolysis of 2-(3-cyclopentenyl)-2-^{14}C-ethyl nosylate; the radioactivity of the C^3 atom in the formed products was found to exceed that of the C^7 atom and this was accepted as an argument in favour of the formation of intermediate, both corner- and edge-protonated, cyclopropanes [56, cf. 57, 58].

The participation of σ-bonds is by no means always weaker than that for π- and n-donors. Schleyer [13] notes that the neighbouring group ability should mirror basicity. The proton affinity of the strained σ-bonds of cyclopropane is 179 kcal/mole which exceeds that of the π-donor, ethylene (59 kcal/mole) and even that of water (165 kcal/mole).

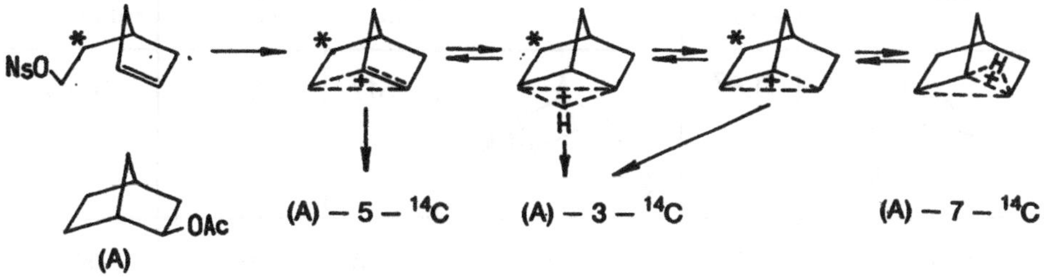

3.3 Criteria for the Formation of Nonclassical Carbocations

According to Sargent, the following criteria indicate the intervention of a bridged ion [2]:

1) enhanced rate of solvolysis relative to a similar model deprived of σ-delocalization capacity;
2) stereospecific formation of reaction products;
3) intensive skeletal rearrangements;
4) loss or retention of optical activity in cases when formation of the classical ion would lead to opposite results;
5) the phenomenon of "ion-pair return" to isomeric starting material.

The well-known American chemist H. C. Brown has come out against the concept of nonclassical ions [3,4,13,20,21]; his main arguments will be discussed on the 2-norbornyl ion.

3.3.1 Kinetic Data and Nature of the 2-Norbornyl Ion

The difference between the solvolysis rates of 2-exo- and 2-endo-norbornyl derivatives can be accounted for by the low reactivity of the endo-isomer; this was attributed to the effect of spatial hindrances since in the transition state the departing ester group passes unusually close to the endo hydrogens at C^5 and C^6. However, there are practically no experimental data corroborating this point. For example, if a CH_3 group is introduced to C^5 the rate of endo-isomer solvolysis does not practically change [60]. Introduction of a CH_3 group to C^6 decreases the rate of solvolysis of endo- and exo-isomers so that the rate ratio remains unchanged [40] (Table 2). For the endo-isomer this fact can be explained by the difficulty of detaching the leaving group due to spatial hindrances (see also p. 21).

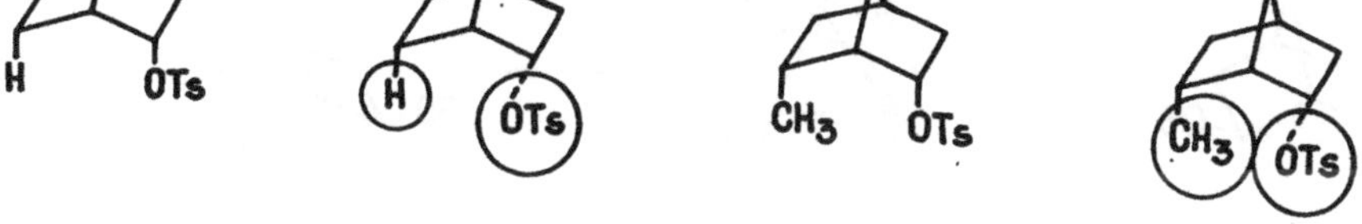

Table 2. Relative Rates of Acetolysis of Arylsulphonates at 25 °C

Compound	Relative rate	Compound	Relative rate	Exo : endo rate ratio	References
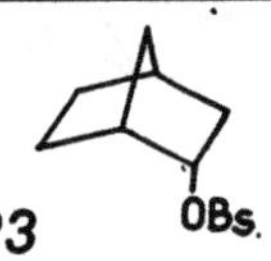 23	1.47	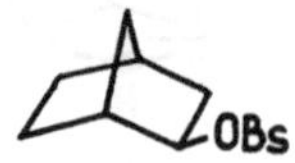 22	517	350	24,25)
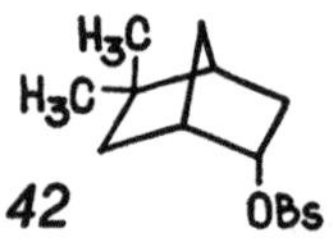 42	1.05	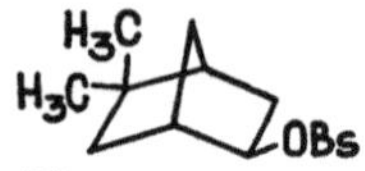 43	153	145	60)
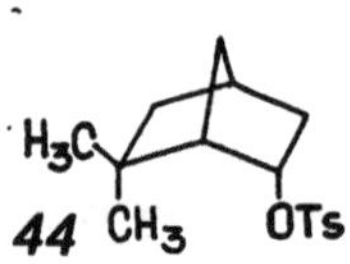 44	0.09	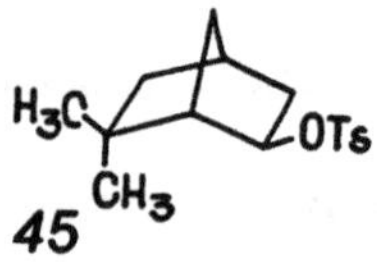 45	25.4	222	46)

It is more difficult to explain the deceleration of solvolysis for exo isomers. If the bulky substituent at C^6 in the endo-position hinders the solvation of the arising classical 2-cation from the endo side, it is difficult to explain the hindrance to the solvolysis of compounds having hydrogens at the endo-position in C^5 and C^6, e.g., *46*, because for these compounds and for *22* there must be no difference in the ease of solvent approach from the endo-side.

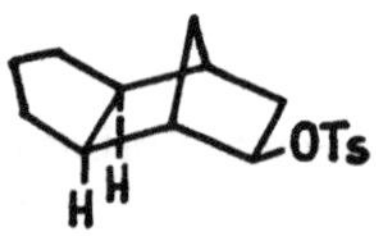

46

$k_{rel} = 2.76$

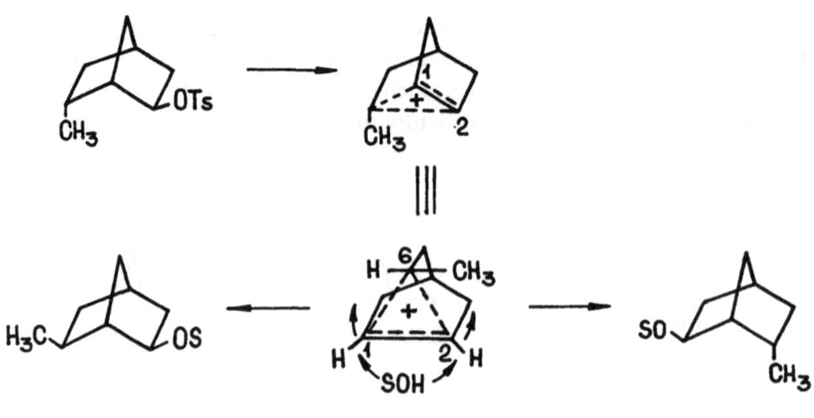

Thus, in the concept of classical ions the steric factors at C^6 fail to account for the deceleration of solvolysis at C^2. At the same time the models show that for 6,6-dimethyl-2-exo-norbornyl brosylate *45* in the transition state leading to a nonclassical ion the non-bonded interactions between the CH_3 groups on C^6 and the hydrogens on C^1 and C^2 are markedly increased. It is clearly seen from the scheme on the page 18. In the transition state of the nonclassical ion formation unfavourable steric effects occur irrespective of whether the CH_3 group at C^6 is in the endo- or exo-position. Instead, the substitution of the hydrogen by the CH_3 group in C^6 decreases the solvolysis rate of endo-tosylate (for steric reasons) but since the solvolysis rate of the exo epimer decreases too this confirms the participation of the $C^1—C^6$ bond in the exo-tosylate ionization.

It remains unclear whether the endo hydrogen atom in C^6 offers, as Brown believes, any steric hindrance to the ionization of 2-endo-sulphonate because the replacement of 6-endo-H by 6-endo-CH_3 not only increases the bulk of 6-endo-substituent, but also alters the position of the "centre of gravity" of the substituent so that it finds itself in the path of the departing sulphonate anion (increase in the bond length from 1.09 to 1.54 Å upon substitution of H by CH_3). Assumably the detachment of the endo-sulphonate anion does not involve any appreciable steric interaction with a comparatively small hydrogen atom closely bonded to C^6, but it may interact with a more bulky group located at a greater distance [2)] (see the above scheme). The insignificance of interaction between the OTs group and the endo-H can be illustrated by data from Table 3. The small acceleration observed in the

Table 3. Kinetic Data on Acetolysis of Arylsulphonates at 25 °C

Compounds	Relative rate	Compounds	Relative rate	Exo : endo rate ratio
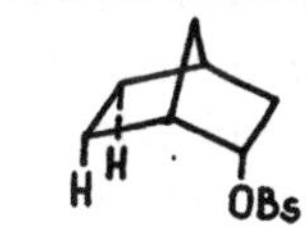	1.47	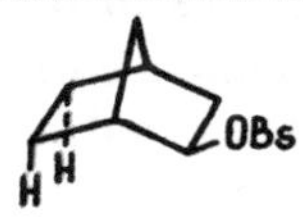	517	350
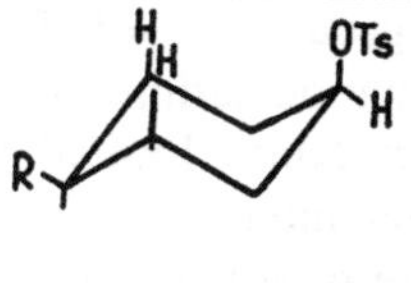	2.61	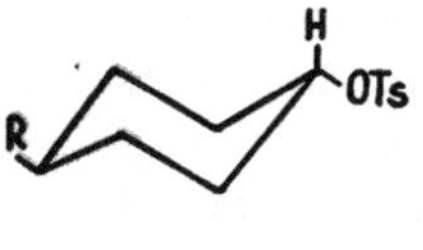	0.8	0.308

case of cyclohexane can hardly be assumed to turn into a 350-fold deceleration, in the norbornyl case by a small change in geometry [61)]. So far there are no precise experimental data proving that the 2-endo-brosylate *22* solvolyzes at an abnormally slow rate; the answer to this question should be evidently based on semiempirical calculations (see p. 22).

3.3.2 Solvolysis Rates of Exo Isomers

To prove his second thesis that exo isomers solvolyze at a normal, rather elevated, rate, H. Brown has:

1) predicted a normal rate of solvolysis of 2-exo-norbornyl derivatives by considering the reactivity of related systems;

2) compared the absolute rate constants of 2-alkyl- and 2-aryl-2-exo-norbornyl p-nitrobenzoates and the respective chlorides with the rate constants of cyclopentyl analogues [62–65];

3) considered the exo:endo solvolysis rate ratio for tertiary norbornyl systems [63, 64, 66]. Let us discuss each of these approaches in more detail.

Since the introduction of a two-carbon bridge into a cyclohexane system increases the rate of acetolysis of the respective brosylate by a factor of *43* [3, 4] one might expect a similar rate increase upon introduction of this bridge into a cyclopentane system as well. Sargent [2], however, showed this reasoning to be wrong. In cyclopentyl derivatives the transformation $sp^3 \rightarrow sp^2$ in the solvolysis destroys four eclipsed torsional interactions and overcomes some increase in the angle strain. In the cyclohexyl system the $sp^3 \rightarrow sp^2$ transformation increases both conformational and angle strains and decelerates the solvolysis rate as compared with the cyclopentyl derivatives.

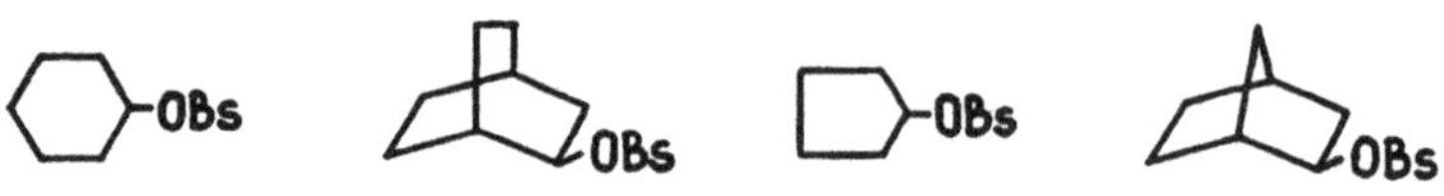

$k_{rel} = 1$ 43 32 ?

3.3.3 Results from Carbonyl Stretching Frequencies

In substituted bicyclo[2,2,2]octanes there are 6 eclipsed interactions two of which disappear upon transformation into a corresponding carbocation, but in this case the angle strain increases. It should be noted in this connection that $\nu_{c=0}$ for bicyclo[2,2,2]-octanone is 1731 cm^{-1} [67]. As shown by Halford [68], a deviation of the carbonyl stretching frequency from the normal value of 1718 cm^{-1} [67] in the absence of polar effects is a measure of the deviation of the internal angle of the group $>C=0$ from its optimum of 116.3°. According to Schleyer [69] this frequency deviation can be used as a measure of the internal angle strain from a cation centre located the carbonyl group of a model compound. Thus, the relief of torsional interactions in the solvolysis of bicyclo[2,2,2]octyl-2-tosylate must increase the rate of solvolysis while an increase of the angle strain decelerates the solvolysis; just as in cyclopentyl tosylate the resulting effect gives accidentally a weak (43-fold) increase in solvolysis rate in comparison with cyclohexyl tosylate.

The bridging of cyclopentane to form a norbornane system which, according to H. Brown, is also accelerating the solvolysis leads in fact to other results. An increase in angle strain upon the $sp^3 \rightarrow sp^2$ transformation in both a cyclopentane and a norbornane system must be essential but nearly the same (for cyclopentanone $\nu_{c=0}$ = 1750 cm^{-1} [67], for 2-norbornanone $\nu_{c=0}$ = 1751 cm^{-1} [70]). At the same time the torsional interactions, in passing from cyclopentane to norbornane, change in a way different from that in passing from cyclohexane to bicyclo[2,2,2]octane. In cyclopentane four of ten eclipsed interactions disappear upon the $sp^3 \rightarrow sp^2$ transformation; in norbornane two of four eclipsed interactions disappear. One should therefore expect an essential decrease rather than an increase in the solvolysis rate when passing from cyclopentane to norbornane. This decrease amounts to the difference in the free energy 2 kcal/mole (the energy barrier in ethane where there are three eclipsed interactions amounts to ~3 kcal/mole), i.e., cyclopentyl brosylate should solvolyze (at 25 °C) about 33 times as fast as 2-norbornyl brosylate. As a matter of fact, for the exo isomer *22* the solvolysis rate is 16.1 times as high as that of cyclopentyl analogue (in acetic acid, at 25 °C, titrimetrically) while the acetolysis rate of endo isomer *23* accounts only for 0.047 of that of cyclopentyl brosylate, i.e., the endo isomer reacts 22 times as slowly. A good agreement between the predicted rate and the experimental data for the classical 2-norbornyl cation formed from the endo-isomer shows that the solvolysis rate of 2-endo-norbornyl brosylate is not abnormally slow. At the same time exo isomers are solvolyzed at a far higher rate ($33 \times 16 = 528$ times) than it could be expected from data on model compounds; an experimentally found rate ratio of solvolysis for exo- and endo isomers equals 350 [24].

Assumably the driving force to ionize the exo isomer is a decrease in non-bonded interactions when passing from the ground state to the transition state. The model shows that the distance between hydrogen atoms in C^6 and C^2 equals 2.18 Å, while the sum of the Van der Waals radii equals 2.4 Å. This factor, however, seems to be insignificant since substitution of the hydrogens in C^6 by the CH_3 groups not only fails to increase the rate but it even decelerates the reaction by a factor of about 20 in comparison with exo-norbornyl brosylate *22*. Hence the effect of acceleration cannot be attributed to the weakening of non-bonded interactions.

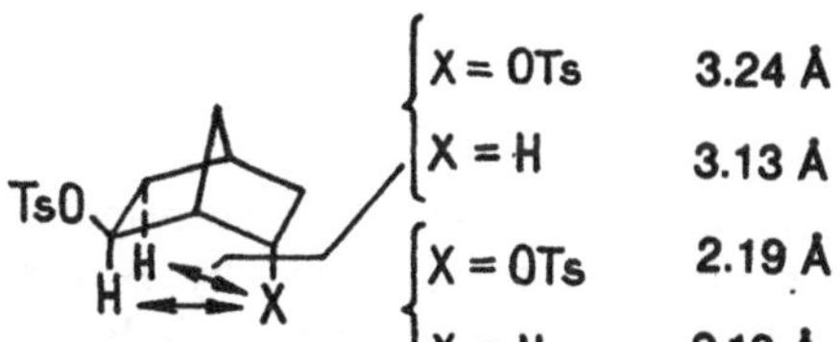

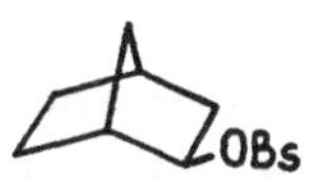

22

$k_{rel} = 517$

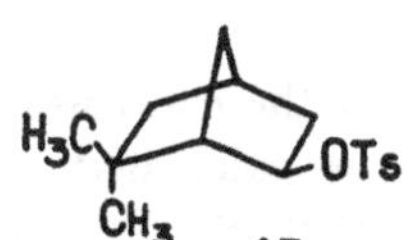

45

$k_{rel} = 25.4$

In endo isomer the strongest non-bonded interaction is that between —OTs and 6-endo-H (2.19 Å, the sum of the Van der Waals radii being 2.6 Å). This interaction decreasing in the transition state should have increased the rate, but the solvolysis rate of endo-brosylate has been shown to be predictable only from the torsional effects. The non-bonded interactions in the ground and the transition state seem to equally destabilize the system and do not, therefore, affect the solvolysis rate of the endo isomer.

This supports Brown's suggestion [59] that 6-endo-hydrogen hinders the detaching of the leaving group from the 2-endo-position; but the magnitude of this effect is far less than postulated by Brown. As a model for calculating the solvolysis rates of secondary arylsulphonates Schleyer [71] used the solvolysis of cyclohexylarylsulphonates in which there is practically no manifestation of anchimeric assistance, angular and torsional strains, changes of non-bonded interactions. The solvolysis rates of secondary arylsulphonates relative to this model have been calculated from an empirical equation on the changes of angular and torsional strains, non-bonded interactions and inductive effects:

$$\log_{10} k_{rel} = 0.125(1715 - \nu_{c=o}) + 1.32(1 - \cos 3\varphi_i) + \\ + (GS - TS)/1.36 + I,$$

where $\nu_{c=0}$ is the carbonyl stretching frequency of the ketone obtained by oxidizing the secondary alcohol of the arylsulphonate under study; φ_i is the avergae torsional angle for each of the C—C bonds adjacent to the leaving aryl-sulphonate group; GS — TS is the difference (in kcal/mole) in the energy of non-bonded interactions for the transition and the ground states: I is the inductive term. The term GS — TS is the most difficult to calculate, but it has the lowest value out of the three steric effects. Calculation from this relationship makes it possible to estimate the solvolysis rate of any secondary arylsulphonate relative to that of the respective arylsulphonate of cyclohexanol in the absence of anchimeric assistance (Table 4).

Table 4. Calculation of the Acetolysis Rates of Arenesulphonates at 25 °C and Estimation of the Magnitude of Anchimeric Assistance [71]

Sulphonate	$\nu_{c=0}$ (cm^{-1})	φ	GS — TS (kcal)	Calc. $\log_{10}$	Obs. $\log_{10}$	$\log_{10} V_{rel}^{exp} - \log_{10} V_{rel}^{cal}$ [a]
Cyclohexyl	1716	60.60	0.0	—0.1	0.00	0.1
2-endo-Norbornyl	1751	0.40	1.3	—0.2	0.18	0.0
2-exo-Norbornyl	1751	0.40	0.3	—0.6	2.71	3.3

[a] Measure of anchimeric assistance [71].

For the endo isomer the experimentally observed rate of solvolysis corresponds to the calculated one, i.e., it is not abnormally low, as suggested by Brown, but that of solvolysis of the exo isomer, contrary to Brown's opinion, is by far higher than the calculated one. Schleyer held that the anchimeric assistance in general can serve as an argument in favour of the nonclassical ion. However, the anchimeric assistance involving the carbon migration would not result in the formation of the nonclassical intermediate, if the cation resulting from the migration is more stable than the ion formed without migration, e.g. [13]:

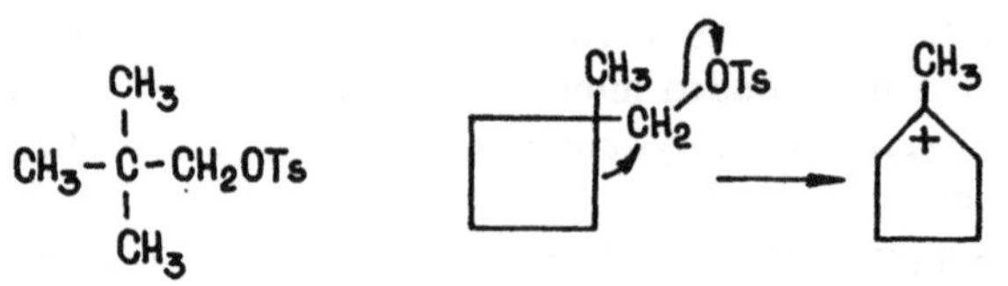

k_{rel} (HOAC, 25°C) = 1.0 24600

This refinement is not relevant to the ionization of 2-exo-norbornyl brosylate because the ion generated upon migration of C^6 is an enantiomer of the initial ion. Thus, in this case the anchimeric assistance indicates the formation of an intermediate more stable than the classical 2-cation. Such a particle can evidently be the 2-norbornyl bridged ion *5*.

In his review [21] Brown has cited examples when the prediction of the solvolysis rates for compounds *44*, *47* and *48* according to Foote-Schleyer yields erroneous results:

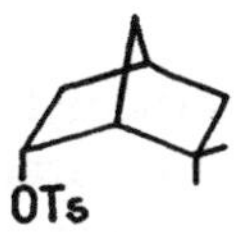

44

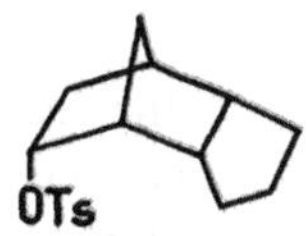

47

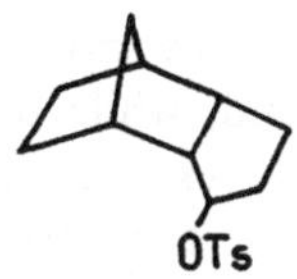

48

Indeed, Schleyer admits [13] that the application of the above equation to sterically crowded system is "dangerous", but this limitation is not extended to 2-exo-norbornyl sulphonates.

3.3.4 Comparison of Cyclopentyl and Norbornyl Derivatives

Having compared the solvolysis rates of tertiary cyclopentyl chlorides and tertiary 2-norbornyl chlorides, Brown found these values close to each other; from this he concluded that the solvolysis of tertiary norbornyl chlorides is not accompanied by the participation of the 1,6 bonding pair of electrons during ionization [62–66, 72–74]. At present this conclusion can be reasoned more strictly. The above example of secondary cyclopentyl and 2-norbornyl derivatives has demonstrated the incorrectness of such a comparison. The changes in the torsional and the angle strain in the tertiary systems can be assumed not to differ markedly from those in the secondary ones. In the case of the second cyclopentyl system the $sp^3 \rightarrow sp^2$ transformation requires energy expenses by about 2 kcal/mole less than for the secondary norbornyl system due to a more considerable decrease in torsional interactions. However, in the ground state of the tertiary 2-exo-norbornyl chlorides there occurs an unfavourable non-bonded interaction between 2-endo-CH_3 and 6-endo-H; it decreases essentially in the transition state of chloride ionization and provides an energy gain of about 3 kcal/mole covering the loss at the expense of torsional interactions. In the cyclopentyl system, due to its nearly planar structure such changes of non-bonded interactions do not occur.

$k_{rel} = 5.6$

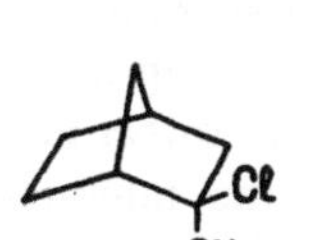

$k_{rel} = 30.2$ (ethanolysis, 25°C)

49

Thus, in the absence of anchimeric assistance tertiary norbornyl compounds would be expected to solvolyze ca. 7 times as fast as the corresponding tertiary cyclopentyl analogues. This is in good agreement with experiments and does confirm the absence of anchimeric assistance due to the participation of the $C^1—C^6$ bonding pair of electrons in the ionization process.

This conclusion agrees with Winstein's point of view: the more the substituents attached to the arising positive centre stabilize the cations the less is the degree of anchimeric assistance [75]. In case the substitute at C^2 exerts so strong a stabilizing influence that the localized classical ion becomes more stable than the nonclassical one no anchimeric assistance should be observed.

On the other hand, while the stability of the secondary norbornyl nonclassical ion essentially decreases (i.e. for sterical reasons) and becomes comparable with that of the classical ion, the exo-brosylate *50* may solvolyse to form a significant amount of endo-acetate [76]. But since exo-brosylate *50* yields about 65% of the

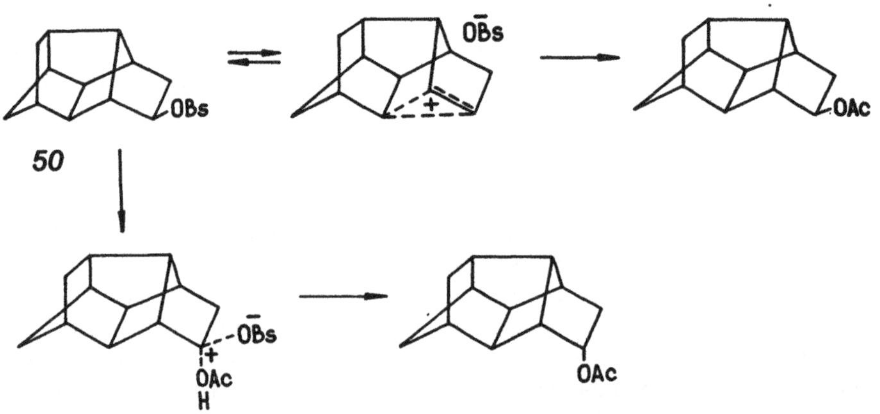

product with preserved configuration (in contrast to the prevailing inversion in the classical solvolysis of secondary substrates) and, in addition, the rate ratio of solvolysis for exo-brosylate *50* and its endo epimer is 45:1 one can infer that the solvolysis of exo-brosylate *50* occurs partially via the non-classical ion.

The anchimeric assistance is a sufficient, but not a necessary sign of formation of the nonclassical ion. Anchimeric assistance is only observed if it is essential for the transition state of the rate-determining step for solvolysis. Absence of anchimeric assistance does not preclude formation of the less stable nonclassical ion at stages succeeding the primary formation of the classical carbocation at the limiting stage [77].

By analogy with A. Streitwieser's [78] views on the solvolysis of tertiary alkylhalogenides the neighbouring group in intramolecular nucleophilic substitution can participate like a nucleophile in direct nucleophilic substitution. The only difference is that steric hindrances to the interaction of the neighbouring group with the developing cation centre will be somewhat smaller than in the attack by an external reagent since the participating group is covalently bound to a substituent attached to the reaction centre. Drawing an analogy between the participating neighbouring group in intramolecular nucleophilic substitution and the solvent molecule in unassisted solvolysis one should replace the molecule of the substituent S in structures *B* and *C* by the neighbouring group. As seen from the graph the considered nonclassical ion *C'* is more stable than the classical 2-methyl-2-norbornyl cation; nevertheless the intermediate arising at the limiting stage is a classical ion, and no anchimeric assistance will be observed.

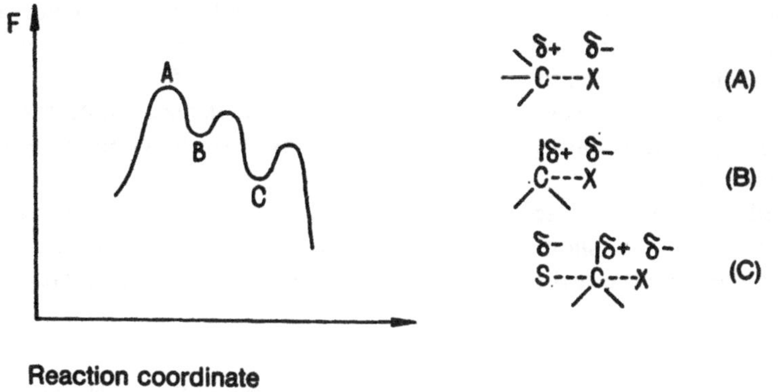

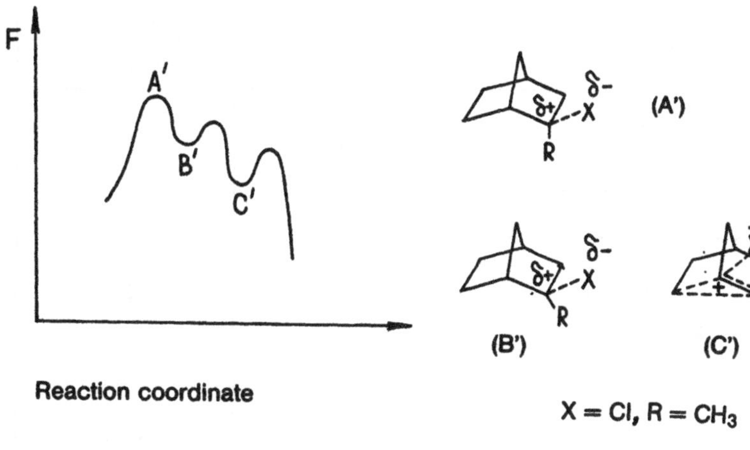

3.3.5 Exo:Endo Rate Ratio for the Solvolysis of Sec. and Tert. Norbornyl Derivatives

The solution of the problem whether the nonclassical ion takes part in solvolysis if there occurs no anchimeric assistance should be based on other criteria, such as the stereochemistry of reaction products (see p. 47).

Brown suggested that since in the solvolysis of tertiary norbornyl derivatives there is no σ-participation of the 1,6-bond, the exo:endo rate ratio for them must be far lower than for the secondary ones in which it is allegedly present [3,4]. But since for the tertiary norbornyl derivatives this ratio has the same order of magnitude as for the secondary ones Brown asserts that the participation of the 1,6-bond cannot be the main reason for a high exo:endo rate ratio for the secondary norbornyl derivatives. Brown denies the possibility that the observed exo:endo rate ratio for tertiary systems could arise from the relief of steric strain upon solvolysis of exo-epimers.

Table 5 shows that for tertiary systems the exo:endo rate ratio predicted only with steric factors agrees with experiments; in the case of secondary systems there is no such accordance. Thus the difference in activation energy for the solvolysis of secondary exo- and endo-derivatives which is due to the σ-participation upon solvolysis of exo isomer coincides by chance with the difference for tertiary systems due to a steric strain decrease in the transition state for exo isomers owing to the removal of 2-endo-alkyl from 6-endo-hydrogen [2]. The relief of the steric strain in the transition state of solvolysis of 2-exo-esters makes it necessary to correct for the steric accelerating effect in tertiary systems in comparing tertiary and secondary substrates.

Finally, for a more accurate comparison of kinetic data for secondary with tertiary systems one should use the exo:endo ratio of polarimetrically found rates because these are the real rates of ionization; in tertiary systems there usually occurs

Table 5. Exo:Endo Rate Ratio for the Solvolysis of Secondary and Tertiary Norbornyl Derivatives [2]

Compound	Exo : endo rate ratio		References
	Found*	Calc.	
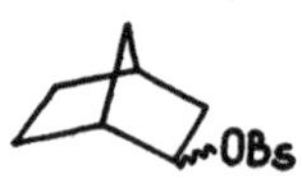	350	1	25)
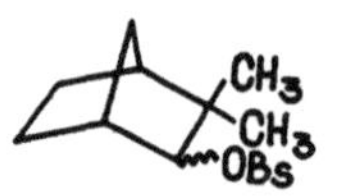	1230	1	60)
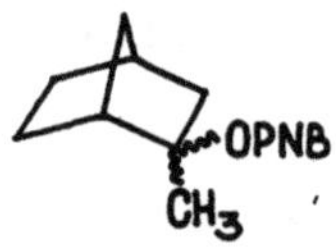	183	160	66)
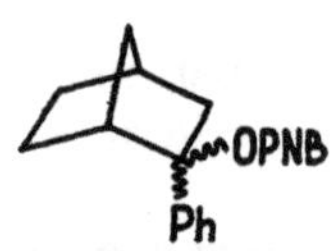	140	115	66)
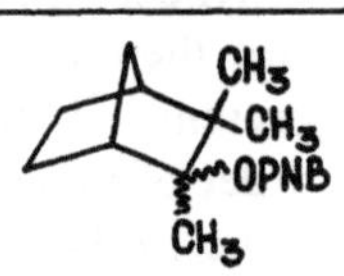	593	550	66)

*Obtained titrimetrically.

non "internal return" in ion pairs; and the titrimetric constants are almost a true reflection of ionization rate. After these corrections (Table 6) it follows that the exo:endo rate ratio for tertiary systems is lower than that for secondary ones, and that those for tertiary esters are rather close to one another.

Table 6. Exo:Endo Rate Ratio for the Solvolysis of Secondary and Tertiary Norbornyl Derivatives (after Corrections for Steric Effects, "Internal Return")

Compound	Exo : endo rate ratio		References
	Found	Calc.	
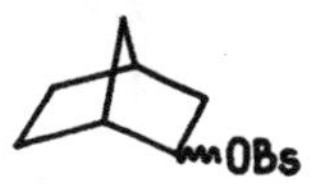	1600*	1600	25)
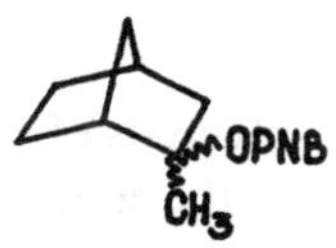	193**	27.5	66)
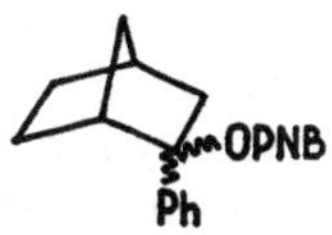	140**	23.8	66)
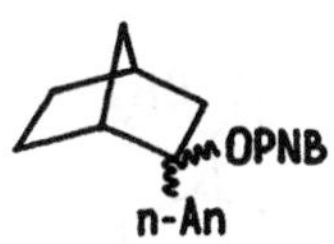	284**	47.4	72)

*Polarimetrically **Titrimetrically

In recent years H. Brown has done a great deal in the research on the solvolysis of 2-tert-norbornyl-p-nitrobenzoates attached at C^2 to a para-substituted phenyl group [20,21,72,78-82]. The p-anisyl group is capable of levelling the participation of adjacent groups even though it is very large (10^{11}) [83]. So this group would undoubtedly remove a much smaller σ-participation in the case of solvolysis of 2-exo-norbornyl esters (the exo:endo rate ratio for secondary p-nitrobenzoates equals 280); in reality, however, the exo:endo rate ratio remains practically unchanged. There is no doubt that in the case of tertiary substrates *52* and *53* the difference in solvolysis rates is due to steric factors, but the question is: which? H. Brown considers the reason for the high exo:endo ratios to be the steric hindrances in ionization of endo isomer *53*. However, comparison of the values $\Delta F^{\neq}$ for the solvolysis of secondary and tertiary substrates shows both epimers *52* and *53* to react faster than their secondary analogues due to the steric strain relief. Steric hindrances, if any in the ionization of endo isomers *53* and *54* must be identical, but at the same time compound *53* reacts much faster.

In his recent survey [22] H. Brown points out that the high exo:endo rate ratio (885) in the solvolysis of epimers *55* and *56* cannot be due to the relief of steric strain in ionization of exo epimer since the corresponding alcohols are comparably stable [63]. Neglecting whether it is correct to substitute the OPNB group by OH when considering steric effects the similarity of the ground state energies of epimers does not imply the equality of the steric strain relief in transition states from different non-bonded interactions.

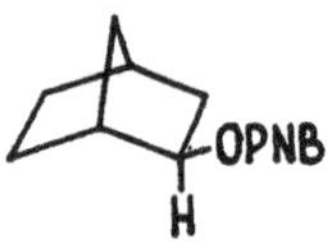

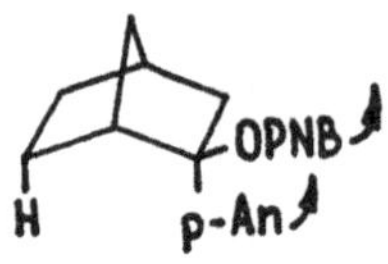

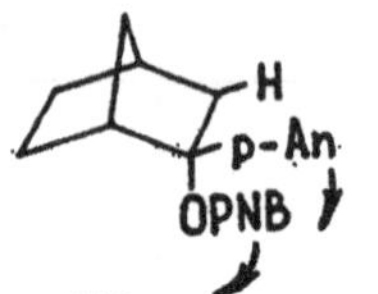

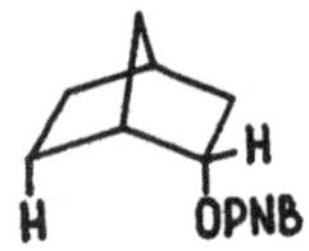

51 52 53 54

$\Delta F^{\neq}$ 22.6 20.1 23.4 27.1

According to H. Brown, the high exo:endo rate ratio in the solvolysis of epimers *55* and *56* is due to hindrances by solvating the leaving anion from the endo side [79,80]. If that is true for the epimers *57* and *58*, one expects a decrease in the exo:endo solvolysis rate ratio (from 885 to 6) because of the hindrances to anion solvation in the ionization of compound *57*. But the comparison of the $\Delta F^{\neq}$ values shows that the main role is not played by this factor, but by an increase in the solvolysis rate of endo isomer *58* as compared with endo isomer *56* owing to the steric strain relief in the transition state of ionization of the former:

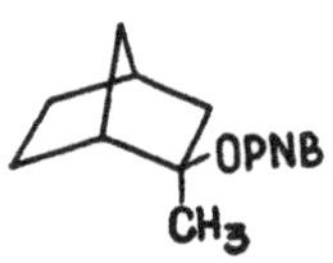

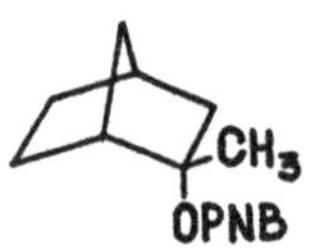

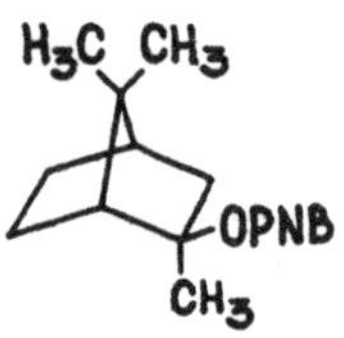

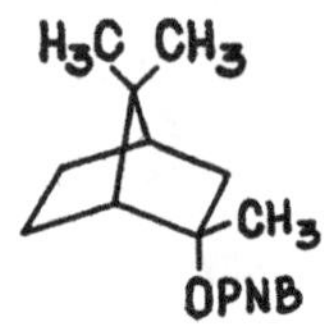

55 56 57 58

$\Delta F^{\neq}$ 28.4 32.4 27.5 28.6

Thus steric effects are really the main factor determining the exo:endo solvolysis rate ratio for tertiary substrates; but this cause is not the steric hindrance to ionization of endo isomers, but the steric strain relief in the transition state of ionization of exo and endo epimers.

Schleyer has recently [13] proposed a different explanation of the high exo:endo rate ratio for tertiary 2-norbornyl derivatives — steric hindrance to ionization of endo epimer due to a bulky 2-R group. Such hindrances fail to occur with secondary substrates due to a smaller volume of 2-H than that of 2-R.

As seen from Fig. 1, there is a great resemblance between the Goering-Schewene diagrams for the solvolysis of secondary and tertiary 2-norbornyl esters [13]. Thereby, H. Brown has shown that the high exo:endo ratio of rates and products does not yet

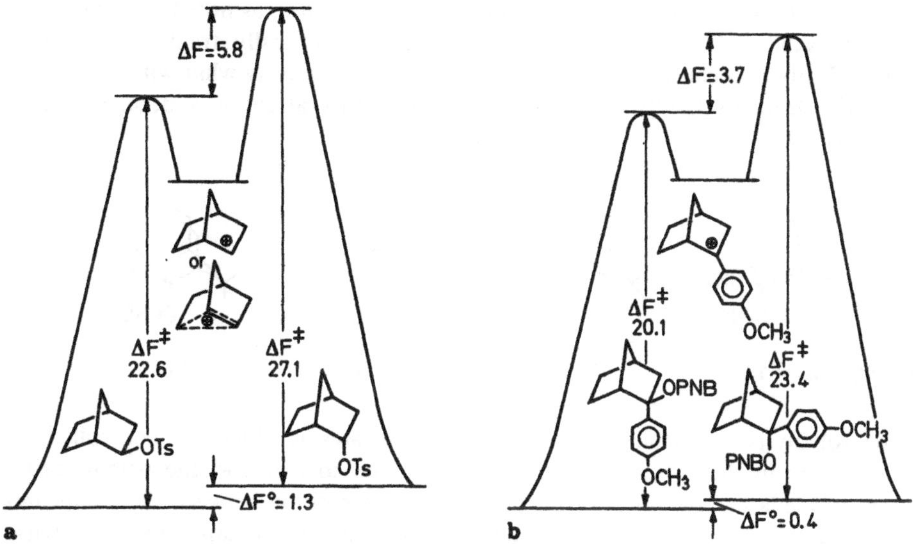

Fig. 1. Free energy diagram for: **a** the acetolysis of exo- and endo-norbornyl tosylate; **b** the solvolysis of 2-p-anisyl-2-norbornyl p-nitrobenzoates in 80% aqueous acetone at 25 °C [13]

prove the σ-participation in the solvolysis of the exo epimer. But the problem is whether the resemblance of diagrams is due to similar reasons or if it is an accidental coincidence though the basic reasons are different. In other words: can the tertiary substrates be models for describing the behaviour of secondary analogues in solvolysis reactions?

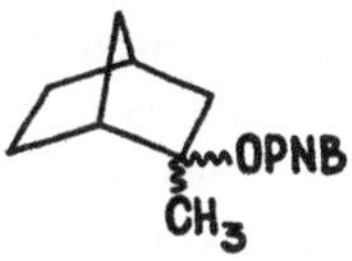

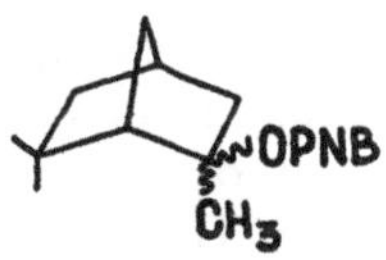

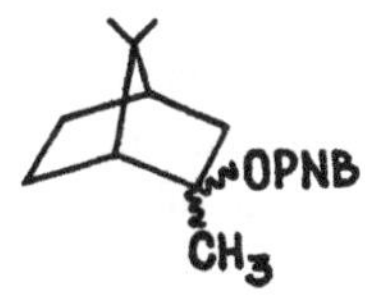

Exo : endo	885	3630000	6.1

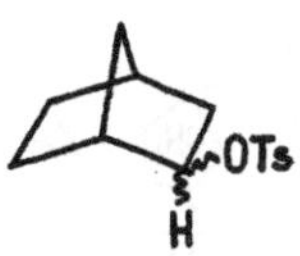

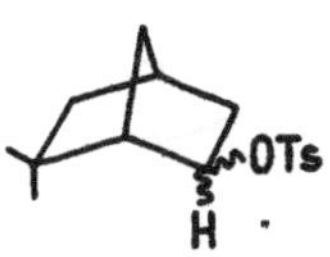

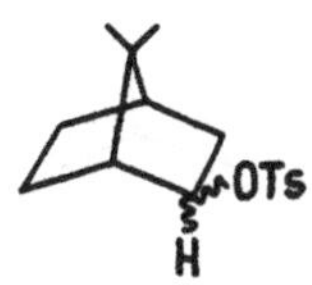

Exo : endo	280	206	4000

This is not the case. As follows from the above examples, the tertiary substrates are far more sensitive to changes of steric factors than their secondary analogues; neither do the data on the influence of steric factors on the solvolysis rate of tertiary substrates allow to predict even qualitatively, in which direction in the solvolysis rates of the secondary analogues change. Besides, as noted by Schleyer [13] H. Brown's theory [13] of strain relief in the solvolysis of sterically strained structures is rather "flexible" because one cannot predict even qualitatively what will result from the steric strain of the parent molecule — an acceleration or a deceleration of solvolysis, e.g.:

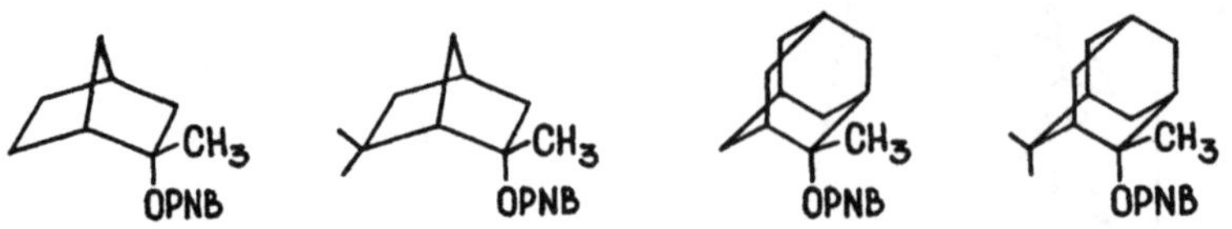

(25 °C) $k_{rel} = 1.0$ 0.18 1.0 14.0

The solvolysis of tertiary and secondary substrates is obviously controlled by steric factors in the former case, assistance of the solvent and the neighbouring groups in the latter. In this case must occur the competition between the bonds and the medium, a decrease in solvent nucleophility being accompanied by an increase in the exo: endo rate ratio; this is really observed [15).1]

	Medium	Exo : endo
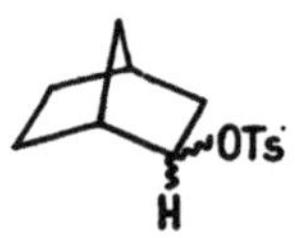	AcOH	280
	CF_3COOH	1120
	$CF_3COOH + SO_2FCl$	5200

The high exo:endo rate ratio in these examples is due to σ-participation in the ionization of exo epimer; accordingly in literature there seems to be not a single high rate ratio of an epimeric pair of secondary substrates for which a neighbouring group could not participate. The examples below confirm well this point particularly when compared with the tertiary analogues.

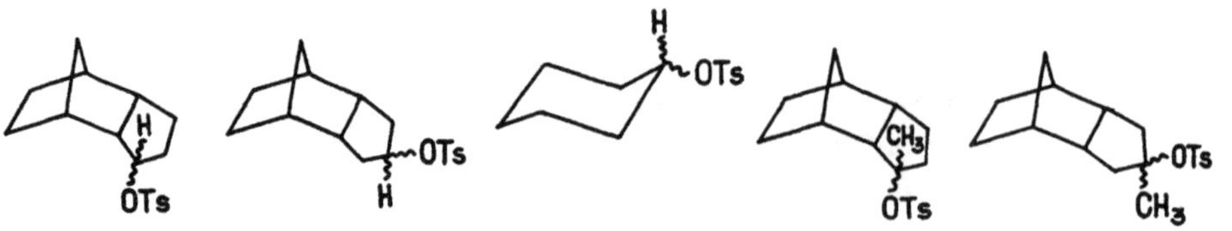

1.6/1.0 5.7/1.0 2...4/1 4300/1 11000/1

1 Winstein's choice of acetic acid as a "standard" medium for comparing the rates of solvolysis of tosylates was a failure, since it did not reveal the true amount of anchimeric assistance. A better judgement is obtained by the trifluoroacetolysis.

3.3.6 σ- or π-Participation in Solvolysis

Gassman and Fentiman proposed the so-called tool of increasing electron demand of the developing cation centre [84]. For their example of the solvolysis of 7-syn-aryl-7-anti-norbornenyl-p-nitrobenzoates they confirmed the π-participation by three criteria:

1) the amplification of acceptor properties in aryl is accompanied by a growing rate ratio of unsaturated and saturated analogues;

2) there is a "break" on the Hammett-Brown plot for the unsaturated series (see Fig. 2);

3) for unsaturated substrates in which aryl is a stronger acceptor than the anisyl group the value of ϱ^+ is relatively low (—2.30 cf. with ϱ^+ of saturated substrates —5.27).

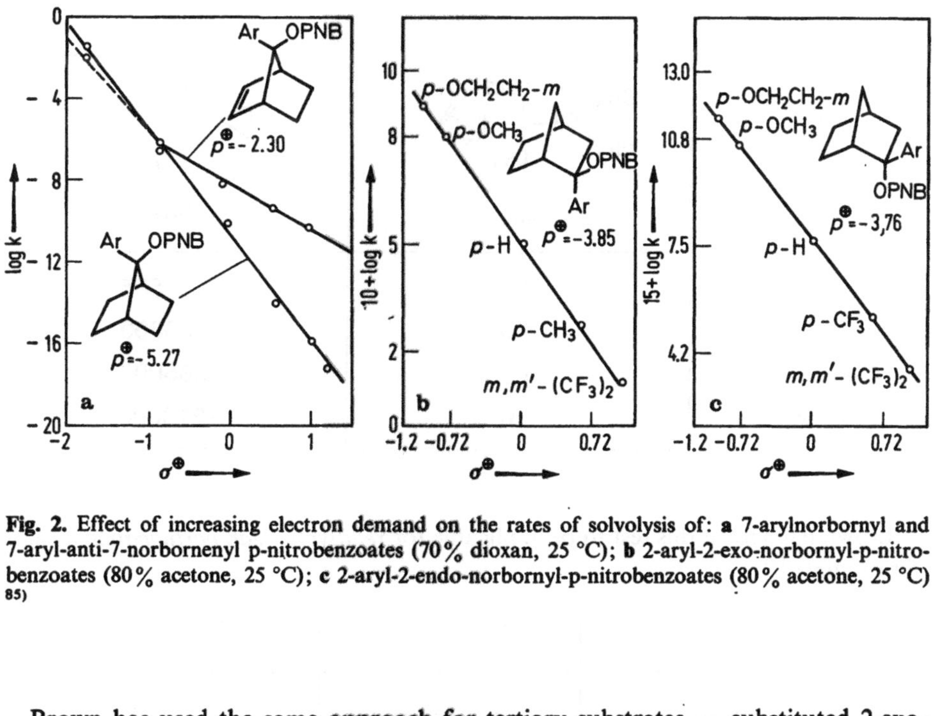

Fig. 2. Effect of increasing electron demand on the rates of solvolysis of: **a** 7-arylnorbornyl and 7-aryl-anti-7-norbornenyl p-nitrobenzoates (70% dioxan, 25 °C); **b** 2-aryl-2-exo-norbornyl-p-nitrobenzoates (80% acetone, 25 °C); **c** 2-aryl-2-endo-norbornyl-p-nitrobenzoates (80% acetone, 25 °C) [85]

Brown has used the same approach for tertiary substrates — substituted 2-exo- and 2-endo-norbornanes [81]. The exo:endo solvolysis rate ratio does not change essentially with increasing electron demand of the developing cationic centre; the value of ϱ^+ for the exo and endo series are similar (3.85 and 3.76, respectively), and there is no "break" on the curve for the exo series in passing from donors to acceptors, i.e. where one can expect σ-participation (Fig. 2, b, c) [85]. On this ground Brown has concluded that both series of epimers are solvolysed the same mechanism with the same demand of the developing cation centre without σ-participation in the exo series.

Perhaps the Gassman-Fentiman criterion is non-operative for σ-participation? — inquires Brown. No, the following examples show, it is not the case [86].

	 k_{rel}	 k_{rel}	 k_{rel}	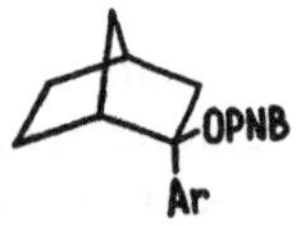 k_{rel}
4 – CH_3O	1.0	3.4	1.0	264
4 – CH_3	–	–	1.0	232
H	1.0	41.5	1.0	127
4 – CF_3	1.0	3400	1.0	187
3, 5 – $(CF_3)_2$	–	–	1.0	176
3, 5 – $(CF_3)_2$	1.0	25500	–	–

X	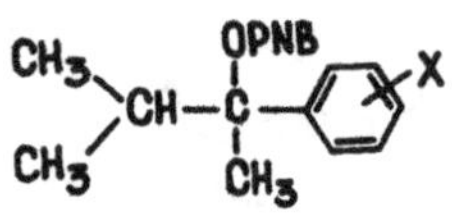 k_{rel}	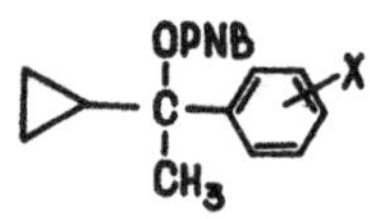 k_{rel}
4 – CH_3O	1.0	505
H	1.0	25300
4 – CF_3	1.0	285000
3, 5 – $(CF_3)_2$	1.0	1210000

Brown has shown this test to be applicable for *relatively weak participation as well, for example* [86]:

	X	k_{rel}
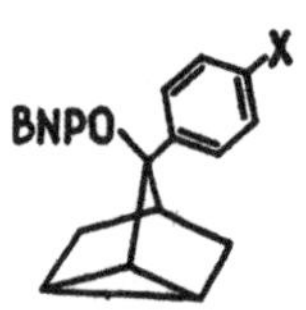	4 – CH_3O H CF_3	2.36 38.40 270.00

Finally, Peters has demonstrated the applicability of this approach not only for detecting the σ-participation of "specific" cyclopropane bonds, but also the more normal C—C bonds of the cyclobutane ring [87]:

Y	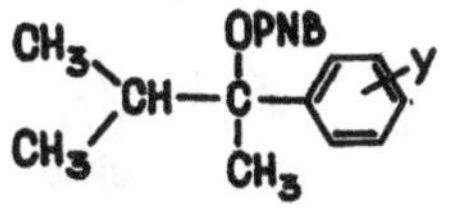 k_{rel}	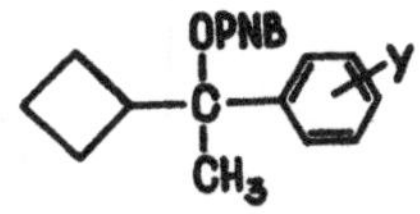 k_{rel}
4 – CH_3O	1.0	1.0
H	1.0	5.7
4 – CF_3	1.0	9.0
3, 5 – $(CF_3)_2$	1.0	17.6

From all these facts Brown concludes that σ-participation cannot be essential to determine the exo:endo rate ratio for 2-norbornyl esters. Consider, however, Brown's arguments more closely. Battiste and Fiato [88] have shown that if the solvolysis rates of exo and endo substrates are compared not with each other but with those of compounds incapable of σ-participation, then an increasing demand results in an increase of the solvolysis rate of both exo and endo epimers; this accounts for the small changes of exo:endo ratios. In endo epimers, the authors [88] observe the σ-participation of the 1–7 bond. When mentioning Battiste and Fiato's work [88], Brown did not reply to their objections [89].

Can the closeness of the values of ϱ^+ for exo and endo series lead to the conclusion that the mechanisms of their solvolysis are identical? Obviously it cannot,

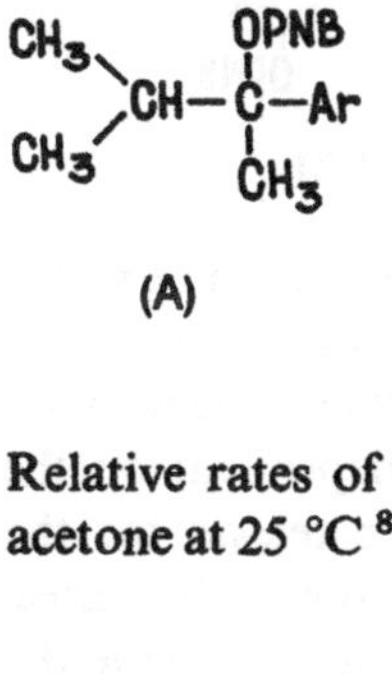
(A)
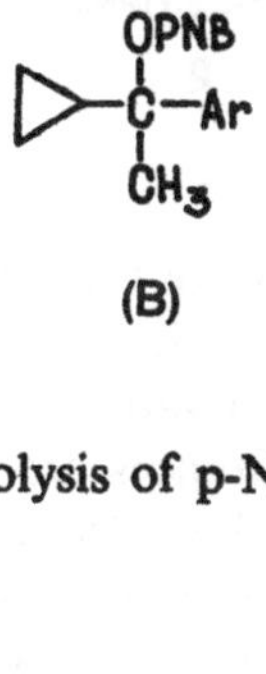
(B)
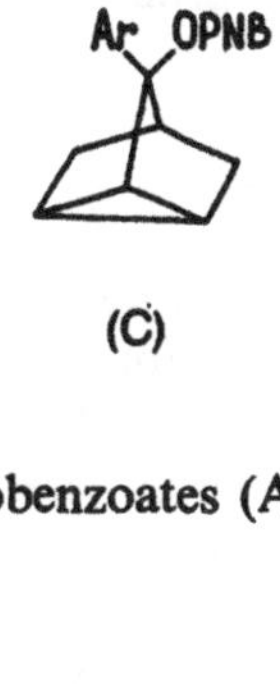
(C)
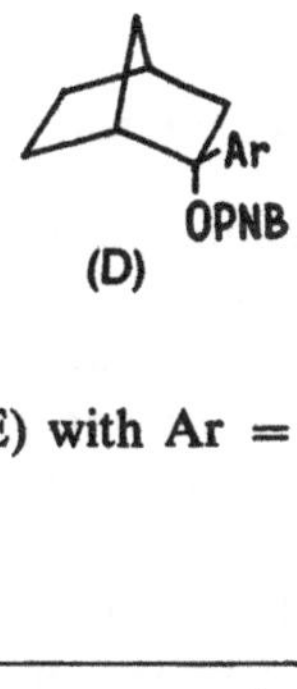
(D)
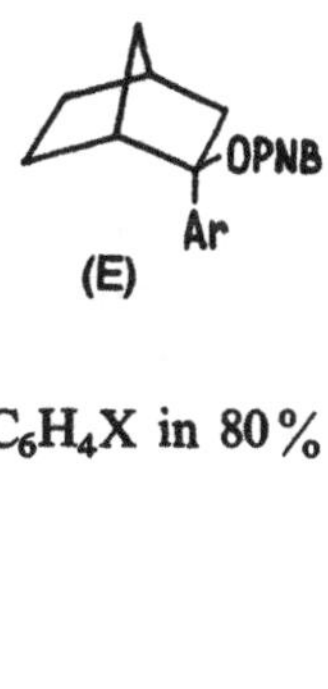
(E)

Relative rates of solvolysis of p-Nitrobenzoates (A–E) with Ar = C_6H_4X in 80% acetone at 25 °C [88].

x	*A*	*B*	*C*	*D*	*E*	exo:endo
4-CH_3O	1.0	505	2.36	0.62	175	284
H	1.0	25000	38.40	5.57	795	143
4-CF_3	1.0	285000	270.00	16.70	3120	188

if we do not know all the factors affecting the value of ϱ^+ in each series. To confirm this conclusion the following examples show close values of ϱ^+, but widely different in reactivity and vice versa [90–92]:

			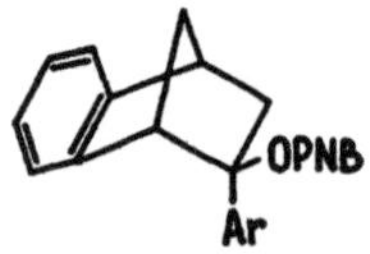
ϱ^+	−3.27	−3.47	−4.50
k_{rel} (Ar = Ph)	10^4	1	1

Recently McManus and Harris [93] have proposed to establish the influence of steric factors on the ϱ^+ value. The method consists in calculating the rate constant of the methyl derivative of the *A* series from the data on ϱ^+ for the series of aryl-substituted analogues of *B* and from the value of γ^+CH_3.

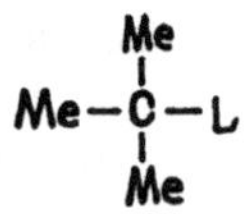

(A)

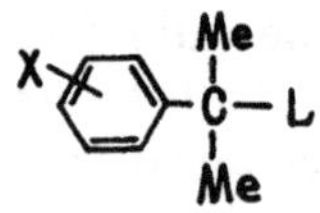

(B)

If the solvolysis of the *B* series is accompanied by specific steric effects — relief of steric strain or steric hindrance to ionization — there will be a discrepancy between experiment and calculation.

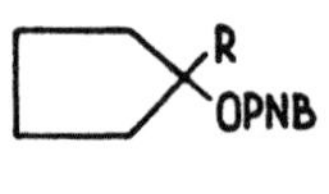

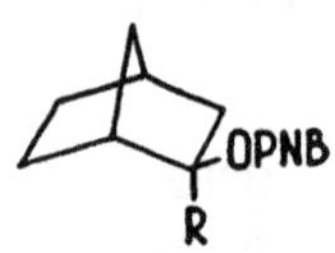

$\lg k_{calc} - \lg k_{exp} = 0.68$ 0.48 1.31

From these examples it is clear that the solvolysis of the exo series does not reveal any complicating steric effects while for the endo series they are rather great.

Brown suggested [13] that all 2-endo-norbornyl derivatives react slowly due to steric hindrances for the leaving group. But such hindrances are not dependent on the nature of the radical R! One should therefore look for another reason for the discrepancies between experimental and calculation results for the endo series. McManus and Harris suggest as a possible explanation the relief of steric strain in the transition state of ionization of endo isomers due to the weakening of non-bonded interaction of aryl ortho-hydrogens with H^1 and $H^{3\text{-exo}}$. There is no such effect for $R = CH_3$.

Thus, steric factors affect the ϱ^+ value only for the endo series; the observed similarity of the ϱ^+ values for the exo and endo series is an incidental result of simultaneous influence of steric and electronic factors. It is wrong to conclude from this similarity of ϱ^+ values that the solvolysis mechanisms in either series are similar as well. [2]

2 The theoretical calculation of the ϱ^+ values from the PMO theory was developed by Jorgensen [94].

Thus, it is not the ϱ^+ value, but the "break" on the Hammett-Brown curve which testifies the changing mechanism and the onset of σ- or π-participation. This criterion, however, is of low sensitivity, it is only applicable to systems with very strong anchimeric assistance. Thus for the series below there is no break on the respective curves though π-participation in the solvolysis of the respective secondary analogues has been established:

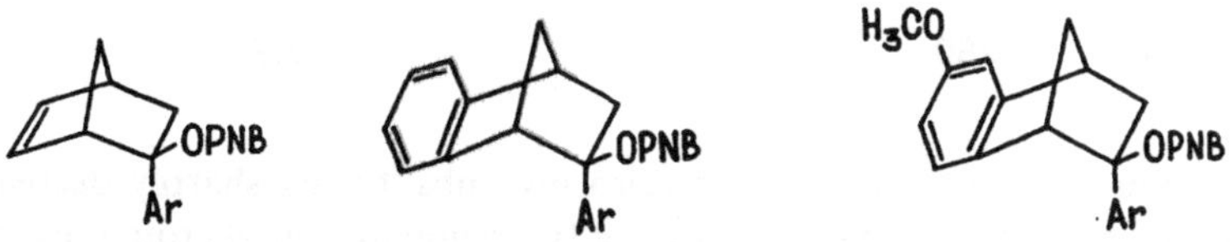

Apparently all aryl groups, even the most electron-seeking, are π-donors and so they decrease the demand of the cation centre for the σ- and π-participation.

In 1975 Peters suggested compare the data on solvolysis of tertiary 2-aryl-2-norbornyl substrates and secondary analogues with the use of so-called group constants γ^+ [95]. With $\gamma^+_H = 2.53$, Brown has shown the data on the solvolysis of secondary and tertiary 2-exo- and 2-endo-substituted norbornanes to correlate well with log k (γ^+ for aryl groups is equivalent to σ^+) without any "break" of the curves (Fig. 3). From this fact Brown has excluded a σ-participation in the solvolysis of secondary 2-exo-norbornanes. Criticizing Peters' method Schleyer notes [13] that the accuracy of determining the rate constants of secondary substrates from the data on tertiary analogues is rather low and the error may amount to 10^2.

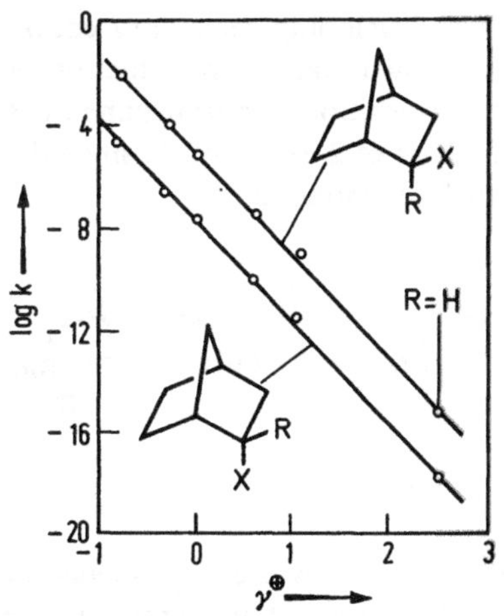

Fig. 3. Correlation of the tertiary 2-norbornyl derivatives with the secondary [13]

All the above approaches have the disadvantage that the demand of the tertiary cation centre in them varies while the σ- and π-participations must manifest themselves to a much greater extent upon formation of secondary and primary carbocations. In the examples below (not too numerous so far) the demand of the secondary cation centre C^2 is varied by substituents at C^3 located symmetrically relative to the bonds C^2-exo-O(Ts) and C^2-endo-O(Ts) [96,97]:

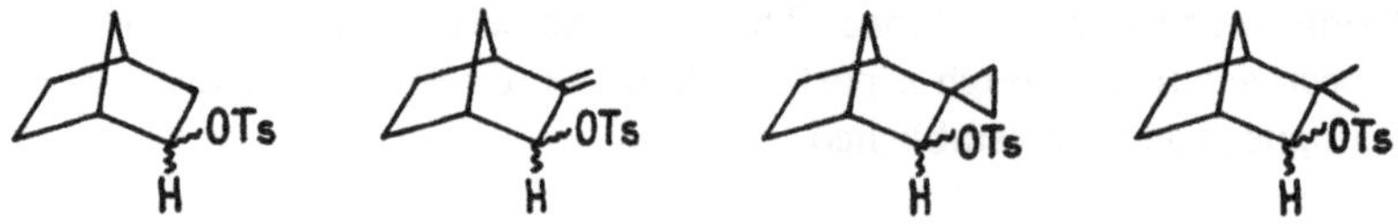

Exo : endo 280 3.9 3.2 1240

Introduction to C^3 of electron-releasing substituents sharply decreases the exo: endo rate ratio; there is no such effect if two geminal methyl groups are introduced to C^3. The value ~3 for the exo:endo rate ratio resembles greatly the usual epimeric rate ratio for secondary substrates without anchimeric assistance (see above). Introduction of an electron-releasing substituent to C^3 levels the σ-participation in the solvolysis of the secondary 2-exo-epimer, but it does not affect the rate ratio of the tertiary analogues for which the solvolysis rate is predominantly determined by steric factors, for example [98,99]:

Exo : endo 885 831 127 357

A different method of varying the demand of the developing secondary cation centre has been propsed by Lambert [100–102]. He has shown that the introduction of an electron-seeking tosyloxy- or acetoxy group to the 3-position in norbornane does not lead to an increase of σ-participation in the solvolysis of 2-exo-tosylate since the epimeric rate ratio for diesters is even lower than that for monoesters:

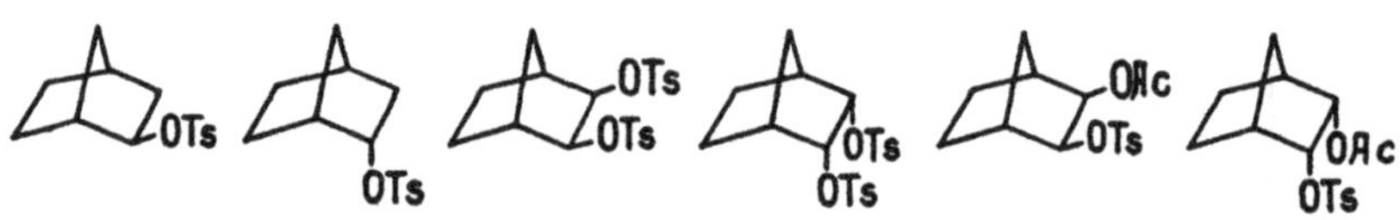

Exo : endo 280 48 11

Using Lambert's data Brown [13] has pictured Fig. 4 to show the dependence of log (k_{exo}/k_{endo}) on the log k endo-norbornyl. This graph shows that an increase in the demand of the developing cation centre at C^2 does not result in an increase in σ-participation. Reproducing Brown's graph Lambert [102] is of opinion that the graph implies a complete absence of σ-participation in the solvolysis of unsubstituted 2-exo-norbornyl tosylate.

Criticizing Lambert's approach, Schleyer [13] introduces acyloxy groups to the 3-position to bring in different steric effects for the C^2-exo-X and C^2-endo-X bonds.

Also, the dipolar effect of the 3-tosyloxy group adjacent to the reaction site introduces serious complications [103]. As for Fig. 4, it would have been greatly changed (and so would the conclusions drawn from it), if one had included the data for compounds *59* and *60* and other similar systems and excluded those for *61* and *62*.

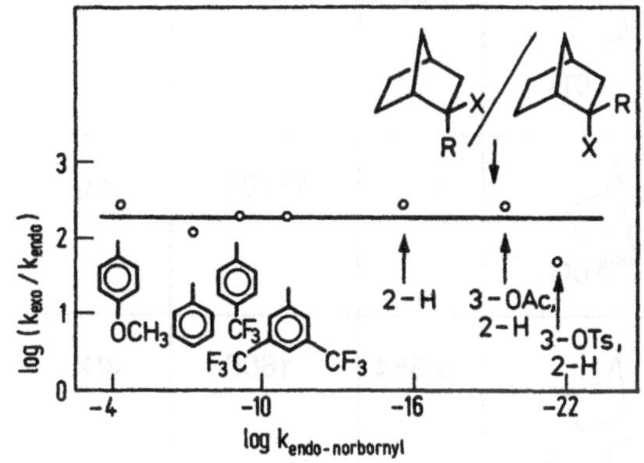

Fig. 4. Effect of enhanced electron demand in the 2-norbornyl system on exo:endo rate ratio [13]

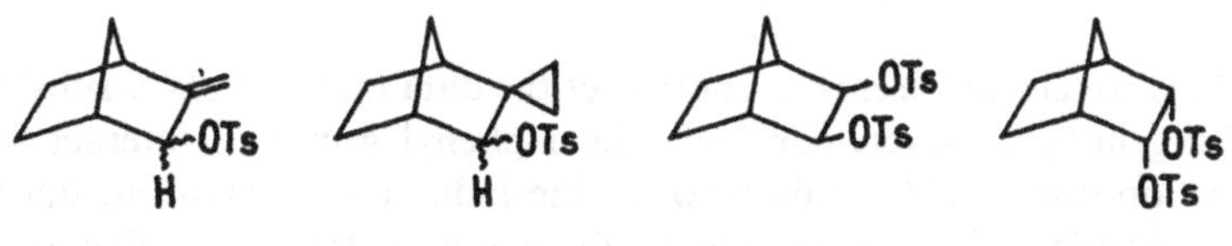

	59	*60*	*61*	*62*
Exo : endo	3.9	3.2		

To elucidate the formation of the nonclassical 2-norbornyl ion *5* in secondary systems the effect of the substituents at C^1, C^6 and C^7 on the solvolysis rate of 2-norbornyl derivatives [2,12,104–111] was widely studied. Since on solvolysis of 2-exo- (and not 2-endo-)norbornyl brosylate *22* the transition state will resemble the nonclassical ion *5* the addition of electron-releasing substituents to the above atoms must enhance the exo:endo rate ratio. As seen from Table 7, the introduction of a methyl group into the 1-position results in a drastic increase of this ratio from 280 to 12700.

Traylor has shown [111] that the addition to C^1 of the substituent $CH_2Sn(CH_3)_3$ isoinductive with the methyl group, but capable of σ, π-conjugation sharply accelerates solvolysis of exo-epimer (by a factor of 10^5), while the exo:endo rate ratio reaches the tremendous value of 10^9! According to Traylor the facts observed agree with the delocalization of the σ-electrons from the C^1—C^6 bond in the transition state of ionization of 2-exo-isomers and does by no means support Brown's concept of steric hindrance to ionization.

The solvolysis rate of 1-phenyl-2-exo-norbornyl tosylate *65* is only 3.9 times as high as that of compound *63*. This has been interpreted to indicate that only a small part of the positive charge in the transition state of ionization is transferred to C^1 [4].

Table 7. Solvolysis Rates of 1-Substituted 2-Norbornyl Tosylates

Compound	Relative rate	Compound	Relative rate	Exo : endo rate ratio	References
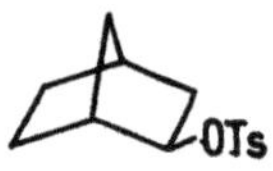 63	1.00	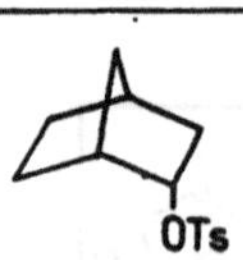	1.00	280	40)
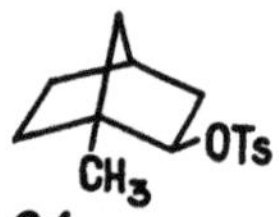 64	51.2		1.13	12700	104)
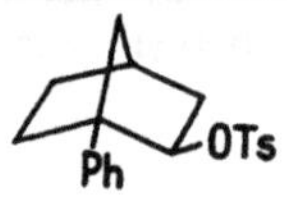 65	3.91	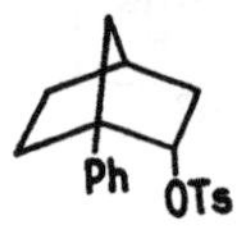	0.684	1600	104)

This conclusion seems to contradict the data on the 1-alkyl-substituted norbornanes. Sargent [2], however, considers the 1-phenyl group to interact with the developing cation centre of C^1 only through the inductive mechanism, but without any mesomeric effect. The phenyl group fails to exert a stabilizing effect on the transition state of ionization of 2-exo-norbornyl derivatives [65, 72, 112]. A similar weak effect of phenyl groups on the kinetics of the "π-route" to form norbornyl compounds has been discovered by Wu and Thornton [113].

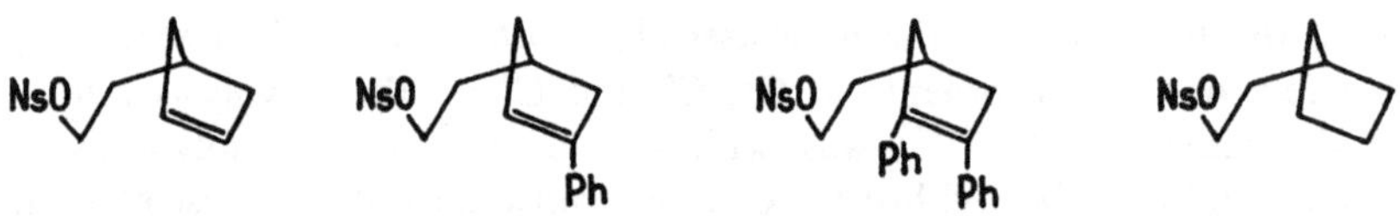

$k_{rel} = 1.0$ 1.70 0.87 0.011

The considerable accelerating effect of the double bond as a consequence of the π-participation is by no means appreciably enhanced by introducing one or two phenyl groups into conjugation with a double bond. The authors believe this to preclude both the formation of an intermediate classical cation and the specific steric effects of phenyl groups.

There is an opinion widespread among organic chemists that the stabilization of a cation centre by a substituent is largely proportional to the charge developing at the site of substitution; therefore, the above examples led to the conclusion about the absence of any significant charge on C^1 in the transition state of exo epimer ionization [4]. In reality, however, as convincingly shown by Wilcox and co-workers [112], these examples are not only far from contradicting the σ-participation and formation of bridged ions, but are evidently typical for nonclassical ions. In the latter the resonance effect of substituents amounts to no more than 10% of the value observed in π-delocalized cations from alternant hydrocarbons, e.g., the allylic ion. In the PMO theory this effect can be accounted for by both a decreasing resonance integral between $HOMO_{Ar}$ — $LUMO_{C^+}$ overlap in σ-delocalized ions and an increasing energy gap between these orbitals.[3]

The effects of substituents in nonclassical ions point to another peculiarity. Since σ- and π-participations are associated with the motion of nuclei in the transition state (as distinct from hyperconjugation), the introduction of substituents even into electroneutral positions of the substrate can sterically inhibit these effects. This is clearly seem from the following examples [115,116]:

(25 °C)
Exo : endo 280 9.5 885 420

Introduction of the cyclopentane bridge to C^5 and C^6 does not affect the exo:endo rate ratio of tertiaty substrate to the extent of the secondary pair of epimers; this is due to an increase in the non-bonded interactions in the transition state of solvolysis of exo epimer as C^6 moves to C^2. Here the introduction of the cyclopentane ring does not involve any steric complications, polar effects or hindrances for departing of the substituent from the 2-exo- or 2-endo-positions. Nor do the calculations by the molecular mechanics reveal any unusual strains from the introduction of a trimethylene bridge.

In 1978 Harris [117] and Brown [118], independently concluded that the solvolysis of 2-endo-norbornyl brosylate proceeds mainly not by route k_s with nucleophilic solvent

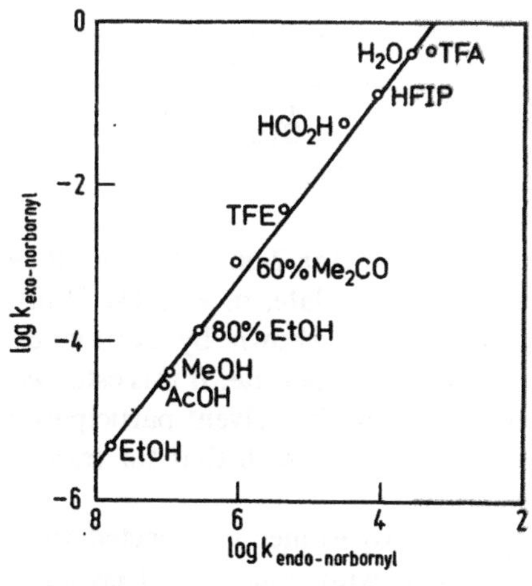

Fig. 5. Linear free energy relationship between the rates of solvolysis in various solvents of exo- and endo-norbornyl derivatives [118]

3 Dewar notes [114] that phenyl groups in basal positions of π-complexes must not exhibit any mesomeric stabilizing effect.

assistance, but by route k_c, i.e. without any nucleophilic assistance of solvent or neighbouring group. Brown [118] also revealed a linear correlation between the solvolysis rate constants of 2-exo- and 2-endo-norbornyl tosylates in a wide range of media (Fig. 5); this allows the conclusion that the solvolysis of exo epimer occurs by route k_c as well.

Lately Brown investigated the exo:endo solvolysis rate ratio of 2-norbornyl chlorides in aqueous acetone with widely varied water content [119]. This work included linear correlations of log k_{exo} on log k_{endo}, as well as those of log k_{exo} and log k_{endo} on log k (tert-BuCl); the latter is a model k_c process. The data obtained agree with the assumptions that the solvolysis of exo and endo epimers are either k_Δ and k_c processes, respectively, or they are both processes of type k_c; the results were found to be independent of the leaving group and hence of F-strain (cf. [120]). According to Brown [119], introducing the α-CH_3 group to epimeric 2-norbornyl chlorides results in about the same kinetic effect; the solvolysis of exo- and endo-norbornyl derivatives, both secondary and tertiary, proceeds by route k_c.

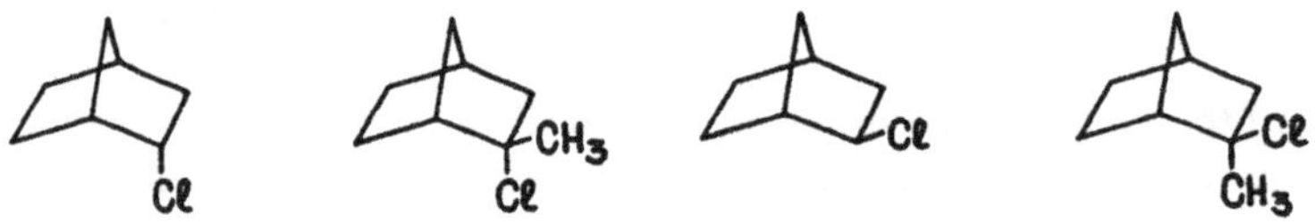

k_{rel} = 1.0 53000 1.0 55000

However, the relatively weak nucleophilic solvent assistance in the solvolysis of 2-endo-norbornyl sulphonate is corroborated, firstly, by the formation of about 8% of optically active 2-exo-acetate *66*; secondly, by a significant decrease in the solvolysis rate of 2-endo-tosylate on introducing 3-exo-substituents shielding the backside approach of solvent molecules [103].

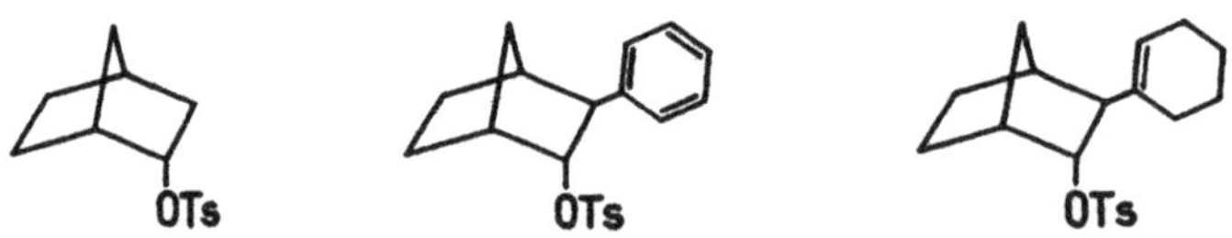

k_{rel} (25 °C) = 1.0 1/260 1/11.5

The value of m from the Winstein-Grünwald equation (for 2-endo-tosylate 0.69, i.e. between 2-adamantyl tosylate, 0.91, and isopropyl tosylate, m = 0.44) [15] also points to some participation of the solvent. Finally, as shown by Nordlander et al. [121], $k_s:k_c$ = 30 for the solvolysis of 2-endo-norbornyl tosylate. If this ester was solvolyzed with considerable steric hindrance to ionization the solvent participation would be unusually great while the ratio observed coincides with that for trans-2-methylcyclopentyl tosylate ($k_s:k_c$ = 30).

For the linear correlation log k_{exo} — log k_{endo} the exo isomer is expected to be solvolyzed by route k_Δ and the endo isomer — by route k_c with weak solvent assistance since either process depends mainly on the ionizing power of the medium rather than on its nucleophility.

3.4 Conclusions from the Kinetic Data [21)]

3.4.1 Solvolysis of 2-Endo-Norbornyl Derivatives

The rate solvolysis cannot be considered, as does Brown, abnormally low for the following reasons:

a) replacing the endo-5-H by a bulkier CH_3 group does not essentially affect the rate of acetolysis (it remains unclear, however, why the replacing the endo-6-H by a CH_3 group reduces the rate of solvolysis of 2-endo-norbornyl tosylate 16-fold);

b) the rate of acetolysis of 2-endo-norbornyl brosylate can be calculated in a fairly good approximation by semiempirical methods only from steric effects;

c) the exo : endo rate ratio for tertiary norbornyl systems is well predicted only from the long-range interactions in the transition state.

3.4.2 Solvolysis of 2-Exo-Norbornyl Derivatives

In contrast to Brown's assertions and in accord with Winstein's and Trifan's assumption, the solvolysis of these secondary systems proceeds with anchimeric acceleration. This is concluded from the following facts: a) the exo : endo rate ratio for 2-norbornyl systems is $\sim 10^2$–10^3; as the reaction rate of the endo isomer is not anomalous (see above), hende the exo isomer reacts at an elevated rate; b) the rate of solvolysis of exo isomers is 10^2 to 10^3 times as high as that calculated according to the semiempirical scheme from only steric effects; c) the ratio of the reaction rate of secondary 2-exo-norbornyl systems to the solvolysis rate of secondary cyclopentyl analogues is 100 times as great as that of tert-2-exo-norbornyl derivatives and tert-cyclopentyl analogues; since tert-2-norbornyl derivatives are solvolyzed without anchimeric assistance, the factor of 100 characterizes tentatively the amount of anchimeric assistance in the secondary 2-exo-norbornyl systems; d) exo- and endo-6-substituents decrease the solvolysis rate of 2-exo-norbornyl tosylate; this cannot be accounted for without participation of the electrons of the 1,6 bond; in the transition state their participation increases the non-bonded interaction due to a decrease in the C^2–C^6 distance.

3.4.3 Solvolysis of Tertiary 2-Norbornyl Systems

The exo : endo rate ratio can be correctly predicted without due regard for anchimeric assistance; this points to its absence. Hence the initial product to ionize these systems is the classical tertiary 2-norbornyl cation. This conclusion, however, does not preclude the possibility of the subsequent rearrangement of the classical cation into a nonclassical one, nor does it give any evidence of the relative stability of classical and nonclassical tertiary 2-norbornyl cations.

3.4.4 Structure of the 2-Norbornyl Cation

Anchimeric assistance to ionization is only observed when the adjacent group participates by formation of an intermediate more stable than the one without such participation. Consequently, the fact of anchimeric assistance in the solvolysis of 2-exo-norbornyl derivatives is a convincing evidence for the formation of the nonclassical ion *5* as an intermediate in this reaction.

Jensen and Smart [122] showed the benzoylation of exo-phenylnorbornane *67* to occur faster than that of endo isomer *68*:

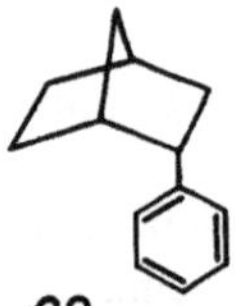

68

$k_{rel} = 1.00$

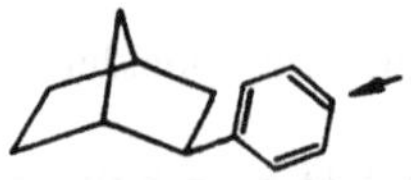

67

1.57

A similar effect has been discovered by Brown [123] who has recorded a higher rate of solvolysis of p-exo-norbornyl-tert-cumyl chloride *69* in comparison with p-endo isomer *70*:

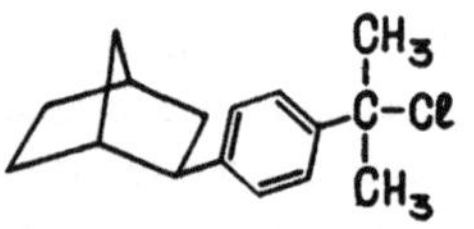

69

$k_{rel} = 25.2$

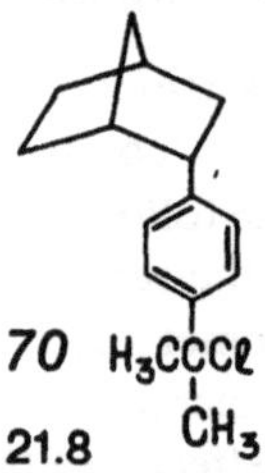

70

21.8

According to Jensen and Smart, the elevated p-character of the C—C bonds of norbornane increases their capacity for hyperconjugation with the developing p-orbital. On solvolysis of 2-exo derivatives the latter is located parallel to the C^1—C^6 bond which is an optimum one for hyperconjugation [124]. At the same time on solvolysis of 2-endo isomers the developing p-orbital is located less favourably for the C—C hyperconjugation with the C^1—C^7 bond so no overlap with the C^1—C^6 bond is possible. For this reason the exo isomer must react faster than the endo isomer.

Similar views are developed by Traylor and co-workers [125–127]. According to them the developing cation centre in the transition state is stabilized via strained σ-bonds without changing the substrate geometry, i.e., by the vertical route *IV*. Such an effect is greatly different both from σ-participation in the nonclassical ion theory where the participating group moves towards the reaction centre as the transition state *III* is reached and from the theory of frangomeric acceleration in which the participating group moves in the same situation from the reaction centre *II* [128]. Both kinds of stabilization are "nonvertical"; at the same time situations *II–IV* are quite different from the solvolysis process without any participation in the transition state *I* (Brown [3,4]).

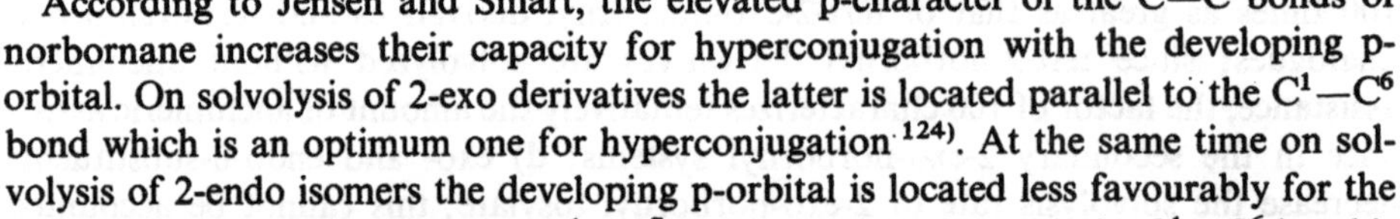

(I) (II) (III) (IV)

Traylor and co-workers have obtained a good correlation between the solvolysis rates of R—CH_2—X compounds where R is bicycloalkyl (Eq. 1) and the position of the "charge-transfer" bands in the electron spectra of the R—C_6H_5 · TCNE complexes (Eq. 2).

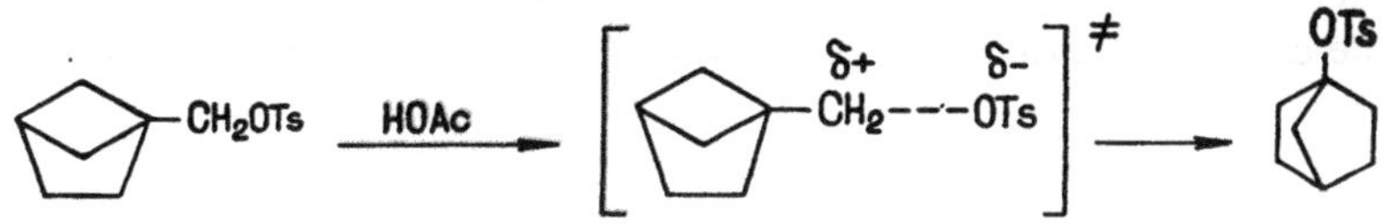

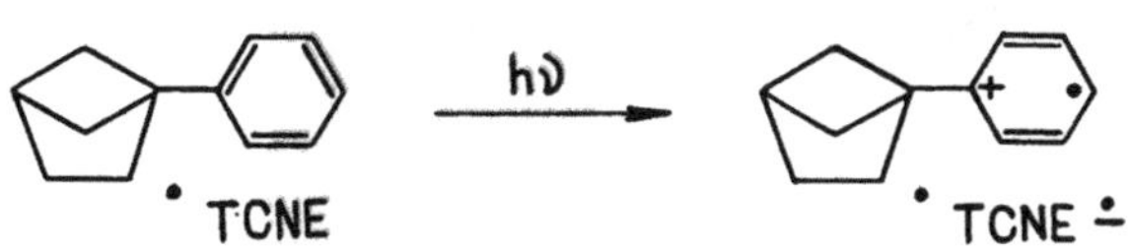

From such a correlation the authors conclude the acceleration in the solfolysis to be mainly due to a vertical electronic effect: the σ, π-conjugation of one or several strained bonds without changing their lengths and location. The authors do not deny the role of nonclassical ions and the possibility of equilibrium of the type

In 1976 Kochi et al. [128a] determined the ionization potential of exo- and endo-2-norbornyl methylmercury and showed the exo isomer to be ionized much easier (by 4.4 kcal/mole) than the endo isomer. Photoionization is known to be a vertical process insensitive to steric effects. The authors believe the data to corroborate well the σ-participation in the transition states of cation reactions of 2-exo-norbornyl derivatives.

3.4.5 Conclusion from the Discussion

Comparing Traylor's ideas with those of Winstein and Olah, on the one hand, and with those of Brown, on the other, we are led to the following conclusions.

1. *According to Traylor and Winstein* the σ-bond participating in the transition state must be coplanar and opposed to the direction of the bond to be broken. Solvolysis of 2-exo-norbornyl derivatives occurs at a rate higher than predicted from steric effects only. *According to Brown,* the solvolysis of 2-exo-norbornyl tosylate occurs at a normal rate without any resonance stabilization [3,4] both in the bicyclic cations themselves and in the preceding transition states.

2. Both Winstein's ion *5* and Traylor's ion *71* can for stereoelectronic reasons be attacked on C^2 only from the exo side. In ion *71* this is due to the microscopic reversibility; the addition of a nucleophile anti to hyperconjugated σ-bond is by far more favourable than the syn-addition in conformity with the principle of trans-addition to multiple bonds. Both ions *5* and *71* are nonclassical since the positive charge is

delocalized 2-electron 3-centre bonding (2e 3C); the carbon atoms participating in the delocalization of the positive charge are penta- and tetra-coordinated. Ion *5* differs from ion *71* in its geometry; secondly, ion *5* is homoaromatic (see p. 98) while ion *71* is not. Finally, in ion *5* atomic orbitals interact with atoms uncombined with the σ-bond (C^2 and C^6), in ion *71* this is not the case. In Brown's ion *6* the nucleophile can attack either lobe of the vacant p-orbital, the ratio of the products must

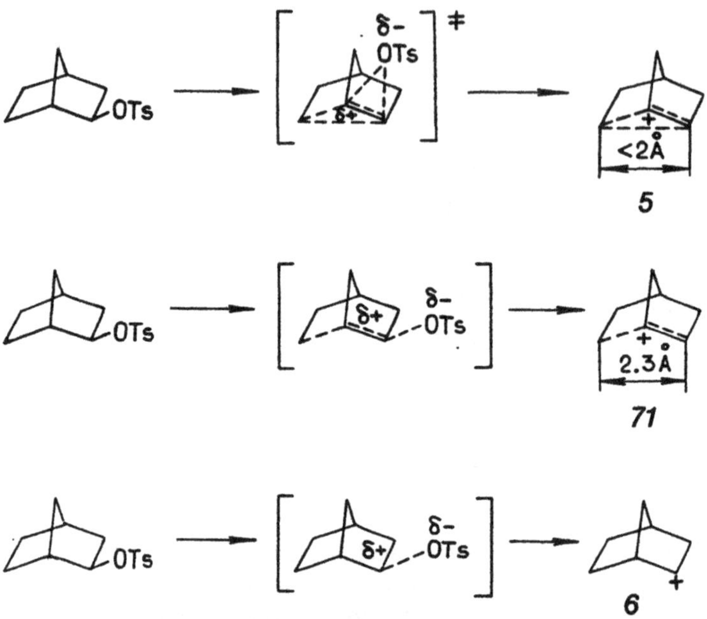

be determined by steric (not stereoelectronic) factors. To explain the exceptional exo-addition to ion *6* Brown advanced a hypothesis about a rapid migration of the σ-bond C^1—C^6 (see below).

3. Traylor's ion *71* does not correspond to the evidence of the complete equivalence of the C^1 and C^2 atoms in the intermediate under solvolysis conditions. To explain the equivalent of the C^1 and C^2 positions in Traylor's ion it is necessary to assume a very rapid equilibrium of ions (Traylor himself writes nothing about it):

This equilibrium is in a way similar to that of classical ions suggested by Brown:

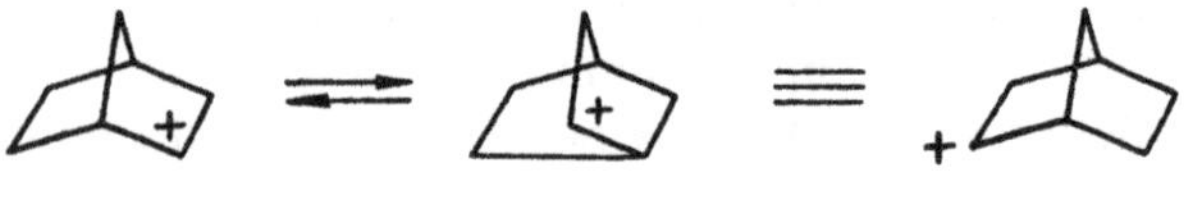

The difference is that in the latter the idea of the very rapid equilibrium is necessary to account for both the equivalence of the C^1- and C^2-positions and the exclusive exo-stereospecificity in the nucleophile attack.

^1H, ^{13}C NMR, Raman and ESCA spectra the long-lived 2-norbornyl ion corroborate reliably its structure as that of corner-protonated cyclopropane (see Chapter 3.8).

Finally, the *ab initio* calculations made in 1971–72 by Pople and co-workers [129] clearly confirm the ions of type *5* to be more stable than those of type *71* since the former are homoaromatic, the latter are not; also most of the strain energy of the norbornane molecule is due to the distortion of normal angles. In the classical ion *6* and in Traylor's ion *71* with unchanged geometry this strain is mainly retained; whereas ion *5* does not preclude the relief of angular strain owing to rather long (1.803 Å) C—C bonds in corner-protonated cyclopropane.

Traylor's data make it possible to judge which groups can participate in stabilizing the cation entre by the "vertical" route. As for the correlation with solvolysis rates it may be (see Ref. 10 in [126]) a consequence of a casual proportionality between the nonvertical stabilization in the solvolysis process and the vertical stabilization in the charge transfer process. It is all the more impossible to deny — and the author himself admits it — that the acceleration of 2-exo-norbornyl tosylate solvolysis is due to both vertical and nonvertical processes.

Hoffman's [130] theoretical investigations show that there is no contradiction between σ-participation and hyperconjugation. The great variation in nuclear motion depends on the systems. With 3-nortricyclyl ions Olah has proved the nonvertical stabilization to be more significant than the vertical hyperconjugation; this is cleary seen from the ^{13}C NMR spectra [131]

$\delta_{\overset{+}{C}} = 263.0$

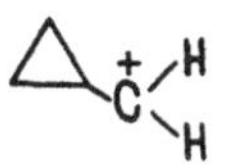

$\delta_{\overset{+}{C}} = 54.9$

Due to the skeleton rigidity in the 3-nortricyclyl cations the σ-delocalization between the C—C bonds of the cyclopropane ring and the vacant p-orbital is highly restricted. This example can be regarded as charge delocalization with minimum motion of nuclei.

Olah [15] emphasizes that genuinely nonclassical and classical ions are nothing but "extremes of a continuum". The movement of nuclei accompanying the bond delocalization can be small (or completely absent: hyperconjugation), partial (σ-participation) or more intense (bridging) leading to the formation of σ-delocalized ions.

3.4.6 Deuterium Isotope Effects

Many works are devoted to revealing and discussing secondary α-, β- and γ-isotope effects of deuterium in solvolysis of the norbornyl cation [132]. Murr and Lee [133–136] have shown that on acetolysis of 2-d-exo-norbornyl brosylate *72* the α-isotope effect observed is lower than that with the 2-d-endo-norbornyl brosylate. The authors attributed it to the fact that the exo-isomer is acetolyzed with anchimeric assistance imparting to the reaction on C^2 a partial S_N2 character. The S_N2 reactions are known to display a negligible α-deuterium isotope effect:

These data, as well as the equality of α-isotope effects in the acetolysis and ethanolysis are considered by the authors to clearly testify to the formation of an intermediate nonclassical norbornyl ion in these reactions. The value of k_H/k_D (25 °C) for the endo isomer is independent of the solvent and equal to 1.20 which is typical for the solvolysis of simple secondary sulphonates. This argument favours the absence of steric hindrance to ionization of 2-endo brosylate *23*.

Collins and Bowman [137], however, have pointed out that the low value of the α-effect of the exo isomer is due to intramolecular rearrangement resulting in the distribution of deuterium between the 1- and 2-positions. If the isotope effect is determined at the moment when the reaction with compound *72* has only passed 3–5% of the way it turns out to be practically the same (1.208 ± 0.09; 25 °C) as for

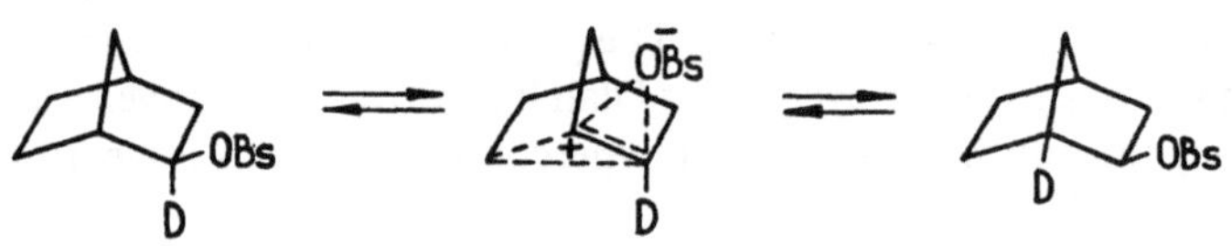

72

the endo isomer; but if the isotopic effect is measured at the 30–50% stage of the reaction, then, according to Lee's data it is far lower — 1.106 at 25 °C. Thus according to Collins and Bowman, the value of the α-effect does not depend on the stereochemistry of the leaving group. They consider the $C^1—C^6$-bond to participate from so large a distance that this does not significantly disturb the C—H(C—D) bond oscillations.

On the other hand, when the 2-norbornyl cation is formed from its Δ^3-cyclopentenyl-ethyl precursor ("π-route"), the more "flexible" structure of the precursor makes it possible for the participating group (double bond) to move closer to the reaction centre which results in a decreasing α-effect for the exo isomer.

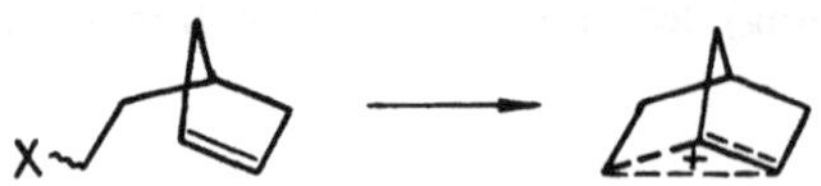

As can be seen from the data below [138–141], the β-isotope effect is rather small on solvolysis of the secondary exo brosylate *73* i.e. just when the nonclassical ion is formed; for tertiary exo and endo epimers the β-isotope effect is practically the same. Of the two deuterium β-atoms the 3-endo-D atom exerts no isotope effect on solvolysis of 2-exo-norbornyl brosylate while the 3-exo-D atom produces a significant effect $k_H/k_D = 1.11$. In the nonclassical ion the 3-exo-H(D) atom is situated ideally for interacting with the vacant p-orbital of C^2 while 3-endo-H(D) has a dihedral angle of 60° with the p-orbital axis. If the ionization of 2-exo-norbornyl brosylate resulted in a classical ion, then the effects of 3-exo- and 3-endo-deuterium would be equal.

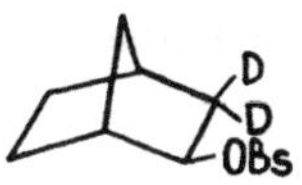

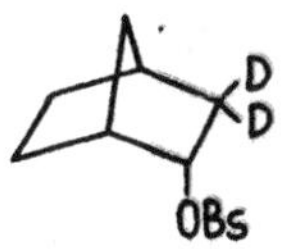

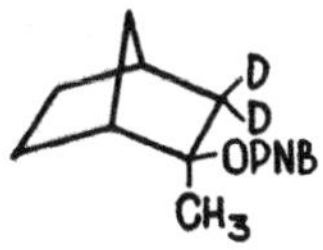

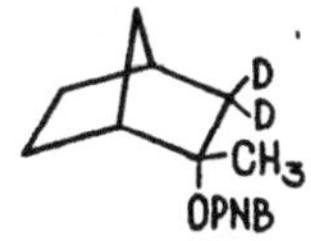

73

$k_H/k_D = 1.10$ 1.26 1.33 1.31

Murr has noted that the γ-isotope effect for compounds *74* and *75* is practically the same [142]; even if ionization of 2-endo-norbornyl brosylate does involve steric hindrances they are not effective enough to affect the value of k_H/k_D during solvolysis of compound *75*.

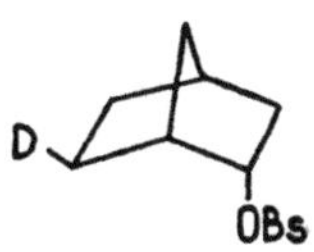

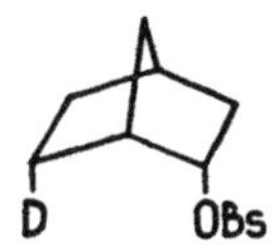

74 *75*

$k_H/k_D = 0.98$ 0.99

3.5 Stereochemical Data

Brown has questioned whether the spatial structure of the solvolysis products of 2-exo-norbornyl brosylate is a proof of the intermediate nonclassical ion [59]. He believes the classical 2-norbornyl cation to yield, when attacked by the nucleophile, only exo compounds for two resons:

a) steric hindrances to the endo attack due to the peculiarities of the norbornane skeleton;

b) rapid equilibrium of enantiomeric classical ions due to the oscillation of C^6 and its bonding pair of electrons between C^1 and C^2 prevents the accumulation of the solvent molecules from the endo side and thereby favours an exo attack. This effect was termed "windshield wiper effect".

3.5.1 Steric Hindrances to the Endo Attack

In the unsubstituted classical 2-norbornyl cation the steric hindrances to the endo attack must yield the same exo products as the attack from the backside of a nonclassical cation 5; the formation of exo products cannot, therefore, serve as a key to the problem of the intermediates for an unsubstituted 2-norbornyl cation. In 1976 Brown has shown the high exo-stereospecificity in the most diverse reactions of norbornane compounds, among them radical reactions, deuterium exchange, cycloadditions, etc. Berson [6)] has drawn the attention to other systems where this ambiguity is excluded. With the reduction of norbornanones-2 he has shown that a methyl group at C^7 directs the nucleophilic attack on the sp^2—C^2 atom into its opposite: steric reasons make the endo direction more favourable. Hence a nucleophilic attack

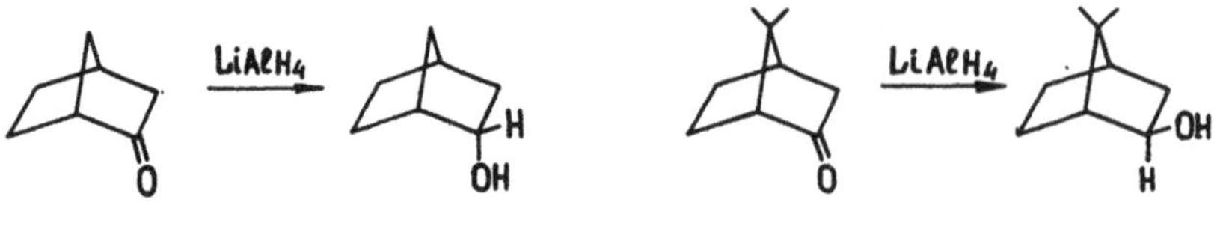

on a classical cation must yield endo products. In reality, however, solvolysis results only in exo products [60, 143–147)], for example:

These data support the assumption about an intermediate nonclassical ion.

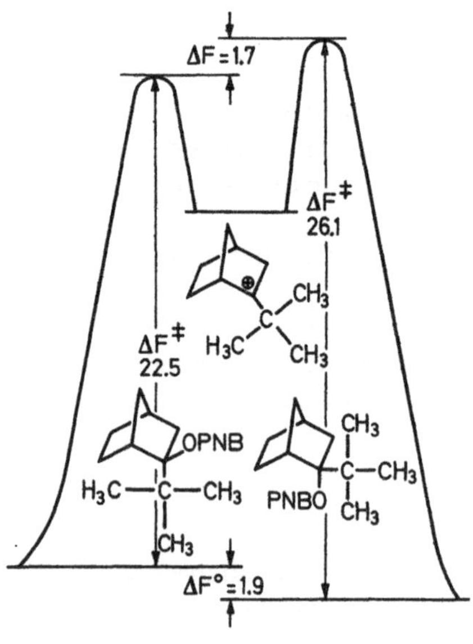

Fig. 6. Free energy diagram for the solvolysis of 2-tert-butyl-2-norbornyl p-nitrobenzoates in 80% aqueous acetone at 25 °C [13)]

An uncommonly high content of endo isomer has been recorded by Brown on solvolysis of 2-tert-butyl-2-exo-norbornyl-p-nitrobenzoate (exo:endo product ratio is 95:5)[148]. The Goering-Schewene diagram (Fig. 6) allows to recognize a higher stability of the ground state of the endo isomer (by 1.9 kcal/mole) and a small difference in transition state energies (1.7 kcal/mole).

3.5.2 "The Windshield Wiper Effect"

Schleyer and co-workers have shown[149] that the ions formed from 2-p-anisyl- and 1,2-di-p-anisyl-endo-2-norbornanols have the same ultraviolet spectra (H_2SO_4) which they believe to speak in favour of the rapid equilibration of classical cations *I* ⇄ *II*; if the developing ion had a nonclassical structure *III* then both aromatic nuclei in case (b) would be simultaneously conjugated with cation centres. However, the simultaneous conjugation of the two aromatic nuclei is impossible for steric reaons [2] since in the case of the coplanar arrangement of the two aromatic nuclei the ortho hydrogen of one nucleus must overlap that of the other by 0.4 Å (according to models). Therefore in a nonclassical ion the effective conjugation of the two aromatic rings will be decreased and the electronic absorption spectrum may accidentally coincide withh that of the monoaryl-substituted ion.

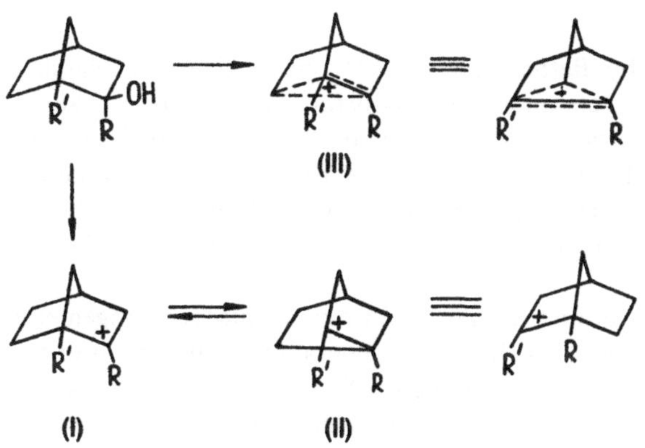

a) R = n – An, R' = H

b) R = R' = n – An

Discussing Schleyer's work Sargent calls attention to the following. The π-conjugation between $C^1(C^2)$ and the aromatic ring for a nonclassical ion is only possible if the latter has the structure of Dewar's π-complex; in Walsh's model, however (see p. 6), implying the sp^3 hybridization of three carbon atoms there are no pure p-orbitals for π-bonding with the aromatic ring. Therefore Sargent[2] holds that in this model neither mono- nor diaryl derivatives can have conjugated nuclei with the positive charge centre; this accounts for the similarity of the electron absorption spectra.

Thus, Schleyer's data cannot convince of the rapid equilibrium of classical 2-norbornyl ions. The tertiary carbocation having a p-anisyl group as a substituent can be assumed to be classical; however, the data reported do not preclude a nonclassical structure either.

Corey and co-workers [150] have shown that solvolysis of optically active 2-exo-norbornyl m-carboxybenzenesulphonate yields a racemic product. If the reaction proceeded via a classical ion, then since the interaction between the carbocation and the carboxylate-anion must be high compared with the Wagner-Meerwein rearrangement (this is not always true, see [151]), some optical activity would be retained. But the solvolysis via a classical ion implies such a fast interconversion of classical ions that Corey, for chemical purposes, takes the ion as symmetrical.

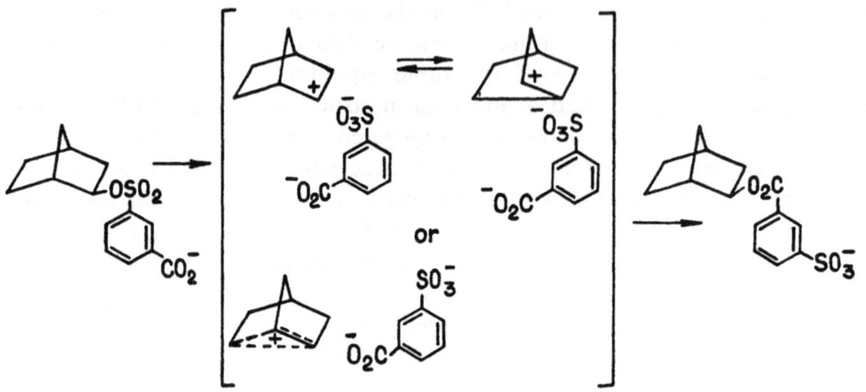

If a reaction, as pointed out by Bethel and Gold [11], proceeds via a classical 2-norbornyl ion, then the rate of conversion of the initially formed ion into an enantiomer must exceed the rate of its interaction with the solvent by more than 2000. The latter interaction is characterized by a constant of about 10^9–10^{10} sec^{-1}; hence the interaction rate constant of classical ions must be of the order of 10^{13} sec^{-1}. Such a rate constant can only belong to reactions whose free activation energy is close to zero.

Having applied the quantum relaxation theory to the intramolecular rearrangements in carbonium ions Fong [152] has concluded that the degenerate rearrangement of the 2-norbornyl cation may occur with a constant of about 10^{12} sec^{-1}. In the 2-norbornyl cation with its almost spherical symmetry the Wagner-Meerwein rearrangement can occur without changing the position of the surrounding solvent molecules, with the minimum motion of the ion nuclei which decreases the activation barrier. The more a substituent causes a deviation from the spherical symmetry the higher becomes the Wagner-Meerwein rearrangement barrier. Brown [21] notes that the value of 10^{12} sec^{-1} is obviously overstated since the solvation of cations must decrease the rearrangement rate as compared to the gas phase. Fong's ideas have been criticized by Olah [15]; his critical remarks will be considered in Section 3.7.

3.5.3 6,2-Hydride Shift

The 6,2(6,1)-hydride shift rarely occurring in the chemistry of acyclic compounds seems to be common in that of carbocations having a bicyclo[2.2.1]heptyl skeleton.

Acetolysis of 7-syn-*76* and 7-anti-chloro-2-exo-norbornyl tosylate *77* yields identical mixtures (1:1):

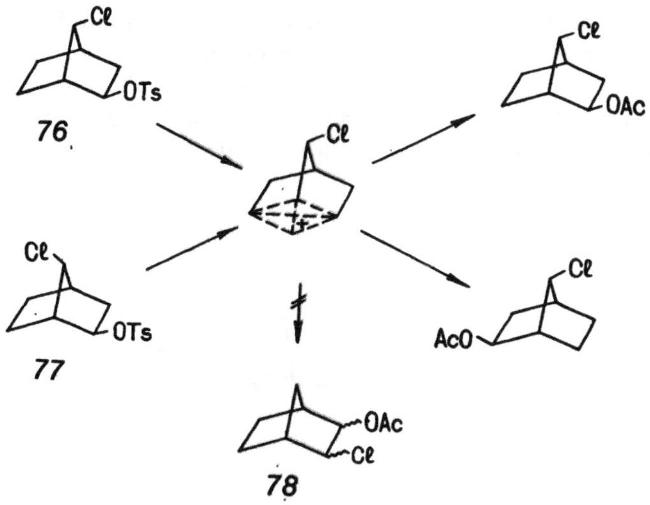

These data have been interpreted by Roberts [153] as an argument to favour the participation of the nonclassical nortricyclonium ion. The absence of 3-chloro-2-norbornyl acetate *78* may result from a smaller positive charge on C^1 than on C^2 and C^6 due to the effect of the adjacent dipole C—Cl. The larger portion of the products formed by the 6,2(6,1)-hydride shift in chloro-substituted, as compared with unsubstituted, 2-norbornyl ion is attributed to a longer lifetime of the chloro-substituted ion due to unfavourable dipole-dipole interactions of C—Cl and the solvent (nucleophile).

Goering [154], however, has shown the ratio of chloroacetates formed from isomeric tosylates *76* and *77* to be different. Therefore, he has suggested, that solvolysis yields isomeric nonclassical ions which turn into one another at a rate comparable to that of capture by the solvent.

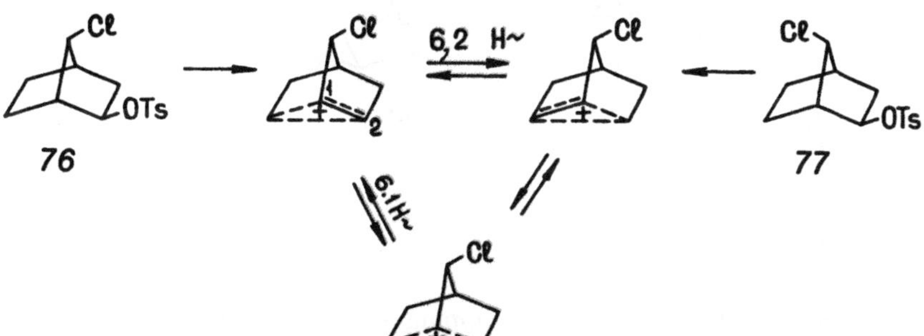

Nortricyclonium ions protonated over the edge and the face cannot be the primary intermediates since otherwise they would be common for both isomers *76* and *77*:

The rate ratio of the 6,2-hydride shift and of the ion capture by the nucleophile depends greatly on the nature of the substrate and on the reaction conditions. Thus, apoisobornyl brosylate *79* and exo-camphenyl brosylate *80* yield the same relative amount of β-phenchocamphoryl product *A* in a wide range of media (~50% in CH_3COOH; somewhat less in more nucleophilic media) [60].

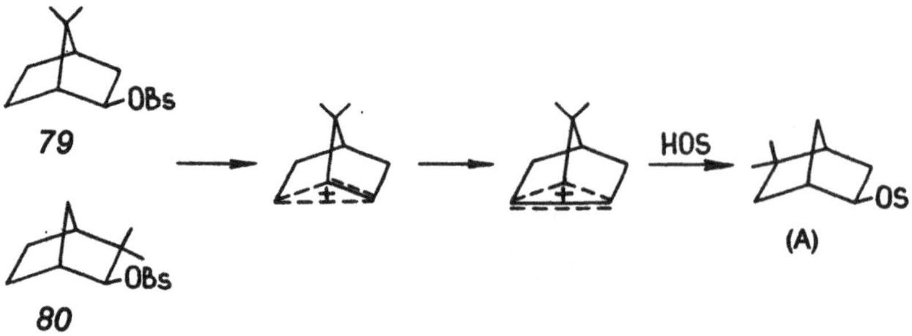

At the same time if alcohols *81* and *82* are treated at 0 °C with thionyl chloride in such a poorly ionizing medium as pentane, they yield practically only apoisobornyl chloride *83* (98%); this results from the Wagner-Meerwein rearrangement and almost completely precludes (<2%) 6,2 hydride shifts.

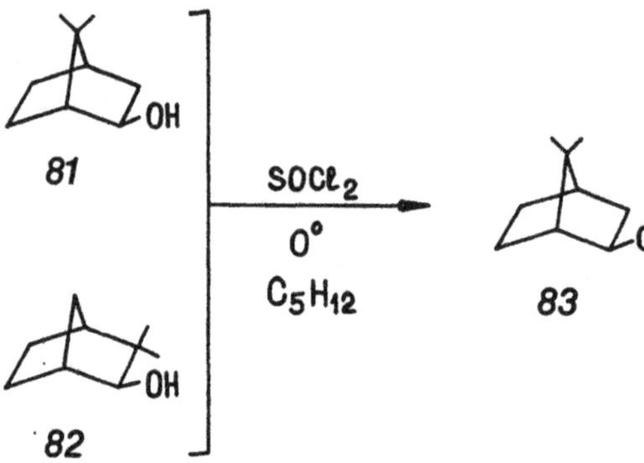

The 6,2-hydride shift in the primarily developing nonclassical carbocations can be assumed to occur with the formation of either a nortricyclonium ion *34* or an edge-protonated nortricyclene *35*. A choice between these two possibilities has been made from the reaction stereochemistry of unsaturated acids *84* and *85* with 50% sulphuric acid. The product in either case turned out to be lactone *86*. The formation of this product clearly indicates the 6,2-hydride shifts in this system to occur by endo,endo-migrations. If the intermediate here were a nortricyclonium ion, then besides lactone *86* the formation of two other isomers of the deuterium position would be equally probable. Suche isomers, however, have not been found [155,156]. This makes it possible to prefer the assumption that the 6,2-hydride shift proceeds via the edge-protonated nortricyclene *35*.

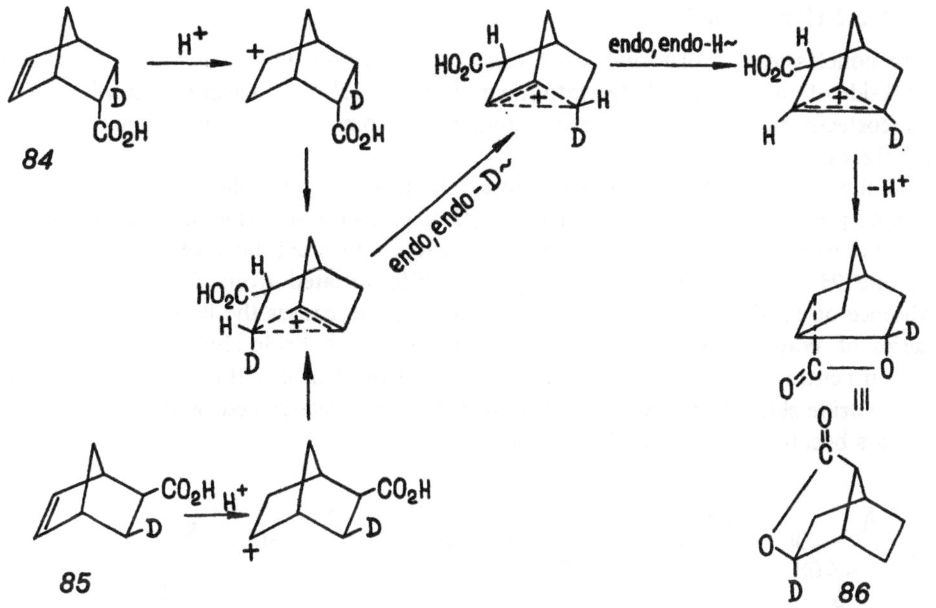

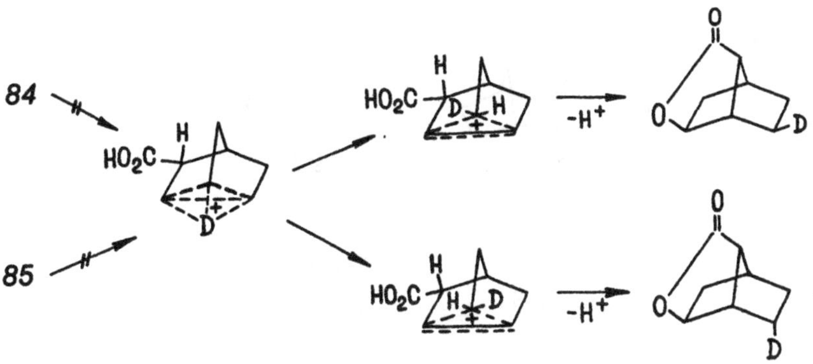

Collins has arrived to a similar conclusion as well[157–159]. By studying the solvolysis of tosylate *87* in aqueous acetone he obtained alcohol *88*, its formation is hard to explain assuming the intermediate nortricyclonium ions *89* and *90*.

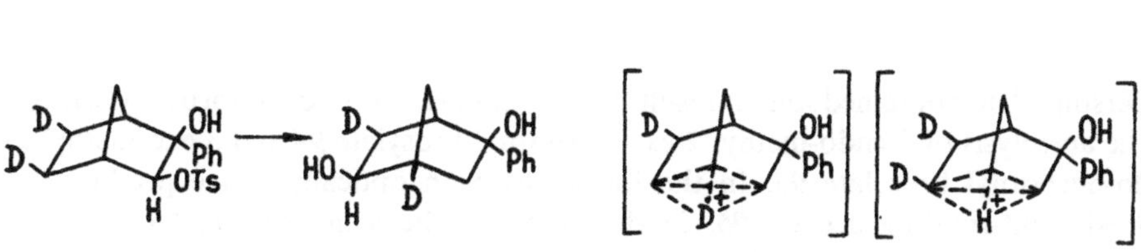

87 *88* *89* *90*

3.5.4 3,2-Hydride Shift

The nonclassical 2-norbornyl ion *5* can be attacked on C^1 and C^2 only from the backside of an orbital included into the multicentre orbital. In accordance with these stereoelectronic demands the group migrating from C^3 to C^2 is to have an exo configuration.

Collins [160] has made a detailed research into the pinacolic rearrangement of 2-endo-phenyl-2,3-cis-exo-norbornandiol *91*. If a 3,2-endo-hydride shift were possible, the diol *91* would yield the ketone *92*. The reaction product, however, is an isomeric ketone *93*, occuring from the 3,2-hydride shift initiated from the exo side. The absence of exo-H in the initially formed carbocation causes the latter to undergo a series of Wagner-Meerwein rearrangements and 6,2-hydride shifts result near the cation centre and at the exo-position of a hydrogen atom; this is subjected to a 3,2-hydride shift. The formation of ketone *93* is not determined by thermodynamic factors because it is less stable than ketone *92*.

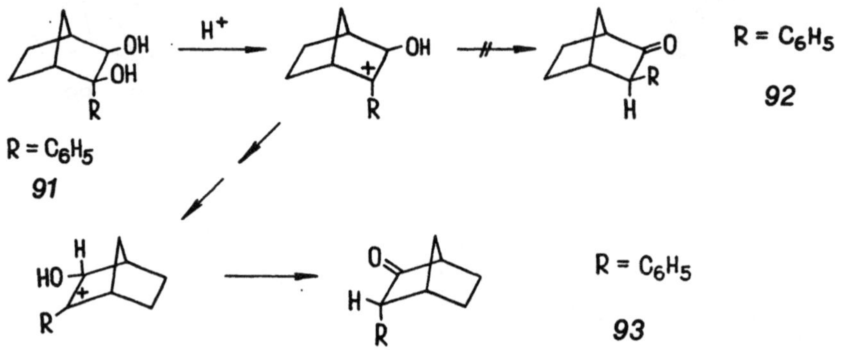

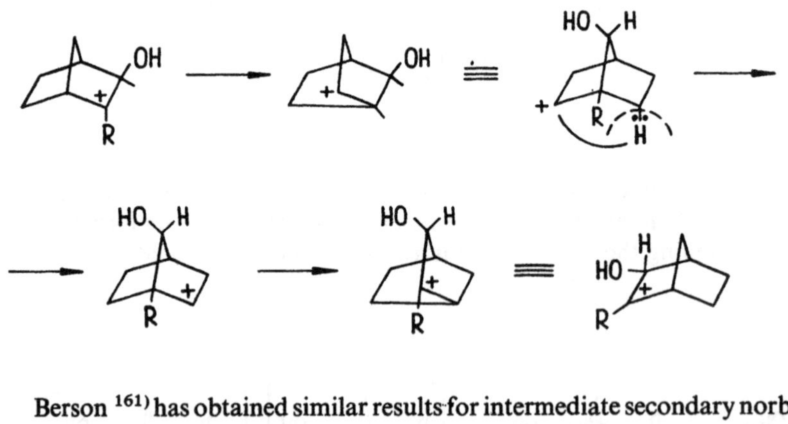

Berson [161] has obtained similar results for intermediate secondary norbornyl cations in the acetolysis of 3-endo-methyl-2-exo-norbornyl brosylate *94* and 3-exo-methyl-2-endo-norbornyl brosylate *95*. In the former case an optically active product has yielded right-rotating acetate *96*, in the latter — the same acetate *96*, but left-handed. According to Berson, these data indicate the rate of the exo-3,2-hydrogen shift to be at least 100 times as high as that of the endo-3,2-hydrogen shift.

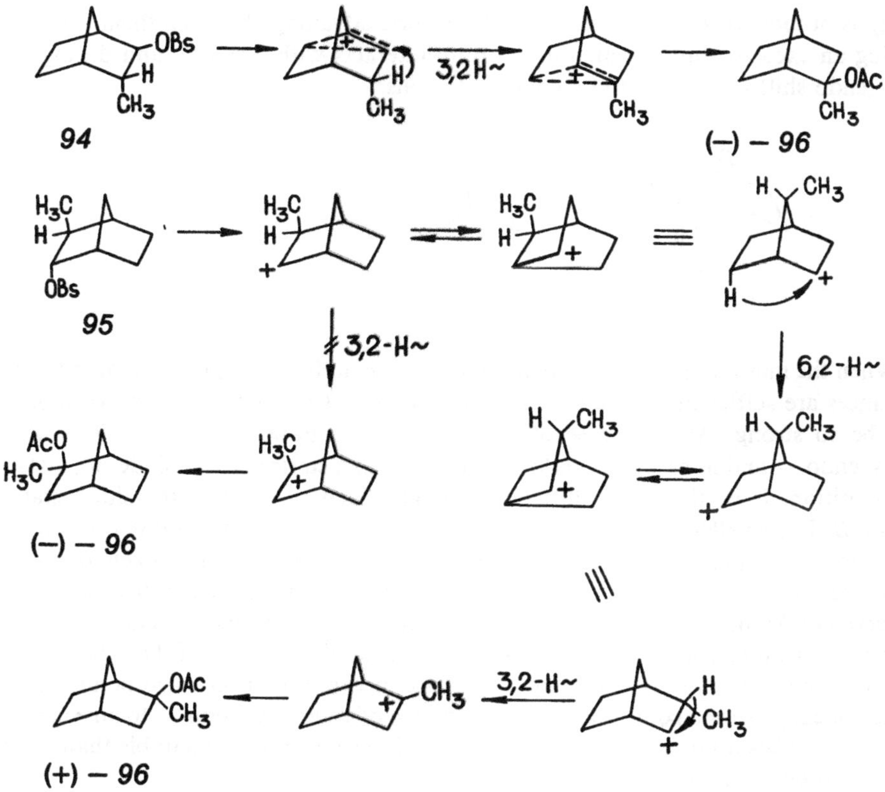

As noted earlier, the most widespread explanation of the high stereospecificity of 3,2-shifts is the assumption of an intermediate nonclassical ion [160-168]. In addition as a result of the σ-participation the extent of transition-state charge delocalization for an exo,exo shift must be greater than that for an endo,endo shift.

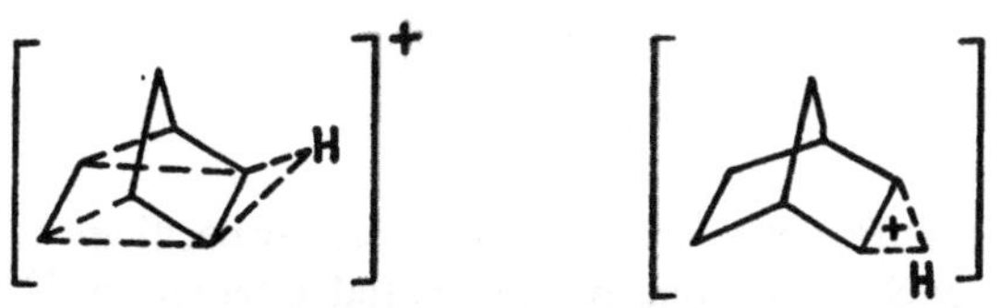

But if one rejects the σ-participation, one sees no other electronic cause providing the exclusiveness of exo,exo migration. One should, however, consider the possibility of the effect of steric factors. As Schleyer has pointed out [169] when a p-anisyl group is stabilizing the cation centre at the C^2 atom and levelling the σ-participation, a considerable amount of exo ketone *92* should appear. But this does not occur, a ketone of the type *93* is formed instead. Consequently, the formation of a non-classical ion does not account for the predominant exo-hydrogen migration. An alternative explanation by Brown [4] is also unsatisfactory. He considers 3,2-shifts to proceed from the exo side since it is less hindering to an attack by a nucleophile. An analysis of models *A* and *B* has revealed no steric advantage between them. Indeed, during an exo,exo shift the migrating group interacts with 7-syn-H, and during an endo,endo shift — with 5- and 6-endo-hydrogens.

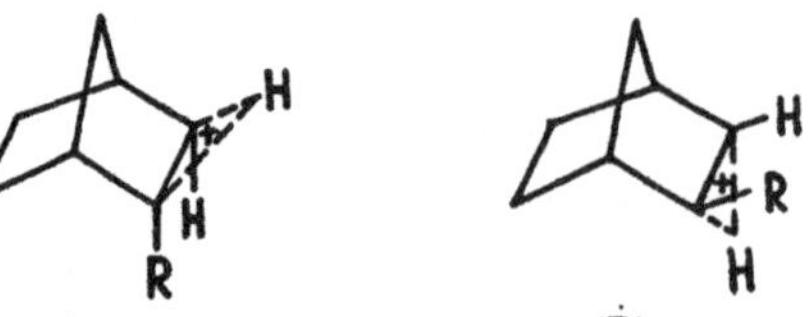

(A) (B)

When the migrating group is a hydrogen atom, then the corresponding interatomic distances are sufficiently large (>2.2 Å) and the effect of spatial hindrances should not be so strong. Also, in case of exo hydrogen migration the group R is at a quasi-endo — and in case if its endo migration — at a sterically less hindering quasi-exo-position. It is, therefore, difficult to judge which of the two transition states (*A* or *B*) has smaller steric strains, but the differences in steric strains can hardly be so great as to provide a 100 fold superiority of exo,exo shift over endo,endo migration (the difference in activation energies is about 3 kcal/mole). Schleyer believes the observed differences to be readily explainable if torsional effects are considered. As seen from the molecular models (norbornene oxides), the location of the substituents at the C^1 and C^2, as well as C^3 and C^4 in case of exo migration corresponds to an almost ideally skewed conformation while in case of endo migration — to an eclipsed one. This makes it possible to expect exo migration to be more favourable than endo migration by 6 kcal/mole.

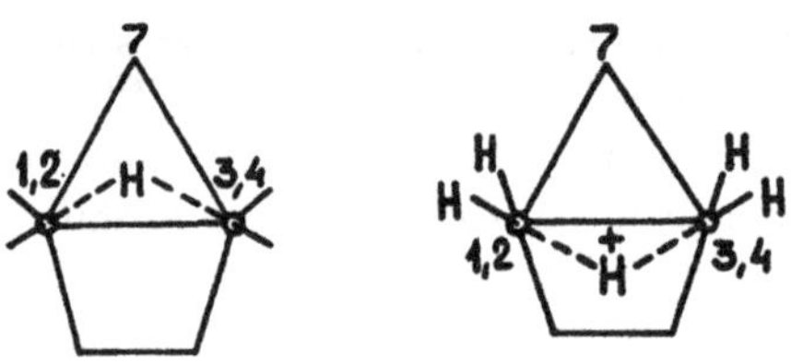

(A) (B)

This value is more than sufficient to account for the stereospecificity of the 3,2-shift without involving the nonclassical concepts or spatial interactions [169]. However, torsional factors can account for a 10–100 fold advantage of exo addition over endo addition to the sp^2-hybridized carbon atom, but they cannot explain a practically absolute exo stereospecificity in the attack of the nucleophile on a 2-norbornyl cation in solvolyses.

Torsional effects, according to Schleyer, can also account for the difference in the rates of solvolysis of 2-exo- and 2-endo-derivatives. In the ground state of the exo isomer the C^1—H and C^2—X bonds are partially eclipsed, this torsional strain is weakened in the transition state when the C^2—X is partly broken. At the same time, the long-range steric interaction between the R group and endo-H is decreased because R moves away from the latter and the partial eclipsing of the C_2—R and

C_6—C_1 bonds diminishes. On the other hand, the break away of the endo-X does not decrease the torsional strain since the C—X bond was not eclipsed in the ground state; but there is an increase in the long-range steric interaction between X and the endo-H-6. Besides, there appears an eclipsed interaction between the C^2—R and C^1—H bonds. These effects are quite sufficient to account for the high exo:endo rate ratio of solvolysis with tertiary norbornyl derivatives and, in conformity with the microscopic reversibility, for the high exo:endo ratio of the products, e.g., on addition of a hydride-ion [170].

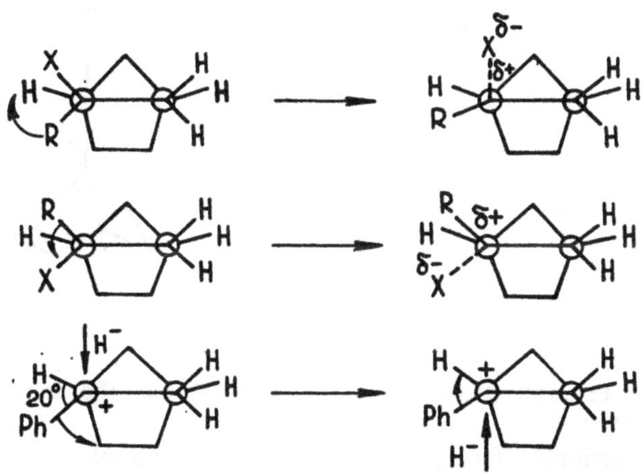

It is of interest that in some, rather rare, cases an endo-3,2-hydride shift occurs as well:

On the other hand, the phenyl group in norbornane derivatives does not migrate even from the exo side; this is likely to be due to an unfavourable steric interaction of phenyl ortho-hydrogen with 7-syn-H [171].

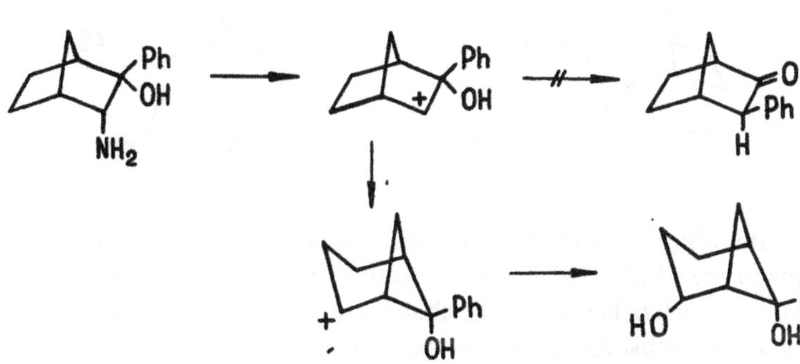

It is also interesting to consider a reaction in which torsional effects and nonclassical interactions are to contribute to the reaction process in different directions. An example of this is the 2-pinanyl-bornyl rearrangement:

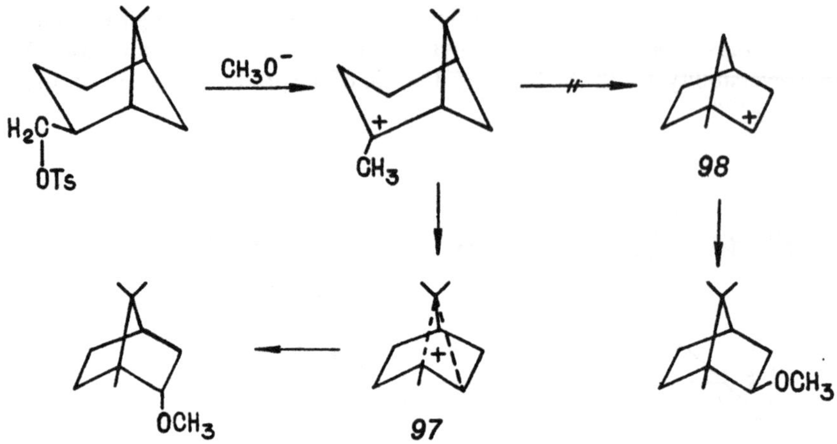

Judging by the reaction product the intermediate has the structure of the nonclassical ion *97*[172].

The concept of torsional effects as the main reason for exo selectivity was used to interpret a large number of data [170,173–178], but all attempts to obtain some additional corroboration of this concept have failed [179,180]. Tidwell [181] has determined the deuterium exchange rate for camphor *99* and 4-methyl-camphor *100* in the basic medium, as well as the hydroboration of 7,7-dimethylnorbornene *101* and 1,4,7,7-tetramethylnorbornene *102*.

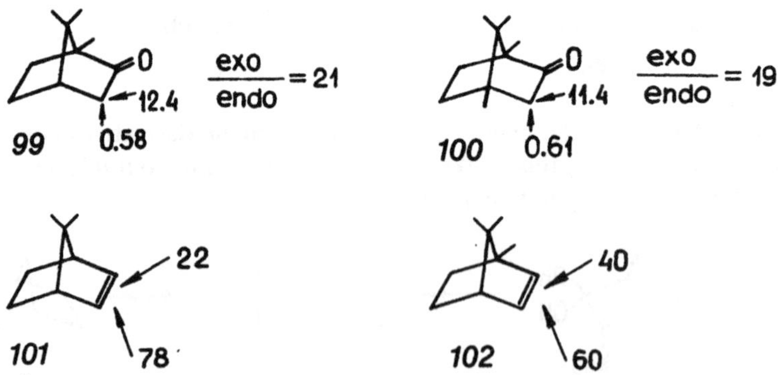

The data obtained indicate the introduction of CH_3 groups into bridgehead positions to exert little effect on exo selectivity though the torsional interactions must be intense. Tidwell believes that the changes in the torsional effects during the reaction are very small. Substituents may distort the geometry of the norbornane ring and, as a consequence, change the value of the torsional effects. Accurate estimation of its contribution, however, is evidently impossible without knowing the true geometry of reagents and transition states.

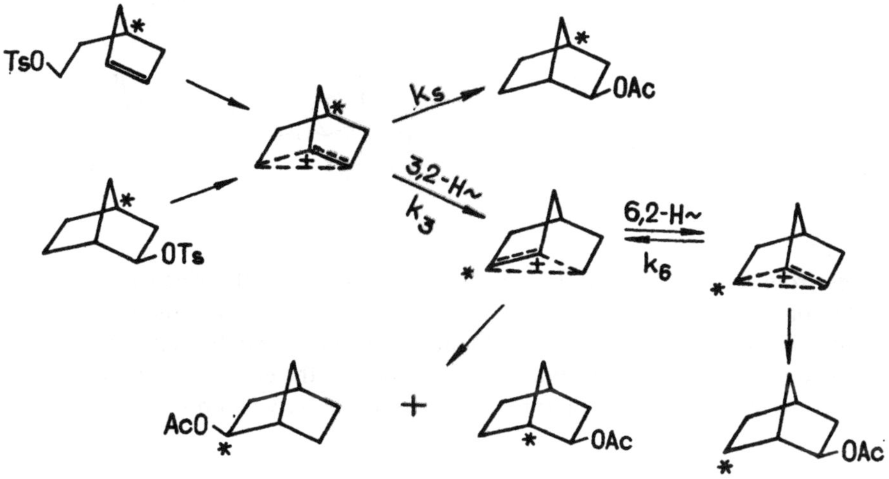

Collins and Harding [182] have estimated the rate ratio of 3,2- and 6,2-hydride shifts, as well as that of capturing the norbornyl ion by a solvent when generated both on the σ- and π-route. It has turned out that $k_s/k_3(\pi) \leqq 115$, $k_s/k_3(\sigma) \leqq 240$, $k_6/k_3 \leqq 100$–200. Olah has calculated for a stable 2-norbornyl ion k_6/k_3 $4 \cdot 10^3$ at 25 °C or $6 \cdot 10^2$ at 100 °C [4]; these data agree fairly well with those of Berson for the solvolysis of 7-methylnorbornyl brosylates.

Lee [186] has shown that in the acetolysis of Δ^3 — [1 — ^{14}C]-cyclopentenyl-ethyl tosylate and 2-exo-[4 — ^{14}C]-norbornyl tosylate k_6/k_3 is <100 and <200, respectively. The distribution of the ^{14}C tag indicates that on solvolysis of exo-norbornyl brosylate the ion pair is internally returned.

3.6 Generation and Conversions of the Classical 2-Norbornyl Cation

As already pointed out, Brown [179] considers the fast oscillation of the C^1—C^2 bond to prevent the solvation from the endo side and to result in exo substitution. Oxer and Wege [186a], however, have shown the solvolysis of isomeric 6-carbomethoxy-2-norbornyl brosylates to be accompanied by both the Wagner-Meerwein rearrangement and the capture of equilibrated ions from the exo and endo side. The low value (4.4) of the exo:endo rate ratio (for unsubstituted 2-norbornyl brosylates 350), the non-stereospecificity in nucleophile addition — all this indicates the ionization to proceed at an intermediate classical carbocation; the participation of the C^1—C^6 is hindered by an electron-attracting carbomethoxy group.

By extending this work Grob [187] has studied the solvolysis of 6-exo-substituted 2-exo- and 2-endo-norbornyl tosylates with severel R. The exo:endo rate ratio

4 The tentative data of paper [183] happen to be erroneous [184].

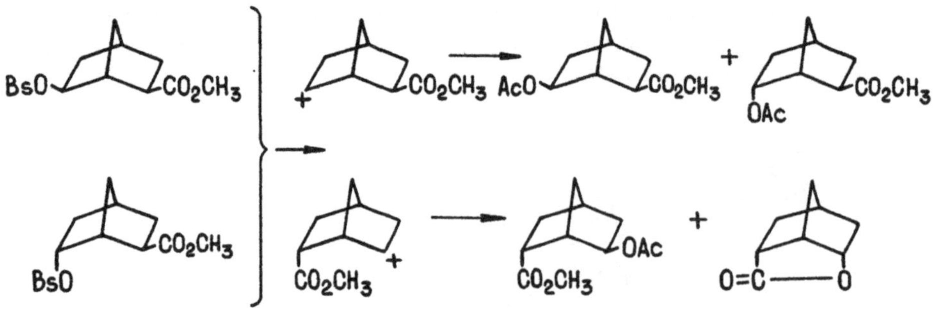

decreases from 2320 [for $R=(CH_3)_2N$] to 0.37 (R=Br) as the electron-attracting properties of the radical R increase. The authors [187] preclude from these results the possibility of explaining the high exo:endo rate ratio by steric hindrances in the ionization of endo isomers (as Brown). But they rather believe the exo:endo rate ratio to be determined by stereoelectronic factors as seen from the Newman projections:

In an exo isomer the $C^1—C^6$ bond is antiperiplanar with respect to the C^2-OTs, resulting in a stronger C—C hyperconjugational effect than in an endo isomer in which the $C^1—C^7$ bond is curved with respect to C^2-OTs and this results in a lower stabilization of the transition state. The electron-attracting substituent at C^6 will destabilize the transition state of ionization of exo isomer to a higher degree than that of endo isomer which may decrease or even invert the exo:endo rate ratio.

The low exo:endo rate ratio of solvolysis of some norbornyl derivatives has been attributed to classical 2-norbornyl ions. Brown asserted, however, that this ignores the selectivity principle; he has shown reactions displaying a low epimeric rate ratio of norbornyl derivatives to give a low rate ratio for a pair of cyclopentyl and cyclohexyl derivatives. Thus, the reaction of acetolysis has a high selectivity [188]:

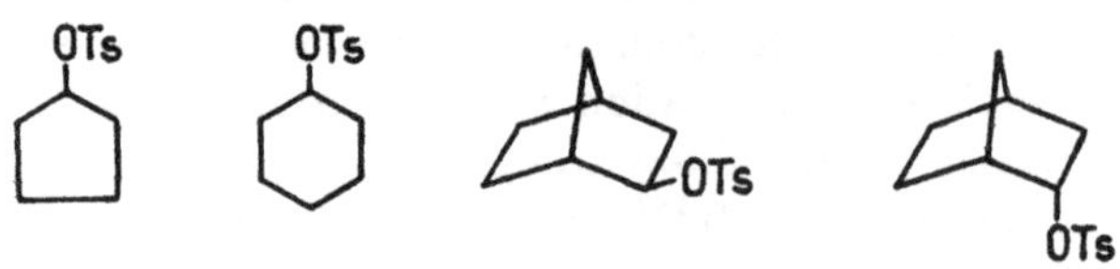

k_{rel} = 32.4 (25 °C) 1.00 280 1.00

On the other hand, the reaction of peracid ester decomposition has a low selectivity in either pair

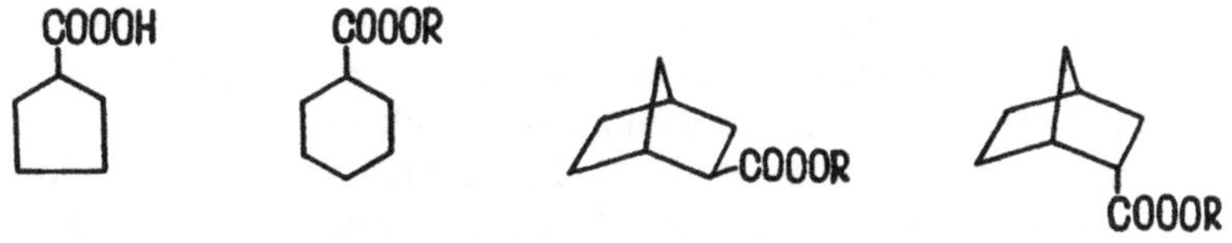

k_{rel} = 0.5 1.00 4.7 1.00
(25 °C)

However, Schleyer [13] refuses to compare data on acetolysis because the contribution of solvent participation is not equal; it is better, therefore, to compare data on trifluoroacetolysis:

Exo : endo 13.8 1120

The exo:endo ratio of norbornyl derivatives proves to be about two orders of magnitude higher than for a model pair. Introduction of a 3-spirocyclopropyl group decreases the exo:endo rate ratio of trifluoroacetolysis from 1120 to 2.8, i.e., by 408; a similar substitution for a cyclopentyl and cyclohexyl pair diminishes the rate ratio of trifluoroacetolysis from 13.8 to 3.2, i.e., only by 4. If the selectivity principle manifested itself in these examples, then its effect would be felt in both secondary and tertiary esters. However, when introducing a 3-spirocyclopropane ring into tertiary 2-norbornyl esters the exo:endo rate ratio of solvolysis remains practically unchanged (see p. 36).

3.6.1 Deamination of 2-Norbornylamines

Corey [150] showed 1963 that 2-exo- *103* and 2-endo-norbornylamines *104* are deaminated in acetic acid to yield practically identical mixtures both in the exo:endo product ratio and in the retention degree of optical activity. He postulated both reactions to the same intermediate classical 2-norbornyl cation since no optically active products were obtained from a symmetrical nonclassical 2-norbornyl ion *5*. Somewhat later Berson studied the composition of reaction mixtures from the deamination of amines *103* and *104* (Table 8) more carefully; three essential differences in the compositions of reaction mixtures [189] were pointed out:

1) endo-amine *104* yields about twice as much endo-acetate *105* as exo-amine *103*;

2) the optical purity of exo acetate *66* obtained from endo amine *104* is higher than that of exo acetate *66* obtained from exo amine *103*;

3) almost no racemization occurs on formation of endo acetate *105* from endo amine *104*. These data preclude the assumption that all the deamination products of amines *103* and *104* are formed from one common intermediate because in this case the extent of racemization for both exo and endo products would be the same.

These data can be interpreted as follows: endo amine *104* gives a classical 2-norbornyl cation *6* which either reacts with a nucleophile or transforms — with a small energy barrier — into a nonclassical ion *5*. The rate ratio of exo and endo addition to the classical ion *6* must equal the ratio of the optical purity of exo acetate *66* to that of endo acetate *105* obtained from endo amine *104*. Deamination of exo amine *103* seems to follow two routes:

1) with participation of the 6,1-bond and formation of the nonclassical ion *5*;

2) without participation of this bond and with formation of the same classical ion *6* from the endo isomer *104*.

Table 8. Products from the Deamination of Optically Active 2-Exo- and 2-Endo-Norbornyl Amines

Products	From exo amine (*103*)		From endo amine (*104*)	
	Yield	Optical purity	Yield	Optical puridy
Exo compounds, total	97 ± 1		95 ± 1	
Endo compounds, total	1.85 ± 0.45		4.8 ± 0.5	
Endo acetate (*105*)	2 ± 0.5		4.7 ± 0.5	85 ± 12
Exo acetate (*66*)	98	11 ± 2	95.3	18 ± 0.6

The percent ratio of the optical purity of exo acetate *66* to the yield of endo acetate is 6. The intermediate product leading to racemates and common for exo and endo amines is a nonclassical ion *5* which has been postulated for the formation of racemic products on solvolysis of arylsulphonates.

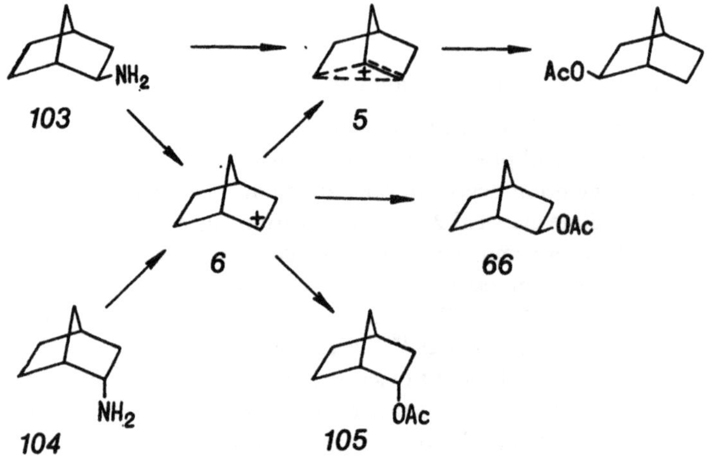

The mechanism proposed can account for facts from Table 8; on deamination of endo amine *104* the purity of endo acetate *105* is higher than that of exo acetate *66*, since optically active products can only be formed from the classical ion *6*. Endo acetate *105* is also obtained from the classical ion *6* only. At the same time exo acetate *66* can be obtained both from the nonclassical ion *5* with indispensable racemization and from the classical ion *6*.

The difference in the optical purity of exo acetate *66* and that in the exo:endo product ratio on deamination of endo *104* and exo amines *103* can be explained as follows. Since endo products and optically active ones can only be formed from the classical ion *6*, their amount on deamination of exo amine *103* must be less because exo amine *103* partially converts directly into the nonclassical ion *5*. Deviation from 100% optical purity of endo acetate *105* seems to be due to a partial equilibrium between the classical cation *6* and its enantiomer which is likely to be established via the non-classical cation *5*.

The mechanism reflects the decrease of σ-participation of the 6,1-bond when forming the 2-norbornyl cation in the deamination reaction as compared to that of solvolysis. On deamination, the leaving group (N_2) is different from halogen- or sulphonate anions:

1) it carries no charge and hence little or no Coulomb interaction occurs in the transition state on its detachment;

2) molecular nitrogen is exceptionally stable, so it is formed with great driving force. The activation energy of deamination reaction is relatively small (~5 kcal/mole), only in the solvolysis of arylsulphonates it amounts to 20–25 kcal/mol [190]. As Streitwieser points out, the decreasing activation energy of carbocation formation results in a decreasing difference between the activation energies of the competing ionization routes ("compression" of the energy scale).

As a result, the rates of ionization with and without the participation of the 6,1 bond become comparable. When the enthalpy term of free activation energy is small the reaction prefers a low activation entropy [191]. The entropy factor favours the

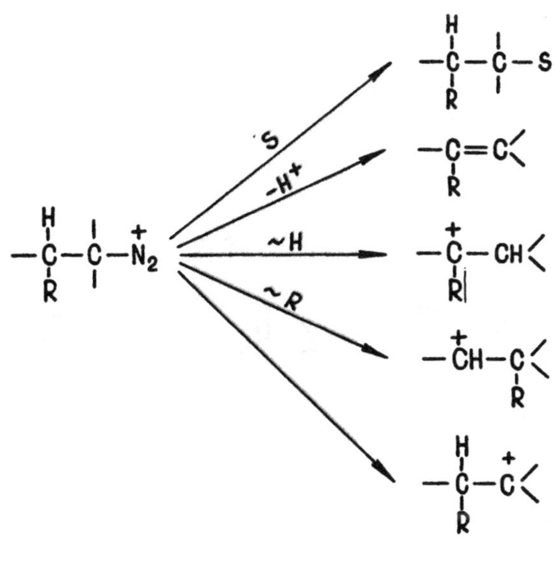

ionization without σ-assistance which requires simultaneous excitation of $C^6—C^1$ and C—N bonds while the simple ionization requires only that of C—N [150].

The deamination shows that from a classical 2-norbornyl ion the exo:endo product ratio does amount to about (4–6):1. This also indicates that the solvolysis products of 2-exo-and 2-endo-brosylates with the ratio of exo and endo acetates of about 330:1 cannot result from a nucleophilic attack on the classical cation *6*.

Decomposition of aliphatic diazonium salts often yields a more complicated mixture of products (see Table 8) than the one formed in solvolysis. Streitwieser explains this difference that in the first case the reaction proceeds not only through carbocations [190,192], but also directly from diazonium ions; the latter react unselectively with a small activation energy.

However, in his numerous works (reviewed in [193]) Collins has shown Streitwieser's assumption to be wrong. Thus he found the 1,2-shifts of aryl groups and hydrogens in the deamination to occur with retained configuration of the atom to which the group shifts, i.e., the nitrogen detachment must precede the 1,2-shift.

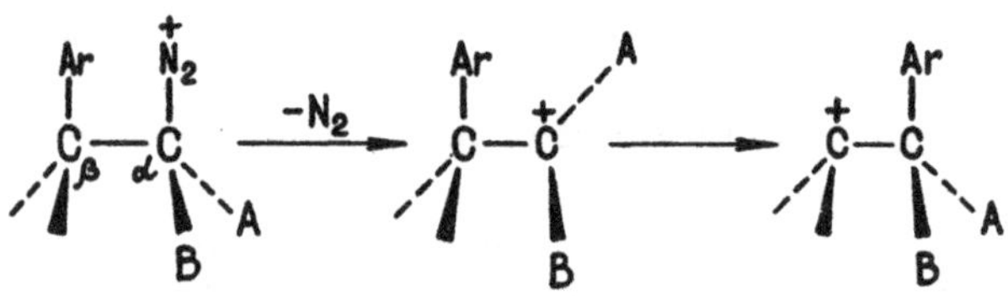

Collins has also criticized an earlier assumption, that nitrogen is detached to yield exclusively reactive, so called "hot" carbocations which, unlike solvolytic "cold" ions, are far less selective and so result in a set of various products [194]. "Hot" ions were assumably formed without anchimeric assistance, the charge being localized. Collins has shown that the unusual properties of cations formed in deamination reactions are retained after several Wagner-Meerwein rearrangements and 6,1,2-shifts [195, cf. 196, 197]. But "hot" carbocations must lose their energy after one 6,2-hydride shift resulting in "normal" carbocations [198] [5] Collins [200–204] considers the intermediate carbocations in deamination reactions to be classical ions, a counterion playing an important role. Indeed, on decomposition of an aliphatic diazonium compound the anion reacts with the cation yielding a product; this situation is sharply different from the one occurring on solvolysis.

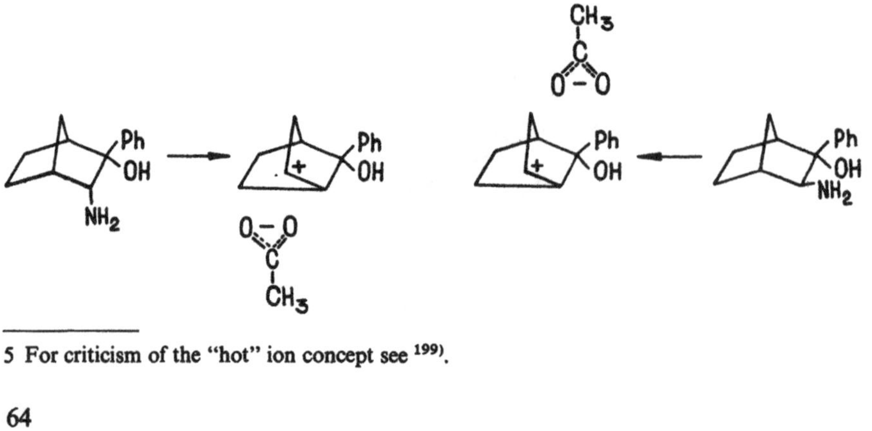

5 For criticism of the "hot" ion concept see [199].

Schleyer [13] has explained Collins's data as a result of forming a bridged ion with an asymmetrically located counterion, e.g.:

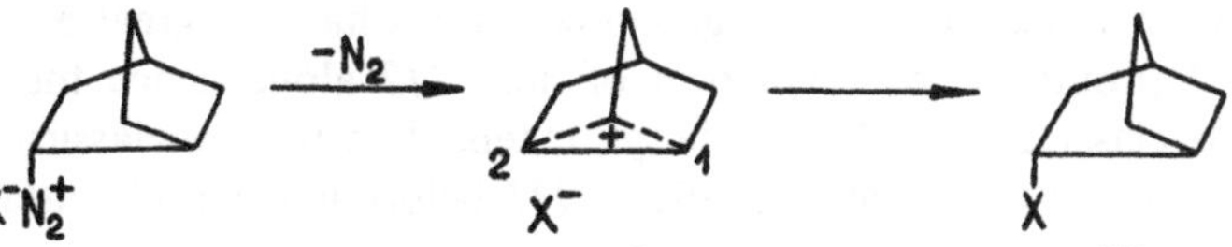

(A) (B)

If in the ion pair *A* the anion interacts with the cation before the counterion occupies a symmetric position, then the product *B* will retain at least a part of its chirality.

According to White, the detached nitrogen separates the counterion and the carbocation; this may lead to various skeletal rearrangements and hydride shifts [205]:

$$RN_2^+X^- \rightarrow R^+N_2X^- \rightarrow \text{products}$$

While carbocations from a deamination are usually observed to be less selective than the same ions from solvolysis, in some cases, there opposite situation occurs, as shown by Berson [206]. He investigated the competing migration of a β-substituent (R_1) and a β-ring atom (G, J) in reactions of compounds of the general formula:

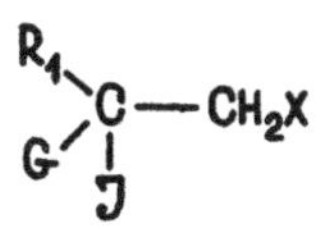

In particular, he has studied reactions of bicyclic compounds of the type

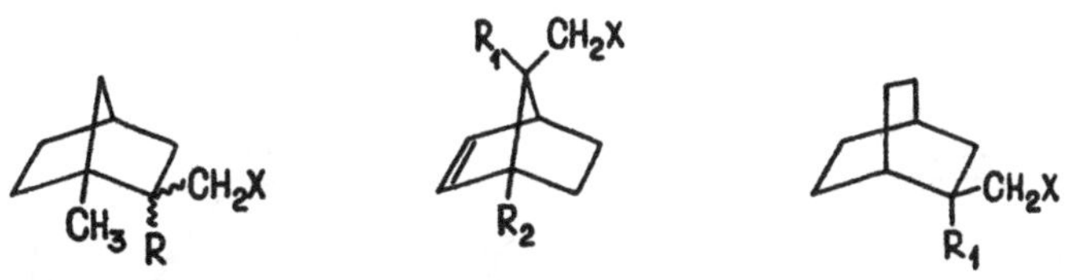

etc.

In a few cases the selectivity of deamination is 100 times as high as that of solvolyses. Berson considers such a selectivity to be due to the small differences between the transition state energy for deamination and the energy barrier for internal rotation; this is why the conformational composition of the ground state turns out to be a significant factor largely determining which group migrates in the deamination.

Yates and Blattel have obtained interesting data on the acid-catalyzed decomposition of 3-diazonorbornanone-2 *106* [207]. Irreversible exo protonation produces an endo diazonium salt *107* which rapidly loses nitrogen yielding a nonclassical ion *108*; this is stereospecifically attacked by water yielding α-*109* and β-ketols *110* or converted into a classical ion *111* and is further opened to form an acid *112*.

With a 5-endo-methyl group no β-ketol is formed because of steric effects; and the amount of acid *112* is reduced due to the unfavourable interaction of the 5-endo-CH_3 and C=O groups in *111*. If acid *112* were formed directly from ion *108* rather than through *111* the 5-endo-CH_3 group could not have so great an effect. A CH_3 group in C^4 promotes the formation of acid *112* alone, since the 4-CH_3 group stabilizes the classical ion *111* to a larger extent than the nonclassical ion *108*; this leads to a complete transformation *108* → *111* before nucleophilic attack.

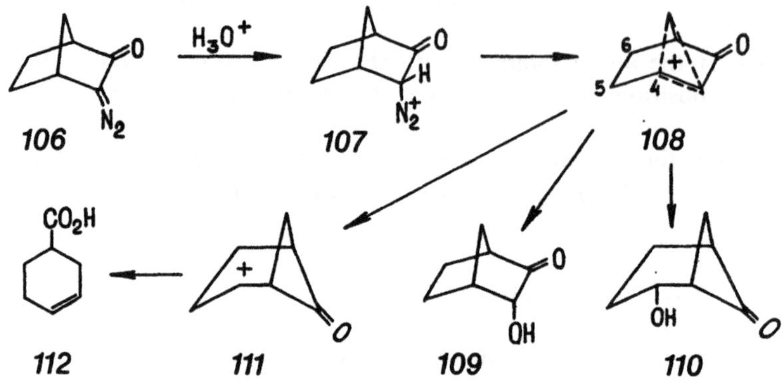

Rearrangement of a norbornane system into a pinane one which, as already noted, rarely occurs in solvolysis, does take place in the photolysis of tosylhydrazones in aqueous NaOH or CH_3OH/CH_3ONa [208, 209].

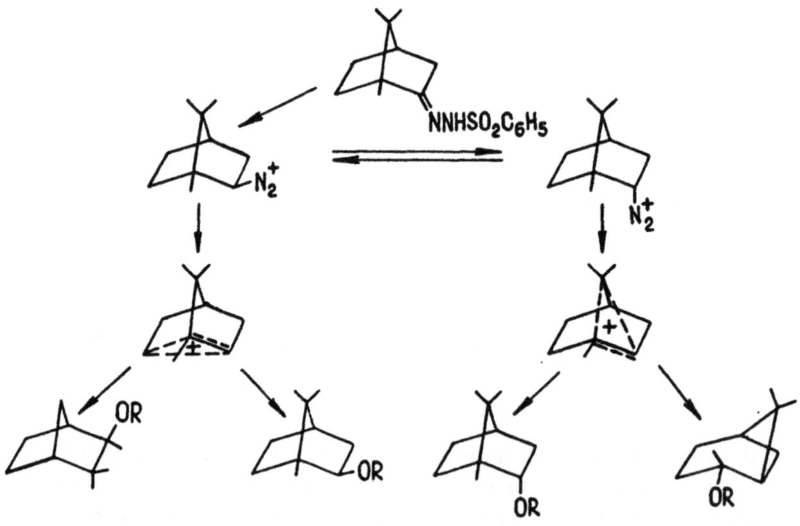

The same rearrangement also takes place on deamination of amines if C^1 has strong donors [210]:

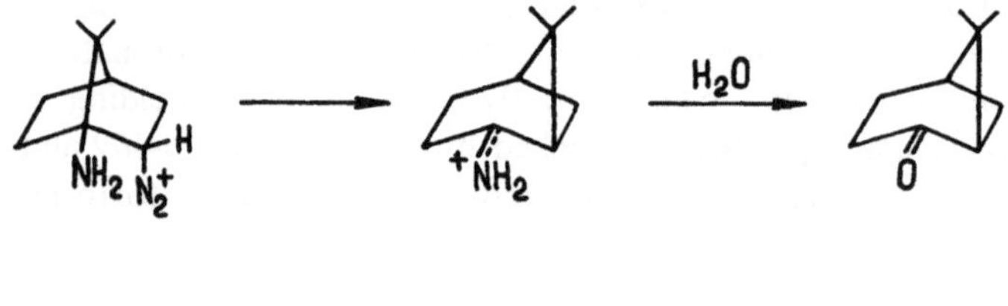

Formation of a tertiary classical ion in solvolyses was clearly shown by Goering [211] in the methanolysis of optically active 1,2-dimethyl-2-exo-norbornyl chloride:

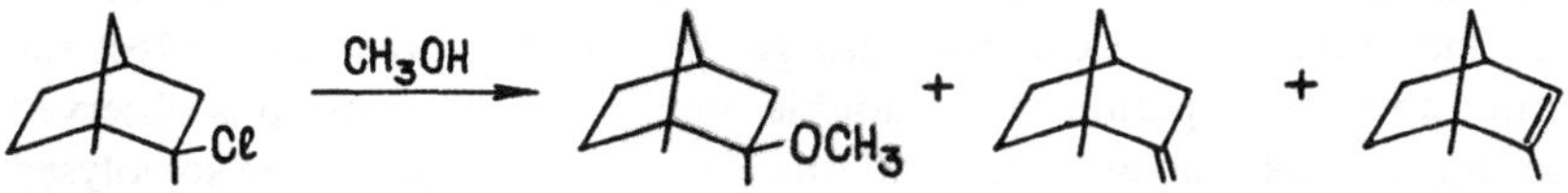

The resulting methyl ether retains 14% of its optical activity. Goering [211] and Brown [21] use this fact to testify the capture of the classical ion before complete equilibration.

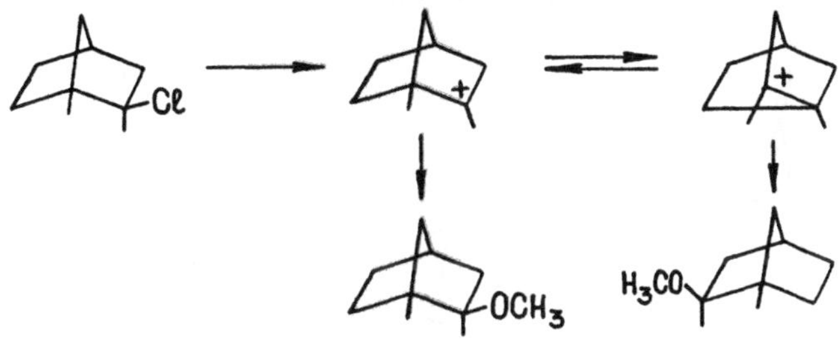

This explanation, however, is not the only one. Another example is:

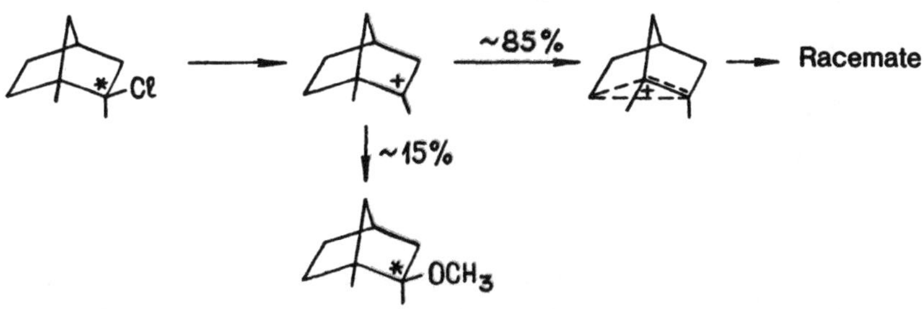

The substitution of hydrogen by methyl at C^1 in the molecules of exo- *55* and endo-epimers *56* practically results in the same accelerating effect which also proves the absence of σ-participation in the transition state of ionization of exo ester *55*.

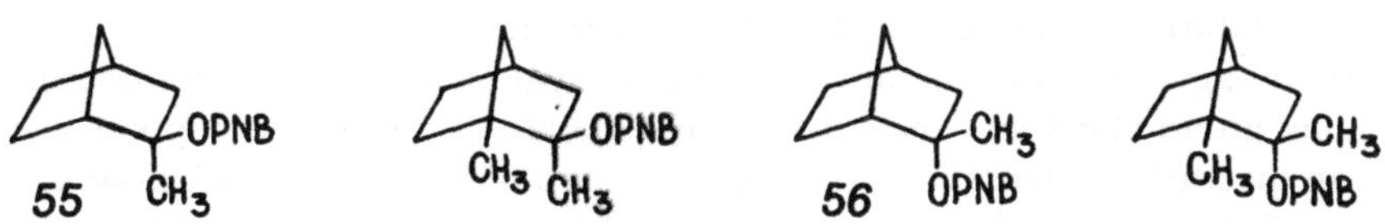

55		56	
k_{rel} = 1.0	8.5	1.0	8.6

3.6.2 Solvolysis of 2-Substituted Norbornanes Containing Electronegative Substituents

Gassman and co-workers [105–109] assumed that such a substituent will hinder the development of the cationic centre at C^1 and that the solvolysis will involve the formation of an intermediate secondary classical 2-norbornyl ion.

Such a classical ion could be formed on solvolysis of tosylates *113* and *114* containing a keto group at the 7-position [105]. The solvolysis rate of endo isomer *114* is 6 times as high as that of exo- *113* but it is the same as that of 2-endo-norbornyl tosylate, i.e., the inductive effect of the keto group does not practically affect the solvolysis rate; no π-participation of the double bond of the keto group is observed either. According to Gassman and Marshall, the exo and endo isomers are solvolysed without anchimeric assistance, via a classical 7-keto-2-norbornyl cation *115*. No electrons of the 6,1-bond seem to take part in the ionization of exo tosylate *113* because the inductive effect of the adjacent keto group hinders the development of the positive charge on C^1 necessary for participation.

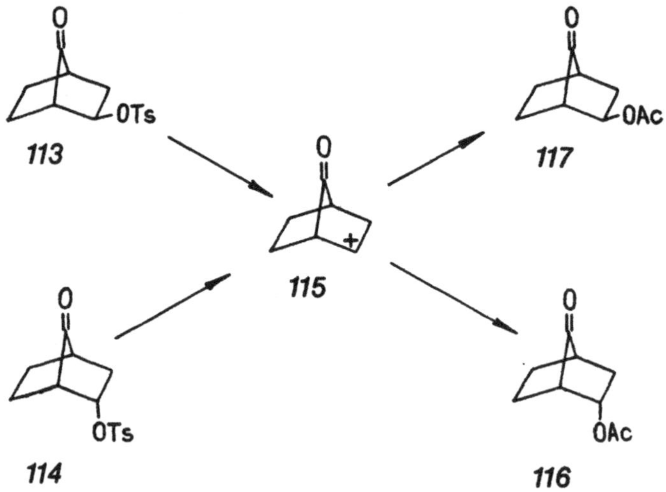

The endo isomer *114* is solvolyzed 6 times as fast as the exo isomer *113*, this agrees well with the value predicted by Schleyer for the steric strain relief in the transition state caused by the interaction of 6-endo-H and 2-endo-OTs in the ground state. Brown, however, predicted a retardation effect for endo tosylate ionization (see p. 17). Gassman and Marshall have shown the exo isomer *113* to yield a mixture of 34% endo *116* and 66% of exo acetates *117*, respectively; endo tosylate *114* has given a mixture containing only 2% of endo acetate *116*.

These results agree with the formation of a classical cation in either case considering the shielding group effect and the solvent backside attack which was supposed to result in an inverted configuration; the predominance of exo acetate *117*, however, indicates the 2-norbornyl cation to be less accessible from the endo side than from the exo side. The formation of 34% of endo isomer *116* from exo tosylate *113* testifies

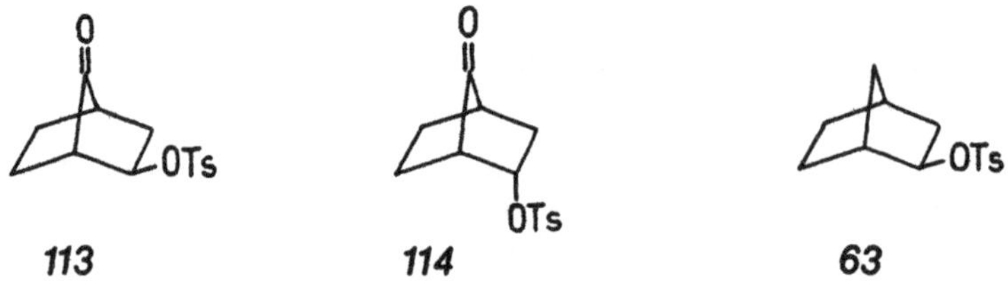

113 114 63

that steric requirements alone are unable to explain the exclusive formation of exo products from the 2-norbornyl cation.

The reactions studied have been the first examples of solvolysis of 2-norbornyl aryl sulphonates without exclusive formation of exo products. But the comparison of solvolyses between compounds *113*, *114* and *63* is complicated by different hybridization of the C^7 atom and by the possible formation of an intermediate hemiketal with *113* and *114*. These complications are avoided by acetolyzing the isomeric dimethyl ketals *118* and *119* having the sp^3-hybridized C^7 atom [106]. The solvolysis of compound *118* yielded a mixture of exo *120* and endo acetates *121* (95.5% and 4.5%, respectively).

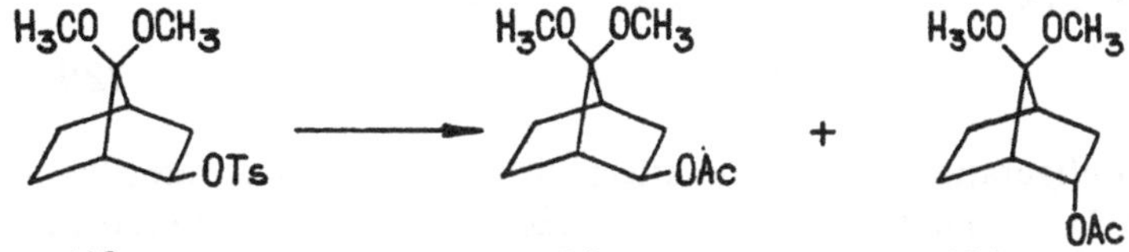

118 120 121

Comparison of the results between compounds *113*, *114* and *118* indicates that the formation of exo and endo products on solvolysis is not influenced by the sp^2-hybridized C^7 atom; on the other hand, if only exo acetate is formed on solvolysis of compound *63* it is apparently formed via an intermediate nonclassical ion *5*.

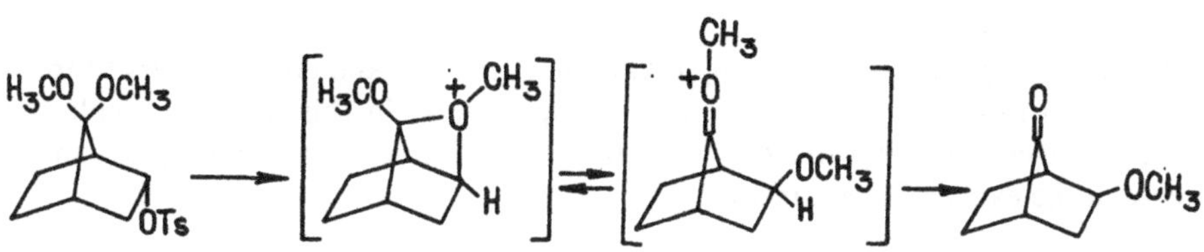

119

Quite differently, a methoxyl-group participates in the solvolysis of endo isomer *119*. This prevents the comparison of the behaviour of exo *118* and endo isomer *119*. To avoid these complications with electronegative substituents, Gassman and Marshall [107] included the heteroatoms of C^7 in the dioxolane ring. The solvolysis of isomeric tosylates *122* and *123* ended in the case of the endo isomer *123* by simultaneously splitting the ring with the oxygen function involved.

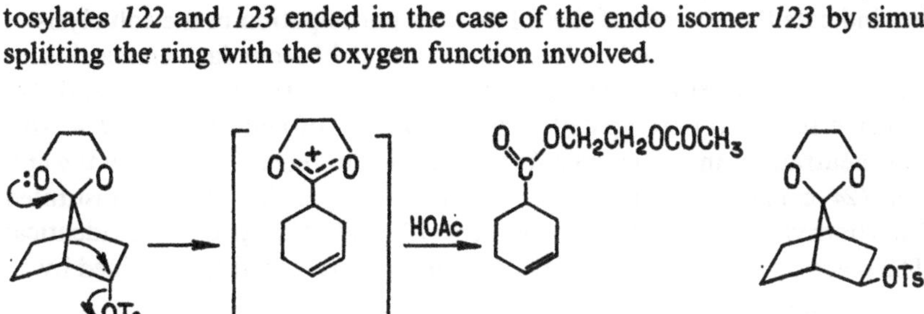

123 122

Thus, in the studied (by Gassman) solvolyses of norbornyl tosylates with oxygen-containing substituents at C^7 the oxygen-containing function has interacted in each epimeric pair in a different and unique way with one of the epimers; hence, these systems cannot be "pure" models of classical norbornyl cations.

To overcome these complications Gassman has investigated [108, 109] the electronegative chlorine substituent capable of destabilizing the developing cationic centre by studying the acetolysis of the four epimeric 7-chloro-2-tosyloxynorbornanes *76*, *77*, *124* and *125*.

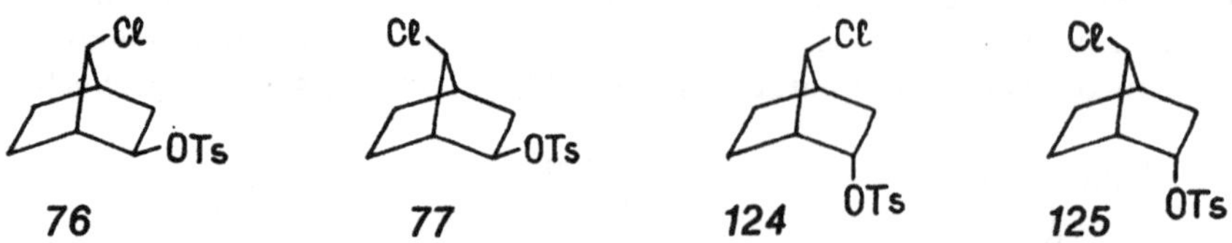

76 77 124 125

The configuration of chlorine affects little the acetolysis rate; the acetolysis rate ratios of *76*, *77* and *124*, *125* are only 2. However the exo:endo rate ratio for the two anti-epimers *77* and *125* and for the two syn-epimers *76* and *124* are 80 and 246, respectively; these values are quite comparable with the exo:endo rate ratio for unsubstituted 2-norbornyl tosylates. The absolute values of rates for the four epimers are essentially lower (about 300 times) than those for the respective unsubstituted 2-norbornyl tosylates as expected from the inductive effect of the chlorine at C^7. All four epimeric tosylates yield qualitatively the same mixtures of products, identical by GLC from anti-chloro tosylates *77* and *125*, quantitatively rather different from syn-isomers *76* and *124*.

Chlorine at C^7 prevents, according to Gassman, the exo and endo epimers to form a nonclassical ion *A* and a classical ion *B* respectively; in this case the 7-chlorine exerts a stronger hindering effect on the solvolysis of exo epimers leading to a sharp decrease in the exo:endo rate ratio.

(A) (B)

These data indicate the absence of the 6,1-bond σ-participation in the solvolysis of 7-chloro-2-tosylates; both exo and endo epimers first yield secondary classical ions. Quantitatively different mixtures are formed from syn-chloro epimers *76* and *124* since syn-chlorine seems to participated in the ionization of endo tosylate *124*; this acceleration is neutralized in exo epimer *76* by steric strain relief. This is why the epimers *76* and *124* cannot serve as "pure" models for studying the unsubstituted classical 2-norbornyl cation; whereas epimers *77* and *125* give quite identical mixtures. The reaction products from the latter compounds is readily explained by an intermediate classical cation *126*.

Gassman's results on the solvolysis of compounds *77* and *125* are of great interest: the high exo:endo rate ratio, the formation of the same products, and the absence of endo products show that the norbornanes in which a cationic centre on C^1 is suppressed by electronic effects behave like unsubstituted 2-norbornyl tosylates.

Although this, as Gassman rightly points out, is no argument that in the latter case a classical 2-norbornyl cation is formed as well, the high exo:endo rate ratio or the exclusive formation of exo products is not yet an absolute criterion for the formation of a nonclassical cation in 2-norbornyl tosylate solvolysis.

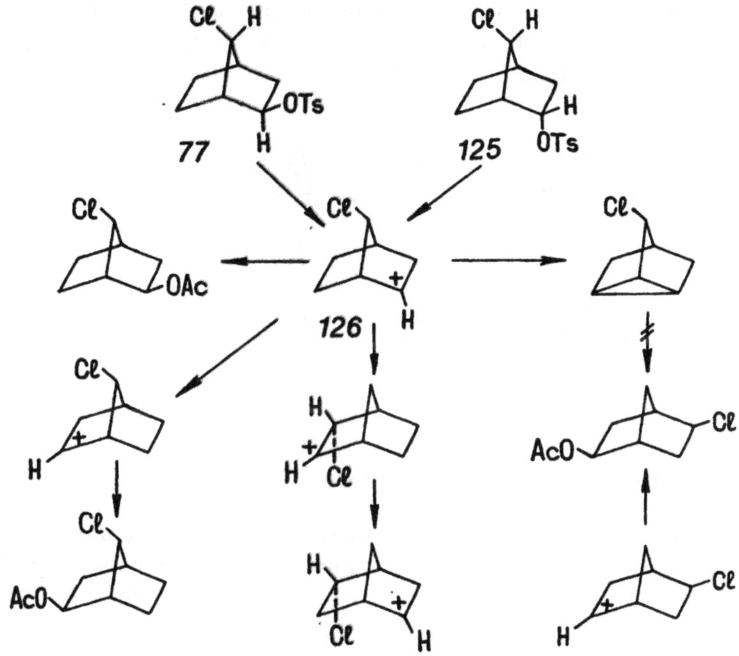

3.6.3 Solvolysis of 2-Norbornyl Brosylates Containing a Cyano Group on C^1

The solvolysis rates of exo brosylates *127* and *79* differ from each other by a factor of $1.7 \cdot 10^{-7}$ while those of endo brosylates *128* and *129* only by $2.4 \cdot 10^{-5}$. Such a difference corroborates the σ-participation of the 6,1-bond in the acetolysis of compound *79* which disappears (or diminishes) on introduction of the 1-cyano group [110].

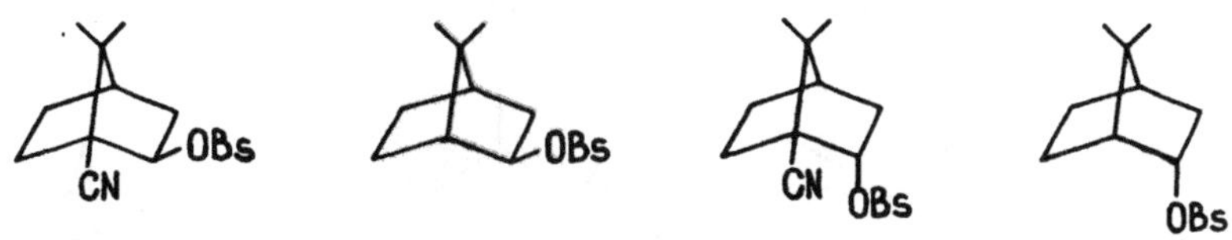

Discussion on the σ-participation and formation of an intermediate nonclassical ion in 2-substituted norbornanes can be expected to have practically no end because this model has many disadvantages. The most essential are:

1) in solvolyses the detachment of both 2-exo and 2-endo substituents results

in the shift of the $C^1—C^6$ bond alone, the $C^1—C^7$ bond does not usually shift due to a rise in the angle strain;

2) the classical ion primarily formed from an endo isomer transforms mainly (~92%) into a nonclassical ion which is formed directly from an exo isomer;

3) the steric effects operating in the transition state of solvolysis of 2-exo and 2-endo epimers (torsional, nonbonded interactions) are different;

4) the steric accessibility of a classical ion and the stereoelectronic accessibility of a nonclassical ion coincide in their direction:

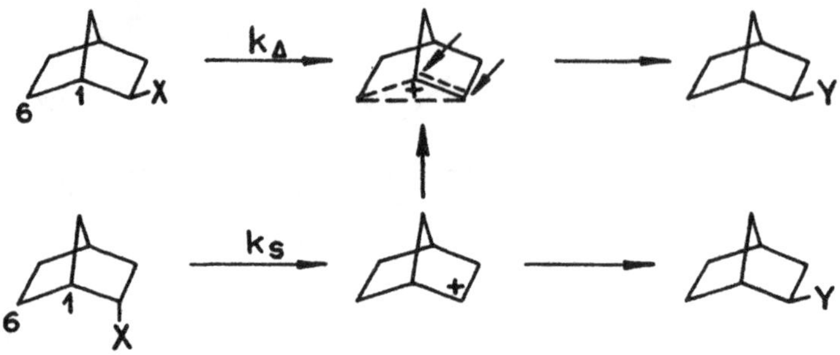

3.7 Structure of the Stable 2-Norbornyl Cation

In recent years Olah and co-workers have researched the structure of the 2-norbornyl cation under so-called "long-life" conditions in exceptional non-nucleophilic, strong-acid media ($SbF_5—SO_2$, $FSO_3H—SbF_5$ — SO_2FCl etc.). In these investigations wide use was made of the ^1H and ^{13}C NMR, laser Raman and ESCA spectra [15,16,183,184,213–221].

A norbornyl cation was generated from various precursors:

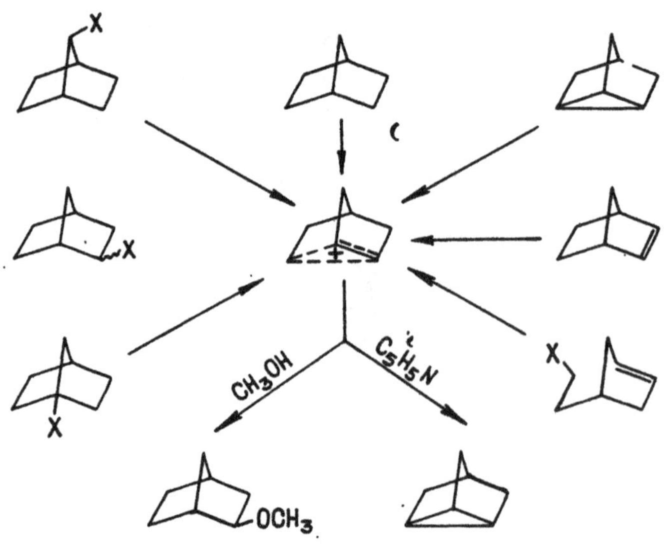

X = F, Cl, Br, OH, $OCOCF_3$

On quenching, different products were obtained depending on the nucleophile and on the method of neutralizing and acid solution.

3.7.1 NMR Spectra

A single peak in the PMR spectra of a 2-norbornyl cation at —23 °C can be accounted for nonclassical (a) or classical (b) cations with simultaneous degenerate rearrangements (see below); they are rather rapid in the NMR time scale.

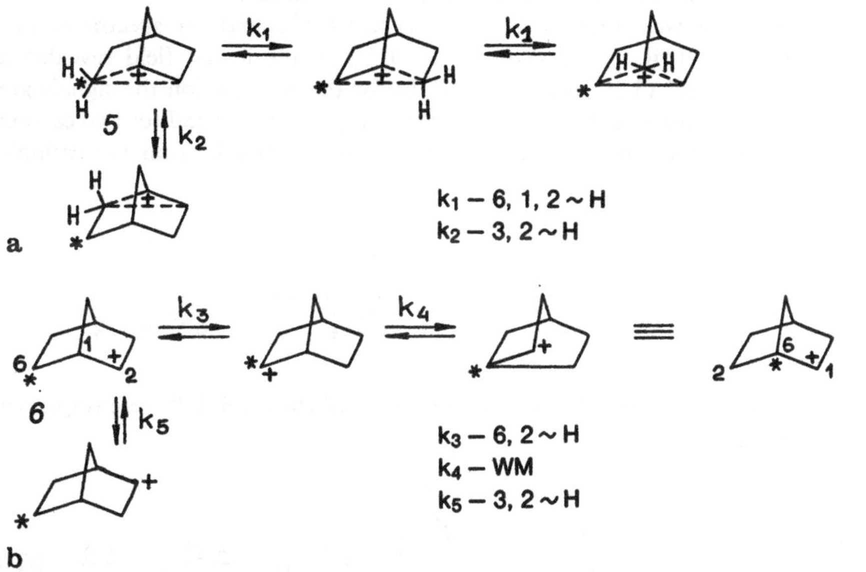

Within —50 to —130 °C there are 3 PMRsignals with an intensity ratio of 4:1:6. This points to the freezing of 3,2-hydride shifts (E_a = 10.8 ± 0.6 kcal/mole; $A = 10^{12.3}\ s^{-1}$). Judging from these data the 3,2-hydride shift rate in a stable 2-norbornyl cation is abnormally low compared with 1,2-hydride shifts in secondary carbocations. Thus the respective activation energies are ~5 kcal/mole for the 1,2-hydride shift in the cyclopentyl cation and 10.8 kcal/mole for the 3,2-hydride shift for the 2-norbornyl cation. This corresponds to the rate ratio 10^{11} at —150 °C and 10^4 at 25 °C. Olah [184] has studied the models of both ions showing that torsional and non-bonded steric effects fail to account for such rate differences. Possibly in the transition state of the 3,2-hydride shift in the 2-norbornyl cation, the σ-delocalization energy is "lost":

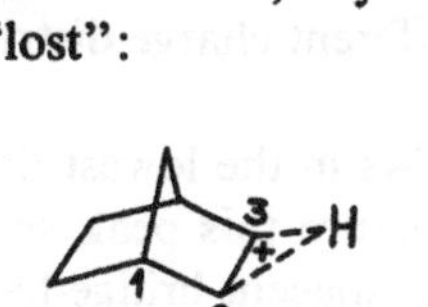

In such a transition state the positive charge is delocalized among C^2, C^3 and exo-H^3 decreasing the "demand" for σ-participation. Hence Olah concluded, that the stabilization amounts to at least 5.8 kcal/mole.

The PMR spectra give no possibility to chose between the nonclassical ion *5* and the rapid equilibrium of a pair of enantiomeric classical ions *6*. The ion structure was therefore studied by Raman laser spectra, because the rates of vibrational transitions are far higher than those of Wagner-Meerwein rearrangements. The spectrum of the norbornyl cation proved to be very similar to that of nortricyclene and quite different from that of norbornane. For rapidly equilibrated classical ions *6*, the RS spectrum would be similar to that of norbornyl derivatives. Besides, the rapid equilibration, according to Olah, would lead to the doublet splitting of the bands of some skeletal deformational vibrations which is not observed.

The ion structure was final established by NMR ^{13}C and ^{1}H spectra at low temperatures. The NMR ^{13}C spectrum at -70 °C in the lowest field revealed a signal at $+101.8$ ppm [6] (3C). In the case of (b) for the classical ion the signals are averaged for the secondary carbon atom (C^2) carrying the charge and for two carbon atoms (C^1 and C^6) located at a distance of one or two bonds from the cationic centre.

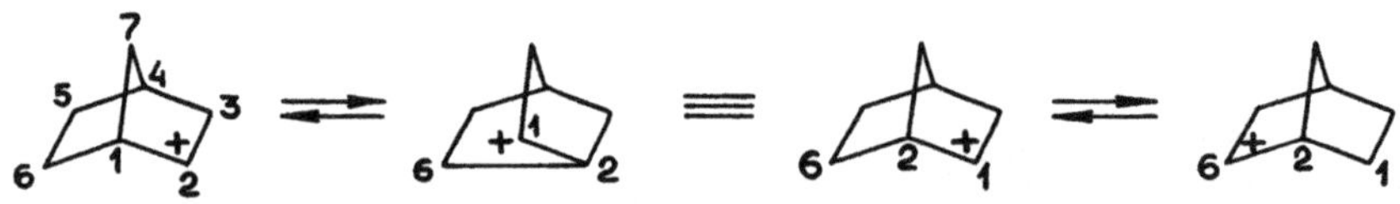

The model compounds used to estimate the averaged chemical shift were isopropyl cation and cyclopentane:

$CH_3\overset{+}{C}_2H\ CH_3$   $\delta^{13}C_{calc.} = \frac{-125 + 150 \cdot 4}{5} = 95.0$

$\delta^{13}C_2 = -125$ $\quad \delta^{13}C = +150$ $\quad \delta^{13}C_{exp.} = +95.4$

The calculation of the averaged chemical shift for the cyclopentyl cation agrees well with a value observed in experiment. However, the same models when used to calculate the averaged shift of C^2, C^1 and C^6 in the classical norbornyl cation $(-125 + 2 \cdot 150/3)$ yield the value 58 ppm quite different from the observed one (+101.8 ppm). To obtain a calculated averaged shift close to the experimental value for the classical ion model one should assume either the chemical shift of the charged C atom to equal -6 ppm or those of either adjacent C atom to equal $+215$ ppm. Both variants (or any other value combinations) are absolutely unacceptable; hence the spectrum supports quite a different structure and a different charge distribution, for instance, like that in the nonclassical ion *5*.

At -154 °C the NMR ^{13}C spectrum of the ion shows in the lowest field a signal of two carbons at $+70$ ppm. In the classical ion structure this peak should be an averaged signal of the secondary cation centre and the adjacent bridge-head carbon. If for the former one takes the chemical shift from the spectrum of an isopropyl cation (-125 ppm) and for the latter — from that of norbornane ($+158$ ppm),

6 Here and further the ^{13}C chemical shifts are given in ppm from CS_2 as external standard.

then the averaged shift will amount to +16.5 ppm which is again widely different from the value observed. This discrepancy points, in Olah's opinion, to the unacceptability of the classical structure for the 2-norbornyl ion.

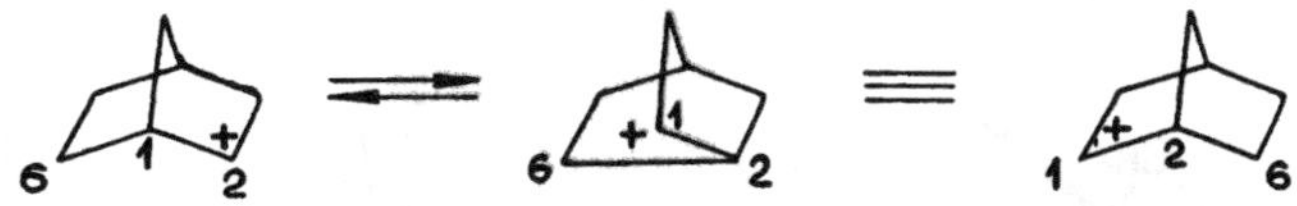

Olah believes that the only structure corresponding to the data of Raman laser and NMR ^{13}C spectra is that of protonated nortricyclene. At —70 °C the "cyclopropyl"carbons C^1, C^2 and C^6 are "averaged" as a result of rapid 6,1,2-hydride shift ($E_a = 5.9 \pm 0.2$ kcal/mole; $A = 10^{12.7}$ s^{-1}). If the protonated nortricyclene structure is corner-protonated, the C^1, C^2 and C^6 are averaged as follows:

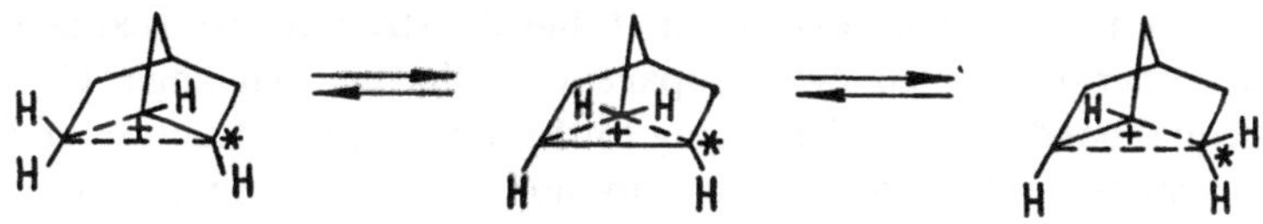

while the edge-protonated nortricyclene is averaged according to the scheme:

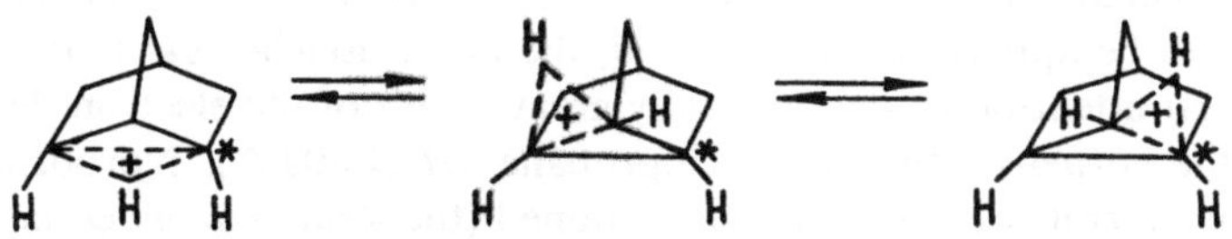

Saunders [222] studied the secondary and tertiary alkyl-carbonium ion rearrangements in superacid media by the NMR method to conclude that the intermediates in these reactions are corner- (but not edge-) protonated cyclopropanes. At the same time Lee et al. [223] studied the trifluoroacetolysis of 1-propyl-1-^{14}C-mercury perchlorate and obtained data indicating that at least part of 1-propyl trifluoroacetate is formed from edge-protonated cyclopropane.

Quantum mechanical calculations made by Klopman [224] have shown the least stable of the norbornyl ions to be the classical ion *6* and the face-protonated nortricyclene *34*; they are by 40 kcal/mole less stable than the edge- *35* or corner-protonated *5* nortricyclenes.

35 34 5 6

However, the difference in the stability of the latter two ions according to calculations, is no more that 3 kcal/mole while the calculation error is 10 kcal/mole (!).

Pople et al.[129] have made ab initio calculations for nine isomeric cations of the common $C_3H_7^+$ formula. They have shown the least stable structures to be those of the face- *I* and the edge-protonated *II* cyclopropanes which are by 10 kcal/mole less stable than the corner-protonated cyclopropane *III*.

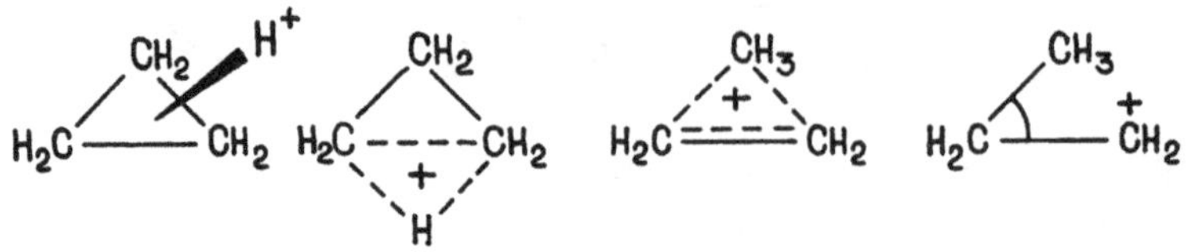

(I) (II) (III) (IV)

The stable primary 1-propyl cation *IV* has the strongly distorted geometry of the opened corner-protonated cyclopropane with the unusual angle C—C—C = 83° and unequal bond lengths. According to the authors, the structure of *IV* illustrates the conventional character of the division of carbocations into classical and nonclassical because there may be classical cations with a geometry sharply different from the accepted one. The 2-norbornyl cation is too large a molecule for ab initio calculations, but the authors consider the bond lengths they found for $C_3H_7^+$ ions to be applicable in describing the other nonclassical ions as well. They call particular attention to the unusually great length of the electron-deficient C—C bonds in the corner-protonated cyclopropane *III* (1.803 Å). Norbornane with its distorted five-member cycles is greatly strained (the strain energy equals 17.55 kcal/mole[225]). Most of this energy is due to deviations of the bond angles from normal. The classical 2-norbornyl ion retains this cycle strain.

Dewar[226] calculated data on the $C_3H_7^+$ ion by the MINDO/2 method; the results differ from those obtained by Pople et al. According to Dewar the most stable particle is the edge-protonated cyclopropane.

Calculations by the MINDO/3 method showed the relative stability of the classical and nonclassical 2-norbornyl cations[227]; consequently the classical ion is more stable than the nonclassical one by 2 kcal/mole. The application of this method in calculating the norbornane formation heat led to an error of 20.8 kcal/mole[228]!

The ab initio calculations taking into account electron correlations have led to the conclusion that the nonclassical 2-norbornyl ion is more stable than the classical one by 8–13 kcal/mole ([229], cf. [230]); similar conclusions have been drawn by the authors of[231]. In[229] the edge-protonated 2-norbornyl cation was shown to be comparable in stability with the corner-protonated one.

Besides certain discrepancies between the results of quantum chemical calculations made by different methods it should be born in mind that they refer to the gas phase and hence take no account of solvation effects.

A choice between the structure of corner- or edge-protonated cyclopropanes for the stable 2-norbornyl ion can be made, according to Olah from PMR spectrum recorded at −150 °C. In the lowest field there are two signals: 3.05 (2H) and 6.6 ppm (2H). These data deny the structure of the edge-protonated cyclopropane since in the latter case there would be three signals in the low field with a 1:2:1

relative area. In the case of classical ion equilibrium the peak 6.6 ppm is an averaged shift of protons at the secondary cation centre and at the bridge-head C atom adjacent to the cation centre. If the isopropyl cation (δ 13 ppm) is taken as a model of the first shift, then the bridge-head proton shift must be 0.2 ppm $\left[\left(\frac{13+0.2}{2}\right)=6.6\right]$ which is quite unacceptable because in norbornane the bridge-head proton shift is equal to δ 2.18 ppm.

The refined parameters of the NMR ^{13}C spectra of the 2-norbornyl ion at —150 °C are [15] presented:

Chemical shifts in ppm relative to TMS; in the parentheses the $J(^{13}C—^{1}H)$ constant values in cps

$C_{1,2}$	C_3	C_4	C_5	C_6	C_7
125 (d, 184.5)	48 (t, 153.5)	33.4 (d, 158.6)	28 (t, 160.2)	22.4 (t, 145.8)	48 (t, 153.5)

On the basis of all the data obtained Olah has ascribed to the 2-norbornyl cation the structure of a corner-protonated nonclassical cation. The bond nature can be best described with a two-electron three-centre molecular orbital (2e3C):

The chemical shift values for $C^{1(2)}$ and C^{6} atoms, as well as those of $^{1}J_{CH}$ constants for these atoms indicate that the charge resides mainly at the "cyclopropane": $C^{1(2)}$ atoms and not on the bridging C^{6} atom which is practically an sp^3-hybridized atom. The resonance theory implies that the main contribution is made by the resonance structures *6* and *7*, and not by *8*; Olah concludes the formula *5a* is a better representation than the usual formula *5*.

Considering Olah's observations one can hardly regard the 2-norbornyl ion, at least under the stable ion conditions, as the usual π-complex (Dewar). For theoretical reasons Winstein considered the charge on C^{6} in the nonclassical 2-norbornyl ion to be rather low [232, 233]. Quantum chemical calculations show the C^{6} atom in the nonclassical ion to be in the sp^3-hybridized state; this increases the coulomb integrals and hence the orbital overlaps of C^6—C^1 and C^6—C^2 and decreases the charge on C^6.

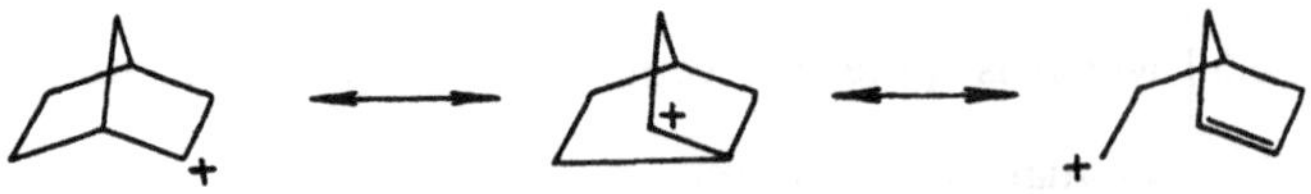

6 7 8

5a 5

Schleyer has come to the same conclusion on the character of charge distribution in the 2-norbornyl ion formed in solvolysis reactions. To estimate the positive charge distribution he introduced an OCH_3 group in the corresponding positions and studies their effect on solvolysis rates [234]. The introduction of OCH_3 group to 5-exo- or 6-exo-positions does not greatly change the exo:endo solvolysis rate ratio (from 300 to 100).

The introduction of the OCH_3 group to the 1-position sharply increases the $k_{exo}:k_{endo}$ ratio (up to 6200); this is accounted for by a considerable decrease in the solvolysis rate of endo isomer (-I-effect). The solvolysis rate of exo isomer changes little; the positive resonance effect of the OCH_3 group added to the positively charged centre in the nonclassical ion is compensated by its -I-effect.

In 1972 Olah revised his ideas of the structure of the norbornyl ion and its analogues under the stable ion conditions. While earlier he practically identified the static 2-norbornyl ion with the corner-protonated nortricyclene *5a* and denoted its structure by the formula *5a* in [34] he considers two possible ways of protonating the cyclopropane ring — on the C—C bond and on C—H bond to form, in the case of protonating the cyclopane itself, the C-cycloproponium (edge-protonated) ion with structure *II* or the H-cycloproponium (corner-protonated) ion with structure *I*.

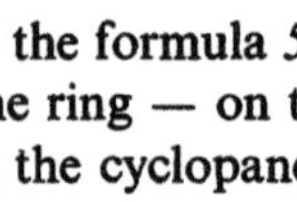

5a

Also, he introduces an additional ion *III* which he calls C-ethenemethonium (methyl-bridged) ion. The latter, he believes, is not formed on cyclopropane protonation but results as intermediate in a rearrangement of the initially formed ions of type *II*. Ion *III*, unlike the two others, contains tetracoordinated methine atoms and a pentacoordinated carbon of the methyl group. Olah gives the following equilibrium scheme of cyclopropane protonation:

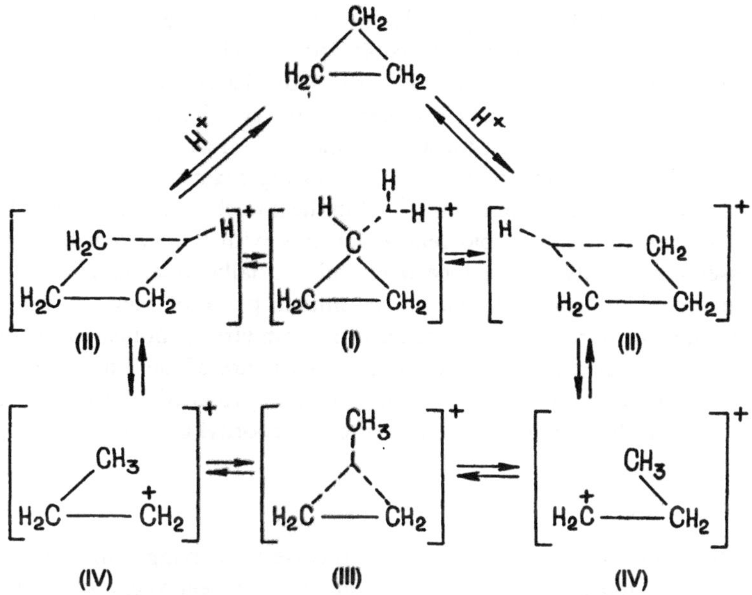

The 1-propenium ion *IV* with the angle C—C—C = 83° has been mentioned on p. 76 in connection with calculations showing this primary carbenium ion to be in the minimum of the potential energy curve for $C_3H_7^+$ ions [129]. In Olah's opinion, the H-cycloproponium ion *I* represents a high-energy transition state, the interconversions of edge-protonated cyclopropanes proceeding not via ion *I* as a transition state, but via ion *III* as an intermediate.

In a similar way the 2-norbornyl ion as a static particle (t ≦ −150 °C) *5a* represents an analogue of ion *III* (methylene-bridged), and the 6,1,2-hydride shifts occur via different forms of protonated nortricyclene; the classical ion *6* is similar to the 1-propenium ion *IV*.

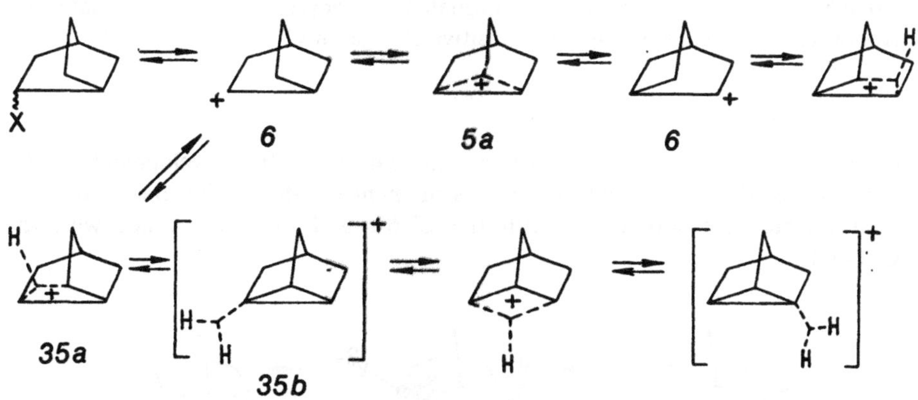

The 6,2,1-hydride shifts can proceed via notricyclonium ions *35a* and *35b* while the Wagner-Meerwein rearrangement — only through delocalization of the C—C bond, i.e. via the norbornonium ion *5a*.

As already noted, Fong postulates a degenerate rearrangement with a rate constant $10^{12}\ s^{-1}$ theoretically possible for the classical 2-norbornyl cation [152]. This viewpoint has been criticized [15] since the barrier is ~1 kcal/mole while its experimentally obtained value (from spectral data) is 5.8 kcal/mol (the barrier of the 6,2,1-shift) which is too high for any "tunneling" effect. MacLean and Mackor had shown C—H bond tunneling to be absent during degenerate rearrangement (the 1,2 hydrogen shift) in benzenium ions [235]. If in the 2-norbornyl ion observed at —150 °C there still occurred superrapid rearrangements with a very low barrier, then such processes could only proceed between substantially σ-delocalized, but somewhat unsymmetrical nonclassical ions [15]. A similar point of view — on the possible equilibrium of partially σ-delocalized unsymmetrical nonclassical ions — was expressed by Sorensen [236] in discussing the structure of the 2-norbornyl cation.

An essential argument against Fong's idea is provided by the data obtained by Olah [237] from the ESCA spectrum of the 2-norbornyl cation.

3.7.2 ESCA Spectra

The ESCA method is applicable for processes occurring within time periods of 10^{-16} sec, so ionic species can be characterized regardless of their intra- and intermolecular interactions, e.g., Wagner-Meerwein rearrangements, hydride shifts, etc. The method permits the carbon 1s binding energy shifts to be measured directly.

In the spectrum of the 2-norbornyl ion there are two lines separated by a barrier of 1.5 eV, of which the one corresponding to a higher electron energy belongs to two methine carbons while the other — to five. The spectrum shows high-energy lines characteristic of localized classical carbocations. Equilibrium of two enantiomeric classical ions should give an electron spectrum identical with a static classical ion even under very rapid equilibration. The data. obtained correspond to the structure of corner- or edge-protonated cyclopropane. Since the latter possibility is precluded by the other spectra the ESCA method corroborates for the 2-norbornyl cation the structure of symmetrical corner-protonated nortricyclene. The ESCA data also corroborate the absence of any large positive charge on C^6.

3.7.3 2-Substituted Norbornyl Cations

One can assume that at the start of the cation centre development at C^2 ionization precedes σ-participation; at this moment the developing p-orbital at C^2 is well oriented for interaction with the C^6-6-exo-H orbital but not with the C^1C^6 bond.

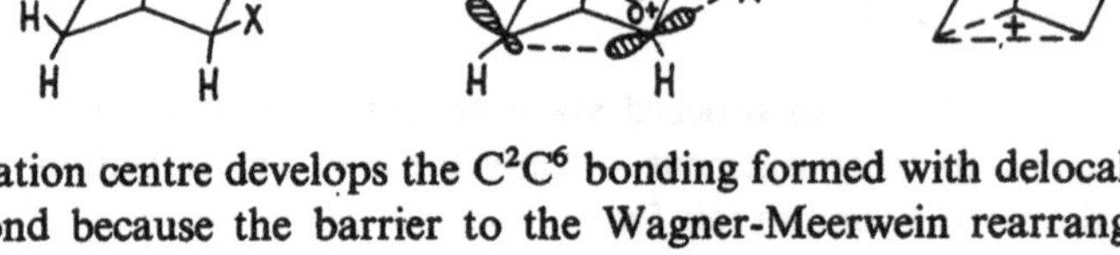

As the cation centre develops the C^2C^6 bonding formed with delocalization of the C^1—C^6 bond because the barrier to the Wagner-Meerwein rearrangement in the secondary-secondary ion systems must be comparatively small. The related barrier in 2-alkylnorbornyl ions should be much higher since it corresponds with the transition of the tertiary ion to the secondary one.

The PMR spectra of 2-methyl- *130* and 2-phenylnorbornyl ions *131* at —60 °C are those of static species where there are no 6,2,1-shifts. In the temperature range —20 to +20 °C such hydride shifts for ion *130*, judging by the PMR spectrum, do occur; the energy barrier is about 15 kcal/mole, by 9 kcal/mole higher than for the 2-norbornyl ion. In ion *131*, even at +35 °C, there are no 6,2,1-hydride shifts. In the NMR ^{13}C spectrum of ion *130* the signal of the C^2 carbon is located at —76 ppm [7], i.e. it is shifted by 66 ppm to the high field relative to the signal of C^1 of the model ion *132*.

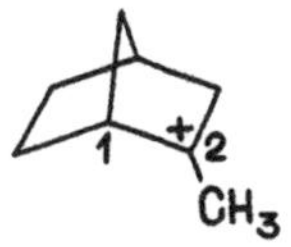

130

$\delta C^2 = -76$ ppm

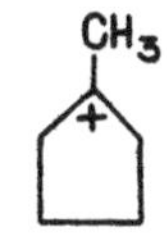

132

$\delta C^1 = -142.2$ ppm

133

$\delta C^1 = -70$ ppm

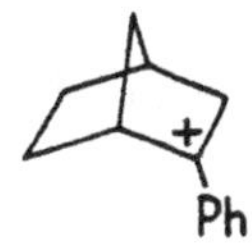

131

$\delta C^2 = -65.8$ ppm

This essential difference may be due to the contribution of σ-delocalization. It will be recalled that the shift for C^2 of the 2-norbornyl ion is +70 ppm; the substitution of hydrogen by methyl at the cation centre usually results in a shift to the low field by about 10 ppm, i.e. the shift of C^2 of ion *130* in its classical structure should be +60 ppm. Consequently the 2-methylnorbornyl ion is an "intermediate" case between limiting classical and nonclassical ions. If the cyclopentyl and norbornyl ions are taken to have 0 and 100% of σ-delocalization, then on this scale the contribution of σ-delocalization in ion *130* amounts to 31%. About the same extent of σ-delocalization has been observed for the 2-fluoronorbornyl ion [238].

In 1972, when studying the PMR spectrum of the 2-methylnorbornyl ion *130* Sorensen [298] discovered at above —80 °C a degenerate rearrangement of this ion which seems to imply a consecutive Wagner-Meerwein rearrangement, a stereospecific endo-6,2-hydride shift and a new Wagner-Meerwein rearrangement; $\Delta F^{\neq}$ is equal to 11.1 kcal/mole.

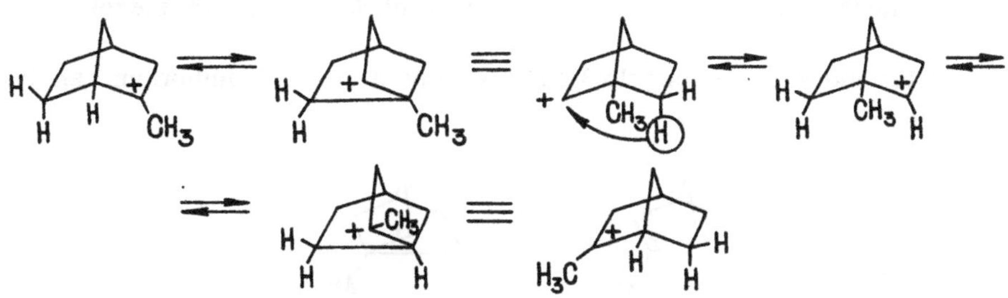

The close similarity of chemical shifts of C^2 in ion *131* and of C^1 in 1-phenylcyclopentyl ion *133* suggests that the σ-delocalization is very small or absent in ion *131*. Thus, the latter under stable ion conditions is a classical ion. This is also

7 Relative to CS_2.

corroborated by the Farnum's study of this ion [239]. Later Farnum studied the PMR spectra of a whole series of 2-aryl-2-norbornyl ions [240] and showed that in these ions the σ-participation of the C^1—C^6 bond displays itself in case the aryl group is a stronger acceptor than the phenyl is.

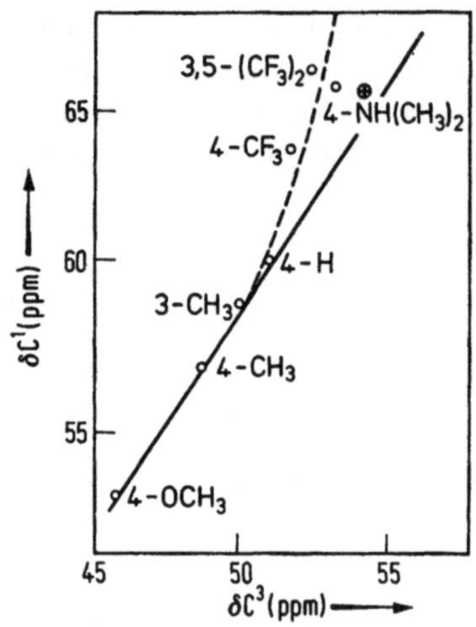

Fig. 7. A plot of δC^1 vs δC^3 chemical shifts of 2-aryl-2-norbornyl cations [241]

In 1977 Olah [241] studied the characteristics of the ^{13}C NMR spectra of the same series of ions. Fig. 7 presents a graph of δC^1 depending on δC^3; the observed break in the slope is accounted for by an increase in the contribution of the C^1—C^6 bond σ-participation to the charge delocalization with increasing electron demand at C^2. Of similar character is the plot of δC^1 vs σ^+. The character of the latter dependence does not change qualitatively when use is made of σ^{++} constants [242] reflecting an increased "demand" of stable ions in comparison with solvolysis transition states.

Similar results were obtained by Farnum et al. [243]. He observed curves which cannot explain the assumptions of equilibrium pairs of classical ions for the following reasons:

1) For the model 6-aryl-bicyclo[3,2,1]octyl ions a similar equilibrium can be assumed as well:

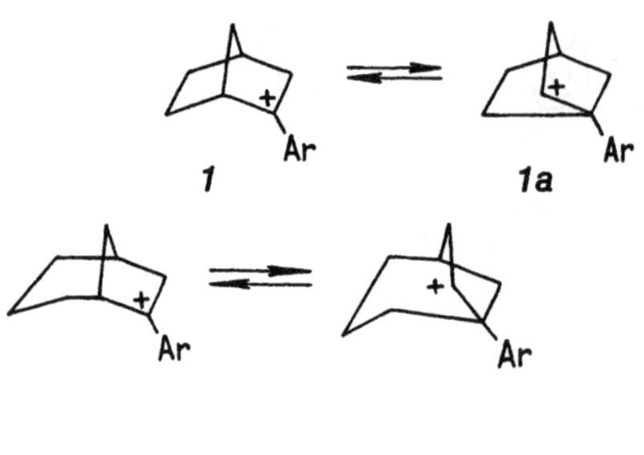

In the last case a deviation from linearity must also occur, just as for norbornyl ions, but in reality it is not observed.

2) The deviation from linearity of the plot δC^1 vs δC^3 for 3,5-bis(trifluoromethyl)-phenyl ion reveals that in the equilibrium between *1* ⇄ *1a* it is necessary to have a *1a*:1 > 10^{-2} concentration ratio. At the same time the analysis of the PMR spectra of 2-arylnorbornyl ions shows that for the same ion one should expect a ratio of *1a*:*1* ~ 10^{-4}–10^{-5} [240], i.e. by far less than required to account for the ^{13}C NMR data.

Applying a similar approach Coxon et al. [244] have shown that in the series of stable 1,7,7-trimethyl-2-arylnorbornyl ions σ-delocalization manifests itself even more effectively than in that of 2-aryl-substituted ions investigated by Olah. Comparison of the C^1 and C^7 shifts in these two series testifies to much larger deshielding in the former case than what one could expect from the effect of substituents.

By observation of symmetrical 1,2-bis-substituted norbornyl cations, Nickon and Lin [245] have shown the 1,2-dimethoxy-2-norbornyl cation to be classical and studied its degenerate rearrangement. Later Olah [246] using the NMR ^{13}C and H^1 spectra, as well as Raman laser spectra studied the 1,2-dimethylnorbornyl ion *134* and showed that in the latter, just as in the 2-methylnorbornyl ion, σ-delocalization of C^6-6-exo-H-C^2 occurs. But while in the 2-methylnorbornyl ion the barrier to the Wagner-Meerwein is rather high (the tertiary-secondary ion transition), in ion *134*

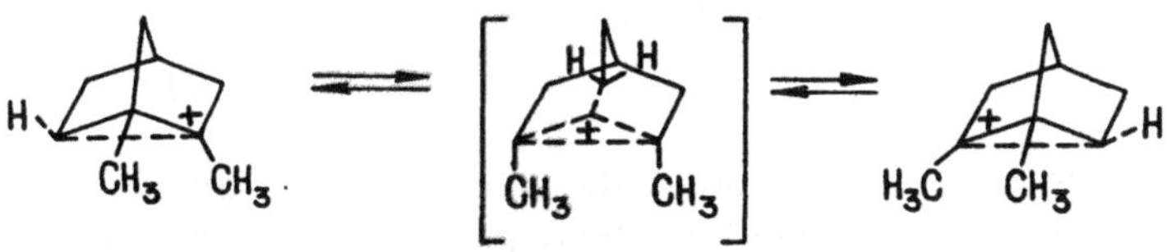

134 *135*

it is very low and even at −130 °C the rearrangement proceeds at a high rate (the degenerate tertiary-tertiary ion rearrangement). The transition state for this rearrangement must resemble the nonclassical 1,2-dimethylnorbornium ion *135* (for Olah's new nomenclature) which obviously has a higher energy than ion *134*.

Subsequently Sorensen [247] critized Olah's method of comparing the observed chemical shifts with those expected on the basis of comparisons with models. To choose between one nonclassical ion *135* and rapid equilibration of classical ions *134* the equilibration of "modified" ions *136* and *137* was studied. Addition of another two CH_3 groups to C^3 should not, the author believes, greatly affect the cation structure. In the PMR spectra even at −130 °C no separate signals of isomers *136* and *137*

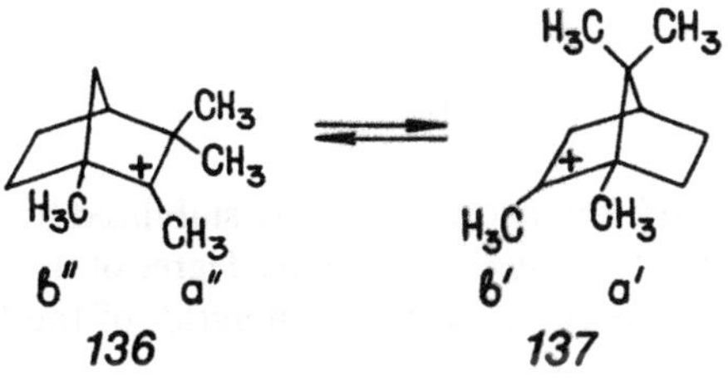

136 *137*

have been observed. However, since structures *136* and *137* are not degenerate the averaged signals of the methyl groups a and b for ions *136* and *137* must have different shifts which has really been observed. The values of averaged chemical shifts depend on the equilibrium position between the ions and should therefore vary with temperature. The author has shown that this dependence does exist and has thereby proved the rapid equilibration of classical ions.

When studying polymethylbornyl cations of type *136* Sorensen [248] found the exo-3,2-methyl shift in these ions to occur much faster (by about 10^8 times) than the endo-3,2-methyl shift. The author offers the following explanation. The presence in tertiary 2-norbornyl ions of some overlap of the back-side part of the sp^3-orbital of C^6-6-exo-H with the empty p-orbital at C^2 causes the change in spatial orientation of the latter. As a result the exo substituent at C^3 turns out to be more coplanar, and the endo one, less coplanar with the vacant p-orbital at C^2 *II*, which does explain the observed difference in the rates of the 3,2-shift. In more nucleophilic solu-

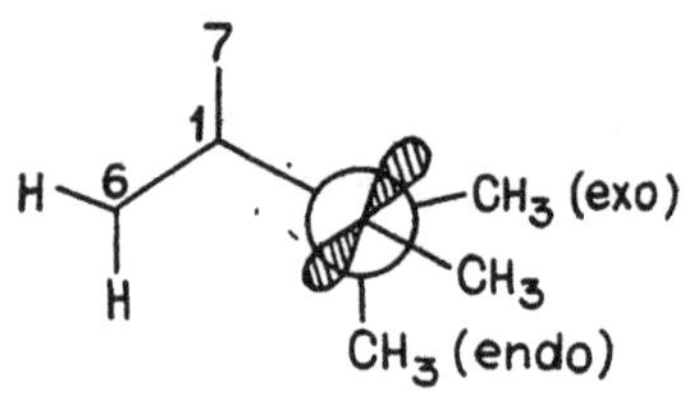

(I)

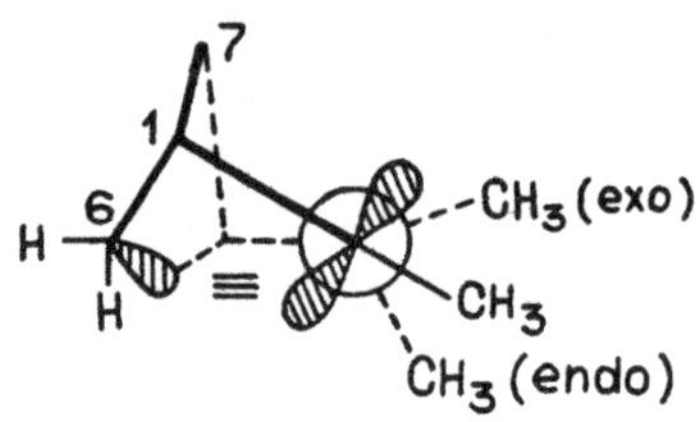

(II)

tions used in works on solvolysis the stabilization of the cation centre by the solvent is higher, the demand for σ-bond participation is less so the ion geometry begins to approach the normal model of classical ion *I* which enormously decreases the exo:endo rate ratio of 3,2-methyl shift — from $\sim 10^8$ in superacids to $\sim 10^3$ under the conditions of acid catalysis.

In 2,3-dimethyl-2-norbornyl cations the rate of the exo-3,2-hydride shift is far higher than that of the endo-3,2-hydride shift (the difference in activation energies ≥ 5.5 kcal/mole [249]). Contrary to Olah's data [250] of the two isomeric ions-2,3-endo-dimethyl-2-norbornyl *138* and 2,3-exo-dimethyl-2-norbornyl *139* the more stable one is the former [249]. As has been noted earlier, the vacant p-orbital at C^2 interacts mostly with the exo substituent at C^3; this hyperconjugative interaction with the C^3—H bond is more effective than with the C^3—CH_3 bond.

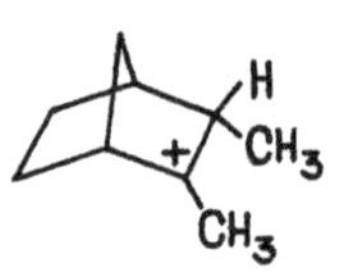

138

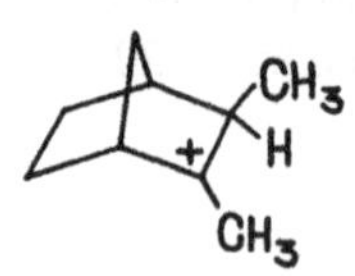

139

Both the secondary and the tertiary 2-norbornyl ions are stabilized by the σ-participation of the C^1—C^6 bond, the only difference being in the degree of this participation [236]; as the measure of the latter Sorensen used the "strength" of the C^2—C^6 bond (Fig. 8).

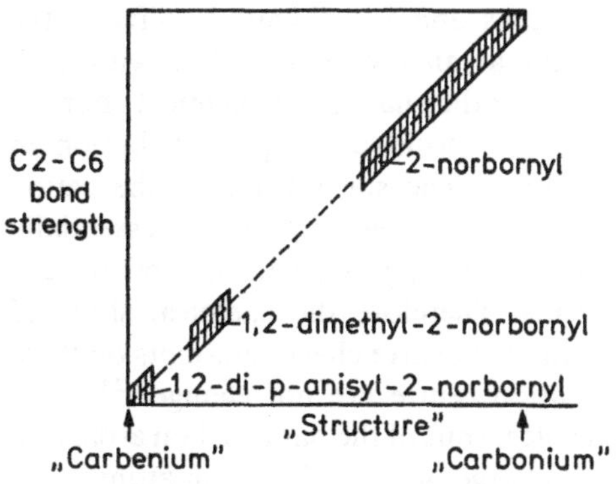

Fig. 8. Sorensen's graph for secondary and tertiary 2-norbornyl cations

Kramer [251] attempted in 1975 to revaluate critically Olah's data on the stable 2-norbornyl ion structure. Let us consider briefly Kramer's objections.

3.7.4 Critical Revaluation

1. *ESCA spectra*: Olah has shown that while in the spectrum of the 2-norbornyl cation the signal of the cation centre is separated from those of the other carbons by a barrier (ΔE_b) of 1.5 eV, for the cyclopentyl cation it amounts to 4.3 eV; this is due to far larger charge localization in classical carbocations. Kramer maintains that the cyclopentyl cation is an erroneous model to compare with the 2-norbornyl cation. However, in the spectrum of the 2-methylnorbornyl cation (which can obviously raise no objections as a model compound) the value of ΔE_b is 3.7 eV, i.e. again much higher than for the 2-norbornyl cation.

Further Kramer has measured the peak area ratio in the ESCA spectrum of the 2-norbornyl cation (from the figures published by Olah) and found it is not 2:5 (as required by the nonclassical ion model) but 4.5:1 or 6:1, depending on the method of measurement. Olah, however, points out that in the ESCA spectrum the main thing is not the ratio of peak areas which is often distorted due to admixtures coming from the vacuum systems and so can somewhat vary from experiment to experiment, but that of the ΔE_b value. Finally, Kramer assumes the reported spectrum to belong to a neutral compound or a mixture of substances rather than to a stable ion. But according to Olah, the NMR spectra of the solutions under study before and after the recording of the ESCA spectrum are perfectly identical and coincide with that of the 2-norbornyl cation.

The theoretical calculations [230, 252] published in 1977 agree with Olah's experimental data and corroborate the conclusion that the 2-norbornyl ions is nonclassical.

2. *The ^{13}C NMR spectra*: To calculate the averaged shift of the C^1 and C^2 atoms for the rapid equilibration of epimeric 2-norbornyl cations Olah used an isopropyl ion as a secondary carbocation model. Kramer observes that the signal of the C^+ atom in the tert-butyl cation spectrum is in a lower field than for an isopropyl cation. Kramer means that the positive charge on the secondary cation centre is *lower* than on the tertiary one; this contradicts the fact that tertiary alkyl carbocations are *more stable* than secondary ones. Hence he concludes that averaged shifts cannot be calculated from the ^{13}C NMR spectrum of the isopropyl cation since it is not a typical secondary carbocation.

Data on the relative stabilities of carbocations and the chemical shifts in the ^{13}C NMR spectra of the same cations do not at all always correlate with each other (for details see [253–255], cf. [256,257]). On the other hand, the chemical shifts of the ^{13}C atoms in carbocation spectra reflect the positive charge distribution in carbons of similar hybridization and substitution. The substitution at the cation centre of the hydrogen by methyl, phenyl, cyclopropyl etc. introduces essentially different effects of the neighbouring groups (α-effect), just as in any other substituents of different sizes and different bonds. Therefore the chemical shifts of cation centres cannot be directly correlated with the positive charge amounts on these atoms except when the substituents at the cation centre are not changed [258–260]. Thus, in benzyl cation spectra it is only the chemical shift of the para-carbon atom that can serve as a relative measure of the positive charge on the C^+ carbon atom.

Moreover, all the changes occurring in the ^{13}C NMR spectra in passing from the isopropyl cation to the tert-butyl one are typical for secondary carbocations in substituting H at C^{α} by the CH_3 group [261] (a tentative analysis of CH_3 group electron effects has been reported by Olah [262]); hence the ^{13}C NMR spectrum of the isopropyl cation can be used to calculate averaged shifts. Indeed, as shown by Olah, the ^{13}C NMR spectrum of the isopropyl cation permits to predict averaged shifts for fast equilibrations of classical carbocations, in particular, for degenerate rearrangement of sec-butyl cation [16]. Irrespective of concrete model compounds in comparing calculated averaged shifts for hypothetic equilibrium of classical ions and experimental shifts of nonclassical ions, as a rule, one can observe a great difference (by tens and even over a hundred ppm) between them, the experimental signal manifesting itself in a higher field; in other words, a considerable part of the deshielding positive charge effect "disappears".

In his chapter-to-chapter comments on Brown's views [13] Schleyer [13] points out that in analyzing the ^{13}C NMR spectra he would like to calculate chemical shifts both for classical and nonclassical ions and to compare these data with experiment. But it is not yet possible, he believes. However, as far back as 1971 Ditchfield and Miller [263] calculated the averaged chemical shift of ^{13}C for equilibration of classical ethyl cations and the chemical shift of ^{13}C for a bridged ethyl cation; the latter signal is shifted to a high field relative to the former by 50 ppm. Secondly, there is a correlation between the chemical shifts of ^{13}C and ^{11}B in the spectra of carbonium ions and the corresponding boranes [264], which was successfully used to determine the chemical shifts of some nonclassical ions [265,266]. Finally, Cheremisin and Schastnev

Table 9. Calculated and Experimental ^{13}C Chemical Shifts and $^{1}J_{CH}$ Values for 2-Norbornyl Cations

Cation		Structure	C^1	C^2	C^1—H	C^2—H
2-Methyl-norbornyl	Calculated	Classical	66	251	153	—
	Experimental		82	273	168	—
2-Norbornyl	Calculated	Averaged classical	151	151	164	164
	Calculated	Nonclassical	122	122	187	187
	Experimental		127	127	185	185

[267] made a direct quantum chemical calculation of the chemical shifts of ^{13}C and the constants of $^{1}J_{CH}$ for isomeric classical and nonclassical ions. The authors conclude from Table 9 that the 2-norbornyl ion has a nonclassical structure and the 2-methylnorbornyl ion — mainly classical.

For the 2-methylnorbornyl cation the calculations suggest a specific interaction of the cation centre C^2 with the C^1—C^6 bond even without any distortion of the structure.

Another variant to identify the classical or the nonclassical nature of carbocations by ^{13}C NMR data has been suggested by Olah, Schleyer et al. [268]. They calculate the difference between the sum of the chemical shifts of all the carbons of an ion and that of all the carbons of the corresponding hydrocarbon formed by adding a hydride-ion. For static and rapidly equilibrated classical carbocations the difference is usually ≥ 350 ppm. For nonclassical ions this value is by hundreds of ppm less: thus for the 2-norbornyl ion $\Delta\delta$ is $+175$ ppm, for the 7-norbornenyl one -1 ppm etc.

Apart from the two "extreme" categories of classical and nonclassical ions there is a large group of ions with intermediate types of structure. Thus, for a 2-methylnorbornyl ion which was earlier, on the basis of the ^{13}C NMR spectrum, assigned the structure of a partially σ-delocalized ion the value of $\Delta\delta$ is 304 ppm: less than the analogous value for the typically tertiary ions — methylcyclopentyl (374 ppm), methylcyclohexyl (362 ppm), 2-methyl-2-bicyclo[2,2,2]octyl (385 ppm) and 2-methyl-2-bicyclo[3,2,1)octyl (355 ppm).

For 2,3- and 1,2-dimethylnorbornyl ions the values of $\Delta\delta$ are 287 and 274 ppm, respectively, lower than for classical 1,2-dimethylcyclopentyl and 1,2-dimethylcyclohexyl ions (~ 350 ppm) being in a rapid degenerate equilibrium. These facts agree with the structure earlier ascribed by Olah to the 1,2-dimethyl-2-norbornyl ion — a structure of partially σ-delocalized, rapidly equilibrated ions (Wagner-Meerwein rearrangement) [246]. The same applies to Saunders's data on the 1,2-dimethylnorbornyl cation in which one CH_3 is replaced by CD_3 (see below) [269, 270].

Referring carbocations to classical or nonclassical on the basis of the value of $\Delta\delta$ (^{13}C) it is also useful to compare these values for the secondary ion in question and for the corresponding methyl-substituted tertiary ion. The comparison of classical secondary and tertiary ions shows small differences (for isopropyl and tert-butyl the difference is -4 ppm, for cyclopentyl and methylcyclopentyl it is -10 ppm), for nonclassical ions these differences are far larger: 2-norbornyl — 2-methylnorbornyl, -129; 2-bicyclo[2,1,1]hexyl — 2-methyl-2-bicyclo[2,1,1]hexyl, -97 ppm etc.

In 1977 Saunders and co-workers [269–271] proposed a new experimental approach for choosing between the structure of the nonclassical ion and the fast equilibrium of classical ions. The substitution of hydrogen by deuterium is accompanied in the first case by an insignificant splitting of signals at the cation centre due to isotopic resonance perturbation, in the second — by a far greater isotopic

140

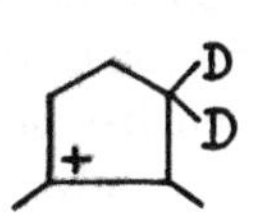

141

effect — equilibrium disturbance. Thus, when studying the influence of the isotopic deuterium effect on the ^{13}C NMR spectrum of the 2-D-norbornyl ion *140* Saunders [272)] showed the isotopic splitting of the peaks for $C^{1,2}$ to be less than 2.3 ppm. Note for comparison that in ion *141* the isotope splitting amounts to 105.3 ppm for $C^{1,2}$; in a undeuterated ion these signals are averaged due to a rapid 1,2-hydrogen shift. The data on ion *140* are in good agreement with a static, symmetrical structure and are not consistent with the postulated rapid Wagner-Meerwein rearrangement.

A similar approach with isotopic effects of deuterium has been developed by Servis and Shue [273)]. They have shown that in static classical carbocations the β-deuterium isotope effect involves a shift of the C^+ signal in the ^{13}C NMR spectrum to a low field (a smaller hyperconjugation effect of CD_3 than CH_3); in delocalized classical ions the same effect does not manifest itself at all, while in nonclassical ions it results in a shift of the C^+ signal to a high field due to isotopic resonance perturbation. Thus in passing from the 2-methylnorbornyl cation to the 2-methyl-d_3-2-norbornyl ion in the ^{13}C NMR spectrum the signals are shifted by (—2.2 ppm C^2), (—0.3 ppm C^3) and (+0.4 ppm C^1); the authors interprete this in favour of the σ-delocalized ion structure, the isotopic substitution increasing the relative content of the canonical form *1b*:

1a *1b*

As a result, the signals of C^2 and C^3 are shifted to a high, and that of C^1 to a low field, respectively. The isotopic signal shift observed agree neither with the fast equilibration of *Ia* and *Ib*, nor with the static structure of the classical ion *Ia*: the latter alternative is also contradicted by the values of the chemical shifts of C^1 (79.9 ppm) and CH_3 (26.6 ppm).

3.7.5 Relative Stability of the 2-Norbornyl Cation

Studying the $R^+ + CO \rightleftarrows RCO^+$ equilibrium in HSO_3F — SbF_5 (1:1) for tert-butyl, 1-adamantyl and 2-norbornyl ions Hogeveen [274–277)] obtained the values of $K = 700$, $2 \cdot 10^4$ and $1 \cdot 10^4$ $l \cdot mole^{-1}$ (20 °C), respectively. These results show the norbornyl and 1-adamantyl ions to be of about the same stability and only 15–30 times less stable than an acyclic tertiary ion. These facts, as well as the reaction rate constant values for the reduction of carbocation salts ($R^+SbF_6^-$) by molecular hydrogen in strong acid media support the nonclassical stabilization of the 2-norbornyl ion.

Several publiscations [17,278,279)] support the stability of the 2-norbornyl cation in the gas phase and in superacid media, this ion is to be by 6–10 kcal/mole more stable than classical secondary cations. Sorensen has shown [273)] that in fluorosulphonic acid the secondary 2-norbornyl cation is only by 7.5 kcal/mole less stable than the tertiary 2-methylnorbornyl ion. This value is much lower than the usual energetic difference between the classical secondary and tertiary ions (~14.5 kcal/mole).

Arnett has determined the heats of formation for 2-norbornyl and 2-methylnorbornyl ions [280, 281] in a low-temperature calorimeter and a chloride ionization reaction in a $SbF_5 + SO_2FCl$ mixture [282]. The value of $\Delta\Delta H_i$ for this pair of ions (7.6 kcal/mole) agrees with Sorensen's data. However, in comparing the ionization heats of secondary and tertiary chloride Arnett thought it necessary to account for different stability. Thus, while tert-butyl chloride is by 5 kcal/mole more stable than sec-butyl chloride, cyclohexyl chloride is by 5 kcal/mole more stable than 1-methylcyclopentyl chloride.

In 1980 Arnett [283] determined the heat of isomerization of the secondary 4-methyl-2-norbornyl cation to the tertiary 2-methyl-2-norbornyl ion in SbF_5—SO_2FCl; in such an experiment the stabilities of neutral molecules are of no significance. Rearrangement of a 4-methyl-2-norbornyl cation into a 2-methyl-2-norbornyl releases 6.57 ± 0.41 kcal/mole; in contrast, rearrangement of sec-butyl to tert-butyl ion releases 14.20 ± 0.60 kcal/mole. Thus, the secondary 2-norbornyl ion is more stable than the "usual" secondary ions by 7.5 kcal/mole; the 4-methyl group is assumed to exert an insignificant effect on the charge at C^2. Also the ionization heats (ΔH_i) of 2-exo-norbornyl chloride and 4-methyl-2-exo-norbornyl chloride into the respective secondary ions are very close to each other: —23.16 ± 0.43 and —22.20 ± 0.49 kcal/mole at —100 °C. All these data indicate a specific stabilizing effect in the secondary 2-norbornyl ions.

In 1981 Farcasiu [284] applied the molecular mechanics technique to calculate the relative stability of secondary methyl-2-norbornyl cations and showed the most stable isomers to be those with methyl in the bridge-head (cf. [285]). Therefore he, as well as Schleyer [286] substituted Arnett's models by others deprived of secondary effects:

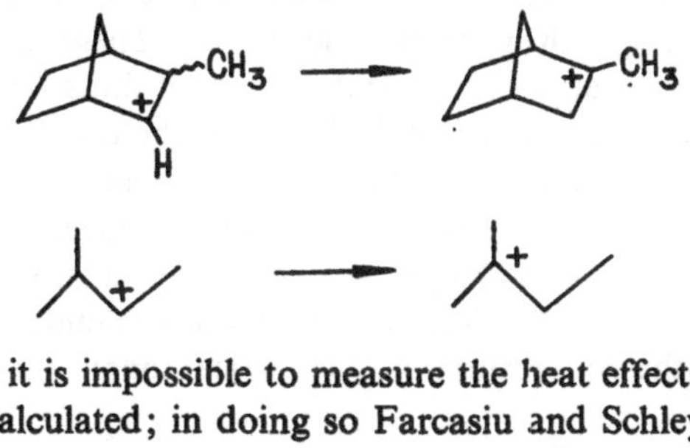

But since it is impossible to measure the heat effects of these isomerizations they should be calculated; in doing so Farcasiu and Schleyer obtained different results. Schleyer found errors in Farcasiu's calculations and determined the heat effects for Eq. (3) and (4) to —8 ± 1 and —14.3 kcal/mole; the extra stabilization energy of the secondary norbornyl ion in superacids amounts to 6 ± 1 kcal/mole. If the secondary 2-norbornyl ion can be assigned the nonclassical structure on the basis of spectral data, the above value is considered the energy of nonclassical stabilization.

Can the data on the structure of a stable 2-norbornyl ion be applied to the intermediates in the solvolysis of 2-substituted norbornanes? Brown says no [13]. Indeed, in superacids the solvation of the cation, though feeble, is unspecific. Nucleophilic media exhibit specific interactions with the counter-ion and one or more solvent molecules, which may alter the structure, energies and properties of intermediates. Nevertheless, if we compare the same substrates in solvolytic media and in superacids, we obtain related information provided we really compare carbocations in different media, and not processes of S_N2 type and stable ions.

4 Bicycloheptyl, Bicyclohexyl, and Bicyclooctyl Cations

4.1 The Bicyclo/2,2,1/hexyl Cation

It is interesting to compare the 2-norbornyl cation with its lower homologue — the 2-bicyclo[2,1,1]hexyl cation. Calculations with the MINDO/3 method show that the steric strain in the bicyclo[2,1,1]hexyl ion makes it a particularly favourable model for exhibiting σ-participation [287].

Meinwald et al. [288,289] assumed from kinetic and stereochemical data that the solvolysis of the 2-endo-tosyloxy-bicyclo[2,1,1]hexane *142* yields an intermediate nonclassical ion *143*:

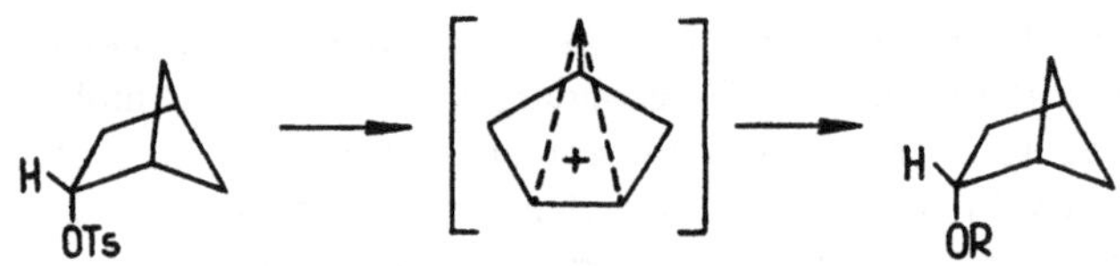

142 *143*

Wiberg [290] studied the structure of the 2-bicyclo[2,1,1]hexyl cation (generated from 2-chlorobicyclo[2,1,1]hexane and SbF_5 in SO_2FCl) under the "stable ion" conditions by PMR. The PMR spectrum consists of three signals at 8.32 (2H), 3.70 (6H) and 2.95 (1H) ppm which are not broadened as the temperature decreases from −90 to −130 °C. If the ion is generated from 2-chlorobicyclo[2,1,1]hexane-2-d_1, then the signal intensity at 8.32 ppm drops to be twice as low. These data show that 6 methylene protons are equivalent at −130 °C while hydrogen shifts do not occur at a marked rate even at −90 °C. If the value of $\Delta S^{\neq}$ is zero the activation enthalpy of "averaging" the methylene groups proves to be less than 7 kcal/mole and that for possible hydrogen shifts — more than 13 kcal/mole. The latter value is far larger than the typical value of the activation enthalpy for hydrogen migration in open carbonium ions (< 5 kcal/mole [291–293]). All these data can be accounted for, according to the authors, to assume the equilibrium of nonclassical ions:

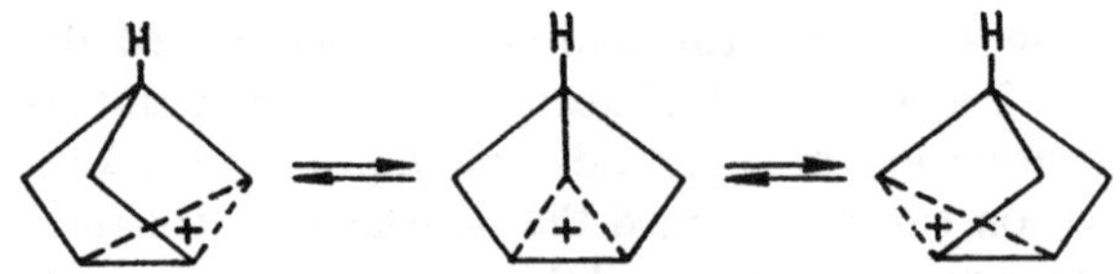

The CNDO calculations show that the activation energy for this rearrangement must be very low [294]. At the same time the nonclassical ions can be expected to permit but a low rate of hydrogen migration from the methylene group to the cation centre.

It is of interest to compare the chemical shifts in the PMR spectra of the 2-norbornyl *5* and the 2-bicyclo[2,1,1]hexyl *143* cations. While the bridge-head protons have similar values of chemical shifts, the other protons of $\geqslant$CH groups in the spectrum of ion *143* exhibit their signals in a much lower field than ion *5*. If the spectra of ion *5* are used to calculate the averaged chemical shift of CH_2 group signal of ion *143*, the predicted value $(2 \cdot 1.70 + 3.05)/3 = 2.2$ is much less than the experimental one (3.70). The authors allocate this discrepancy between the two ions to the difference in the geometry of the protonated cyclopropane fragment.

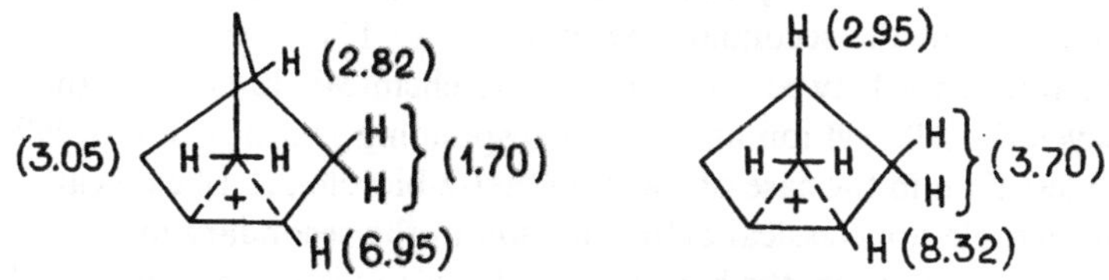

5 143

The solvolysis of tosylate obtained from bicyclo[2,1,1]hexane-2-ol-2-d_1 resulted in bicyclo[2,1,1]hexyl-2-acetate in which deuterium is equally distributed between the "alcoholic" carbon and the neighbouring bridge-head carbon which is in good agreement with the NMR spectra data.

Later Olah et al. [295] studied the ^{13}C NMR spectra of 2-bicyclo[2,1,1]hexyl cations and concluded these ions to be classical; for the secondary ion they suggested the equalibrium of the 6 equivalent ions

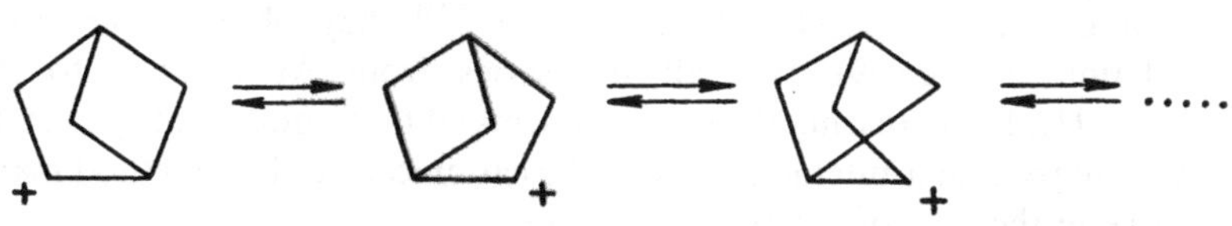

These data also agree with Brown's results on the solvolysis of 2-aryl-2-bicyclo-[2,1,1]hexyl-p-nitrobenzoates [296] for which the value of ϱ^+ was found to be —4.31, i.e. still more negative than that of 2-aryl-2-endo-norbornyl-p-nitrobenzoates ($\varrho^+ = -3.72$). The acetolysis rate of the secondary tosylate *142* is 3 times as low as that of 2-endo-norbornyl tosylate and 1000 times as low as that of 2-exo-epimer; i.e. there are no data in favour of kinetic acceleration due to σ-participation.

Thus, the data reported by Brown and Olah, on the one hand, and those by Wiberg and-Dewar, on the other, contradict each other and fail to provide definite conclusions on the structure of ion *143*.

Sorensen tried to clarify this question [297]; the bicyclo[2,1,1]hexyl cation exhibits thermodynamic and spectral properties "intermediate" between those for typically nonclassical (2-norbornyl cation) and classical ions (cyclopentyl, aliphatic ions). This thesis is favoured by following reasons:

1) The difference in the free energies for the secondary and the tertiary 2-methyl-bicyclo[2,1,1]hexyl cations is 7.0–9.8 kcal/mole while that for 2-norbornyl ions it is 5.5 kcal/mole [236, 298] and for cyclopentyl it is ≤ 12.5 kcal/mole [299].

2) The isotopic effect of the CD_3-for-CH_3 substitution on the equilibration of 1,2-dimethylbicyclo[2,1,1]hexyl cations has an intermediate value as compared to 1,2-dimethylnorbornyl and 1,2-dimethylcyclopentyl ions [270].

3) The chemical shift of the C^+ atom in the ^{13}C NMR spectrum of the 2-methylnorbornyl ion is equal to 270.2 ppm [300], in that of the 2-methyl-2 bicyclo[2,1,1]hexyl ion — to 322.0 ppm [295], of the methylcyclopentyl ion — to 336.7 ppm [301].

4) The averaged chemical shift of the C^1 and C^2 atoms in the ^{13}C NMR spectrum of the 2-norbornyl ion is equal to 125.0 ppm [221], of the 2-bicyclo[2,1,1]hexyl ion — to 157 ppm [290], of the secondary butyl ion — to 171.6 ppm [302].

5) The difference between the sum of the chemical shifts of all the ^{13}C atoms of the 2-bicyclo[2,1,1]hexyl ion and the corresponding hydrocarbon is 297 ppm, i.e. by 100 ppm less than in the case of the 2-methyl-2-bicyclo[2,1,1]hexyl cation; this points to a considerable nonclassical delocalization in the secondary ion [268].

According to Sorensen, the bicyclo[2,1,1]hexyl ion is a nonsymmetrical nonclassical ion:

The partial σ-bridging seems to be a general phenomenon in organic chemistry and is suggested by quantum chemical calculations [129].

Olah et al. tried to record the σ-participation in stable bicyclo[2,1,1]hexyl ions by increasing the electron demand of the cation centre. When they studied the ^{13}C NMR spectra of 2-aryl-2-bicyclo[2,1,1]hexyl ions [303] they observed an excellent linear correlation (r = 0.999) over all the series from Ar = anisyl to Ar = 3,5-bis-$(CF_3)_2C_6H_3$. By increasing the electron demand they coud identify in stable ions only very strong σ- and π-interactions with the cation centre. This method may be even less effective in the case of solvolytic reactions.

4.2 The 7-Norbornyl Cation

When acetolysing the tosylate *144* Gassman and Miles obtained the acetate *146* (~90%); consequently, the reaction proceeds with retained configuration. This is explained by a nonclassical ion *145* [304–307]. Another explanation for the solvolysis of 7-norbornyl tosylate is the fission of the S—O bond rather than that of C—O.

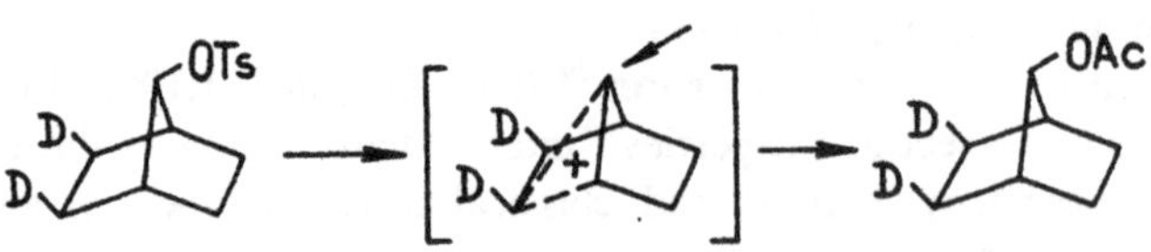

144 *145* *146*

To verify this assumption Gassman solvolyzed the tagged tosylate *147* [308]. With very strong nucleophiles (an inactive substrate) the fission of the S—O bond is really feasible, but under the usual conditions of secondary tosylate acetolysis the C—O bond is split.

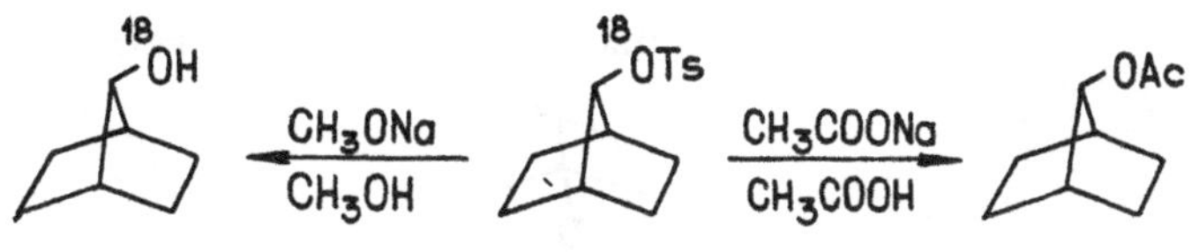

147

Miles suggested that the retention of configuration on acetolysis of 7-norbornyl tosylate can be accounted for by stereochemical control [301,307]. To verify this assumption Dilling and Alford [309] studied the solvolysis of secondary *148*, *150* and tertiary *149*, *151* tosylates. The secondary tosylates *148* and *150* retained their configuration to a very high degree (>95%). Both isomeric tertiary tosylates *149*

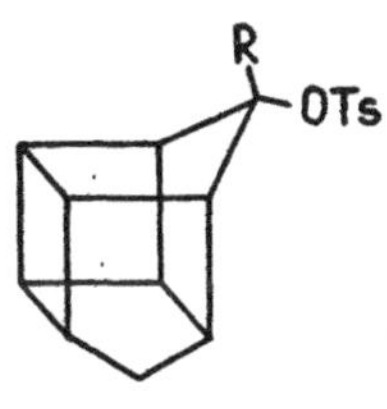

148, R = H

149, R = CH_3

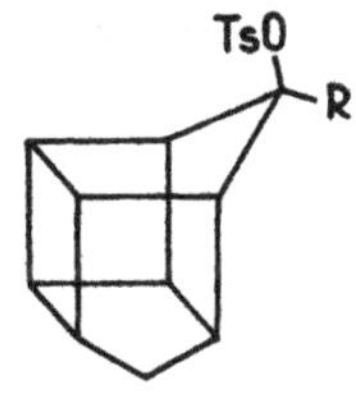

150, R = H

151, R = CH_3

and *151* make up practically an equivalent mixture of acetates. Thus, the high stereospecificity at R = H is not explained by steric effects which in the tertiary systems would manifest themselves still stronger, but the σ-participation which is effective when a secondary ion is formed but is not essential for a tertiary ion (see also [310–312]).

Funke and Winstein [313] obtained, from the solvolysis of α-deutero-2-bicyclo[3,2,0]-heptyl brosylate *152* a mixture of acetates *153–156*. An important assumption is the bridge-flipping in the molecule of the 7-norbornyl ion — *152'* ⇄ *153'* (cf. [314]).

In the ^{13}C NMR spectra of stable 7-aryl-7-norbornyl ions Olah has formed an excellent linear correlation (r = 0.998) between δC^7 of these ions and δC^1 of the respective 1-aryl-1-cyclopentyl ions — the model classical cations [315]. These data testify to the classical nature of 7-aryl-7-norbornyl ions.

4.3 Bicyclooctyl Cations

The bicyclo[2,2,2]octyl and bicyclo[3,2,1]octyl systems have advantages over the norbornyl system in studying the problem of nonclassical carbocations. A higher symmetry of the skeleton, ease of detecting skeletal isomerizations and smaller steric effects — all this helps one to obtain more definite answers to questions on the classical or nonclassical nature of carbocations. Unfortunately, these systems have been investigated neither in depth nor in breadth, but all attention has been focussed on its lower homologue (2-norbornyl ion) [2]. When studying the solvolysis of 2-bicyclo[2,2,2]octyl brosylate *157* Walborsky found the following peculiarities [316].

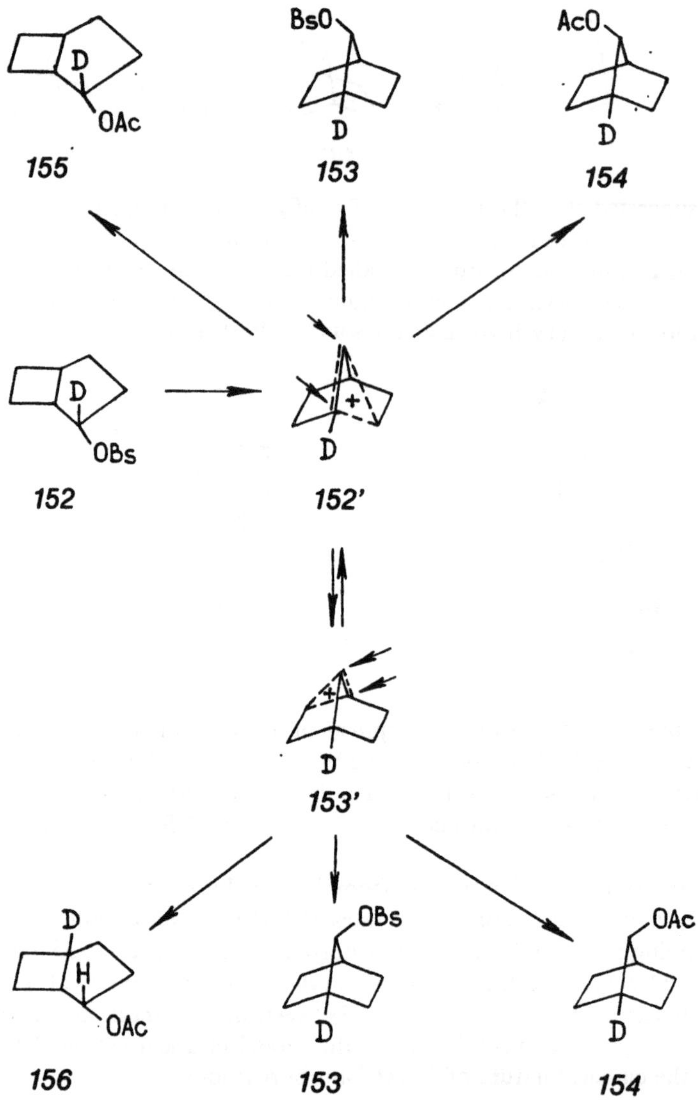

1. The nonclassical ion *158* in contrast to the 2-norbornyl cation *5* has no plane of symmetry and so, if formed from an optically active precursor it gives rise solely to optically active products.

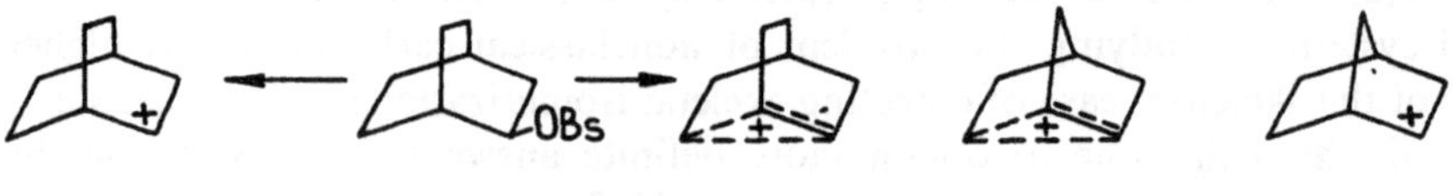

159 157 158 5 6

2. The classical bicyclo[2,2,2]octyl ion *159*, unlike classical 2-norbornyl ion *6*, has a symmetry plane so that products derived from it must be racemic.

3. The skeletal rearrangements occurring via an intermediate nonclassical ion *158* are readily detected since they lead to a different, bicyclo[3,2,1]octyl system (cf. with the rearrangement of norbornyl derivatives to norbornyl ones).

4. The stereochemical interaction of the classical ion *159* with the solvent should not be determined by the ion geometry since it is symmetrical. The participation of the solvent and the shielding by the leaving group should lead to some net inversion of the configuration if the solvolysis proceeds via the classical ion *159*.

The acetolysis of optically active brosylate *157* [316] results in an optically active bicyclo[2,2,2]acetate *160* with 85 ± 15% retention of configuration; the second product — the axial 2-bicyclo[3,2,1]octyl acetate *161* (63:35) — turned out to be optically active as well. Not even any traces were found of the more stable [317] equatorial isomer *162*. From these data the authors [316] concluded an intervention of the nonclassical ion *158*. The degree of retention (85 ± 15%), however, does not preclude the classical ion *159* to play a minor role as well.

If in the solvolysis of tosylates *163* and *164* the reaction is interrupted after 50% of the initial compound has been spent [318, 319], the same mixture of products results; but the mixture of tosylates in either case consists of 70% *164* and 30% *163*, no equatorial 2-bicyclo[3,2,1]octyl tosylate being practically formed. These results indicate that tosylates *163* and *164* are first ionized to form an ion pair whose cation part is certain to be the nonclassical ion *158*. As a result of "internal return" this ion pair may give a mixture of tosylates *163* and *164* and also react with the solvent to yield the acetates *160* and *161*. The rate of ionization exceeds that of solvolysis about fivefold.

To prove the anchimeric assistance in the solvolysis of tosylate *163* Goering and Sloan compared it with 2-endo-norbornyl tosylate having no anchimeric assistance. Without this assistance, the authors believe, the solvolysis rate of *163* must be lower than that of 2-endo-norbornyl tosylate; due to greater flexibility of the bicyclo[2,2,2]-octyl skeleton the steric interactions will not weaken so much in the transition state. Indeed the solvolysis of tosylate *163* proceeds 100 times as fast as that of the model compound which corroborates the anchimeric assistance in the solvolysis of *163*.

Sargent [2] points out, however, that Goering's reasoning does not consider the transition from the sp^3- to the sp^2-hybridized state in the norbornane and the bicyclo[2,2,2]octane systems connected with various angle strain changes. By employing Foote's correlation [320] Sargent shows that if the angle strain effects are corrected, "corrected" value of the solvolysis rate for tosylate *163* is even somewhat less than for 2-endo-norbornyl tosylate. These data also agree with the calculations made by Schleyer [71] who has predicted but a small anchimeric acceleration for tosylate *163*.

Later Walborsky et al. [321] showed that solvolysis of brosylate *157* yields 1.7% of equatorial acetate *162*. The authors conclude that the nonclassical ion *158* is not the only product-forming ion, that the initially forming ion *158* is in equilibrium with an optically inactive classical ion *159* and an optically active classical ion *165*.

Berson has come to a similar conclusion [322–325].

In 1968, detailed kinetic and stereochemical research showed Goering that while 2-bicyclo[2,2,2]octyl esters and 2-exo-bicyclo[3,2,1]octyl esters form an intermediate unsymmetrical nonclassical ion *158* maintaining the optical activity and retaining the configuration), 2-endo-bicyclo[3,2,1]octyl esters form a symmetrical nonclassical ion *166*; thereby the optical activity is completely lost. The equilibrium cannot account for the stereoselectivity in the nucleophilic attack since the classical ion *165* generated from a deamination yields exo:endo products in a 4:1 ratio [326-328].

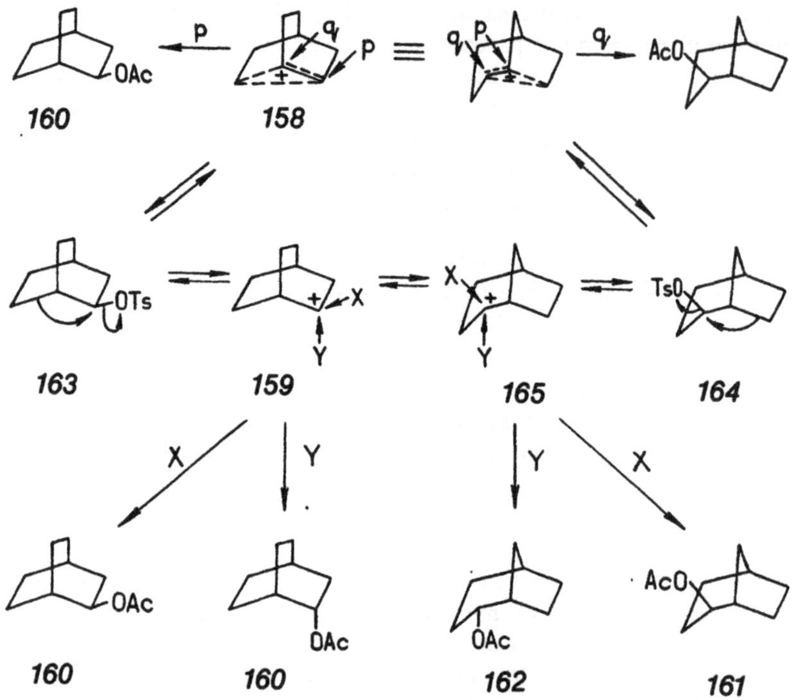

The bicyclo[2,2,2]octyl cation, as shown by calculations, is less stable than the 2-norbornyl one both in the gas phase (by about 6 kcal/mole) [329] and in solutions [330]. At the same time the angle strain in the 2-norbornyl ion is higher, and the torsional and long-range nonbonded effects are comparable with those for the bicyclo[2,2,2]-octyl ion; the difference seems to be due to the norbornyl cation geometry more favourable for σ-delocalization as compared with the bicyclo[2,2,2]octyl ion.

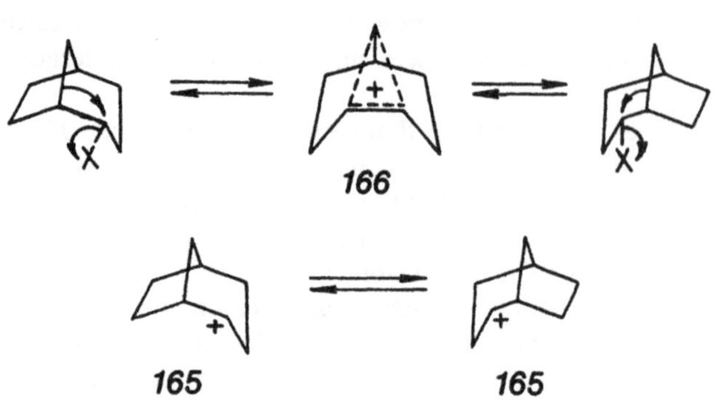

4.4 Nonclassical Ions with Hydrogen Bridged

Sorensen et al. [331, 332] succeeded in generating the first representatives of a new type of nonclassical ions with a hydrogen-bridged atom (μ-hydrido bridging); the unusual spectral characteristics of these carbocations are surprisingly similar to the parameters of diborane:

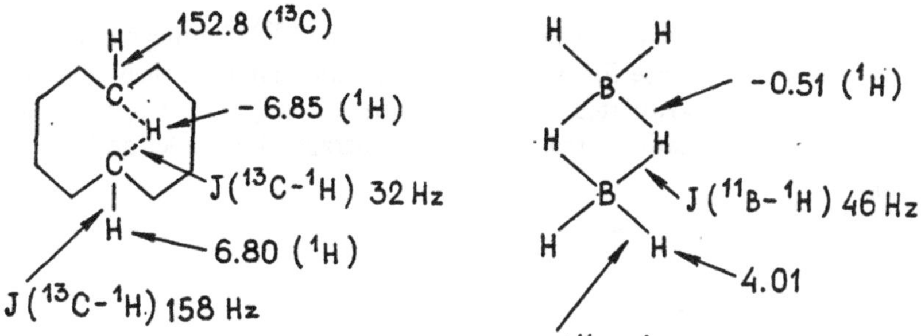

The secondary cyclodecyl ion *167* was generated from 1-chlorocyclodecane in SbF_5—SO_2FCl—SO_2F_2 at −140 °C; at −120 °C this ion decomposes to liberate hydrogen and form a decalyl ion *168*:

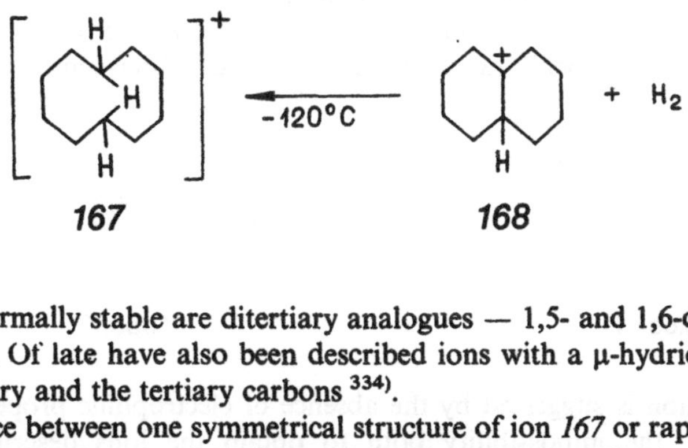

167 *168*

More thermally stable are ditertiary analogues — 1,5- and 1,6-dimethylcyclodecyl cations [333]. Of late have also been described ions with a μ-hydrido bridge between the secondary and the tertiary carbons [334].

The choice between one symmetrical structure of ion *167* or rapid equilibration of cyclodecyl ion in favour of the former is made from the following data:

1) the constant J(^{13}C—^{1}H) between the bridged hydrogen and the neighbouring carbons (32 ± 5 cps); the value of the above constant for the equilibrium should be ~65 cps [(130 + 0)/2];

2) the unusual high-field chemical shift of the hydrogen signal [−6.85 ppm (!) for the secondary ion *167*] in the ^{1}H NMR spectrum; for equilibrium this signal should be expected at ~1.5 ppm;

3) the very insignificant isotopic perturbation of the ^{13}C NMR spectrum when the Saunders method is applied [269]. Thus, for ion *169* the most low-field signal at 142.0 ppm is splitt into a doublet with 0.6 ppm; this agrees well with the isotopic perturbation of "resonance" and the temperature independence of this "splitting".

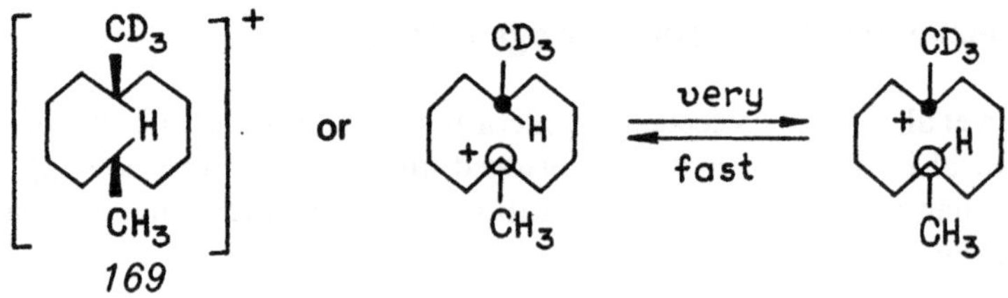

This kind of ions with a bridged hydrogen can be generated [335] from cycloalkanes with 8–11 carbon atoms, i.e. those for which direct transannular hydrogen shifts in carbocation reactions was postulated [336]. Ions with C_8, C_9 and C_{11}-rings as distinct from a cyclodecyl cation, rearrange to contract the cycle and form tertiary ions, e.g.:

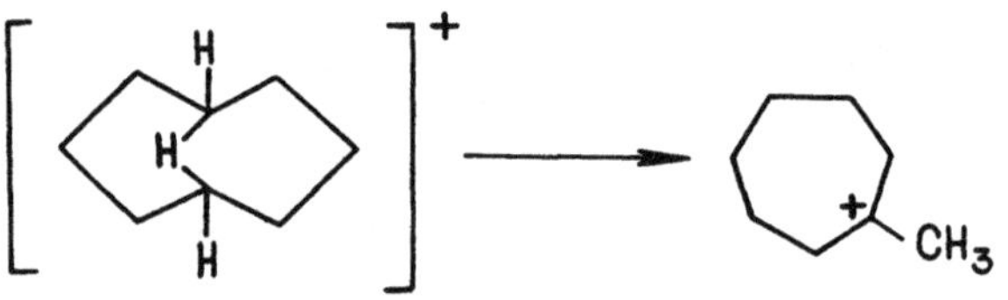

Theoretical calculations of similar ions by the ab initio method and by MINDO/3 show the negative charge on the bridging hydrogen increasing as the separation of the two bridge-head carbons is increased; correspondingly, the terminal protons increase their positive charge. Judging both by calculations and by the values of $J(^{13}C—^{1}H) \sim 150$ cps, the bridge-head carbons have a hybridization intermediate between sp^3 and sp^2. The electronic structure of μ-hydrido ions can be best represented [333] by an "open" rather than a "closed" three-centre two-electron system:

"closed" "open"

This representation is suggested by the absence of electrophilic properties of the bridged hydrogen, the impossibility both to obtain the ions described by the protonation of the central bond in decalins and to abstract the μ-hydrogen as a proton.

5 Homoallylic and Homobenzylic Cations

This section deals with the participation of the electrons of the adjacent π-bond (the olefin double bond or the aromatic ring) at the stage of the cation centre development, as well as with the formation of intermediate nonclassical homoallylic and homobenzylic ions.

Just as in the case of σ-participation, the electrons of the neighbouring π-bond interact covalently with a carbon from which a halogen, a tosylate or other "leaving" group detaches, contributing thereby to the bond fission. This interaction (intramolecular nucleophilic substitution) occurs at the backside of the leaving group.

The term "homoallylic ion" was introduced by Winstein when he studied the so-called "cholesteryl-i-cholesteryl" ion *170* [337–340]. This term stands for carbocations with a β-olefin double bond in which there occurs a "homoconjugation" — a 1,3-overlap of orbitals bypassing the CH_2 group. As is known, in the usual

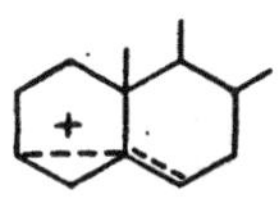

170

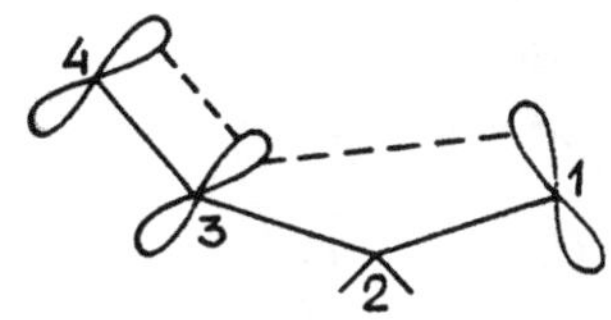

p, π-conjugation the electron delocalization proceeds only via the neighbouring carbons (just as in an allylic ion).

Later on Winstein extended the concept of homoconjugation for a nonclassical MO joining the atoms taking part in its formation into a cyclic system. A particular systems is that of homoaromatic compounds [233, 341–344]. Both for aromaticity and for homoaromaticity the two essential criteria are:

1) increased stability of the substrate in comparison with the similar acyclic conjugate π-system;

2) ability to sustain the induced ring-current which shields the protons located under (above), and deshields the ones located in the ring plane. Examples of acyclic, aromatic and homoaromatic particles are given below together with the respective values of delocalization energy (DE) in β units:

0.828 2.000 1.372

$\beta_1 = 0.5\beta_0$

As seen from these data the DE of the aromatic cyclopropenyl cation is far larger than that of the acyclic allyl cation because for the cyclopropenyl ion with one split σ-bond the 1,3-resonance interaction (β_1) is still significant. Such particles are called homoaromatic. The σ-skeleton of an aromatic particle can be broken at one, two or three sides, and such ions are called, respectively, mono-, bis- and tris-homoaromatic. The prefixes "mono-", "bis" and "tris"-stand here for the number of the sides where the σ-bond is broken or stretched, but not for the number of CH_2 groups introduced in place of one σ-bond.

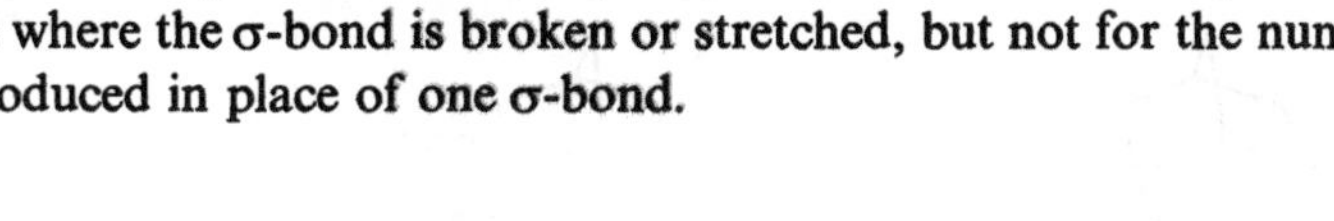

Hehre[345] and Haddon[346] proposed a different approach to the estimation of homo- or antihomoaromaticity than Winstein. The "external" CH_2 group is not considered to be outside the conjugation system and is taken into account together with the adjacent CH groups as part of the cyclopropane ring. In this case the number of π-electrons "delocalized" in the cycle is increased by 2, but thereby the system changes from Hückel to Möbius ring structure. The homotropylium ion (8π) is an aromatic Möbius compound and the bicyclo[3,1,0]hexenyl cation (6π) is an

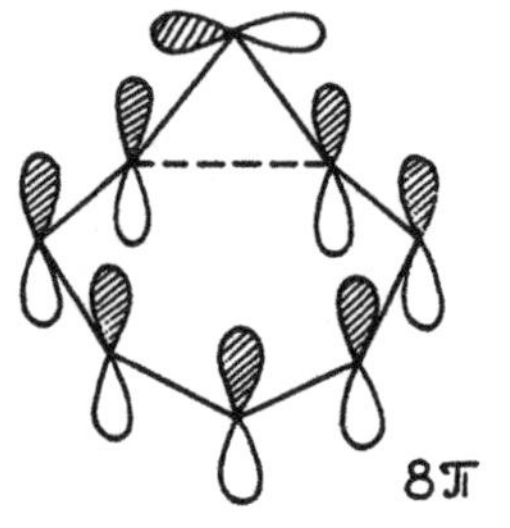

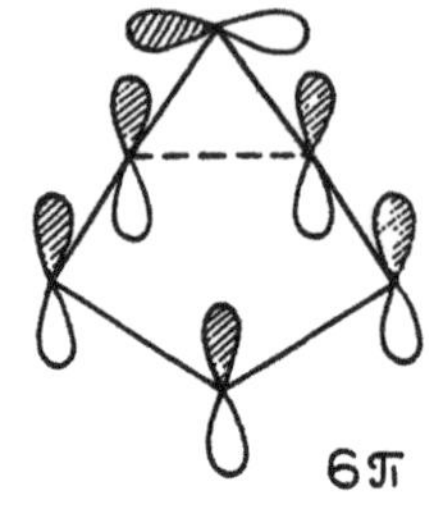

antiaromatic Möbius system (cf., however, [347]). Consequently, the approach of Winstein and that of Hehre and Haddon lead to the same conclusion. In a real ion both mechanisms of electron delocalization — Hückel and Möbius — occur simultaneously.

Jorgensen has utilized extended Hückel, MINDO/3 and perturbation theory calculations of the energy of homoaromatic interaction for some systems[348].

5.1 Stereochemical and Kinetic Data on π-Participation and Formation of Intermediate Homoallylic and Homobenzylic Ions

5.1.1 The 2-Norbornenyl Ion

Much interest has been shown for the solvolysis of esters of 2-exo-norbornenol and its homoallylic isomer — 3-nortricyclenol[337, 339, 340, 349–352].

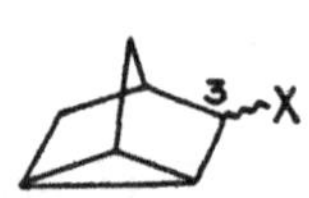

Due to the negative inductive effect the introduction of a double bond decreases the acetolysis rates of both 2-exo- and 2-endo-norbornenyl brosylates as compared with saturated analogues (the numbers stand for the relative acetolysis rates of brosylates):

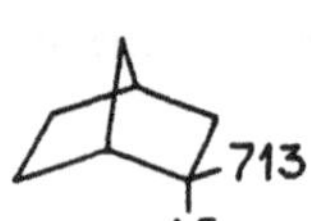

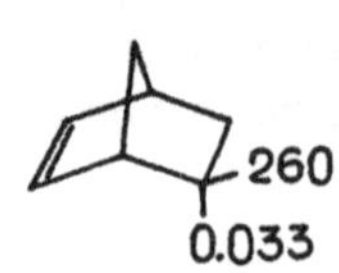

The solvolysis rate of unsaturated exo-sulphonate is 7900 times as high as that of endo isomer, i.e. by far higher than that for saturated analogues (~475). This is explained, since in the unsaturated exo isomer the p-orbital arising in the solvolysis is accessible for π-interaction with the transannular double bond. The orbital overlap is maximal owing to the rising angle strain; the "compromise" provides the energy of stabilization of about 4.2 kcal/mole for the homoallylic ion *171* in comparison with the classical ion *172*.

172 *171* *173*

The major products of solvolysis of 2-exo- and 2-endo-norbornenyl or nortricyclyl halides or sulphonates are nortricyclyl derivatives; 2-exo-norbornenyl derivatives are formed with a small yield. The predominant formation of nortricyclyl derivatives is associated with its higher stability compared to the norbornene one [353]. Even in the direct equilibrium between norbornene and nortricyclene the tricyclic isomer is predominant (3 : 1) at reflux temperature. This difference in the stability of products can be partially reflected on the activation energy for partitioning the cationic intermediates into bicyclic and tricyclic solvolysis products.

Solvolysis of 2-exo-norbornenyl sulphonates tagged with ^{14}C on C^2 and C^3 shows [354], that the process is accompanied by skeleton rearrangements (30–50%). Judging by the extent of tag "entanglement", part of the norbornenyl products (0–40%) is formed from the homoallylic ion *171*, but the precursor of the main part is the symmetrical ion *173* in which the C^1 and C^2 atoms are equivalent. The tag entanglement, however, is less than if the symmetrical ion alone were formed.

The solvolysis of norbornenyl derivatives can be explained in two ways:

1) the solvolysis is accelerated by homoallylic participation alone, but the initially produced ion *171* is not only converted into the final product, but also rearranged into ion *173*;

2) the solvolysis is accelerated both by homoallylic interaction and by σ-participation; so the ions *171* and *173* are formed competitevely from the starting material.

The 2-exo-norbornenyl isomer can be assumed to yield first ion *171* while the nortricyclyl isomers are likely to yield directly the symmetrical cation *173*.

But in 1966, when acetolysing the brosylate *174*, Cristol et al. [355] concluded: the equal distribution of deuterium in the forming norbornenyl acetates between the 3- and 7-positions indicate that the ionization of compound *174* is accompanied by the formation of a symmetrical ion *173* or rapid equilibration of unsymmetrical ions

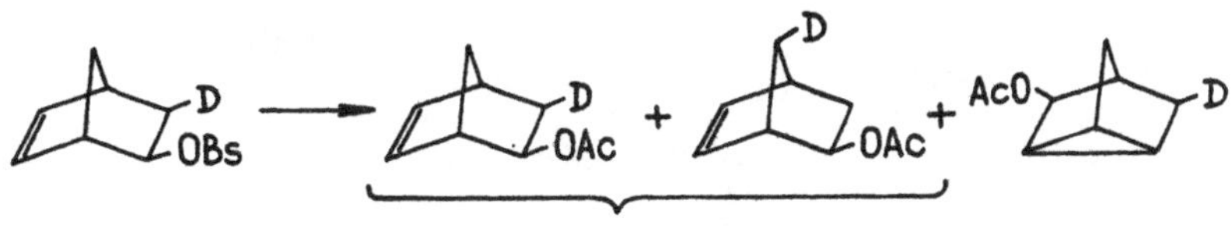

174 7%, 1 : 1 93%

171 and *171a*; this contradicts the initial formation of the homoallylic ion *171* which slowly isomerizes into ion *173*.

173

171

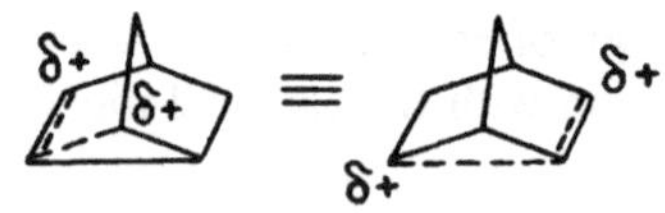

171a

Thus, at first sight the data of Cristol do not agree with those of Roberts [354] on the ^{14}C tag distribution in the solvolysis of the same brosylate in acetic acid. But Lee [356] shows this seeming contradiction to result from different temperatures at which Roberts and Cristol carried out their experiments. At a lower temperature (14–24 °C) the reaction results in the ion pair *A* while at higher temperatures — in two ion pairs — *A* and *B*.

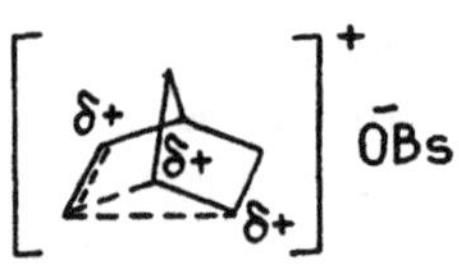

(A)

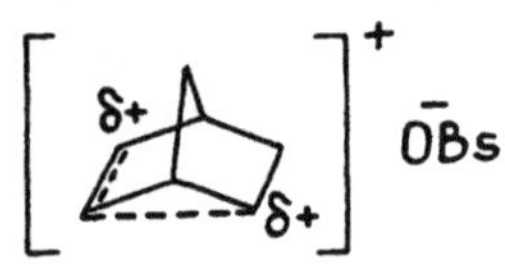

(B)

Later on Cristol [357] repeated Lee's experiments [356] on the solvolysis of brosylate *174*, but the results were different: reactions at temperatures from 24 to 65 °C yield equal amounts of 3-exo-deuterio-2-exo-norbornenyl acetate and of 7-syn-deuterio-2-exo-norbornenyl acetate. It does not seem possible, Cristol remarks, to explain the discrepancy between the results obtained in [356] and [357].

According to Story [358], a symmetrical homoallylic ion *173* is scarcely probable; all the available data can be accounted for by equilibration of enantiomeric cations *171* and *171a*; such a rearrangement must proceed readily enough since the 1,4-interaction is negligible.

Vogel et al. [359] studied the acetolysis of 2-endo-norbornenyl brosylate *175* and observed no processes with C^1—C^7 bond migration. Subsequently, deuterium distribution in the acetolysis products were compared between the epimers *175* and *176* [360]. The exo epimer *176* yields acetates *177*, *178*, *179* and *180* 3.8, 3.8, 46.2 and 46.2%, respectively. The endo epimer leads to: 17.86, 5.34, 38.4 and 38.4% i.e. there are less rearranged products. The authors [360] explain all these data by assuming an intermediate nortricyclyl ion *181*:

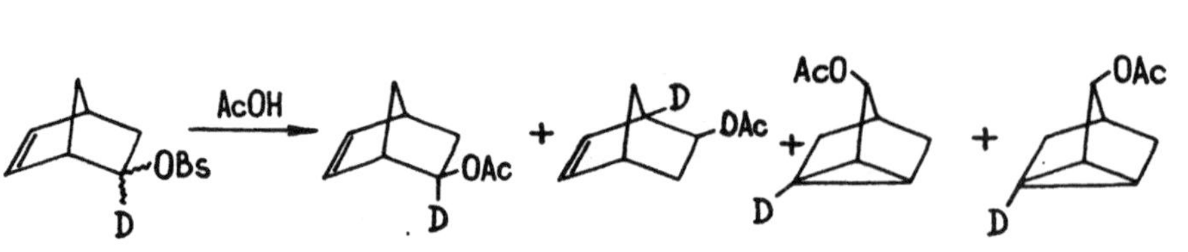

175, 176 177 178 179 180

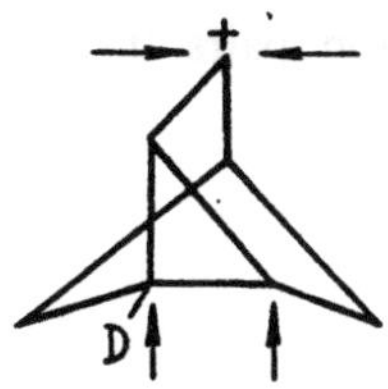

181

Exo brosylate *176* forms the ion *181* which is attacked by a nucleophile on four sides. The less active endo epimer *175* also forms the ion *181*, but with a parallel S_N2-substitution which reduces the amount of nortricyclyl compounds and the fraction of rearranged norbornenyl derivatives. The authors maintain that the concept of equilibrated homoallylic cations is not necessary at all.

The same authors have shown [361] that the solvolytic behaviour of 2-exo- *182* and 2-endo-5,6-bismethylenenorbornyl brosylates *183* is very similar to that of the pair *175* and *176*: the composition of acetolysis products and the tag (deuterium) distribution are well accounted for by assuming an intermediate symmetrical ion *184*.

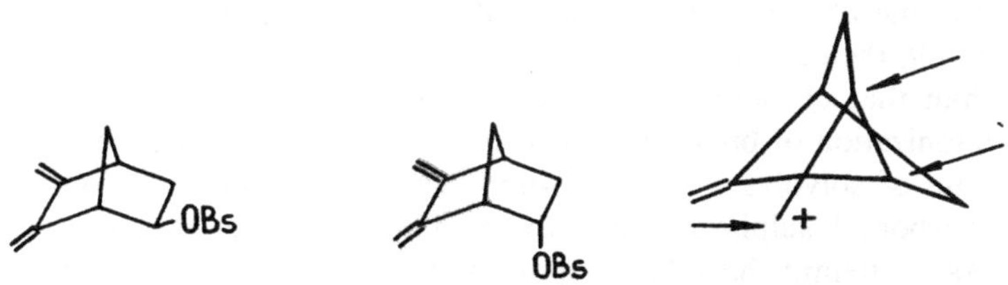

182 *183* *184*

2-exo- or 2-endo-norbornenyl (-nortricyclyl) halogenides or sulphonates are solvolyzed to thermodynamically more stable nortricyclyl derivatives. Different results were obtained by Spurlock in the thermal isomerization of the corresponding thiocyanates [362]. Formed intermediates are ion pairs in which the counterion is a

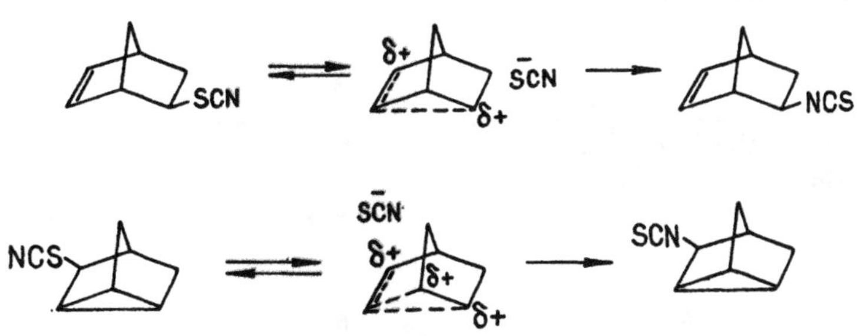

reagent shortening the lifetime of cations due to a rapid reaction. Thus the homoallylic ion *171* is trapped before it is rearranged into ion *173*. In the reaction mixtures no thio- or isothiocyanates have been found with the endo configuration of the substituent; this indicates that among the formed ion pairs there are no pairs containing classical cations:

Cristol [357] did not agree with the interpretation of the data on the thiocyanates isomerization suggested by Spurlock. He showed the ionization of thiocyanate *185* to yield isothiocyanates *186* and *187* (with the latter slightly predominant) along with thiocyanate *188*.

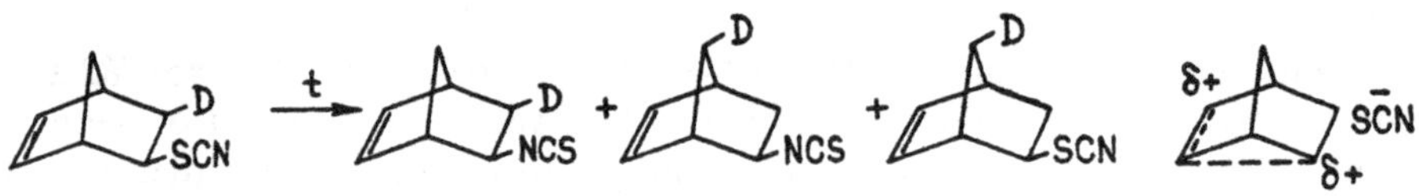

185 186 187 188 189

If the isomerization of the compound *185* resulted, as Spurlock believed, in the ion pair *189*, then the amount of compound *186* in the reaction products would be larger than that of isothiocyanate *187*. This corroborates indirectly Cristol's idea that the ionization of brosylate *174* does not yield the nonsymmetrical ion *171*.

The rate of solvolysis of 2-endo-norbornenyl sulphonate is lower than that of 2-endo-norbornyl sulphonate due to the polar effect of the transannular double bond. Many attempts have been reported to determine the deceleration of solvolysis caused only by the inductive effect of the homoallylic double bond [363]. This problem, however, is complicated by simultaneous action of the opposite π-participation effect. According to Wilt, a good way of determining the inductive effect of the homoallylic double bond may be to compare the reactivities of tosylates *190* and *191*. In compound *190* the system geometry "forbids" the homoallylic delocalization and, hence, the tosylate *191* reacts 5 times as fast as *190*; this can serve to measure the inductive effect of the homoallylic double bond [364].

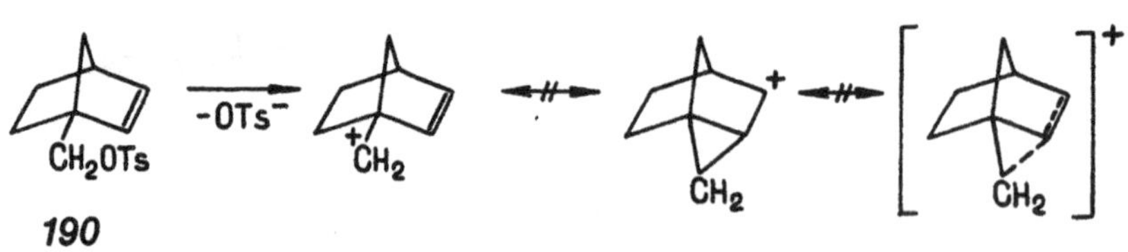

190

191 192

Cis-exo-ditosylate *193* acetolyzes 505 times as fast as its saturated analogue *61* [365].

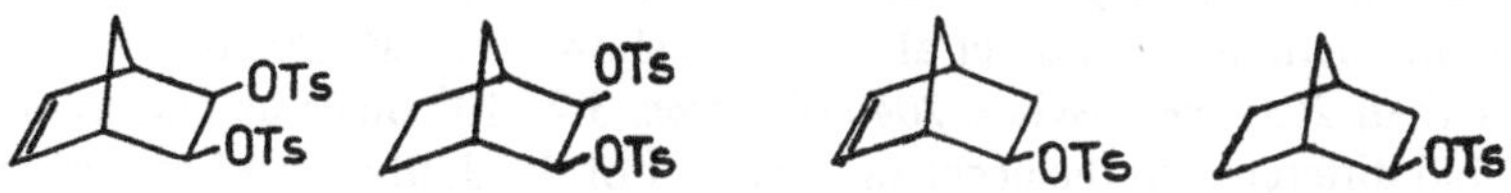

193 | **61** | **194** | **63**

$k_{rel} = 1.00$ | $1.98 \cdot 10^{-3}$ | $2.03 \cdot 10^{2}$ | $6.85 \cdot 10^{2}$

Since the double bond at the homoallylic position, as shown above, decelerates solvolysis, the rate ratio of solvolysis of tosylates *63* and *194* is 3.37; but this fact requires a special discussion. Two explanations can be proposed. One of them assumes diester *193* to detatch simultaneously both tosylate groups to form a dication *195* which further reacts with the solvent to form a homoallylic ion *196* and finally — diacetates.

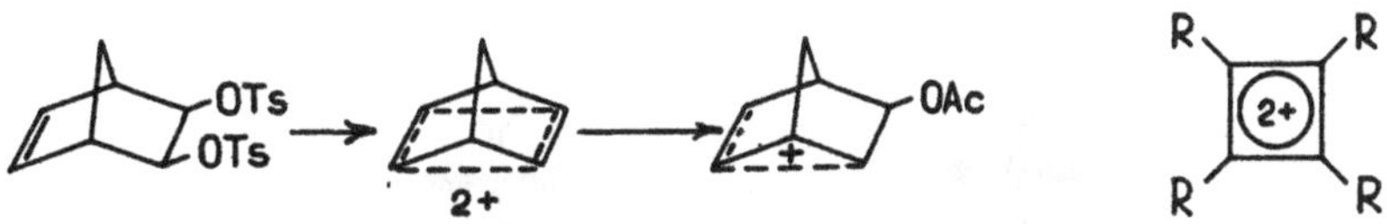

193 | **195** | **196** | **197**

The formation of cyclobutenium dications *197* which are analogues of ion *195* has been proved by spectral methods [366]. Among the products of *193* acetolysis interrupted after half of the parent compound has been spent there are no acetoxytosylates; this seems at first sight to corroborate the above mechanism. However, the unsaturated trans-ditosylate *198* and cis-exo-acetoxytosylate *199* which have only one readily detachable group (under the reaction conditions) solvolyze at about the same rate as ditosylate *193* and that these data preclude the possibility of dication formation.

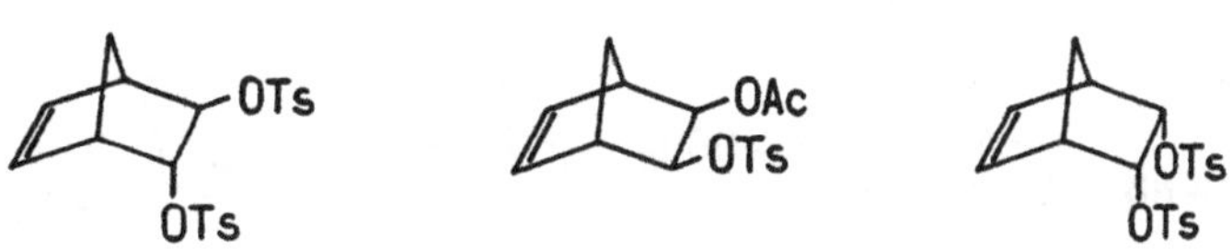

198 | **199** | **200**

$k_{rel} = 1.08$ | 1.84 | $1.28 \cdot 10^{-5}$

The alternative explanation is that the role of homoallylic participation increases when near the leaving group there is an electronegative substituent. This appears to refer only to π- rather than to σ-participation; indeed, the rate ratio of acetolysis for tosylates *63* and *61* is 336,000 while for *194* and *193* it is only 203. The increase in the homoallylic participation is suggested by the fact that the acetolysis of cis-endo-tosylate *200* in whose solvolysis no homoallylic participation is possible proceeds 100 times more slowly than that of its saturated analogue *61*.

Homoallylic participation can also be intensified by the effects of nonbonded steric interactions. De Puy et al. [367] have shown tosylate *202* to solvolyze 270 times faster than *203*; the tosylate *204* solvolyzes $3.2 \cdot 10^4$ times as fast as *205* does. The 100-fold difference in the accelerating effect of the double bond can be accounted for

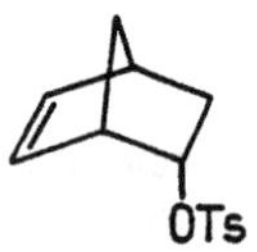

201

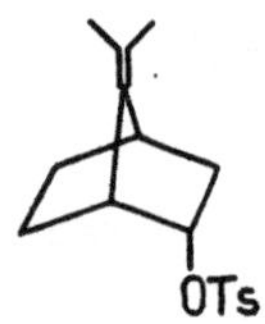

202

203

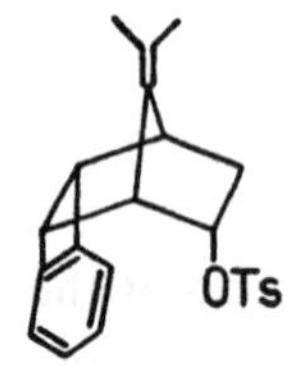

204

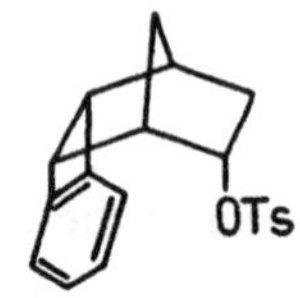

205

by significant nonbonded interactions in the ionization of *205* owing to which the developing p-orbital finds itself in a more favourable position for overlapping with π-electrons [368].

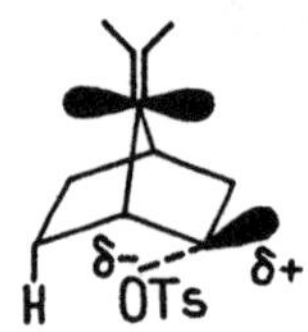

Brown has shown that on introduction of a double bond to isomeric p-nitrobenzoates *52* and *53* the rate of solvolysis in either case decreases equally (about fivefold) which is due to the inductive double-bond effect [369] and clearly points to the lacking anchimeric assistance in the solvolysis of *206*; this is also evidenced by the absence of tricyclic compounds among the reaction products. The isomers *206* and *207* are characterized by a high exo:endo rate ratio and predominant formation of exo alcohol.

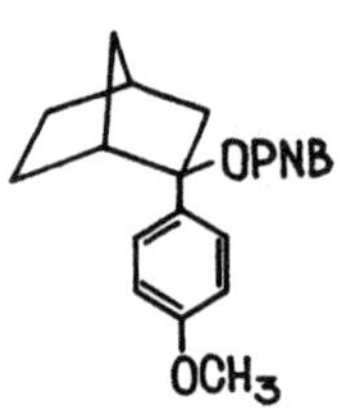

52

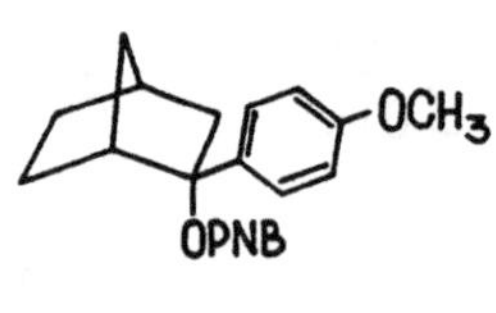

53

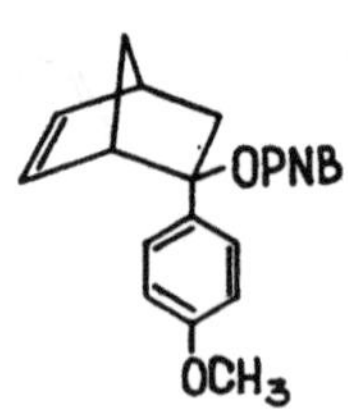

206

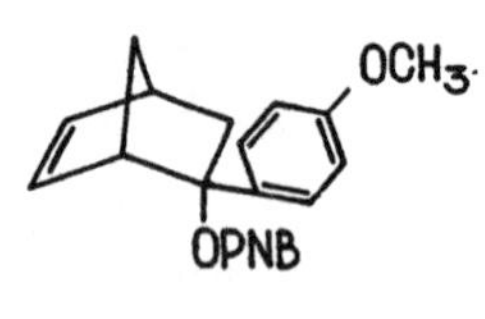

207

Introduction of a homoallylic double bond to the molecules of 1-phenylcyclopentyl-, 2-exo- and 2-endo-phenyl-2-norbornanyl-p-nitrobenzoates was shown [370, 371] to lead in all cases to an equal deceleration of solvolysis (80% acetone). On the other hand,

Brown has found that the exo:endo rate ratio of solvolysis of 5-aryl-5-norbornenyl-p-nitrobenzoates remains practically unchanged (~300) while the electron nature of the substituent in phenyl varies in a wide range (from p-CF_3 to p-OCH_3-group). Consequently, the high exo:endo rate-ratio of solvolysis of tertiary esters is due not to the π-participation in the solvolysis of exo isomers, but to other, possibly steric, factors. These factors are, according to Brown, unfavourable interactions between the π-electron cloud and the detaching anion in the transition state forming from an endo isomer.

In epimeric 2-(3,5-bistrifluoromethylphenyl)-2-norbornenyl-p-nitrobenzoates the exo:endo rate ratio increases (447; 80% acetone, 25 °C), as well as the product is rearranged [372]:

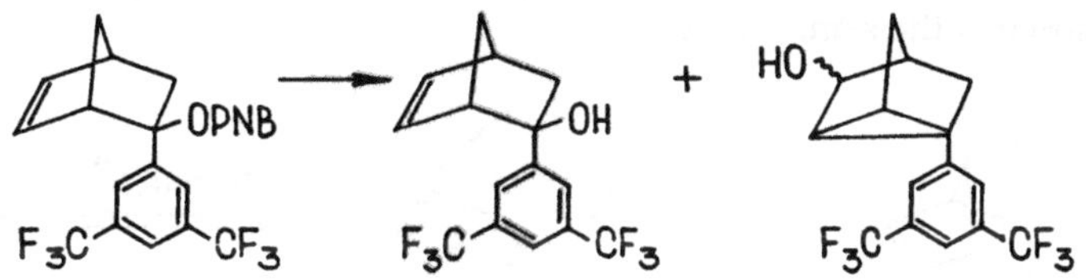

208 23% 77%

These facts testify to homoallylic assistance in the transition state of ionization of *208*.

The introduction of a methyl group to the C^5 atom increases the homoallylic assistance effect. Thus, the exo:endo rate ratio of solvolysis of 2-aryl-5-methyl-2-norbornenyl-p-nitrobenzoates increases markedly with increasing electron demand by varying the substitute in the 2-aryl group; for p-OCH_3 this ratio equals 354 (80% aqueous acetone, 25 °C), for p-H — 1260, for p-CF_3 — 6700.

By the ab initio method Hehre and Hiberty [374] have calculated different conformations of the homoallylic ions $C_4H_7^+$ and the isomeric bisected cyclopropylcarbinyl cation to show all the homoallylic ions *I–IV* being energetically less stable than ion *V* by 20–30 kcal/mole. The solvolysis products whose composition points to the intermediate formation of homoallylic cations can be attributed either to the fast trapping of unstable ions by counterions or to the change of the stability ratio of the isomeric ions $C_4H_7^+$ by solvation effects.

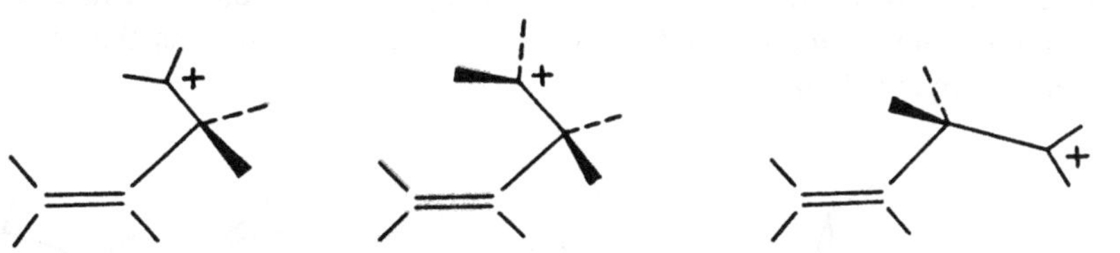

(I) (II) (III)

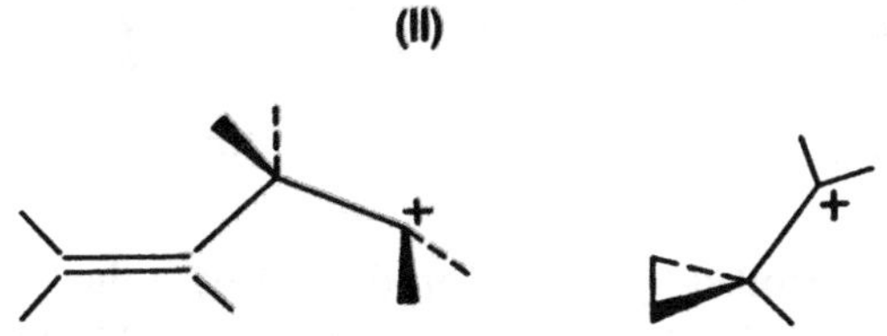

(IV) (V)

5.1.2 7-Norbornenyl and 7-Norbornadienyl Ions

For systems in which the homoallylic double bond is located unsymmetrically relative to the leaving group, like in 2-substituted norbornenes there is a comparatively small rate increase. At the same time in 7-anti-substituted norbornenes where a symmetrical position of the double bond contributes to intermediate formation of the 7-anti-norbornenyl (bishomocyclopropenyl) ion *19* the π-participation is strong [153]. Such an ion, according to quantum chemical calculations, is by 4 kcal/mole more stable than the 2-norbornenyl ion *171* [340].

The 7-norbornenyl ion *19* can be generated by the solvolysis of 7-anti-norbornenyl derivatives ("π-route") [352, 375, 376] as well as from tricyclic isomers ("σ-route") [377–380]. In fluorosulphonic acid both types of compounds yield (according to the PMR spectra) the same stable ion *19*.

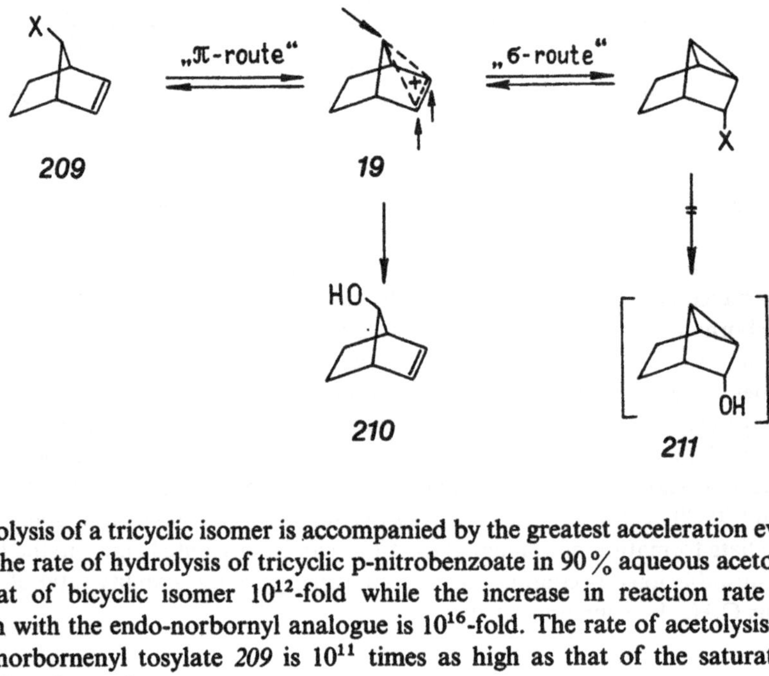

209 19 210 211

The solvolysis of a tricyclic isomer is accompanied by the greatest acceleration ever observed; the rate of hydrolysis of tricyclic p-nitrobenzoate in 90% aqueous acetone exceeds that of bicyclic isomer 10^{12}-fold while the increase in reaction rate in comparison with the endo-norbornyl analogue is 10^{16}-fold. The rate of acetolysis of the 7-anti-norbornenyl tosylate *209* is 10^{11} times as high as that of the saturated 7-norbornyl analogue *4*.

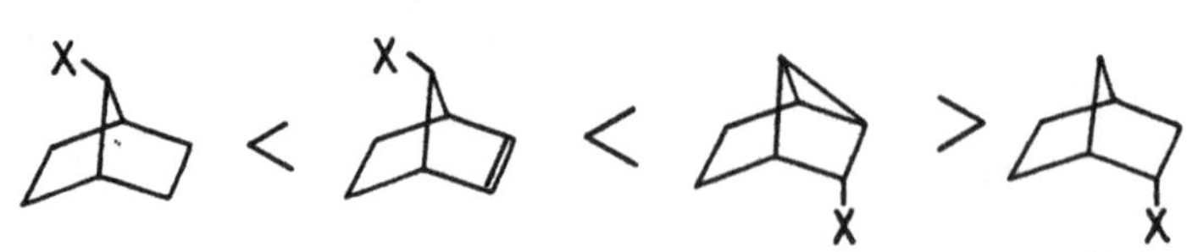

$k_{rel} = 1$ 10^{11} 10^{23} 10^{7}

The kinetically controlled hydrolysis product of both the "π-" and the "σ-route" is the 7-anti-alcohol *210*, and not an alcohol with a cyclopropane ring *211*, i.e. under solvolytic conditions, too, where the ion lifetime is rather short in comparison with that required for observation by the NMR method both the "σ-" and the "π-route" result in the same cation. This seems to be the symmetrical ion *19*.

In principle a nucleophile can attack this ion both on C^7 and on $C^2(C^3)$; in this case, as shown by experiment, under kinetically controlled conditions the ratio of C^7- and C^2-products changes essentially as the nature of the nucleophile changes. Thus, the solvolysis of tosylate *209* with sodium methylate in methanol results in a mixture of about equal amounts of tricyclic *212* and bicyclic ethers *213*:

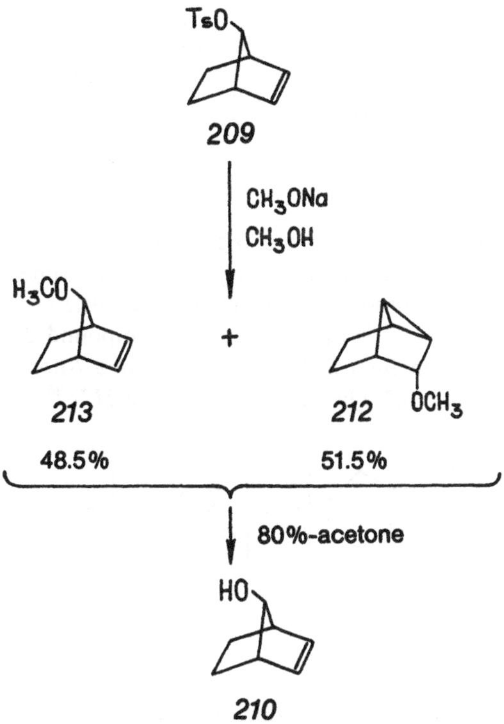

while an acetate buffer yields only 0.3% of tricyclic product. The latter is not a product of an nucleophile S_N2' reaction because its formation rate is independent of the concentration. According to Winstein, strong nucleophiles react with a carbocation at its most electron-deficient centre (the transition state reflects the cation structure) while weaker nucleophiles attack the sites where the reaction results in the most stable product [381].

Hydrolysis of the tricyclic ether *212* proceeds 10^6 times as fast as that of the bicyclic ether *213*; this is due to the higher free energy of the ground state of *212* [382, 383]. These data are supported by Spurlock [384] who has shown that the thermal isomerization of 7-anti-norbornenyl- and 7-anti-norbornadienyl thiocyanates in polar aprotic media results only in isothiocyanates with retained structure, and no tricyclic products. The bicyclic compound has a higher stability then a tricyclic one.

Interesting results have been obtained by Tufariello [385] on similar models:

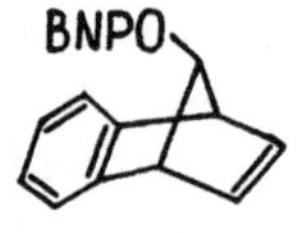

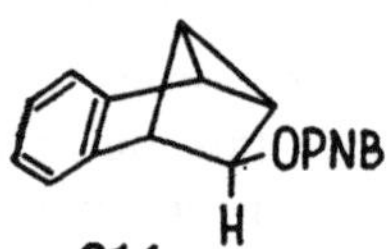

214

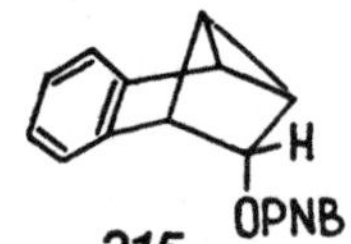

215

$k_{rel} = 10^8$ 10^{15} 10^{21}

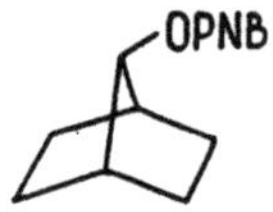

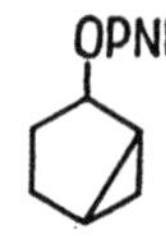

216

$k_{rel} = 1$ 10^{15} 10^{22}

As can be seen from the above data, the exo-epimer *214* solvolyzes at a rate typical for secondary cyclopropylcarbinyl systems e.g., *216*; at the same time the endo epimer *215* is by 6 orders of magnitude more reactive. This difference is explained since the participation of the $C^3—C^4$ is only possible in compound *215*.

Another "π-route" leading to the intermediate formation of a 7-norbornenyl ion is the solvolysis of the exo-2-bicyclo[3,2,0]hepta-6-enyl tosylate *217* [386]:

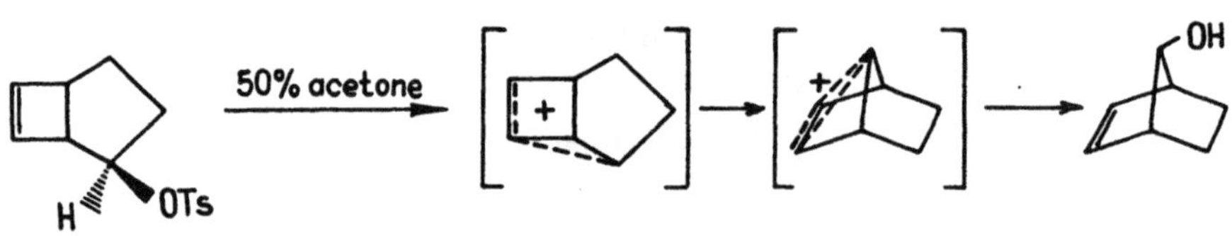

217 218 19 210

The exo:endo rate ratio amounts to 2400 which points to the primary formation of the homoallylic ion *218* with the parent skeleton further rearranged into ion *19*.

From the above data Brown [387] assumed the intermediate carbocation not to be the nonclassical ion *19* but to have the structure of the classical tricyclic carbocation *219* in the rapid equilibration with its epimer. As noted above, the solvolysis of tosylate *209* under weak-alkaline conditions yields only alcohol *210*.

In this connection Brown suggested that in this reaction the unstable, kinetically controlled product is the tricyclic alcohol *211* which is further isomerized to compound *210*. However, Tufariello has shown alcohol *211* to be quite stable under alkaline

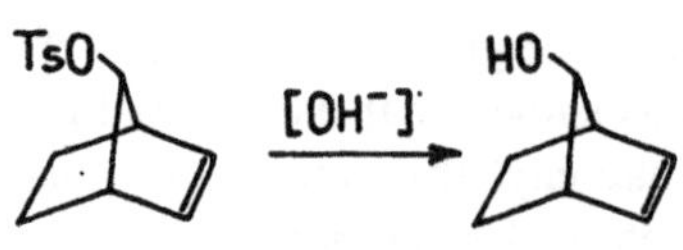

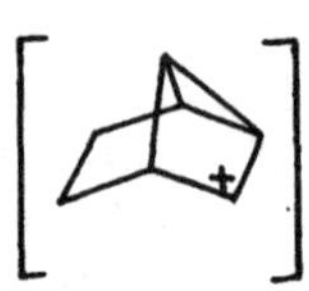

209 210 211 219

hydrolysis [378]. Winstein noted the formation of only the endo tricyclic product is well supported by the nonclassical structure of the intermediate ion *19* [377]; for steric reasons one should expect mainly exo products from the classical ion *219*.

To choose finally between the nonclassical and classical structure of the 7-norbornenyl ion Gassman [388] studied the solvolysis of benzoates *220–222*. If an intermediate nonclassical ion is formed, the effect of the CH_3 groups on the solvolysis rate will be cumulative, but if classical ions are in an equilibrium, the effect of the second CH_3 group should be much smaller than that of the first. Indeed, the first CH_3 group shift the equilibrium between classical ions essentially towards the tertiary ion *223*; so the second CH_3 group produces a weak effect since ions *223* and *224* are very similar. The introduction of the first CH_3 group increases the solvolysis rate by 13.3, that of the second by 11.1; i.e. the effect is cumulative.

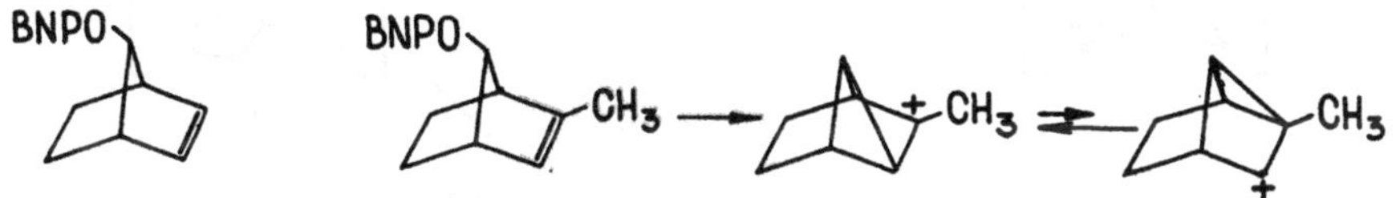

220 *221* *223*

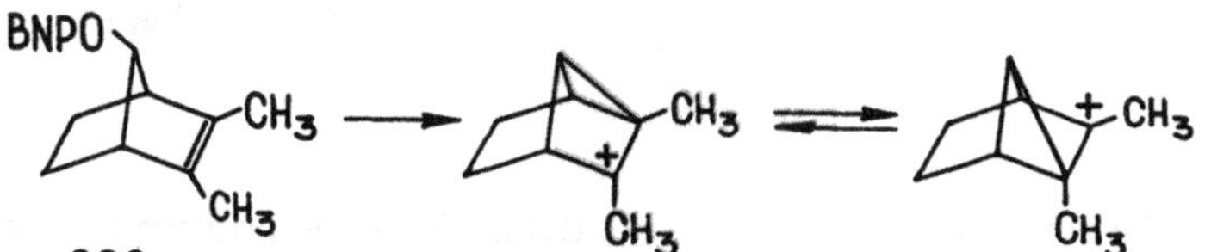

222 *224* *224*

Earlier Bartlett and Sargent [51] had pointed out that the cumulative effect of the CH_3 on solvolysis rates manifested itself in the formation of symmetrical transition states leading to symmetrical intermediate nonclassical ions.

Even the allegedly "classical" ions *219* which are in equilibrium have the structure of the nonclassical secondary α-cyclopropylcarbinyl cations [389].

The solvolysis of tertiary 7-norbornenyl esters *225* and *226* yields only the products with the retained configuration; the relatively low rate ratio α-CH_3:α-H (10^2 at 25 °C), the negative isotopic effect for the solvolysis of ester *225* as compared with that of *226* — all this testifies to an effective π-participation of the double bond and to a relatively weak hyperconjugative stabilization of the cation centre by the 7-methyl group [390].

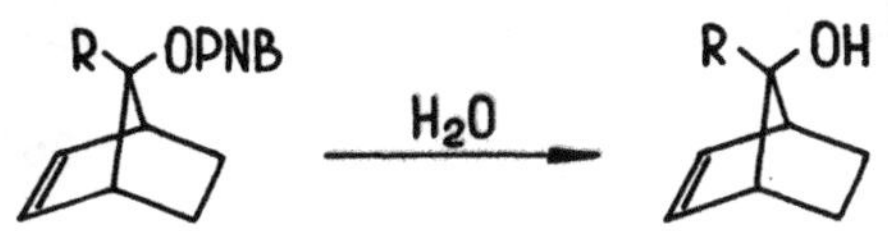

225, R = CH_3

226, R = CD_3

A far more complicated reaction mixture is formed in the solvolysis of epimeric substrates with a syn-position (relative to the double bond) of the leaving group, e.g.:

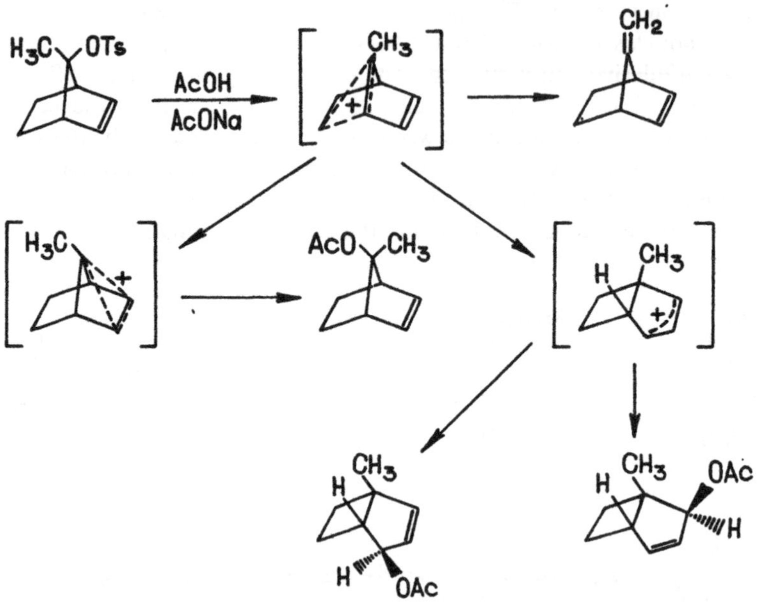

In order to evaluate the correlation of the effects of delocalizing the positive charge from the double bond participation as well as of conjugation of the carbonium centre with the aryl group Gassman et al. measured the rates of solvolysis of the following p-nitrobenzoates [391]:

1)

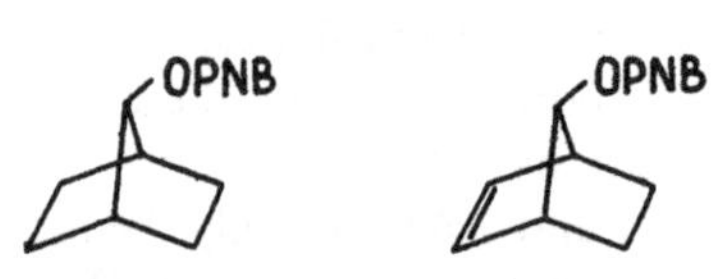

2)

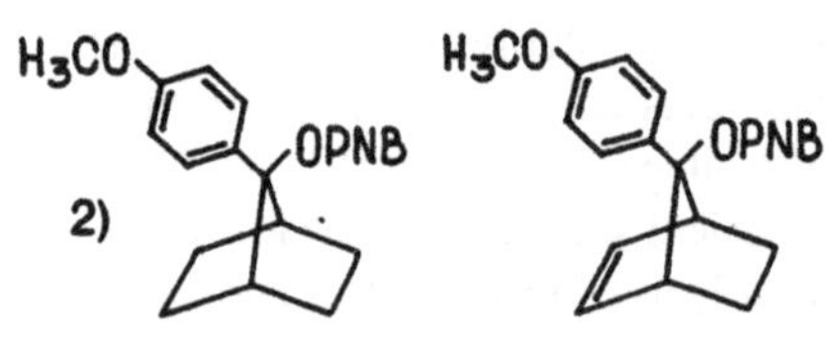

$k_{rel} = 1$ 10^{11} 1   3

3)

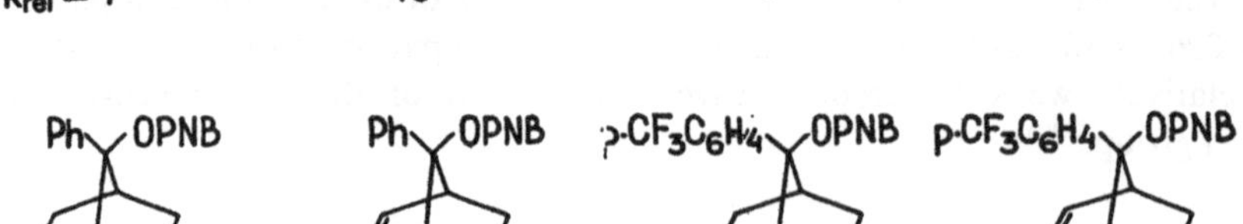

$k_{rel} = 1$ 41   1   35000

From the data cited the p-anisyl group is clearly seen to "level" the anchimeric acceleration of about 10^{10}, the solvolysis proceeding, as shown by Gassman, non-stereospecifically:

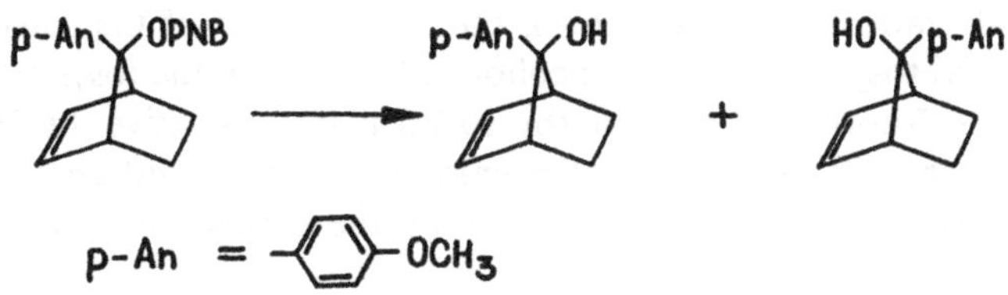

The phenyl group "levels" the acceleration of 10^9, the p-trifluoromethylphenyl one — 10^6; the weaker the stabilization of the developing cation centre, the stronger the effect of the π-participation, i.e. the latter effect is the function of the electron demand of the arising cation centre.

Subsequently Gassman showed [84], that while thecorrelation of solvolysis rates of the compounds *A* with the σ^+ constants gives a straight line ($\varrho^+ = -5.27$) while a similar dependence for *B* has a "break" at a point corresponding to a p-anisyl compound. In this case for X = 4-OCH_3, H, 4-CF_3, 3,5-$(CF_3)_2$ there is a linear dependence with $\varrho^+ = -2.30$ while the points for X = OCH_3 and — $N(CH_3)_2$ lie on the straight line of the compounds *A*.

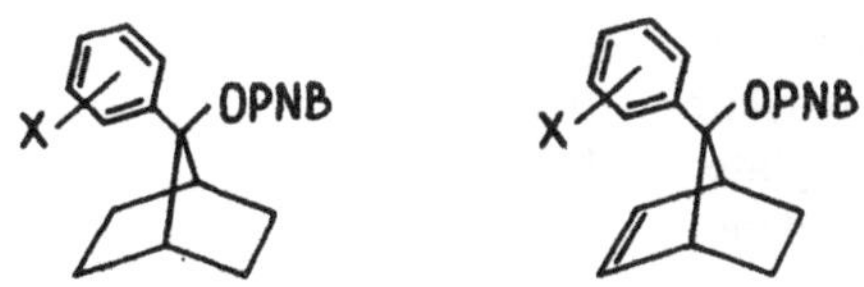

(A) (B)

The authors recognise a change in the solvolysis mechanism in the curve "break". For the substituents stabilizing the cation centre to a lower extent than the anisyl group does, the participation of the double bond is very important. The p-dimethylamino-phenyl and anisyl group effect no such anchimeric participation, the stabilization of the cation centre by the anisyl group being just sufficient to "level" the double bond participation. These considerations are also well supported by the study on the stereochemistry of reaction products.

In his research on compounds *A* and *B* Lustgarten [392] has shown the substitution of 7-H by 7-CH_3 to lead in the case of saturated compounds *A* to an enormous acceleration ($10^{8.2}$), while in the case of monoenes *B* the acceleration is as low as $10^{2.6}$ and for dienes *C* it is only $10^{2.2}$.

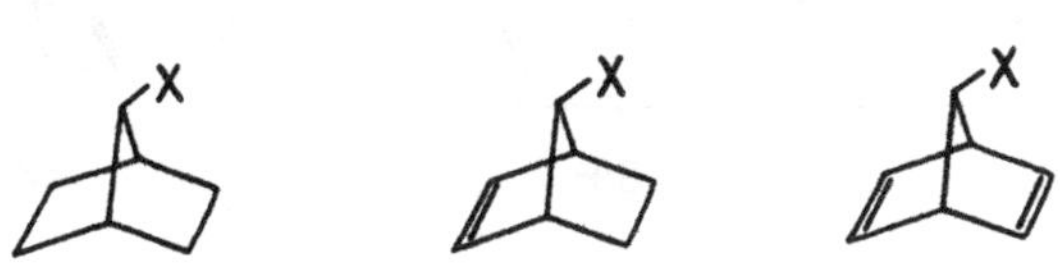

(A) (B) (C)

As a rule, an external nucleophile attacks the C^7 atom of the intermediate π-delocalized 7-norbornenyl ion *19* from the anti-side; this also applies to the intramolecular migrations of the 7-substituent to C^7 [393]. The only example of such a migration from the syn-side [394] seems to be due to steric factors.

Winstein and Stafford [395] have shown, the only product of the solvolysis of the 7-syn-norbornenyl tosylate *227* is the alcohol *228*. Although the tosylate *227* is more effective than the 7-norbornyl tosylate (by 10^4) it is far less active (by 10^7) than the anti-isomer *209*; the different course of reaction induced the authors to account for the accelerated solvolysis of the syn-isomer by the formation of a stable allylic ion.

227 228

Listed below are the relative rates of acetolysis of some norbornyl tosylates at 25 °C borrowed from Winstein [352, 375]

Cyclohexyl tosylate	1
7-Norbornyl tosylate	10^{-7}
5-exo-Norbornenyl tosylate	10^3
5-endo-Norbornenyl tosylate	10^{-1}
7-anti-Norbornenyl tosylate	10^4
7-syn-Norbornenyl tosylate	10^{-3}
2-endo-Norbornyl tosylate	1

To differentiate between a retained configuration and the formation of an endo product Tanida [396] has pointed out three peculiarities in the deamination of the isomeric 7-norbornenyl amines *229* and *230*:

1) a complete retention of the configuration in the deamination of the anti-amine *229*;

2) an usually high yield of intramolecular reaction products (chloride, alcohol) in comparison with that of intermolecular reaction ones (acetate) 78:22 from the anti-amine *229*;

3) the reaction of the syn-amine *230* proceeds with a complete rearrangement into bicyclo[3,2,0]heptene.

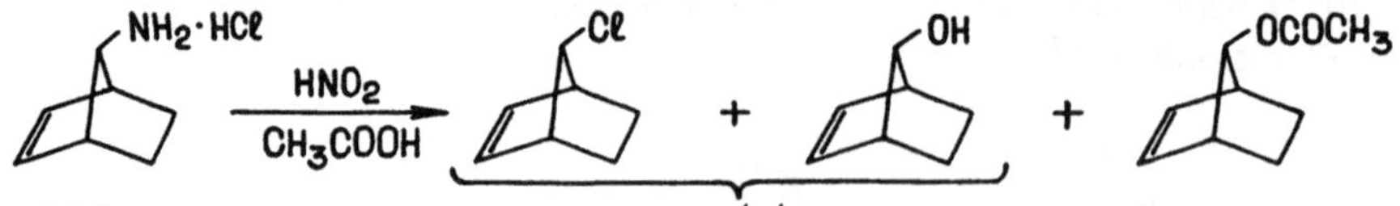

229 78% 22%

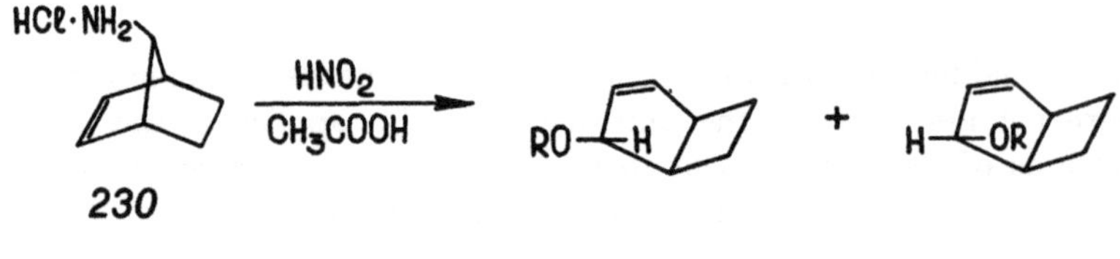

230

The simultaneous mechanism S_N2 with the substitution of a diazonium group by the solvent for deaminating the anti-amine *229* is rejected because it would result in configuration inversion:

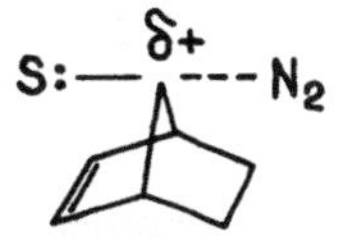

Since the driving force of alkyl diazonium salt decomposition is the detachment of stable nitrogen molecule, the activation energy must be much lower than for the solvolysis of halogenides or sulphonates. The participation of the neighbouring group in such processes is hardly necessary and possible because of a low positive charge appearing on the C atom in the transition state [397,398]. Another viewpoint is quite opposite: the neighbouring group participates in such transition states because it contributes not only to pushing out the leaving group, but is also pulled by the positive charge formed on detachment of nitrogen [190,192]. In case the latter viewpoint is accepted the configuration retention on deaminating *229* can be attributed to the 2,3-π-participation in the dissociation of the C—N bond when the positive charge is still mainly located at the nitrogen.

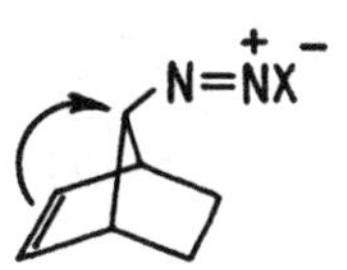

In the case of syn-amine *230* the driving force is the formation of a stable allylic cation.

Ohorodnyk [399] calculated the geometric and electronic structure of the 7-norbornenyl anion *231*, radical *232* and cation *19*. As early as 1947 Coulson and Dewar [400] had attempted to compare the energies of isomeric ions in which the atom positions are different are quite intenable. So the earlier published calculations by the Hückel method [340] are very inaccurate.

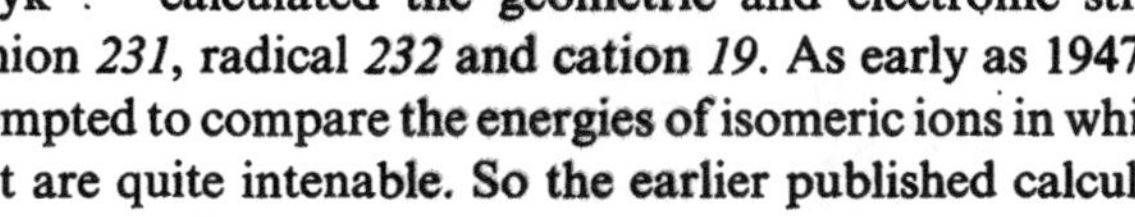

231 *232* *19*

In the anion *231* the unshared electron pair is almost completely localized on C^7; the electron distribution here is similar to their distribution in norbornene. In the case of the radical and particularly of cation *19* the unpaired electron or charge is delocalized between C^2, C^3 and C^7 (Table 10).

Hoffmann suggested the following diagram for the interaction of the vacant 2p-orbital of C^7 in ion *19* with the π- and π*-orbitals of the double bond [401]. The plane of symmetry passes through C^7 and the middle points of the 2–3 and 5–6 bonds.

Table 10. Distribution of Charges

Atom	Anion	Radical	Cation	Atom	Anion	Radical	Cation
C^1	0.083	−0.045	0.09	C^5	0.014	0.020	0.010
C^2	−0.037	−0.078	0.095	C^7	−0.407	0.242	0.191

Since interaction is possible only between the levels of equal symmetry the level of π^* (A) remains unperturbed. The levels of π (S) and 2p (S) are mixed up in accordance with the perturbation theory principles. As a result the π-orbital is stabilized. According to the perturbation theory, the upper component of the interacting pair is more destabilized than the lower one is stabilized. Therefore the four-electron system (the 7-norbornenyl anion) must on the whole be destabilized, while the stabilization of the three-electron system (the 7-norbornenyl radical) is problematic (cf. [402]).

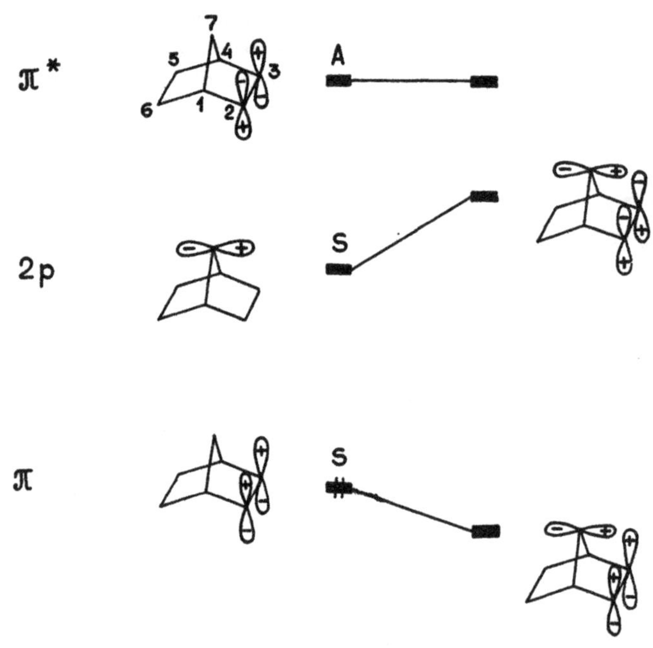

The interactions C^2—C^7 and C^3—C^7 in the cation can be increased on condition the C^7-bridge is inclined to the double bond — within the limits admitted by an increase in steric strain.

If two double bonds are introduced at the anti-2- and anti-3-positions of the molecule of the 7-substituted norbornane, then the accelerating effect in the solvolysis reaction turns out to be by 7 orders of magnitude weaker than from one 2,3-double bond [403]. The observed effect of acceleration relative to the saturated analogue, as well as the structure of acetolysis products is well accounted for by the formation of the nonsymmetrical ion *233*.

Although the 3-cyclopentenyl derivatives formally resemble the 7-anti-norbornenyl ones, their solvolysis has until recently been considered to proceed without any double-bond participation due to the practically planar conformation of the cyclopentenyl ring and an increased angle strain[404]. Lambert, however, has shown[405] that 3-cyclopentenyl tosylate *234* is formolyzed under practically complete (>99%)

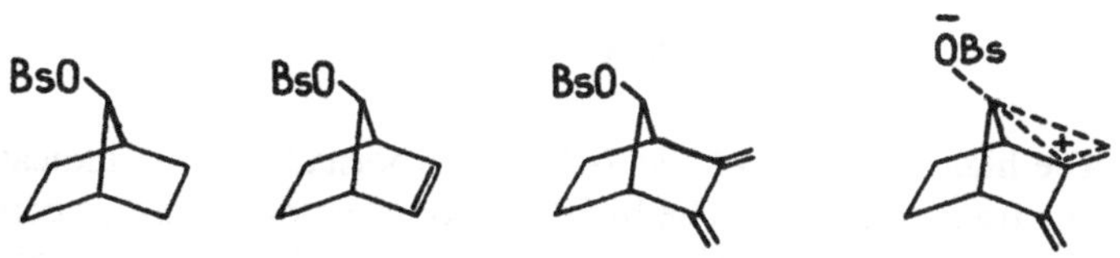

k_{rel} (25 °C) = 1 10^{11} 10^{4} *233*

retention of configuration, while the respective saturated analogue — cyclopentyl brosylate — is solvolyzed with complete reversion[406]; thus, a double bond changes the ion-paired mechanism (k_s) for the process with π-participation (k_Δ). The 3-cyclopentenyl system represents an example of π-participation without rate increase (cf. [43]). According to the latest data, the cyclopentene ring is non-planar[407] and a substituent at C^1 must increase this effect. The participating double bond forms the symmetrical bishomocyclopropenyl cation:

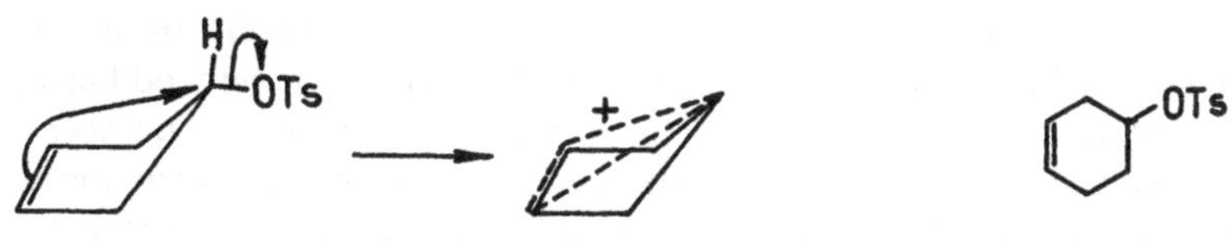

234 *235*

Under similar conditions the 3-cyclohexenyl tosylate *235* is formolyzed with only 40% retention of configuration[408], i.e. the π-participation is more effective if a symmetrical cation centre is developing. In tertiary cyclopentenyl derivatives — 1-aryl-3-cyclopentenyl p-nitribenzoates — no increase is observed in the relative rate of solvolysis with increasing electron demand[409].

The 7-norbornadienyl ion *236* is formed on solvolysis of 7-chloro-norbornadiene *237*; this solvolysis proceeds 10^{14} times as fast as that of the saturated analogue *238* and 800 times as fast as that of the monoene *239*[410]. The sole product of the reaction is 7-norbornadienol.

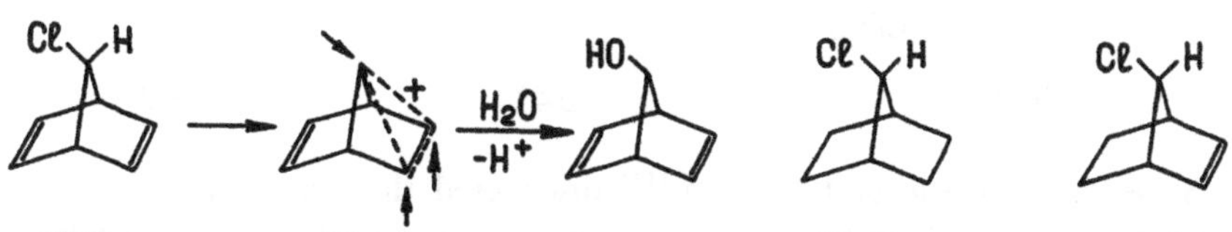

237 *236* *238* *239*

10^{14} 1

The reduction of chloride *237* by $LiALD_4$, as well as its methanolysis with sodium methylate result in a mixture of tricyclic (mainly) and bicyclic products [383,411]:

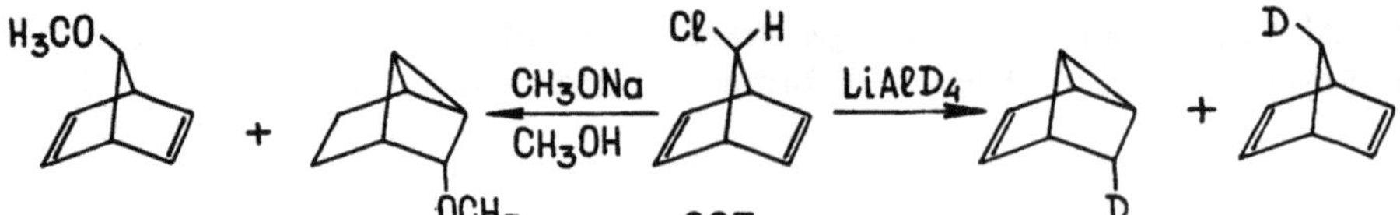

237

The high reactivity of the 7-chloronorbornadiene *237* indicates high stabilization of the transition state. The high activity of chloride *237* as compared with chloride *239* may be due to the "bicycloaromaticity" of the 7-norbornadienyl cation which can be regarded as an aromatic Möbius structure (4π-electrons):

The relative stability of 7-norbornenyl and 7-norbornadienyl ions serves to discuss the relative role of homoaromatic and bicycloaromatic stabilization. Goldstein has expanded the concept of homoaromatic ions (two π-bridges) into that of bicycloaromatic ions (three π-bridges) [412,413]. He considered the orbital interaction to stabilize the bridge system if the sum of π-electrons of the only odd bridge and of the longest even bridge equals $4m + 2$ ($m = 0, 1, 2 \ldots$) and the total number of π-electrons in the system is equal to 4n. If a bicyclic three — bridge system contains $(4n + 2)$ electrons, it is antibicycloaromatic. In passing from the 7-norbornyl to the 7-norbornenyl cation the homoaromatic stabilization increases the solvolysis rate by eleven orders of magnitude; the second double bond results only in a small increase. It is obvious that the homoaromatic stabilization plays a main role for 7-norbornadienyl cation [414].

The rate of solvolysis of p-nitrobenzoate *240* exceed that for ester *220* under similar conditions by three orders of magnitude. Allred and Hinshaw [415] attributed this effect to the formation of an intermediate "laticyclic" (2 + 2 + 0) ion *241*:

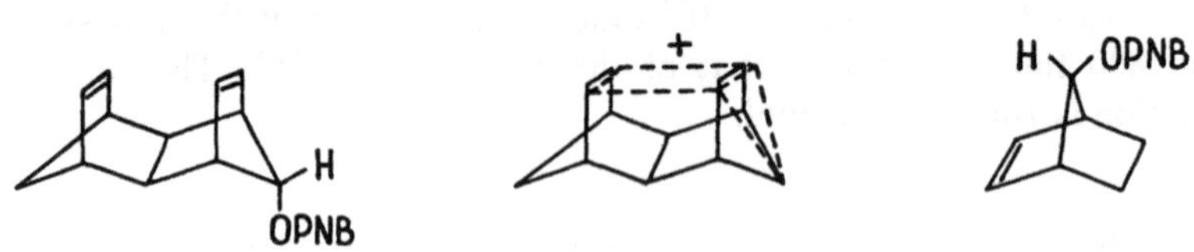

240 241 220

Later, however, Paquette and Dunkin [416] discovered that the solvolysis of dinitrobenzoates *242* and *243* in 80% aqueous acetone occurs practically at the same rate; this precludes the special role of laticyclic stabilization (i.e. π-participation of the remote double bond) as compared with homoaromatic stabilization taking place in 7-anti-norbornenyl derivatives. Paquette and Dunkin maintain that the homo-aromatic stabilization of the developing cation centre on solvolysis of esters *240*, *242*, *243* "levels" the demand for the assistance of the other π-bonds.

Allred et al. reverted to the problem under discussion[417]. Using the secondary deuterium isotopic effect they showed that a similar reactivity of dinitrobenzoates *244* and *245* exceeding that of ester *240* by 3 orders of magnitude is due to π-bond participation in *244*, to the steric strain resulting in the convergence of the π-electrons of the double bond and of the developing cation centre.

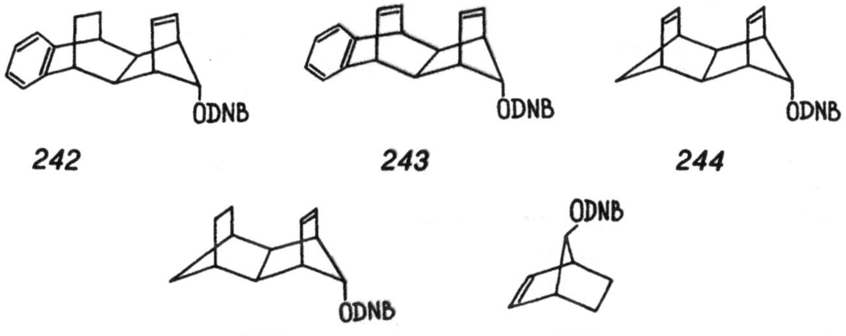

242 *243* *244* *245* *246*

5.1.3 Homobenzylic Ions

As already remarked, two explanations are advanced for the acceleration of the solvolysis process of exo(anti)-derivatives in comparison with endo (syn)-isomers. One of them rests on the concepts of anchimeric participation of σ- or π-electrons in the transition state. The alternative explanation is the different influence of steric factors on the ground and the transition state for exo:endo (syn:anti) isomers.

At first Bartlett and Giddings[418] showed the 9-anti-benzonorbornenyl brosylate *247* to be acetolyzed yielding acetate *248* and retaining the skeleton and configuration.

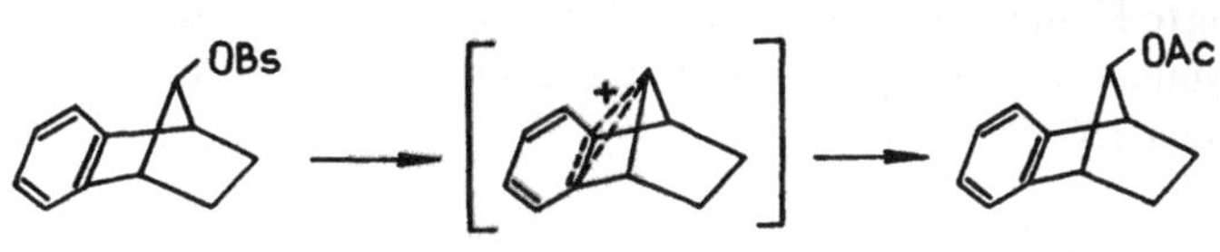

247 *249* *248*

The benzene ring just as the double bond brings about anchimeric acceleration, but far less than the latter. The solvolysis rate of brosylate *247* is $6 \cdot 10^5$ times higher than that of 7-norbornyl brosylate but it is 10^5 times lower than that of 7-anti-norbornenyl brosylate. The authors consider this to be due to some destabilization of the aromatic ring in the intermediate ion *249*.

In 1963–1964 Tanida et al.[419,420] synthesized 6-substituted 9-anti-brosyloxy-benzonorbornenes. Their solvolysis is a first-order reaction yielding only anti-acetates under complete retention of configuration. Introduction of the OCH_3 group in the 6-position increases the rate of acetolysis 53.7 times, CH_3 — 5.7 times; Cl, Br, NO_2 decrease it down to 0.045; 0.030; $1.4 \cdot 10^{-4}$, respectively (the rate of acetolysis of the nonsubstituted brosylate *247* is 1). The factor k_{OCH_3}/k_{NO_2} for the indanyl series equals only 8 which is in accord with purely inductive effects attenuating when passed through saturated bonds [421]. Consequently, the transition state of solvolysis of the 9-anti-derivatives must involve benzene ring participation. The Hammett linear correlation for the rate constants of acetolysis of 6-substituted brosylates can be obtained in two ways: by using as an abscissa the value $1/2(\sigma_p^+ + \sigma_p)$, $\varrho = -4.80$ or, respectively, $1/2(\sigma_p^+ + \sigma_m^+)$, $\varrho = -5.10$. The first variant implies nonsymmetrical

$k_{OCH_3}/k_{NO_2} = 386000$ $k_{OCH_3}/k_{NO_2} = 8$

homobenzylic participation; the second method of data treatment corresponds to the simultaneous participation of meta- and para-positions of the benzene ring involving a symmetrical nonclassical transition state.

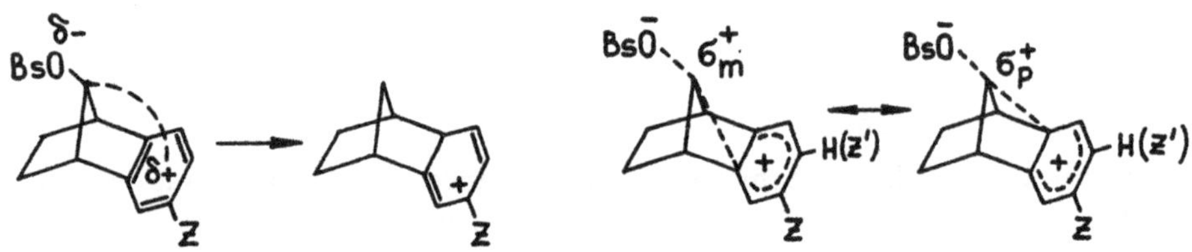

A forcible argument favouring the second type of transition state is the increase in the rate on introduction of the second OCH_3 group (factor 56) being the same as on introduction of the first OCH_3 group into the 6-position (factor 54 [422]). More justified, therefore, the linear correlation with $1/2(\sigma_p^+ + \sigma_m^+)$.

According to Hammond's postulate the intermediate carbocation can also be assumed to be symmetrical. Story considers the cation structure to be best represented by the formula *250* [358].

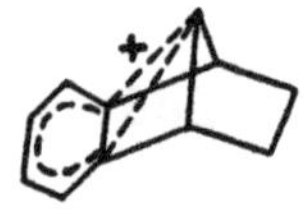

250

The value $\varrho = -5.10$ obtained by Tanida is the largest for all the β-arylalkyl systems studied and falls within the interval of values of ϱ (from -4 to -12) typical for the electrophilic substitution in aromatics [423]. The value of ϱ for the 2-indanyl system is equal to -1.1.

The efficiency of π-participation for different aromatic rings in one molecule was compared by Tanida et al. [424]. They showed the rate of solvolysis of brosylate *251* to be higher than that of isomer *252*; in either case only acetates are formed with complete retention of structure and configuration. The authors consider the solvolysis to proceed with the formation of the isomeric nonclassical ions *253* and *254* which are "separated" by a considerable energetic barrier preventing the less stable ion *254* from converting into the more stable ion *253*; the formation of the intermediate symmetrical ion *255* is denied.

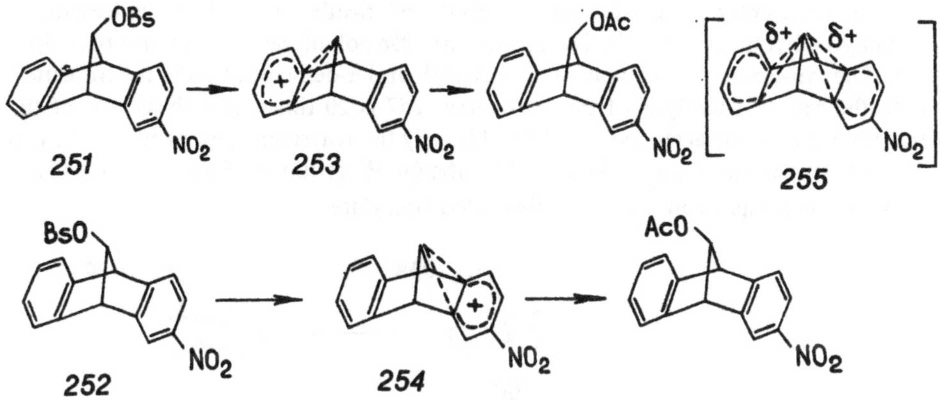

The solvolysis of the 9-syn-benzonorbornenyl brosylates [421] proceeds under complete rearrangement and stereospecific formation of 2-anti- and 2-syn-3,4-benzobicyclo[3,2,0]heptenols:

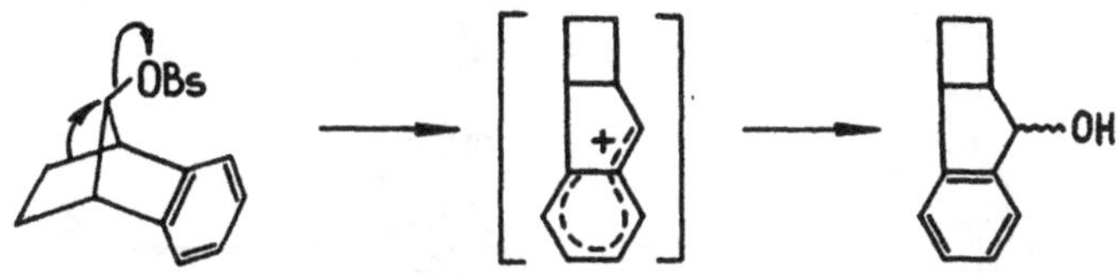

The rates of acetolysis of 7-norbornyl brosylate and 6-nitro-9-syn-benzobornenyl brosylate are practically equal, i.e. there is no π-participation; only the latter compound is completely rearranged under solvolysis reaction. The driving force of rearrangement for 9-syn-brosylates seems to be the tendency to form a stable benzylic carbocation. The $k_{OCH_3}:k_{NO_2}$ ratio for syn-isomers is only 43, i.e. much less than that for anti-isomers (386000); consequently, the π-participation of the homo-

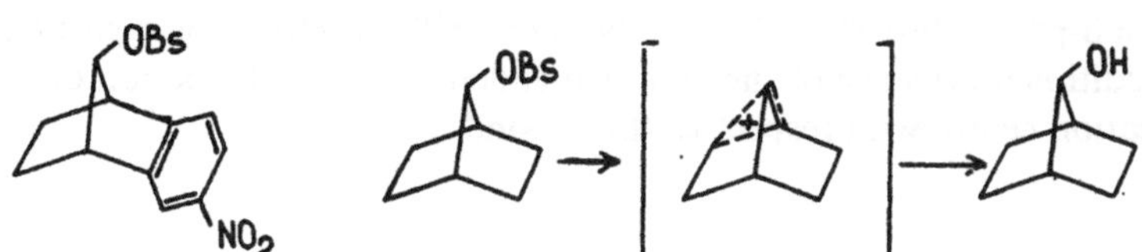

$k_{rel} = 1.05$ 1.00

benzylic type is not observed in the transition state of solvolysis of syn-isomers. The value 43, however, is too high for simple electrostatic interaction. Hence, the factor of 43 indicates that a definite role is played by the polar influence of the π-cloud of the benzene ring ("syn-effect"). Therefore the changes in the rate of solvolysis of different 9-anti-isomers can be assumed to include the influence of two factors: the π-participation of the aromatic system and the "syn-effect". Subtracting the latter we obtain the anti:syn rate ratio attributed only to π-participation.

Indeed, there is a good correlation between lg $k_{anti}:k_{syn}$ and the values of $1/2(\sigma_p^+ + \sigma_m^+)$, $\varrho = -3.57$.

Quite differently proceeds the solvolysis of 6-substituted β-(9-syn-benzonorbornenyl) ethyl brosylates [425]. The anti-isomer *256* solvolyzes without aryl participation. The syn-isomer *257* solvolyzes clearly with aryl π-delocalization in the transition state: the rate of solvolysis of the syn-isomer *257* is 20 times as high as that of the anti-isomer *256*; the acetolysis of *257* yields only the rearrangement products. In this case for syn-isomers $k_{OCH_3} - k_{NO_2} = 146$; the OCH_3 group accelerates the acetolysis 7.3 times in comparison with nonsubstituted brosylate.

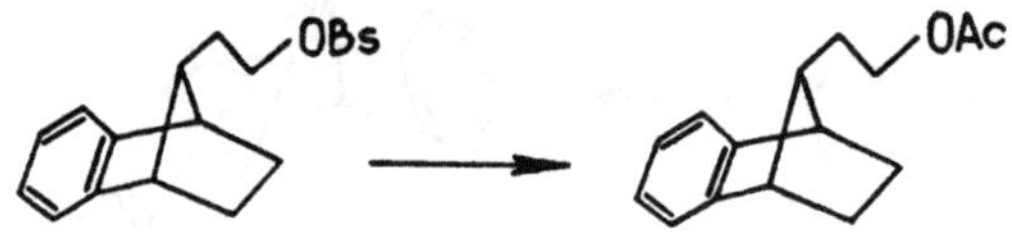

256

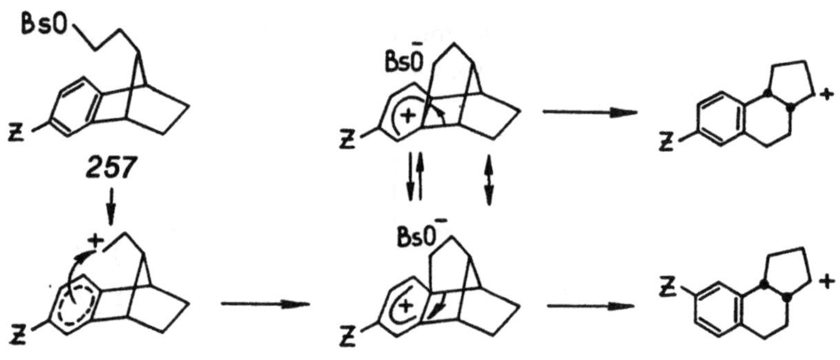

These values cannot be explained by simple inductive effects of the benzene ring. Here the correlation between lg $k_\Delta:k_\Delta^\circ$ and $1/2(\sigma_p^+ + \sigma_m^+)$ is observed as well. In accordance with Winstein's method [426,427] the authors single out the acetolysis with anchimeric participation (k_Δ) and without it (k_s) expecting the latter to result in acetate with the parent structure (as with the anti-isomer).

The β-propyl brosylate *258* is solvolyzed without effects of π-participation; this is an additional evidence of the great dependence of π-participation on the position of the cation centre with respect to the π-bond.

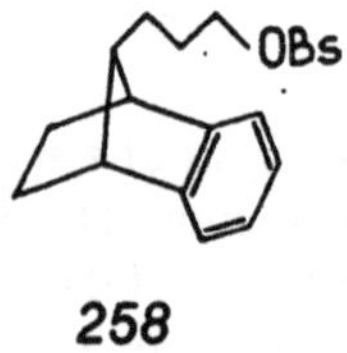

258

Without reliable experimental evidence it was accepted that in passing from the secondary system to the tertiary one, due to the cation centre stabilization, the extent of anchimeric participation and the exo:endo (anti:syn) rate ratio should diminish [63, 78, 428]. The acetolysis of 9-methyl-9-anti-tosylate *259* proceeds with complete retention of configuration, that of the syn-isomer *260* with predominant reversion of configuration. The absence of inversion and the small yield of elimination product from the anti-isomer point to considerable π-participation.

The introduction of a 9-CH_3 group into benzonorbornenyl tosylates decreases the anti:syn rate ratio from 2400 to 493 [421]; this agrees with the above assumption. However, in the presence of a 9-CH_3 group anti-syn rate ratio remains rather high as well. It seems to be due to the fact that the decrease of the ratio resulting from the decreased participation of the benzene ring for the anti-isomer is largely offset by an increase in the reaction rate of this isomer due to the weakening of steric interactions in the transition state in the presence of a 9-CH_3 group.

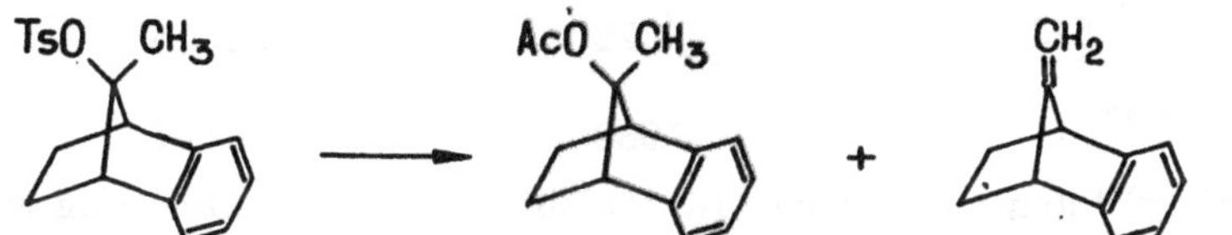

259

$k_{rel} = 493$ 98% 2%

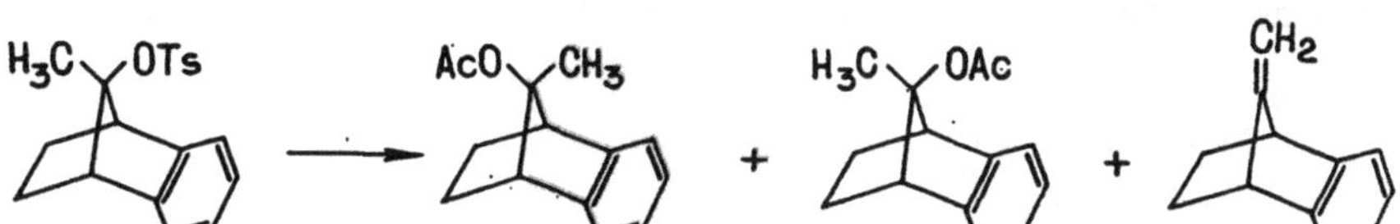

260

$k_{rel} = 1$ 67% 10% 21%

The comparatively small changes of the anti:syn rate ratio in passing from the secondary system to the tertiary one are interpreted by Brown as a insignificant assistance in the secondary systems as well [73, 429]. But the data cited earlier leave no doubts as to the significant π-participation in the 9-anti-benzonorbornenyl (secondary) systems.

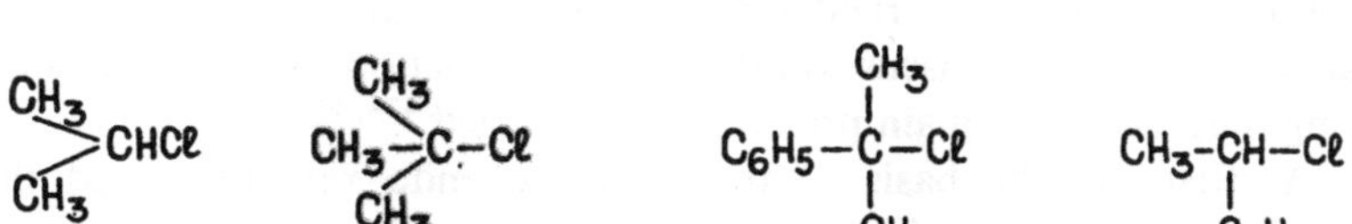

$k_{rel} = 1$ 55000 1800 1

For the tosylates which are 9-benzonorbornenol derivatives the CH_3:H rate ratio for syn-isomers amounts to 86700, and for anti-isomers — to 18000. So the ratio of these values agrees with the π-participation in the anti series. However, the change in the rate of solvolysis in passing from the secondary to the tertiary system is, as a measure of degree of π-participation, of low sensitivity in comparison with the above method of studying the effects of substituents in the benzene ring.

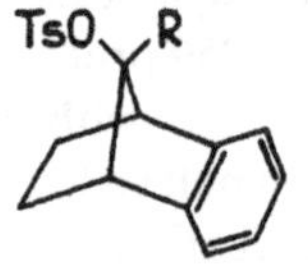

$k_{CH_3}/k_H = 18000$

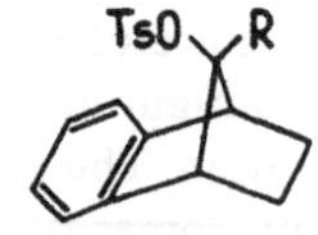

$k_{CH_3}/k_H = 86700$

In 1960 Bartlett and Giddings [418] showed 2-exo-benzonorbornenyl brosylate to be solvolyzed much faster than its endo-isomer. The relative rates of their values for 2-benzonorbornenyl brosylates are lower than for 5-norbornenyl brosylates. Bartlett attributes this to the fact that the greater polarization of the $C^1—C^{8'}$ bond hinders the cation centre development at C^2.

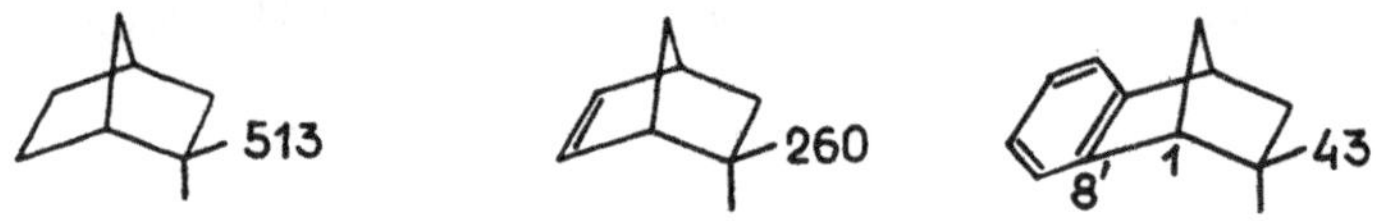

1.47 0.033 0.003

Three teams [430–432] studied simultaneously the influence of the benzene ring substituents on the rate of solvolysis in 2-benzonorbornenyl derivatives. On introducing substituents to the homo-para-position the solvolysis rates of the exo series correlate with the substituent constants σ_p^+ and to the homo-meta-position — with σ_m^+, i.e. just as for the usual reactions of electrophilic aromatic substitution.

The OCH$_3$ group introduced to the homo-para-position accelerates the solvolysis of 2-exo-sulphonates 178-fold while the homo-meta-OCH$_3$ group decelerates it by a factor of 0.4. In the endo series the 6-OCH$_3$ group increases little, and the 7-OCH$_3$ does not practically affect the rate of solvolysis. These results indicate that the effect of π-participation and the exo:endo rate ratio for 2-benzonorbornenyl derivatives are essentially increased by the activating substituents in the benzene ring. Brown's alternative explanation [433] that the high exo:endo rate ratio is due to unusual steric demands of the rigid bicyclic system or to unfavourable electrostatic interactions requires no discussion since on introduction of substituents into the benzene ring these factors remain practically unchanged [434,435].

At first, on the basis of the high exo:endo rate ratio and the retention of configuration in the solvolysis of 2-exo epimers, Bartlett suggested a nonsymmetrical homobenzylic structure of the intermediate ion *261* [418].

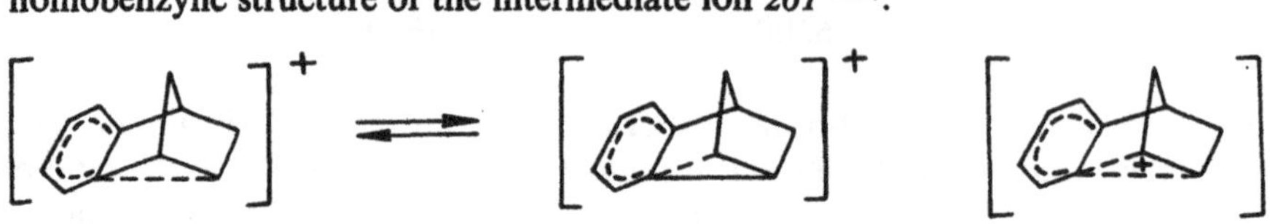

261 *262* *263*

Then Giddings[436] prefered formula *263* to account for the data he had obtained on acetolysis of the optically active 2-exo-benzonorbornenyl brosylate *264*; ion *263* is formed either with *simultaneous* anchimeric participation of σ- and π-electrons at the limiting stage or as a result of the *consecutive* involvement of π- and σ-electrons in the interaction with the formation of an intermediate ion *261*. Later Bartlett assumed ions *261* and *262* to be in equilibrium looking upon ion *263* as a transition state of rearrangement[1]. It should be remarked that ion *261* involves only homobensylic interaction between the π-electrons of the ring and the p-orbital of C^2.

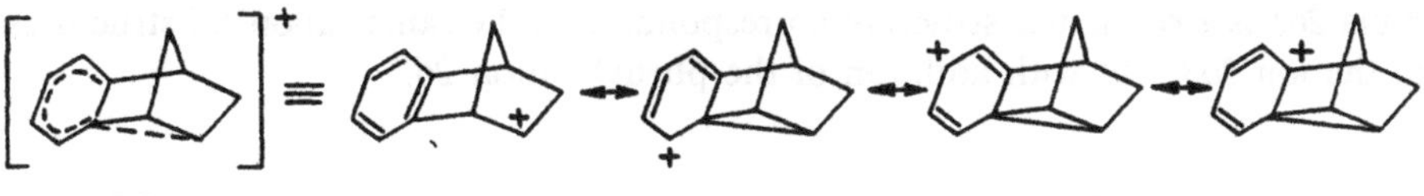

261

In ion *263* both the ring π-electrons and the σ-bond C^1—$C^{8'}$ are delocalized; the charge distribution can be represented by the same equilibrium structures with another one added:

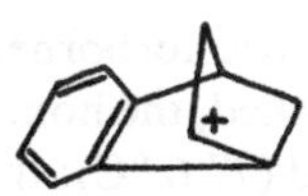

Lancelot, Schleyer and Cram suggest 1972 the following structures for the intermediate ions[437]:

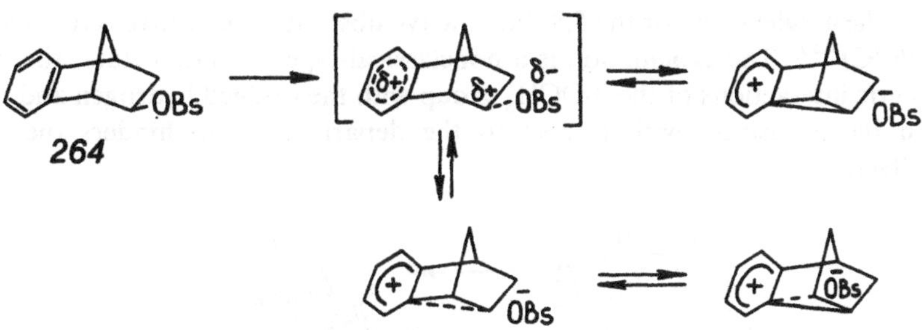

In 1969 Winstein[438] suggested for the 2-benzonorbornenyl ion a symmetrical structure *265*:

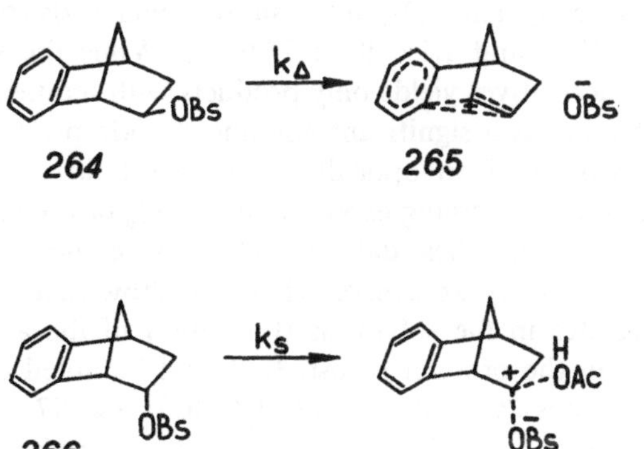

$$\frac{k_\alpha}{k_t} \sim 4.1$$

$$\frac{k_\alpha}{k'_\alpha} = 6200\rho$$

$$\frac{k'_\alpha}{k'_t} \sim 1$$

For the exo isomer *264* the ratio of rate constants determined polari- and titrimetrically ($k_\alpha : k_t$) was shown by him to equal about 4. This means that the formation of racemic brosylate from a symmetrical ion pair as a result of internal return is essential for the solvolysis of the exo isomer; for the endo isomer *266* the ratio $k_\alpha : k_t = 1.0$. In using the polarimetrical constants the exo:endo rate ratio is equal to 62000. Consequently, the exo isomer *264* is ionized with anchimeric participation (k_Δ); the resulting ion pair yields racemic (99.9 ± 0.1%) 2-exo acetate, significant internal return simultaneously taking place. The ionization of the endo brosylate *266* proceeds with the solvent assistance (k_s) and results in a "classical" ion pair. Winstein's formula *265* is a resonance structure corresponding to the same canonical structures as in the ion *263*, but with addition of the phenyl cation *267*.

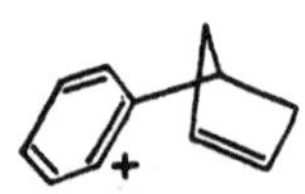

267

Tanida et al. have studied 6- and 7-substituted 2-exo-benzonorbornenyl brosylates with highly deactivating substituents [439]. As already remarked, the homo-para-OCH_3 group accelerates the solvolysis of exo isomer *264* 178-fold (77.6 °C); this effect is the highest for the anchimeric participation of the p-anisyl group in all studied systems. Thus, the benzene ring participation in the C^2—X bond ionization is more effective than in that of the C^9—X bond.

The acceleration for the 5,8-dimethoxy-substituted derivative is relatively low (16 at 25 °C) [440]. The explanation is a nonclassical-type transition state; the unfavourable steric interactions of the 8-OCH_3 group with the bridged hydrogen and the hydrogen in the α-position with respect to the departing group hinders the participation effect.

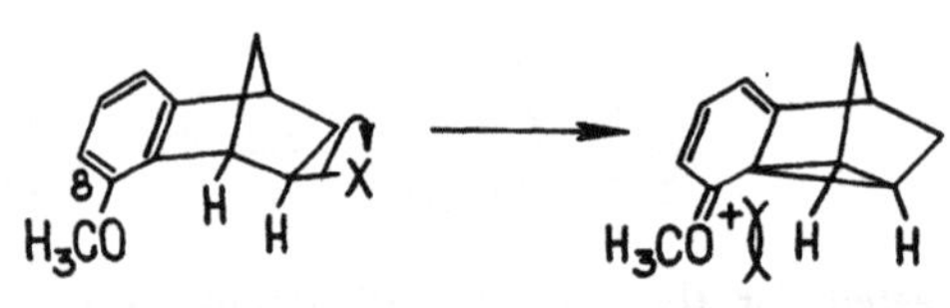

Introduction of 7-OCH_3-6-NO_2- and 6,7-$(NO_2)_2$-substituents leads to a sharp fall in the solvolysis rate ($1.1 \cdot 10^{-3}$ and $1.1 \cdot 10^{-5}$; 77.6 °C). While the solvolysis of 6-methoxy- and 7-methoxy derivatives yields only products with retained configuration, with deactivating substituents a significant amount of endo products and olefines is formed. Upon solvolysis of an optically active 6,7-dinitro derivative the retention of optical activity for the forming exo-acetate is 4.5% of the original value and for the endo-acetate it is 25.7%. These data, as well as those on the acetolysis of 6,7-dinitro-3-exo-deutero-2-exo-brosyloxybenzonorbornene show that the much deactivated benzene ring migrates in the solvolysis (formation of the aryl-C^2 bond) though the stereospecificity in the product is lost. It should be noted that the rate ratios of solvolysis for brosylates *264* and *268* in CH_3COOH and 97% alcohol are very close (respectively 1.3 and 1.5 at 165 °C); these data show that both in the acetolysis of brosylate *264* and of compound *268* (which is surprising) the nucleophilic participation of the solvent is not predominant. Tanida has shown that the rate constants (k_t) and their components — rate constants of the processes with anchimeric participation (k_Δ) are described with good accuracy by the equation $\lg (k/k_0) = \varrho\sigma^+$, where $\varrho = -3.26$.

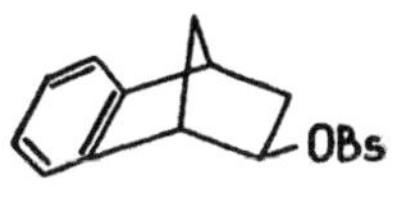

264

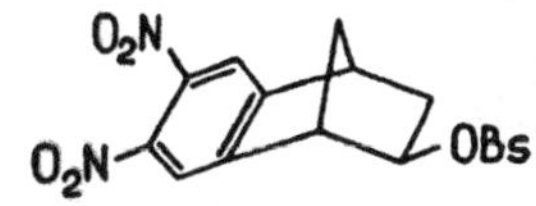

268

Endo-acetate has been formed from the reaction of the brosylate *268* racemized due to internal return with solvent participation (k_s).

Tanida has compared the effect of the same substituents with the solvolysis of 2-endo-brosyloxybenzonorbornenes [441]. While for the solvolysis of exo epimers the ratio of rate constants $k_{6\text{-}CH_3O}/k_{6,7\text{-}(NO_2)_2}$ is equal to $1.6 \cdot 10^7$, for that of 2-endo epimers it is only 230. The stereospecificity of reaction product formation in the exo series falls drastically on introduction of deactivating substituents into the benzene ring; in the endo series the structure of the reaction products depends little on the substituents in the benzene ring. This difference demonstrates the importance of the effect of benzene ring participation on the exo series and its absence in the endo series.

The acetolysis of endo esters results mainly in exo-acetates i.e. it is accompanied by configuration inversion; simultaneously insignificant amounts of olefines and endo-acetates are formed. If in 2-endo brosylates at the 2-exo-position the hydrogen is substituted by deuterium for the brosylate unsubstituted in the ring, as well as of the 7-NO_2- and 6,7-$(NO_2)_2$-brosylates, 44, 25 and 16% of deuterium, respectively, scramble to the C^1 atom of the acetolysis products. This implies that the Wagner-Meerwein rearrangement is decreased by deactivating substituents in the benzene ring. These facts lead to the conclusion that optically active endo brosylates are solvolyzed via a classical ion which then rapidly converts into a nonclassical ion

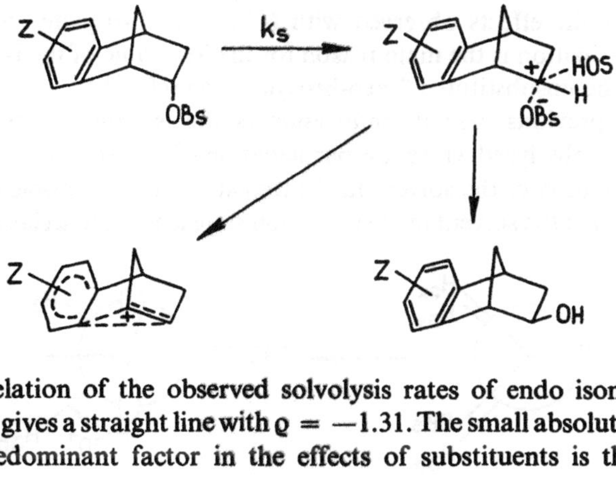

The correlation of the observed solvolysis rates of endo isomers ($k_t = k_s$) with σ-constants gives a straight line with $\varrho = -1.31$. The small absolute value of ϱ suggests that the predominant factor in the effects of substituents is the inductive effect.

In the correlational analysis of the kinetic data on the solvolysis of 1-aryl-2-endo-norbornyl tosylates in which the benzene ring cannot render any anchimeric assistance, ϱ is —1.1 [442] very close to the value —1.3 for 2-endo-benzonorbornenyl brosylates.

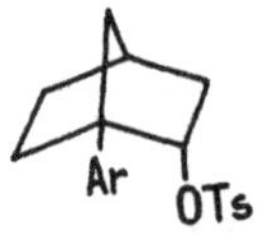

Subsequently Tanida et al. used the secondary β-isotopic effect of deuterium to determine the π-participation in the solvolysis of 2-exo- and 2-endo-brosyloxybenzo-norbornenes with different substituents in the aromatic ring [443]. The isotopic effect increases in the exo epimers as the benzene ring is deactivated. If the steric environment of the reaction centre in the transition state is independent of the substituents in the benzene ring (or if their difference is inessential) then the declining ability of the benzene ring of π-participation is obviously accompanied by the growing hyperconjugative effects of the neighbouring C^3—H bonds in the stabilization of the developing cation centre at C^2.

Equally possible is another explanation: As seen from Newman's projections, the formation of the nonclassical ion (in the k_Δ process) is accompanied by the disappearance of the H-2n and H-3n proton eclipse which results in steric strain relief.

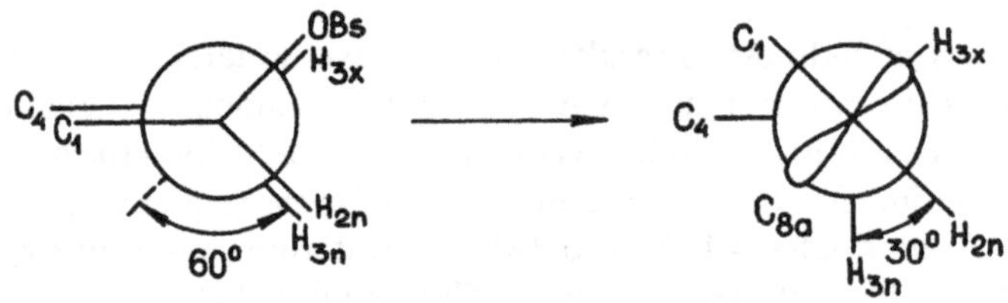

The steric strain relief in the transition state is known to result in a normal kinetic isotopic effect ($k_H > k_D$) [444], so with growing steric strain relief the isotopic effect will also grow.

The authors have failed to establish whether hyperconjugation or steric strain relief is decisive for the effects observed with 6,7-dinitro brosylate *268*; however, the aryl-group participation is the main reason for the low value of the isotopic effect upon solvolysis of the unsubstituted 2-exo-brosylate *264* (cf. [445]).

Contrary to the previous case in endo epimers the isotopic effects are rather large, since the latter solvolyzed without π-participation of the aromatic ring and without effective participation of the solvent in the transition state. Isotopic effects point to weak participation of the solvent in the transition state leading to a classical ion with

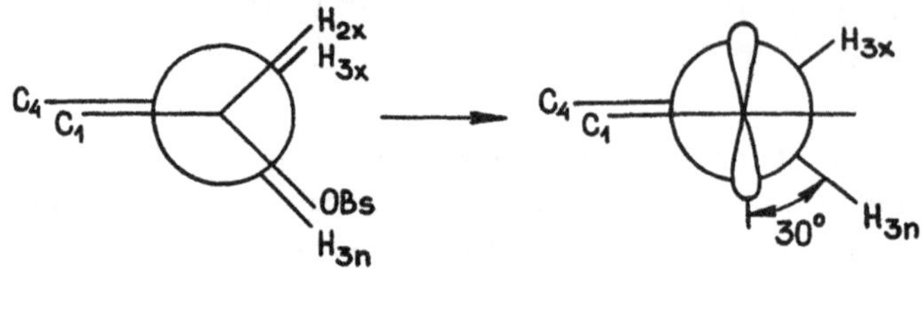

its subsequent conversion into a nonclassical ion; this requires hyperconjugative stabilization at the expense of C^3—H bonds. Another reaon for large isotopic effects in the endo series may be steric strain relief in the transition state. In endo isomers the β-isotopic effects is decreased as the aromatic ring is deactivated; this seems to be due to an increase of solvent participation in the transition state; the latter factor, just as the increase of the ring participation (in the exo series), decreases isotopic effects.

The participation of the ring can also be levelled by introducing into the system a stronger stabilizing group which can accelerate the solvolysis of both exo and endo epimers. Schleyer used in this capacity the α-cyclopropyl group and studied the solvolysis of tosylates *269* and *270* [446].

269 *270*

The cyclopropane ring in both epimers occupies the same position with respect to the departing group; both epimers form a stable bisected cyclopropylcarbinyl cation. The exo:endo rate ratio can reflect the contribution ratio of steric factors on the formation of the unsubstituted 2-benzonorbornenyl cation out of the two epimers. The hydrolysis of compounds *269* and *270* has resulted in an exo:endo rate ratio of 12 and agrees fairly well with the previous value. A comparison of these data with the value of 15000 for the unsubstituted compounds clearly shows the steric and electrostatic effects to have a low value for secondary systems and in the absence of the ring π-participation the exo-isomer solvolyzes faster than its endo epimer only 2- to 10-fold.

According to Paquette [447], the acetolysis of brosylates *264* and *266* as well as of the epimeric pair of 9-exo-brosylates *271* and *272* (5700) have the same order of magnitude in their exo:endo rate ratios; this testifies against the σ-participation in the solvolysis of compound *271* and against the formation of an intermediate ion *273*. The latter case would require an overlap of the unshared pair of oxygen electrons with the developing p-orbital of C^1 (contrary to Bredt's rule); this should have resulted in an extra acceleration as compared with the ionization rate of the brosylate *264*.

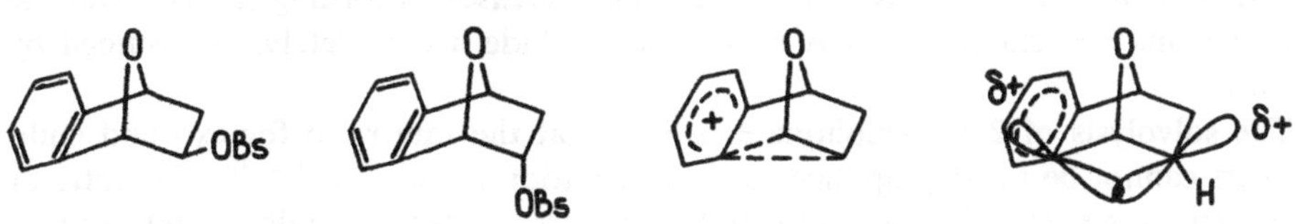

271 *272* *273* *274*

Paquette concludes that the solvolysis of brosylates *264* and *271* involves only the benzene ring π-participation without any positive charge transfer to C^1, with an intermediate *274*. The author, however, says nothing about how to account for the complete racemization upon solvolysis of brosylate *264* pointing to the full equivalence of the C^2 and C^1 atoms.

While in secondary 2-exo-benzonorbornenyl derivatives the π-participation of the aromatic ring raises no doubts, in tertiary analogues the situation is questionable. 2-Endo-methyl-2-benzonorbornenyl-p-nitrobenzoates yielded the following relative rates of solvolysis in 50% aqueous acetone at 125 °C)[448]:

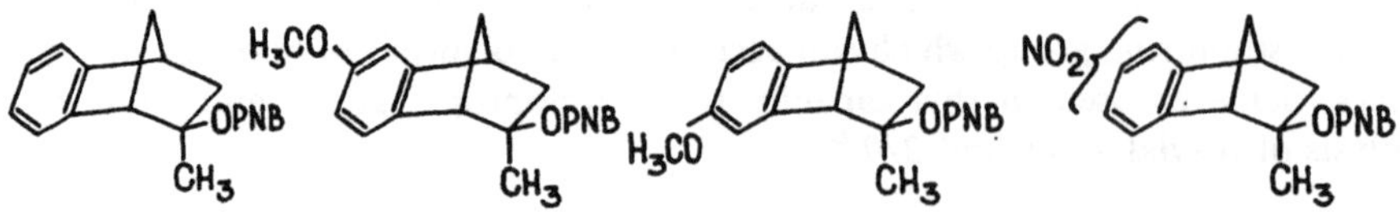

$k_{rel} = 1$ 16 0.74 0.025

These figures make clear that, just as in secondary sulphonates, the homo-para-OCH_3 group accelerates, the homo-meta-OCH_3 group slightly decelerates and the NO_2 group strongly decelerates the solvolysis. The rate constants obtained can be fairly well correlated with the σ^+-constants of substituents where $\varrho = -1.9$ at 125 °C. As has been expected the value of ϱ is less than for the secondary substrates (—3.3 at 78 °C), nevertheless in Winstein's opinion it is too high for the k_s process, since the ionization with anchimeric acceleration (k_Δ) is the main process for the tertiary 2-exo derivatives as well.

The solvolysis of 2-phenyl-2-benzonorbornenyl p-nitrobenzoates yields the factor of 4.7 corresponds to $\varrho = -0.9$[449]; this value shows that at least part of the ionization proceeds with anchimeric acceleration.

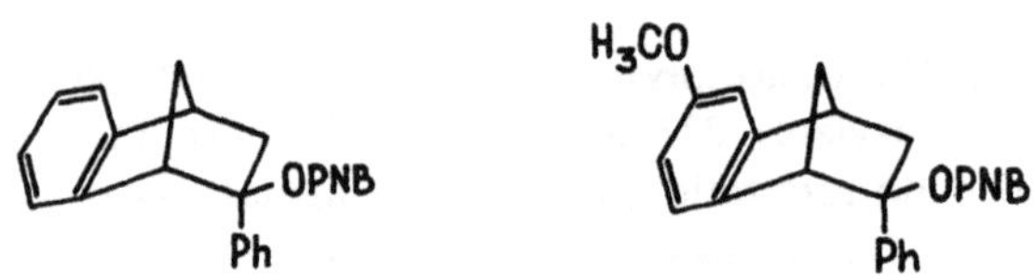

$k_{rel} = 1$ 4.7

The substitution of 2-H by 2-CH_3 and 2-Ph decreases, according to Winstein, the benzene ring π-participation, but it does not exclude it completely, as assumed by Brown.

The solvolysis of 2-exo-sulphonates shows that the rate ratio for exo and endo epimers cannot be reliably applied as a participation criterion (cf. [433]). The ratio of titrimetric rates for the exo and endo epimers of the secondary and tertiary 2-methyl and 2-phenyl derivatives amounts to 15000, 6500 and 3200, respectively

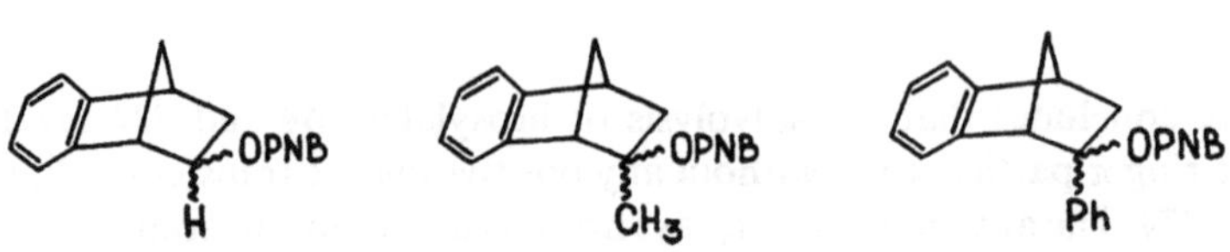

Brown also studied the rates of solvolysis of 2-arylbenzonorbornenyl-2-p-nitrobenzoates and showed that despite a significant change in the nature of substituents in the 2-aryl nucleus the exo:endo rate ratio (as distinct from the 2-aryl-6-methoxy derivatives) remains constant (~3000). The values of ϱ^+ for the series of exo and endo isomers are practically equal, —4.50 and —4.51, respectively [435, 450]. This testifies to the absence of essential π-participation in the solvolysis of the esters of 2-arylbenzonorbornenol.

If π-participation is not essential in esters of 2-arylbenzonorbornenol then what is the reason for their high exo:endo rate ratio observed?

It is very instructive to follow the change in the views of Goering et al. on the solvolysis mechanism of 1,2-dimethyl-2-exo-benzonorbornenyl derivatives. Initially the authors showed the solvolysis of optically active 1,2-dimethyl-2-exo-benzonorbornenyl chloride *275* or -p-nitrobenzoate *276* to form optically active products [451]:

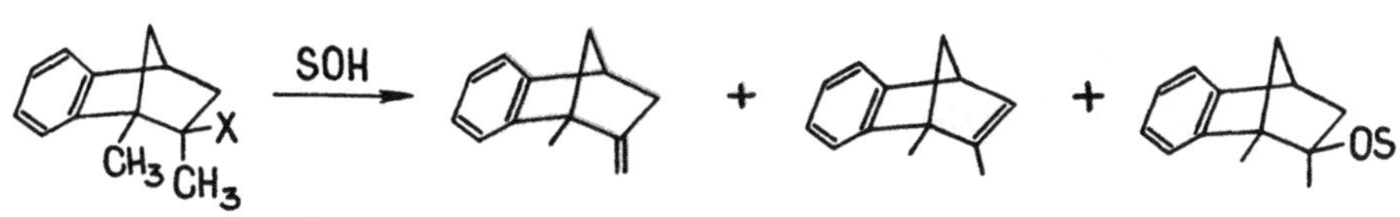

275, X = Cl

276, X = OPNB

This fact is a direct evidence of the formation of the unsymmetrical classical ion *277*.

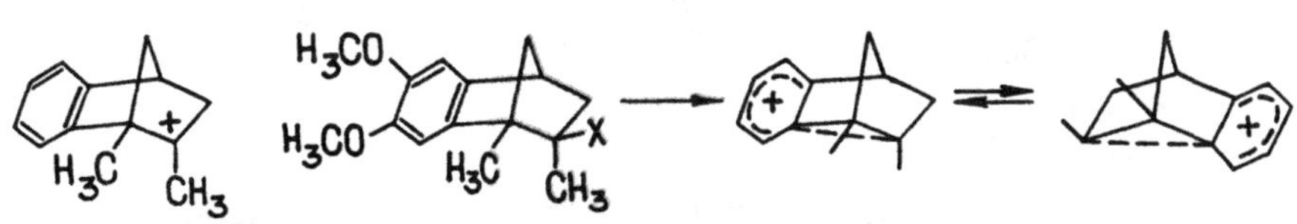

277 X = Cl, OPNB (−) − *278* (+) −*278*

Subsequently the same authors established that the solvolysis of optically active 6,7-dimethoxy-1,2-dimethyl-2-exo-norbornyl chloride or -p-nitrobenzoate is accompanied by anchimeric acceleration as distinct from the unsubstituted analogues *275* and *276*; the resulting products are optically active [452]. Thus, anchimeric assistance leads to the formation of an unsymmetrical cation which may be the homobenzylic ion *278* while the partial loss of activity is due to the interconversion of enantiomeric ions. The authors have also shown that the accelerating effect of the homopara-methoxyl group is practically the same for the solvolysis in methanol, 50% acetone and 90% acetone; as the solvent effects are different for the processes k_s and k_Δ, in [452] contrary to the earlier publication [451], they conclude that compounds *275* and *276* also solvolyze with anchimeric assistance and form an unsymmetrical homobenzylic ion.

Tanida showed that the efficiency of the π-participation of the benzene ring in the stabilization of the developing cation centre depends largely on the position of the vacant p-orbital relative to this ring. Thus, sulphonates *279, 247* and *280* yield on acetolysis only acetates with the same skeleton and with configuration retention indicating a ring π-participation in the transition state [453].

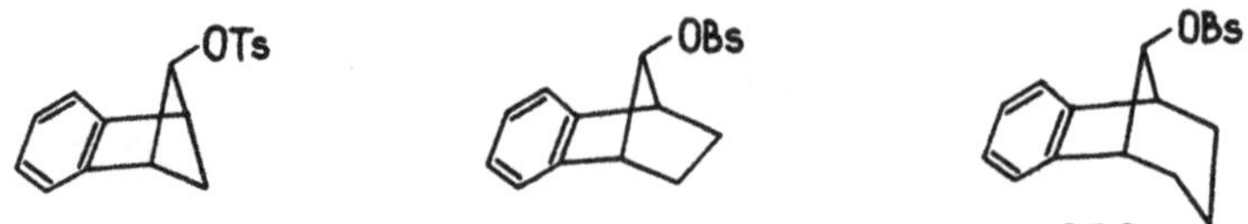

279 *247* *280*

k_{rel} (25°) = 32 1 0.0026

The internal angle at the bridged atom bearing a substituent increases in the sequence *279* → *247* → *280* and the rate of solvolysis falls sharply. These data prove that in passing from C-sp^3 to C-sp^2 the angular effect does not play an essential role. Calculations show, the longer the syn-bridge the longer the distance between the

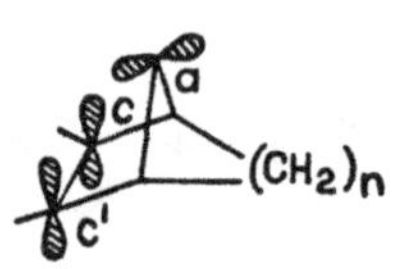

vacant p-orbital and the π-cloud [453]. The overlap integrals ($S_{ac} + S_{ac'}$) for compounds *279, 247* and *280* are 0.14; 0.11 and 0.09, respectively which agrees well with the observed reactivity. In compound *281* where the syn-bridge is still one C atom longer the k_Δ and k_s processes compete with each other [454].

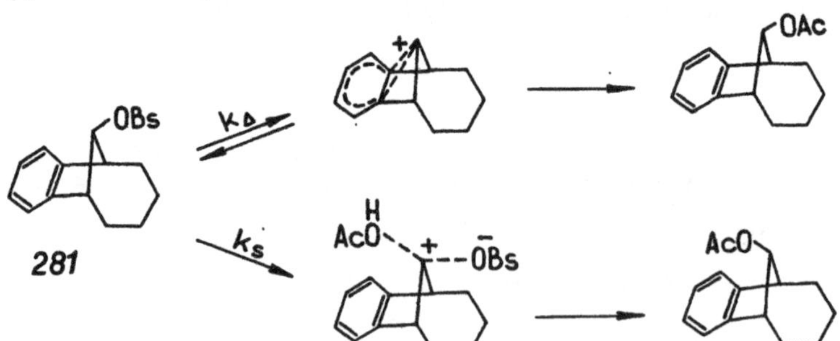

281

5.2 Electrophilic Addition

The stereochemistry of electrophilic addition to norbornenes and norbornadienes is very interesting [455,456]. As these reactions form on intermediate carbocations, there arises, just as for the solvolysis of 2-substituted norbornanes and norbornenes, the question of the nature of these ions — classical or nonclassical. The electrophilic addition to norbornenes and norbornadienes is treated in one place, though in one case it may be only σ-participation while in another -both σ- and π-participation.

The main peculiarities of the addition of electrophilic reagents to norbornene are intensive skeletal rearrangements, formation of cis-adducts, exo-addition of both an electrophile and a nucleophile. The unusual process of cis-exo-addition was explained by nonclassical carbocations or rapid equilibrium of classical ions [457–459], steric [147], torsional [460–462], stereoelectronic effects [463], molecular electrophilic addition [464 to 466], "twist-strain" [455] the difference between the values of exo and endo lobes of π-HOMO [467]. The high stereospecificity of cis-exo addition to norbornene can be due to the steric peculiarities: the endo attack on the sp^2-hybridized C^2 atom is hindered by the hydrogens at C^5 and C^6. If this were the case then the bulky group at the 7-syn-position would lead to the predominance of endo addition. Tidwell [468], however, showed that mercury acetate is added to 1,4,7,7-tetramethylnorbornene and 7-syn-bromnorbornene strictly at the cis-exo-position.

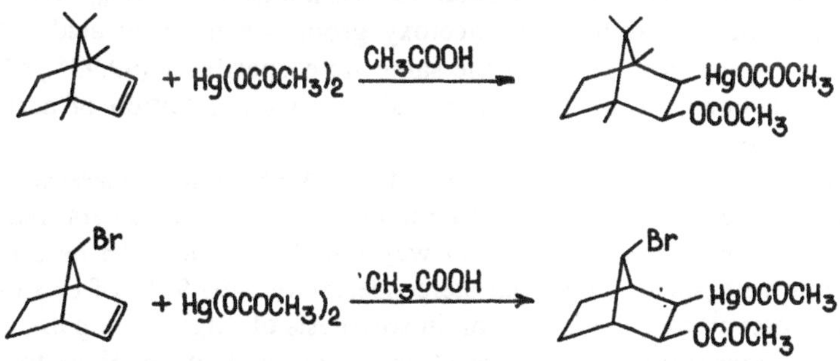

Later on Brown [469,470] compared the stereochemical results of adding various reagents to norbornene and 7,7-dimethylnorbornene. The introduction of a substituent at the 7-position of the norbornene system changes the stereochemical reactions via *cyclic transition states* and intermediates (hydration, hydroboration, epoxidation, complex formation etc.), but affects little the reactions with an "*open*" *transition state* and intermediate (addition of HCl, HBr, CH_3COOH, CF_3COOH etc.). The cis-addition of acids to norbornene is explained by intermediate formation of classical norbornyl cations [471–476]. CD_3COOD and CF_3COOD with norbornene yielded the following product mixtures:

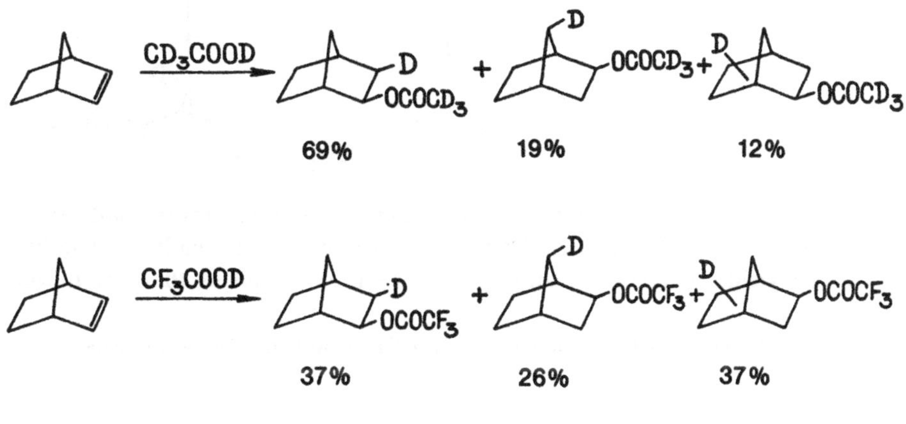

In these reactions Brown pointed out extremely high stereospecificity of formation of exo-norbornyl esters —99.98%[8]. If the sole intermediate were a nonclassical ion the results would be as follows:

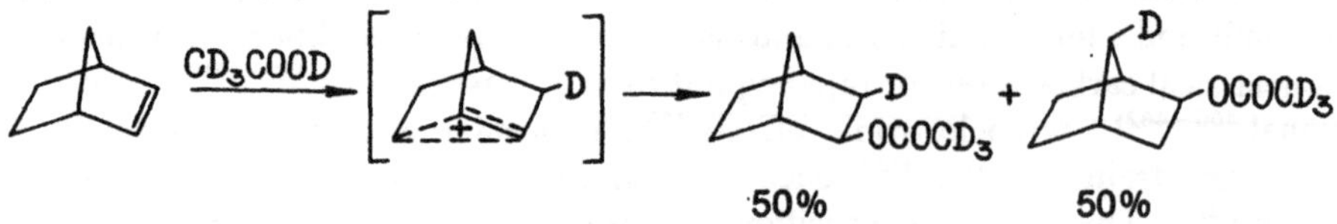

Neither can the experimental results be explained by the assumption of a rapid equilibrium of classical ions imitating a nonclassical ion since in this case the distribution of deuterium between C^3 and C^7 must be equal. The addition of acetic acid can proceed through two mechanisms: ionic addition via a nonclassical ion (~50%) and simultaneous molecular exo-cis-addition (~50%). Brown, however, showed that the addition of the acetoxy group when acetic acid acts on 7,7-dimethylnorbornene proceeds with the same stereospecificity (99.98% of exo isomer) as in the reaction with norbornene i.e. simultaneous addition of acetic acid seem to be highly improbable.

Finally, one can assume the protonation to result in a classical ion which is transformed at comparable rates into a nonclassical ion and is captured by a nucleophile. It is unclear, however, in what way the classical ion can be captured by the nucleophile with such exceptional stereospecificity —99.98% of exo isomer.

Brown notes [21] that the formation in solvolysis of "tight" ion pairs rather than of free carbocations turns the rate of reactions of ions with nucleophiles to be lower than that of reaching an equilibrium between epimers. Therefore in order to "trap" the classical 2-norbornyl cation it is necessary not to use solvolysis, but other reactions in which more "free" carbocations are formed. The electrophilic addition of acids to norbornene belong to these reactions. Brown considers results obtained to be due to the capture of the classical 2-norbornyl cation before a complete equilibrium is established:

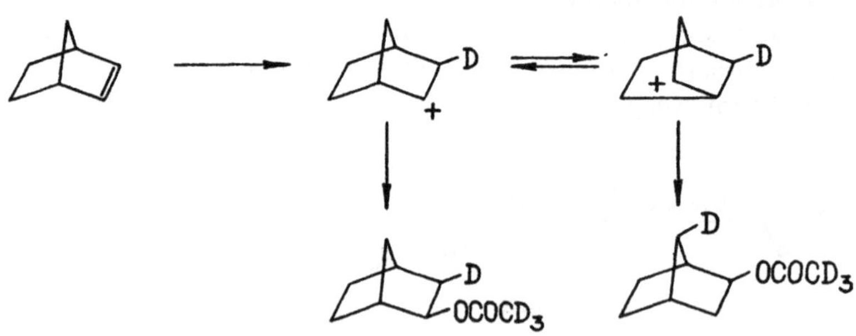

However, his interpretation contradicts the data reported by Cristol and Sallivan [478] on the addition of a CD_3COOD—DCl mixture to benzonorbornadiene. The high ratio of the yields of chlorides and acetates (~15:1) indicates that the nucleo-

8 Cristol obtained similar results on addition of CD_3COOD to norbornadiene and benzonorbornadiene [457, 477].

phility of Cl^- in this medium is far higher than that of AcO^-. Nearly equal amounts of chlorides 282 and 283 from the reaction results while the amount of acetate 284 is by 18% higher than that of acetate 285.

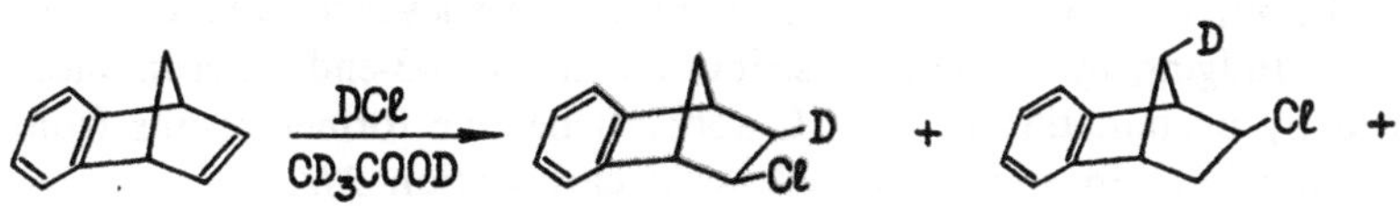

282 283

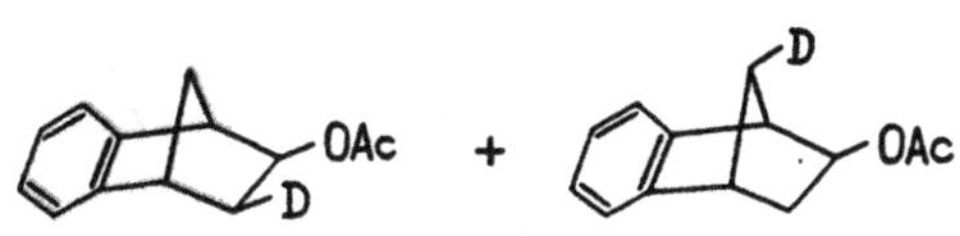

284 285

Consequently, the interconversion of classical cations at a rate comparable with capture by a nucleophile (Brown) is not valid here since it is in chlorides and not in acetates that the less homogeneous deuterium distribution should have been expected. The authors [478] maintain that the HX-type reagents are added to norbornene and some other similar bicyclic olefins both through intermediate formation of nonclassical ions and by simultaneous cyclic processes. As concerns Brown's data on the addition of HCl and CH_3COOH to 7,7-dimethylnorbornene, Cristol believes them to be inapplicable to unsubstituted norbornenes since in the 7,7-dimethylnorbornene the simultaneous processes are strongly hindered. Cristol's objections, however, are hardly convincing because *epoxidation, hydroboration* etc. proceed in the case of 7,7-dimethylnorbornene as well though they only result in endo products.

Norbornene *chlorination* [479,480] results in but a small amount (6%) of trans-dichloride expected from the chloronium ion; but the main products — nortricyclyl chloride and 2,7-di-exo-dichloronorbornene — seem to derive from the nonclassical

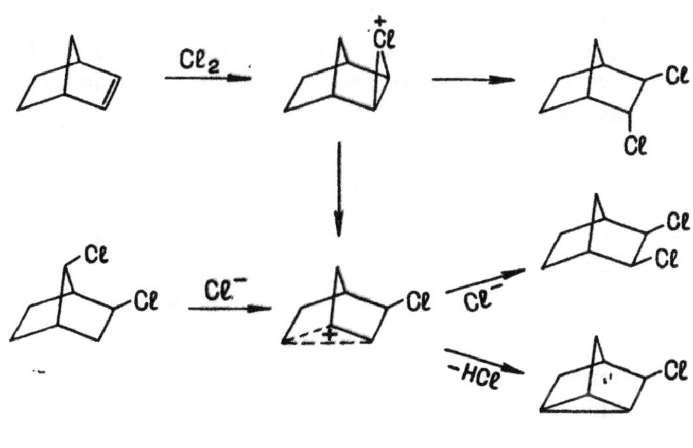

ion, just as a small quantity of cis-dichloride. Thus, in the norbornene system the participation of the σ-bond C^1—C^6 in the stabilization of the developing cation centre proves more effective than that of chlorine at C^3.

Similar results were obtained upon norbornene *bromination*[481,482]. Marshall[483] separated from the mixture five dibromides, a bromonortricyclene and a 2-exo-bromonorbornane. Judging by the label distribution in 2-exo-3-endo-dibromonorbornane obtained by brominating of 5,6-^{14}C-norbornene, the former results from both the bromonium ion *286* and the nonclassical carbocation *287*.

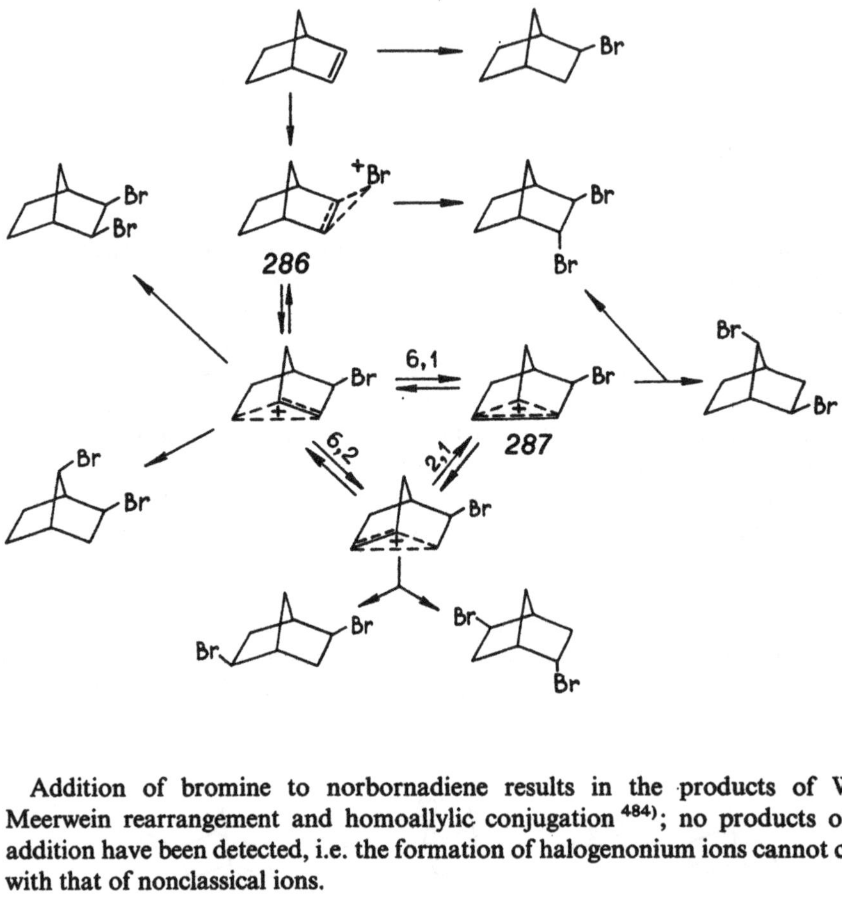

Addition of bromine to norbornadiene results in the products of Wagner-Meerwein rearrangement and homoallylic conjugation[484]; no products of trans-addition have been detected, i.e. the formation of halogenonium ions cannot compete with that of nonclassical ions.

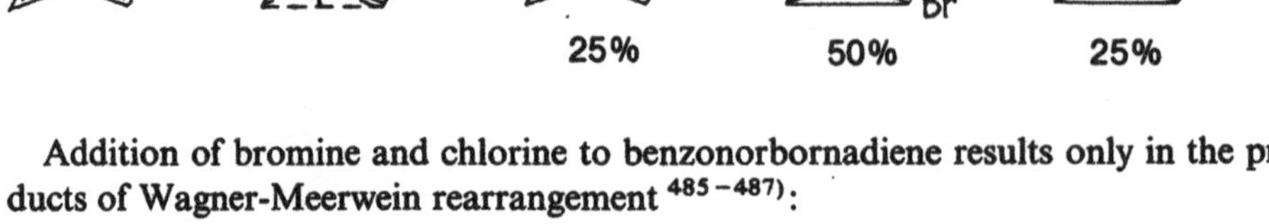

Addition of bromine and chlorine to benzonorbornadiene results only in the products of Wagner-Meerwein rearrangement[485–487]:

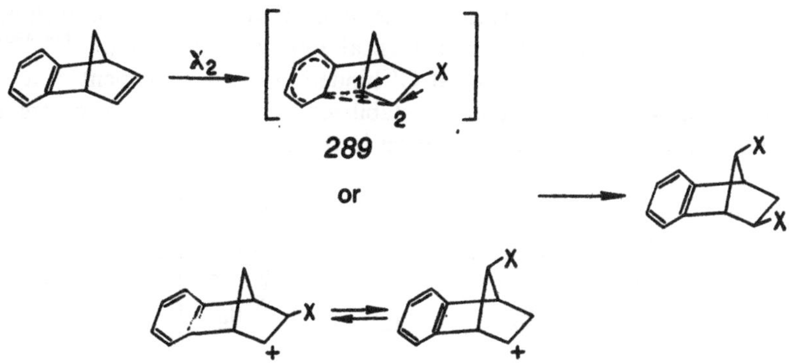

289

290 291

The absence of cis-di-exo-dihalogen derivatives is attributed by Wilt to steric reasons. On the conversion of trans-dichloride *288*, however, Cristol has shown this explanation to be unsatisfactory. The nucleophilic attack at the position nonvicinal to the already introduced substitutent can be accounted for, according to Cristol, either by the higher stability of cation *291* in comparison with cation *290* or by a higher value of the positive charge on C^1 in ion *289*.

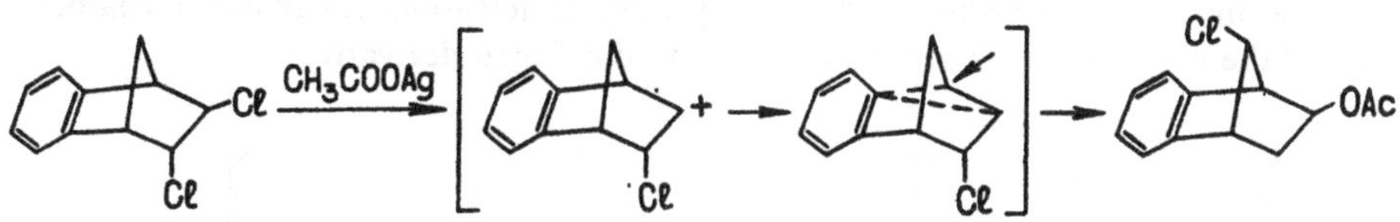

288

To elucidate the steric factors in halogen addition Caple studied the *bromination* of 7-anti-bromobenzonorbornadiene [488]. The bulky substituent at C^7 favours an endo attack on C^5 and C^6; the torsional factors in this system are smaller than in norbornene. Nevertheless the bromine was attached at an exo-position, i.e. irrespective of the steric factors. These results exclude a rearrangement since this would have resulted in a product of the same structure. For the stereoselectivity of the nucleophilic exo attack a nonclassical ion or a rapid equilibration of classical ions is necessary; the reason for the exo attack of the electrophile is not quite clear.

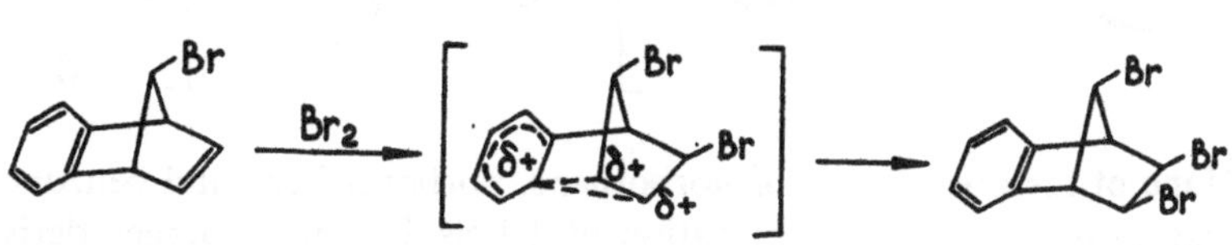

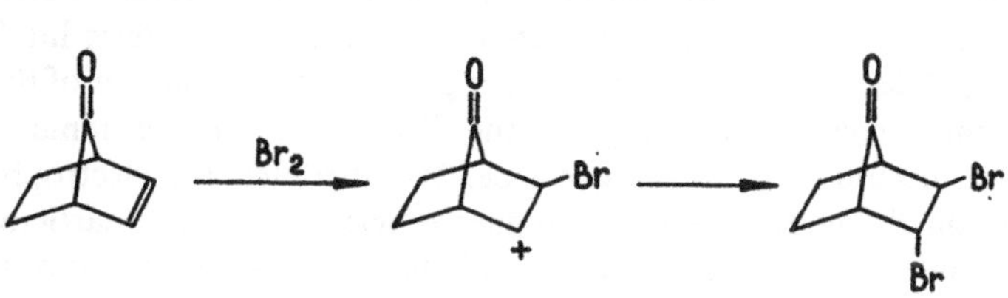

Trans-dibromides are formed as the main products when the anchimeric participation and formation of a nonclassical ion are hindered (cf. [489,490]) either by an electron-seeking substituent at C^7 hindering the development of a cation centre at C^1 or if the substituent at C^2 stabilizes the arising cation centre and thereby decreases the extent of σ-participation in the transition state [491].

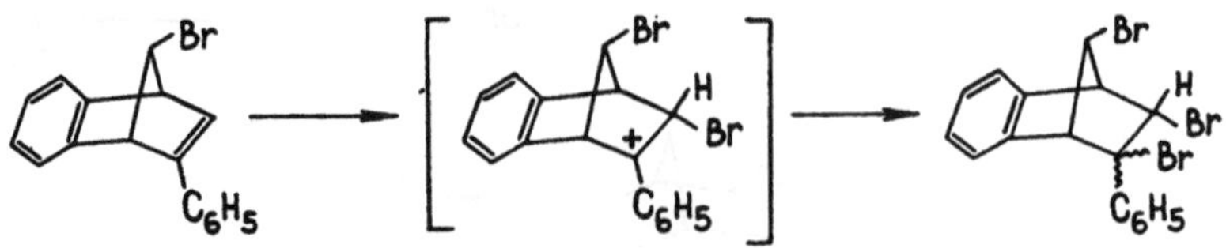

This result indicates that the electrophilic exo-side attack of bromine in 7-anti-brombenzonorbornadiene is due to stabilizing the developing cation centre at the expense of the C^1—C^6 σ-bond. In the electrophilic endo-side attack of bromine the formation of the bridged C^2—C^6 bond is stereoelectronically impossible; the C^1—C^7 bond participation is possible, but its probability is very low due to a considerable increase in angle strain.

As Martin and Koster point out[492], for the electrophilic halogenation of norbornadiene and benzonorbornadiene the endo-side attack is practically null. When brominating and chlorinating the 7-isopropylidentetrafluorobenzonorbornadiene *292* [493] the main products are 2,3-cis-di-endo-dihalogen derivatives.

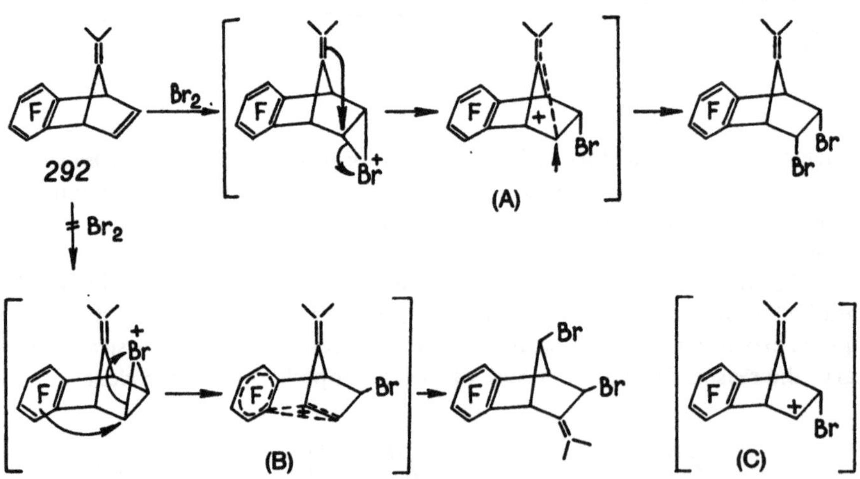

None of the above cases of norbornene, norbornadiene and benzonorbornadiene halogenation involved the formation of 2,3-cis-di-endo-dihalogen derivatives. Such unusual stereochemistry of addition cannot be explained by steric hindrances from the 7-isopropylidene group since even in the presence of bromine or of the CH_3 group (i.e. with far stronger shielding) at the 7-syn-position the ionic addition, as noted above, occurs only on the exo side. We attribute, the electrophilic bromine attack only on the endo side to the higher efficiency of the π-participation of the exocyclic double bond (forming the homoallylic ion *A*) whereas that of the π-participation of the fluorinated aromatic ring could lead to the homobenzylic ion *B* and further to Wagner-Meerwein rearrangement products.

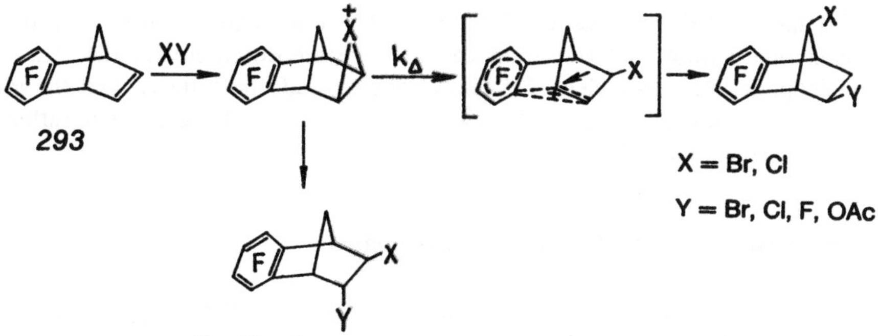

293

X = Br, Cl

Y = Br, Cl, F, OAc

X = Y = J

The latter process is well modelled by the halogenation of the tetrafluorobenzonorbornadiene *293*. The stereospecific endo-side attack of Br^- and the absence of Wagner-Meerwein rearrangement in reactions with 7-isopropylidentetrafluorobenzonorbornadiene are well accounted for by the structure of the homoallylic ion *A* which is not converted into the homobenzylic ion. The fact that the bromine atoms go in to the most sterically unfavourable cis-vincinal position is difficult to explain if the intermediate cation has the classical structure *B*.

The electrophilic addition to norbornenes, norbornadienes and benzonorbornadienes clearly shows that these models make it rather difficult to choose among the three alternative hypotheses assuming intermediate formation of a classical ion, a nonclassical ion or the presence of a rapid equilibrium of classical ions.

6 Bicyclooctenes and Bicyclooctadienes

As far back as 1959 Grob investigated the bromination of dihydrobarrelene [494].

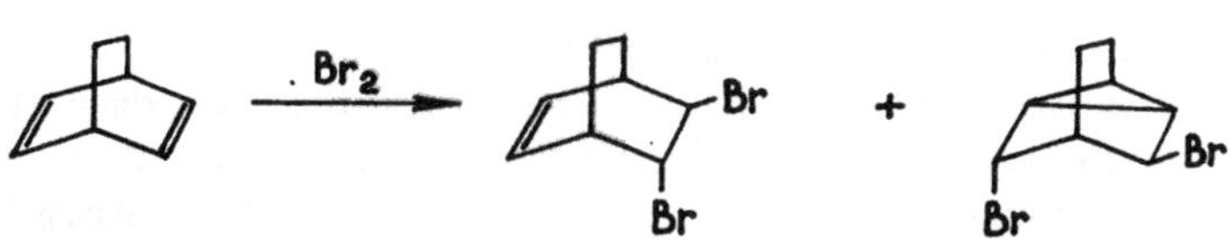

The data obtained, however, solve none of the questions raised above: What does the direction of the electro- and nucleophilic attack depend on? Is there any dependence between the halogenonium ion configuration and the skeletal shift direction?

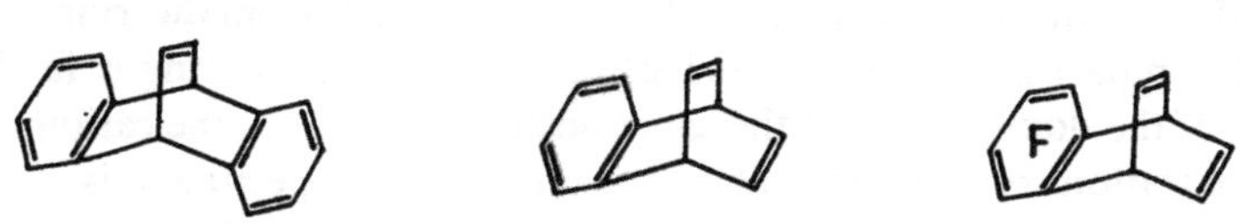

294 295 296

The halogenation of dibenzobarrelene *294*[495,496] is, in our opinion, worse than benzobarrelene *295*: one is unable to study the homoallylic participation; due to the skeleton symmetry it is often impossible to establish which of the bond (C^1—C^6 or C^1—C^2) has shifted as a result of Wagner-Meerwein rearrangement; for the same reason one cannot find out what causes the stereospecificity in the attack by the nucleophile. Benzobarrelene *295* allows one to study both the σ- and the π-participation (homoallylic and homobenzylic) in one model. However, it is rather inaccessible and unexplored[497].

6.1 Halogenation of Tetrafluorobenzobarrelene

A more accessible model is tetrafluorobenzobarrelene *296*[498,499] for the interaction with various halogens[500,501].

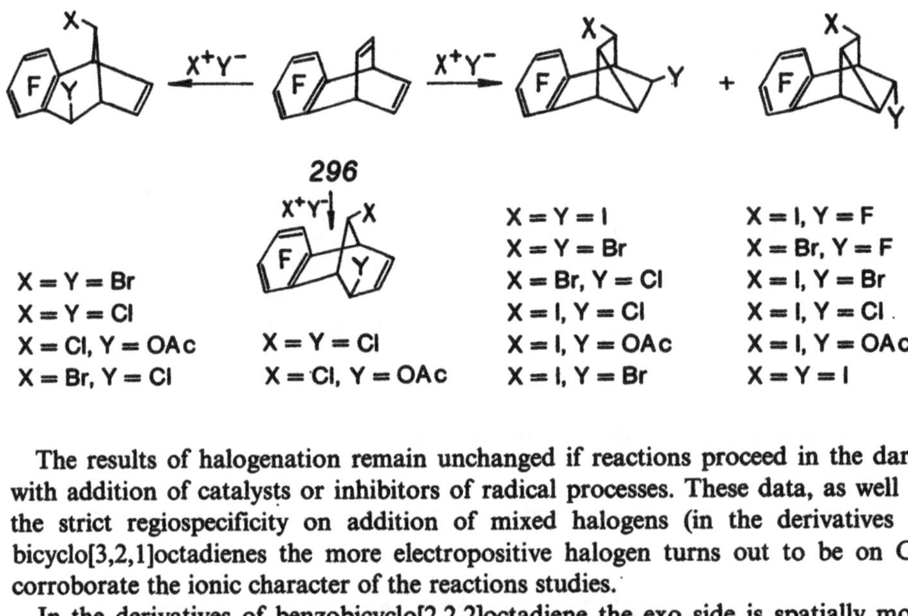

The results of halogenation remain unchanged if reactions proceed in the dark, with addition of catalysts or inhibitors of radical processes. These data, as well as the strict regiospecificity on addition of mixed halogens (in the derivatives of bicyclo[3,2,1]octadienes the more electropositive halogen turns out to be on C^8) corroborate the ionic character of the reactions studies.

In the derivatives of benzobicyclo[2,2,2]octadiene the exo side is spatially more accessible for the attack on the sp^2-hybridized bridge carbon atom than the endo side[502]. In compound *296* the exo side is also sterically more accessible, as found from the results of cis-synchronous reactions[503,504]. Yet the electrophilic attack of bromine and iodine occurs only from the endo side. This endo-stereospecificity in the electrophilic attack can be accounted for by a more effective π-participation of the double bond in the stabilization of the intermediate halogenonium ion *297* as compared with the π-participation of the fluorinated aromatic ring in the stabilization of complex *298*. The π-participation of the aromatic ring is far less effective than that of the double bond in the same position relative to the cation centre arising in the ionization process[418]. Thus the endo stereospecificity is determined by electronic factors, even though the direction of the attack contradicts the steric requirements. The lifetime of complex *297* is likely to be very short, it is not long enough for the complex to be attacked by the nucleophile from the opposite side to form trans-dihalogen derivatives retaining the bicyclo[2,2,2]octadiene skeleton; the complex converts quickly into the homoallylic ion *299*. The latter can either be attacked by the nucleophile to form tricyclooctene derivatives or convert into a still more delocalized ion *300*.

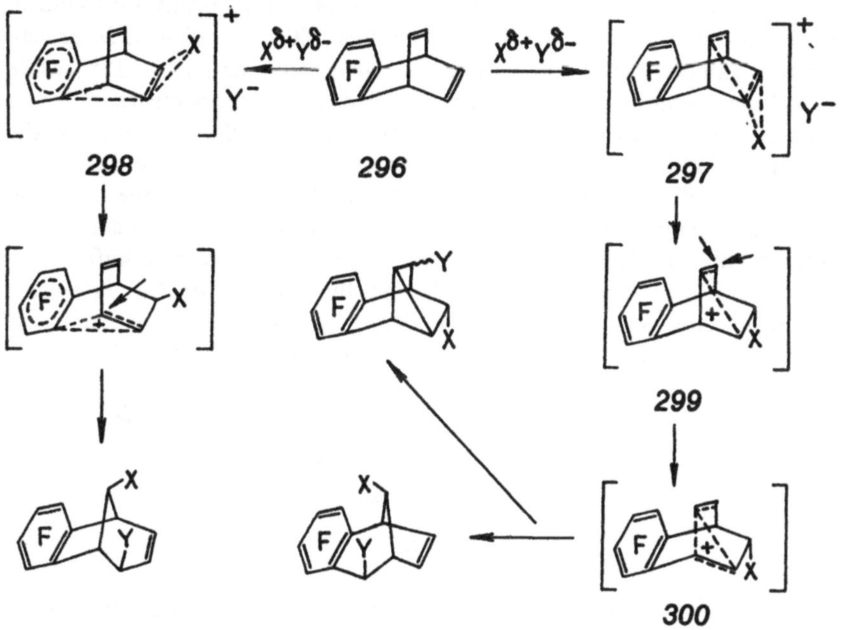

The assumption of the intermediate formation of the latter accounts at once for the two facts: 1) if X^+ has added into the endo-position, then there occurs a 1,2-shift of the C^1—C^7 bond, and not of C^1—C^6; 2) the nucleophile attacks C^1 only from the endo side results in sterically hindered derivatives with 1,3-diaxial arrangement of halogens in the cyclohexene fragment. At the same time in benzo-bicyclo[3,2,1]octenes even in the absence of 8-syn-substituent the spatially less hindered side is known to be the equatorial one [502], so the steric factors should favour the exo-side attack on C^1 in ion *300*. Both facts of stereospecificity are difficult to explain if the opening of the halogenonium ion is assumed to result in the classical ion *301*. The stereospecificity of the nucleophilic attack becomes unaccountable assuming the skeletal shift to occur simultaneously with the splitting of the halogenonium ion and formation of the classical benzyl ion *302*.

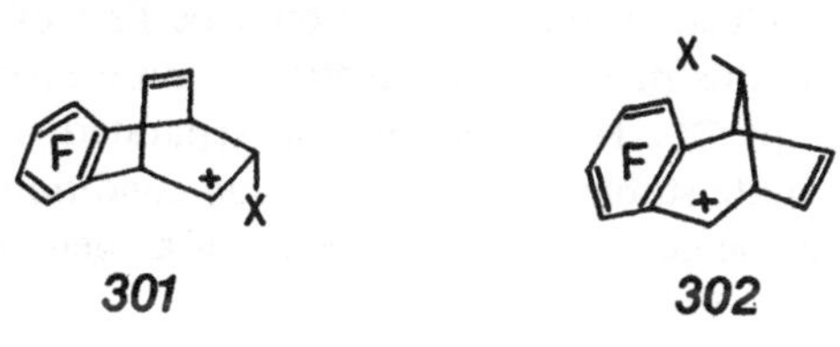

Whether the reaction is accompanied by the Wagner-Meerwein rearrangement (i.e. the rupture of the C-sp^2—C-sp^3 bond) or by the formation of products only due to π-delocalization depends on a variety of factors. If the nucleophility Y^- is relatively large (e.g. J^-), then ion *299* is captured before it converts into a more stable ion *300*. The latter forms and determines the structure of the reaction products with relatively weak nucleophiles (Cl^-, OAc^-). Essential are both the steric volumes X and Y since upon attack on C^1 in ion *300* halogens are found in the 1,3-diaxial position. Indeed, chlorination yields 100% of Wagner-Meerwein rearrangement products, the action of BrCl — 70%, bromination — 25% while iodination and the action of mixed halogens JY yield only tricyclooctenes.

In the case of $X = Cl^+$ there occurs both an exo- and an endo-side attack by the cation apparently due to the high electrophility of Cl^+ for which the differences in the effectiveness of the π-participation of the double bond and the aromatic ring becomes less essential. An exo-side attack yields no tricyclooctene derivatives at all involving only in Wagner-Meerwein rearrangement. There is a shift of the C^1—C^6 bond alone (not the C^1—C^7 bond) and the nucleophile Cl^- is stereospecifically added only on the exo side so that in the resulting compound the halogens are again in a sterically unfavourable 1,3-diaxial position [505,506].

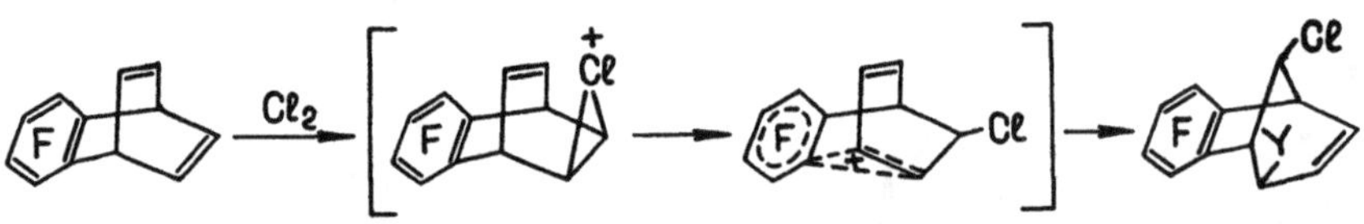

303

The intermediate formation of the homobenzylic ion *303* gives a good account of all the facts of stereospecificity noted above. If the intermediate compound were the classical ion *304* then, while the exo-side attack by a halogen cation occurred, both homoallylic conjugation with formation of tricyclene products and rearrangements would be possible due to migration of the C^1—C^7 and C^1—C^6 bonds as well as addition of a nucleophile from a sterically more accessible equatorial side in ion *305*.

304 *305*

But this does not happen. If the primary intermediates were the classical ions *304* or *301*, then the further products in either case would be first of all those of homoallylic conjugation because this process is energetically more favourable than the 1,2-shift of the C-sp^3—C-sp^2 [364]. However, all the separated derivatives of tricyclooctene are shown to be quite stable under the reaction conditions; hence the formation of bicyclo[3,2,1]octadiene derivatives is the result of a "genuine" 1,2-shift of the C-sp^3—C-sp^2 bond.

As a result one can draw the following conclusions [507,508]:

1) The electrophilic attack on compound *296* proceeds stereospecifically and its direction is determined by the comparative effectiveness of nonclassical assistance in the stabilization of halogenonium ions.

2) The configuration of the intermediate halogenonium ion determines unequivocally the direction of its isomerization into nonclassical carbocations — homoallylic or homobenzylic ions. These are attacked by the nucleophile strictly stereospecifically contrary to the steric factors but in agreement with the steroelectronic requirements of nonclassical ions.

3) The data obtained cannot be accounted for if intermediate classical ions are assumed to be formed both with the initial skeleton of the type *301* and with the bicyclo[3,2,1] octadiene skeleton of the type *302*. The stereospecificity of the attack by a halogen-anion cannot be acounted for by rapid equilibration of classical ions because quite different equilibria have to be assumed for the similar ions *301* and *301a*:

301a

301 302

4) In the reaction no interconversion of nonclassical ions is observed, e.g.:

Thus, the energetic barrier between different isomeric nonclassical ions proves to be higher than the activation energy of the reaction with a halogen-anion.

6.1.1 Comparison between Tetrafluorobenzobarrelene and its Dihydroproduct

The above regularities are mainly retained in the halogenation of tetrafluorobenzodihydrobarrelene *306* as well, but in this case the processes with the participation of the bond (k_{Δ}) and that of the reagent (k_s) compete; the latter route results in trans-dibromides with the same skeleton [509–511]. In ion *307* the skeletal 1,2-shift of the 1–7 bond formating the benzylic ion *308* competes with the capture by a nucleophile; ion *308* yields non-stereospecifically formed products *309* and *310*, the latter being predominant with bromine. In ion *311* the nucleophilic capture precedes the skeletal 1,2-shift which would result in a comparatively unstable ion *312* and finally to *313*.

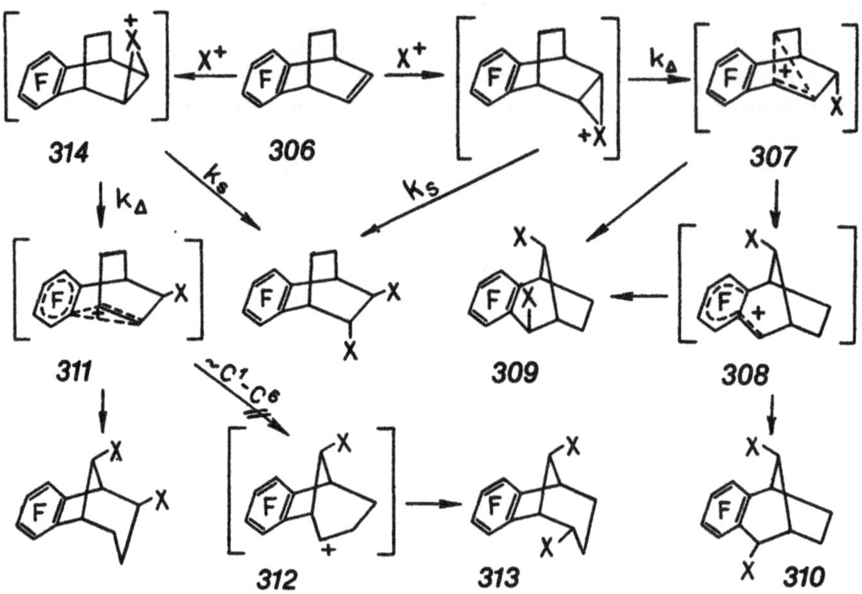

It is of interest to compare the halogenation of compounds *306* and *296*. In the latter, no products at all are formed with the skeleton retained; contrary to steric requirements, the nucleophile is added only stereospecifically. The different behaviour of the exo-halogenonium ions *298* and *314* seems to be due to the C^7—C^8 double bond stabilizing the developing cation centre at C^1 (at the expense of the resonance effect which turns out to be stronger than the inductive destabilization) and thereby favour the k_Δ process. The difference in the behaviour of ions *300* and *307* consists in that in the more stable homoallylic ion *300* the nucleophilic capture forestalls the skeletal shift with the formation of a benzylic cation.

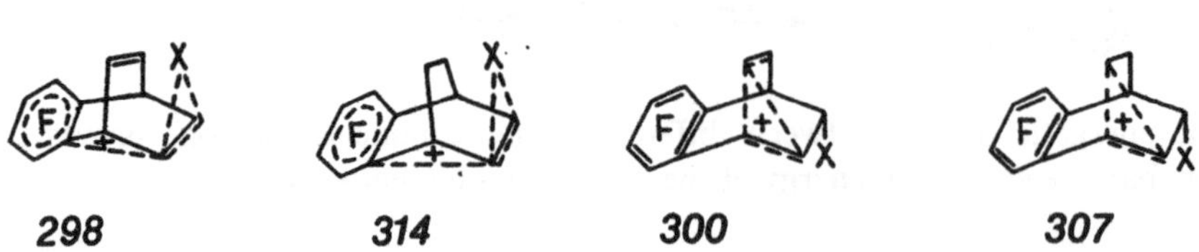

One of the important advantages of using benzobicyclooctene models in comparison with bicycloheptene ones is the relative ease of establishing the intermediate formation of a classical or nonclassical carbocation by the stereochemistry of the products formed. It is clearly seen from the chlorination of benzobicyclo[3,2,1]-octadienes. If the chlorination of derivatives of 6,7-tetrafluorobenzobicyclo[3,2,1]-octadiene first yields an exo-halogenonium ion, then the participation of the aromatic ring and of the σ-bond 1–7 turns out to be effective enough to direct the process along the k_Δ route with intermediate formation of a nonclassical ion and with stereospecific addition of chloro-anion despite the steric hindrances; the process is accompanied by skeletal rearrangement [512]. The assumption that in ion *316* the skeletal 1,2-shift of the $C^1—C^7$ bond forestalls the nucleophilic capture is rejected since in the case of an attack on the sp^2-hybridized atom in ion *317* the endo side is sterically more accessible [513].

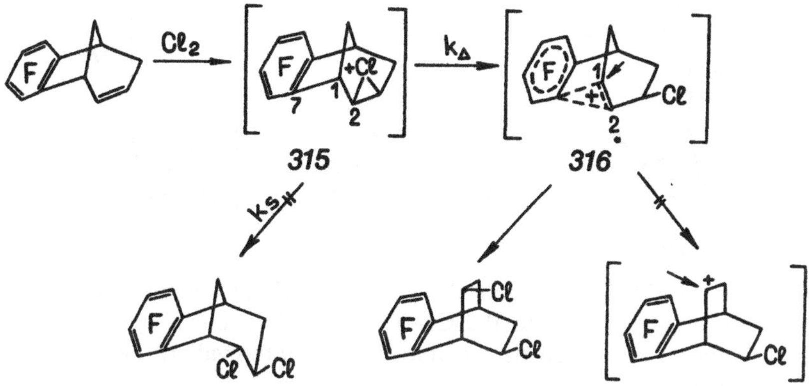

315 316 317

It is interesting to compare the data obtained with those on the solvolysis of 2-exo-bromo-6,7-tetrafluorobenzobicyclo[3,2,1]octadiene, the reaction results in a mixture of acetates with the same skeleton [514].

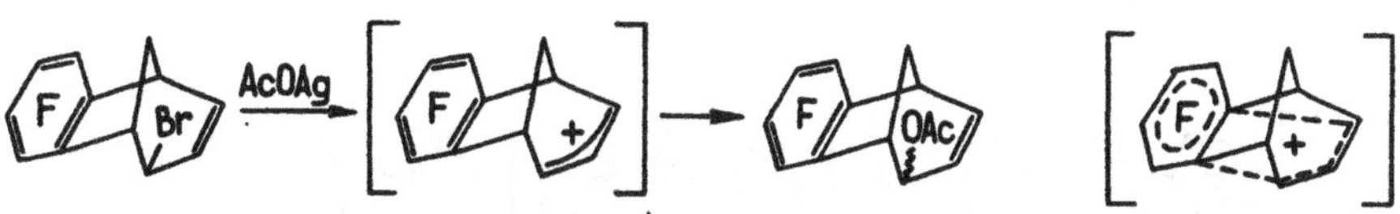

318 319

The $C^2—Br$ bond is ionized without the participation of the $C^1—C^7$ bond to form the classical ion *318*. Thus, if the arising cation centre at C^2 in the bicyclo-[3,2,1]octane system can be stabilized by the classical p,π-interaction, the non-classical participation does not manifest itself (cf. [515]); besides, the homobenzylic participation would result in the antiaromatic (4π-electrons) destabilized cation *319* (cf. [516]). But in the ion *315* the arising secondary cation centre can be effectively stabilized only by the nonclassical σ- or π-participation formating ion *316* of the bishomoaromatic type.

6.1.2 Other Possibilities to form Endo Halogenium Ions

If the derivatives of 6,7-tetrafluorobenzobicyclo[3,2,1]octadiene are first chlorinated to form an endo halogenonium ion (because of steric hindrance from the substituents at C^4 or C^8), then, since the aromtic ring cannot participate and the $C^1—C^8$ bond fails to be realized (seemingly for angular strain reasons [515]), the reaction proceeds via intermediate classical ions and results in chlorine addition products with the same skeleton.

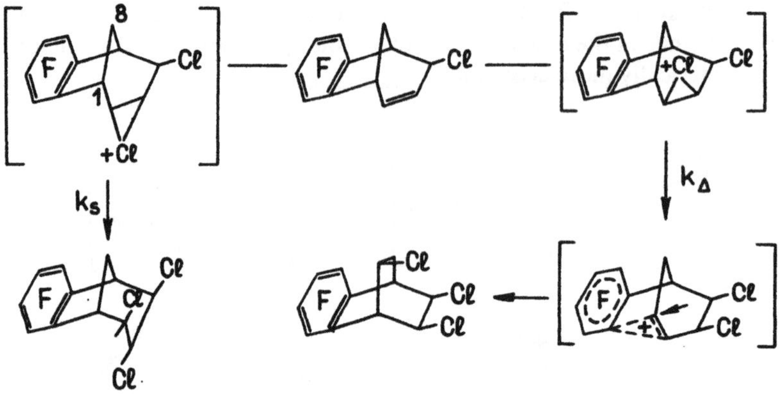

The chlorination of the 3,4-tetrafluorobenzobicyclo[3,2,1]octadiene *320* [517] involves a skeletal rearrangement with the C^4—C^5-bond shift and a stereoselective exo addition of chlorine which can be accounted for by intermediate formation of nonclassical carbocations, e.g.:

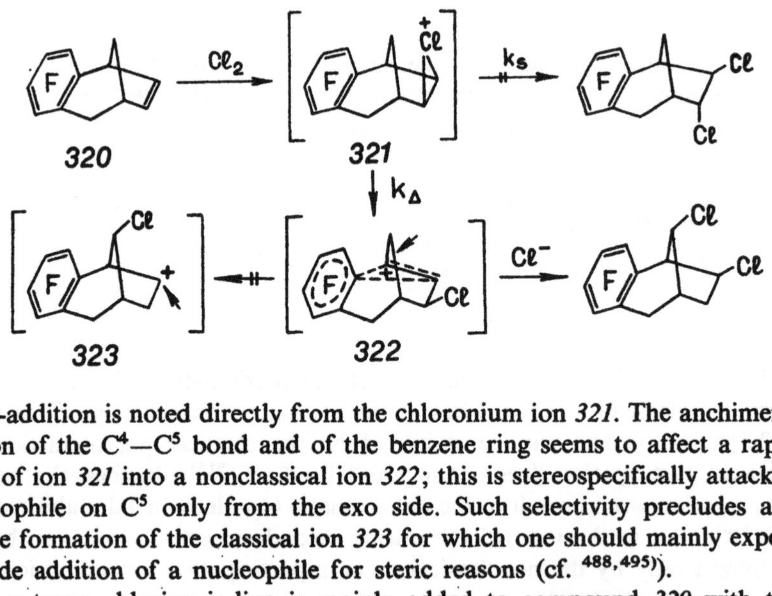

No trans-addition is noted directly from the chloronium ion *321*. The anchimeric participation of the C^4—C^5 bond and of the benzene ring seems to affect a rapid conversion of ion *321* into a nonclassical ion *322*; this is stereospecifically attacked by a nucleophile on C^5 only from the exo side. Such selectivity precludes any intermediate formation of the classical ion *323* for which one should mainly expect the endo-side addition of a nucleophile for steric reasons (cf. [488,495]).

On the contrary, chlorine iodine is mainly added to compound *320* with the participation of the reagent (k_s), because the iodonium ion is more stable than the chloronium one:

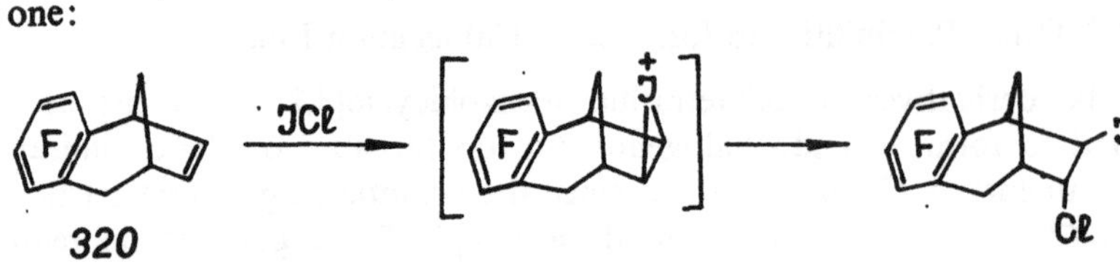

The exceptional regioselectivity in the nucleophilic attack on C^5 when olefine *320* is chlorinated might be due only to the steric effect of the already present chlorine atom. This assumption was verified by allowing the compounds *324* and *325* to react with silver trifluoroacetate[518]; in either reaction the incoming nucleophile finds itself in a non-vicinal position to the present substituent, so the steric factors can hardly exert a decisive effect on the orientation of the nucleophilic attack. These results are better explained by the assumption of a larger value of the positive charge at the non-vicinal positions of the nonclassical ions *322* and *326*.

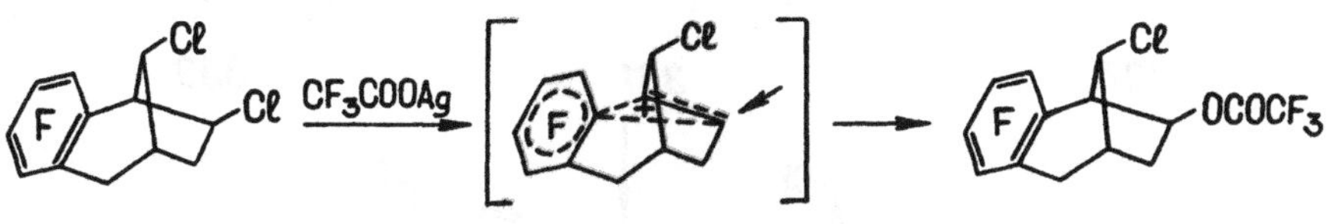

324 *322*

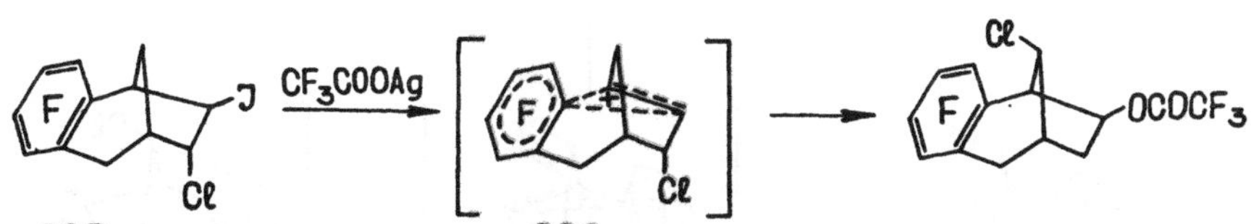

325 *326*

On chlorination of epimeric 8-chloro-3,4-tetrafluorobenzobicyclo[3,2,1]octadienes one could expect that due to the inductive effect of 8-chlorine the σ-participation of the C^4—C^5 bond in the transition state will be weakened (cf. [154]), so the predominant products will be those of opening the chloronium ion without skeletal rearrangment. The reaction, however, will finally yield only the rearrangement products.

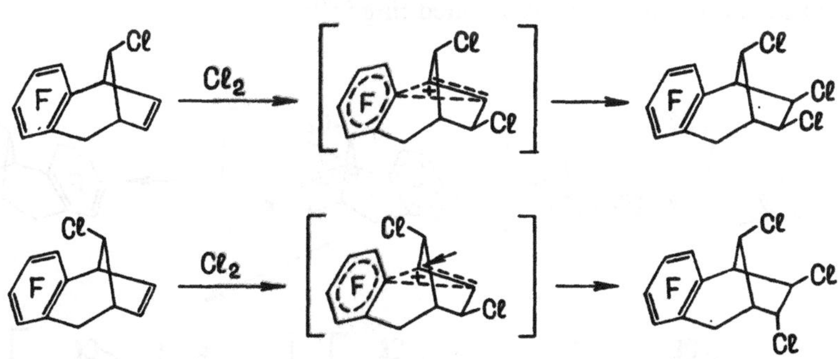

The comparison of the chlorination of 8-chloro-epimers shows that the primary exo addition of a chlorocation cannot be due to steric reasons; the determining factor seems to be only the exo configuration of a chloronium ion providing the π-participation of the benzene ring stabilizing the arising cation centre.

The assumed intermediate formation of *nonclassical ions* is supported by the chlorination of the isomeric chloroacetates *327* and *328*; in either case the reaction yields the same mixture of trichloroacetates *329* and *330* in the same ratio (3:2) The intermediate nonclassical ions *331* and *332* are optical antipodes (contrary to the alternative classical ions *333* and *334*); this explains well the obtained results.

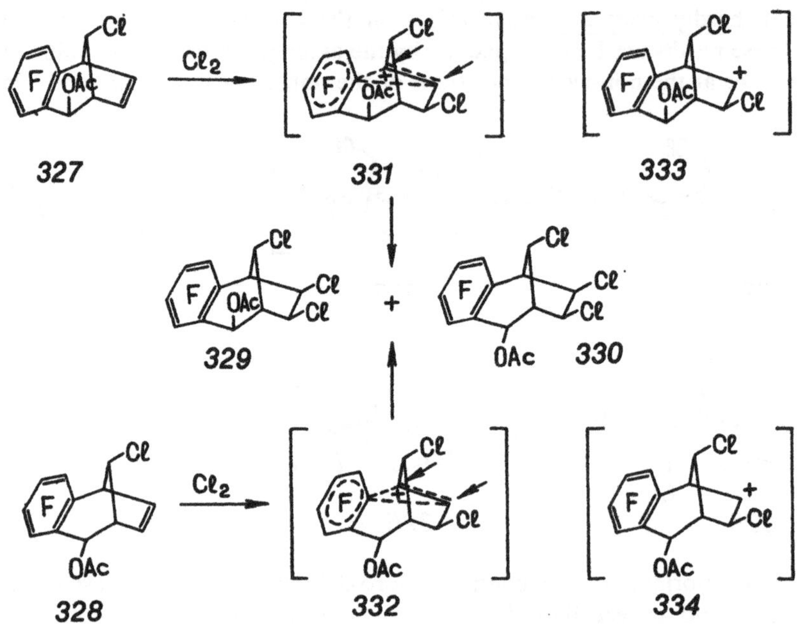

The assumed intermediate formation of *homobenzylic ions* reflects the fact that with rising nucleophility of the anion it attacks not only C^5 and C^6, but also the homo-para position of the fluorinated ring [519]:

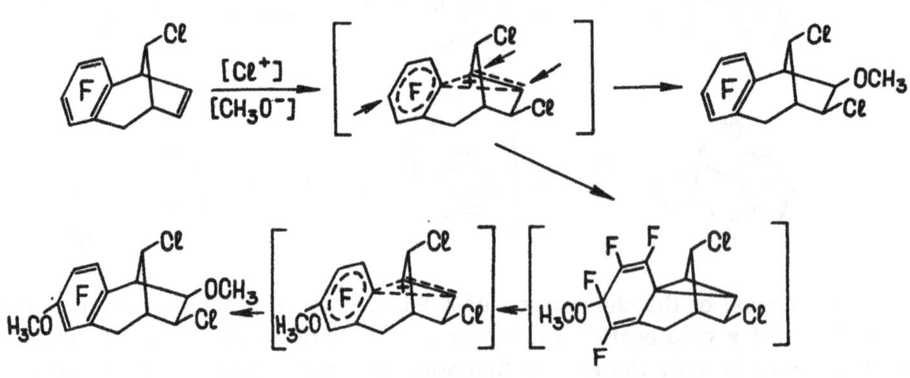

The chlorination of the same substrate in the weakly nucleophilic trifluoroacetic acid makes it possible to study the rearrangements of the homobenzylic ion:

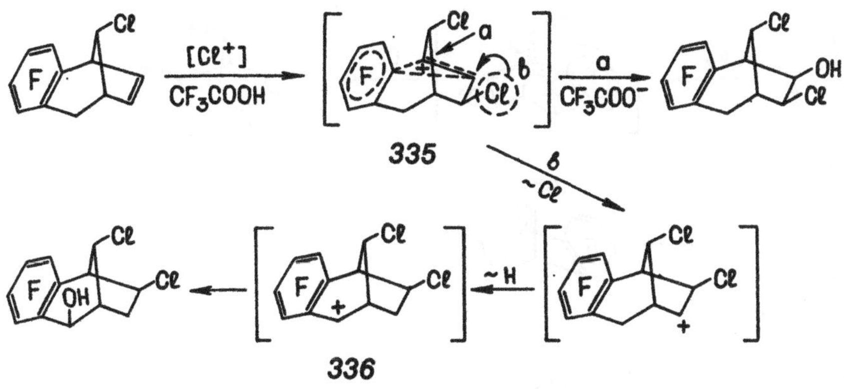

The postulated ions *335* and *336* were generated by chlorinating the same substrate in the mixture $SbF_5 + SO_2FCl$ at —80 °C [519]:

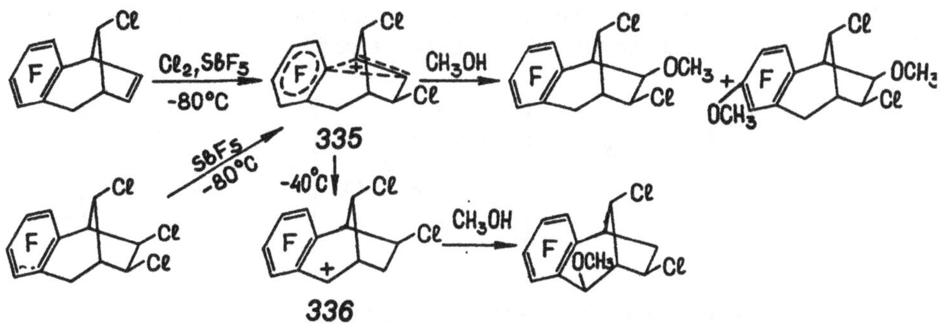

Ion *335* was one of the first representatives of stable nonclassical ions of the homobenzylic type. Its structure was corroborated by NMR ^1H, ^{19}F and ^{13}C spectra; they allow to reject the alternative assumptions of "super-rapid" equilibria of classical ions or unsymmetrical homobenzylic ions:

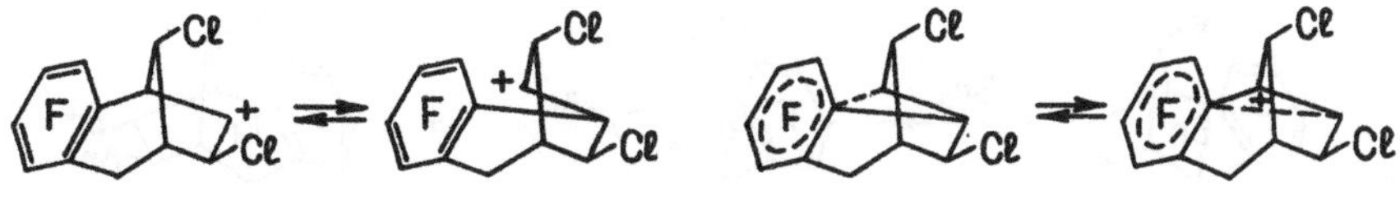

The 2-halo-3,4-tetrafluorobenzobicyclo[3,2,1]octadienes are chlorinated under rearrangement and conversion into derivatives of 3,4-tetrafluorobenzotricyclo-[3,2,1,$0^{2,7}$]octene by benzylic halogen detachment and by further homoallylic conjugation.

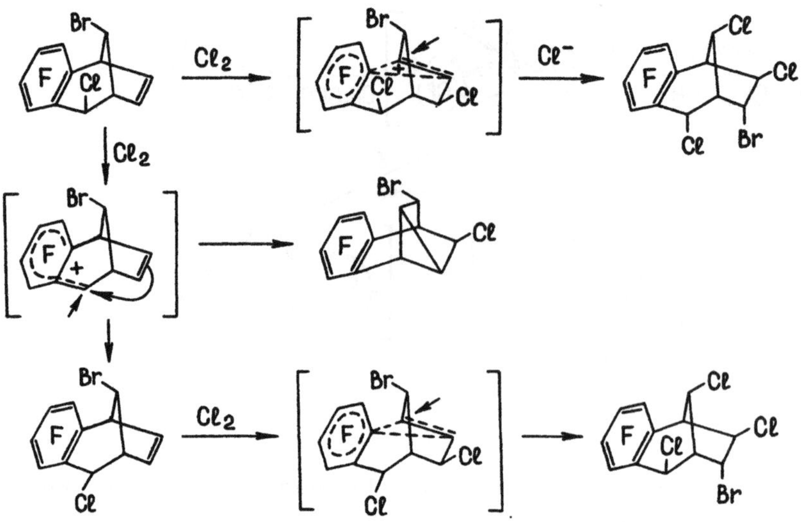

Any other relative arrangement of substituents at C^2 and C^8 (apart from 2-exo-8-endo) introduces no essential steric strain into the molecule. The chlorination of such compounds yields no ions of the benzylic type [518], e.g.:

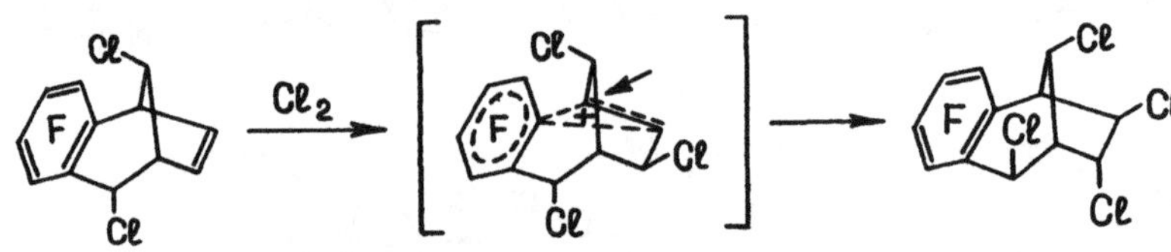

6.1.3 Comparison between Chlorination and Bromination

The scheme depicts that if the addition of chlorine [520] in 3,4-tetrafluorobenzobicyclo[3,2,1]octadienes occurs with the participation of the bond (k_Δ), then that of bromine occurs with the participation of the reagent (k_s) to form trans-dibromides with the skeleton retained. This can be accounted for because the bromonium ion is more stable than the chloronium one and possesses higher nucleophility in $CHCl_3$.

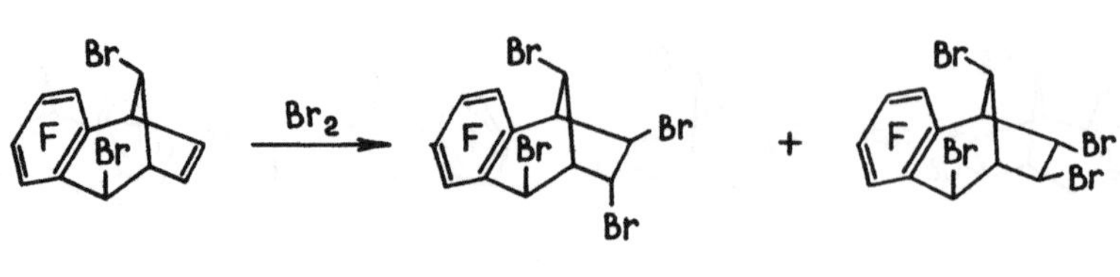

Let us summarize the results of the halogenation. The electrophilic attack on the double bond in 3,4-tetrafluorobenzobicyclo[3,2,1]octadienes always occurs from the exo side; it is favoured by the π-participation of the benzene ring and the π-participation of the C^4—C^5 bond. As a rule, such a participation effects a nonclassical homobenzylic ions. The nucleophilic attack on C^5 and C^6 proceeds stereospecifically — only from the exo-side. The regiospecificity in the nucleophile addition is determined by the relative value of the positive charge on C^5 and C^6 and — if these values are equal — by the polar and torsional effects of already present substituents. The homobenzylic structure of intermediate ions is well corroborated by direct observation under the stable-ion conditions. In case the attacking particle is a positively charged iodine there forms a stable iodonium ion leading to the product of trans-addition without rearrangement.

The nature of the reaction medium effects essentially the structure of the reaction products; in nucleophilic media the double-bond chlorination is accompanied by the nucleophilic substitution of the fluorine atom at the homo-para position while in media with high ionizing force and low nucleophility the irreversible migration of substituents results in the formation of a benzylic ion. The chlorination of compounds containing an exo halogen atom in the benzylic position is accompanied by the detachment of this atom with the formation of benzylic cations which either rearrange with the aid of homoallylic conjugation and in the end yield tricyclooctenes or result in endo isomers on C^2 which undergo further chlorination. Addition of iodine to the reaction medium results in suppression of all the processes, except the electrophilic addition to the double bond.

As shown above, the halogenation of benzobicyclooctenes manifest a high stereospecificity in bond shifting depending on the configuration of initially forming halogenonium ion. Hypothetical halogenonium ions are modelled by oxonium

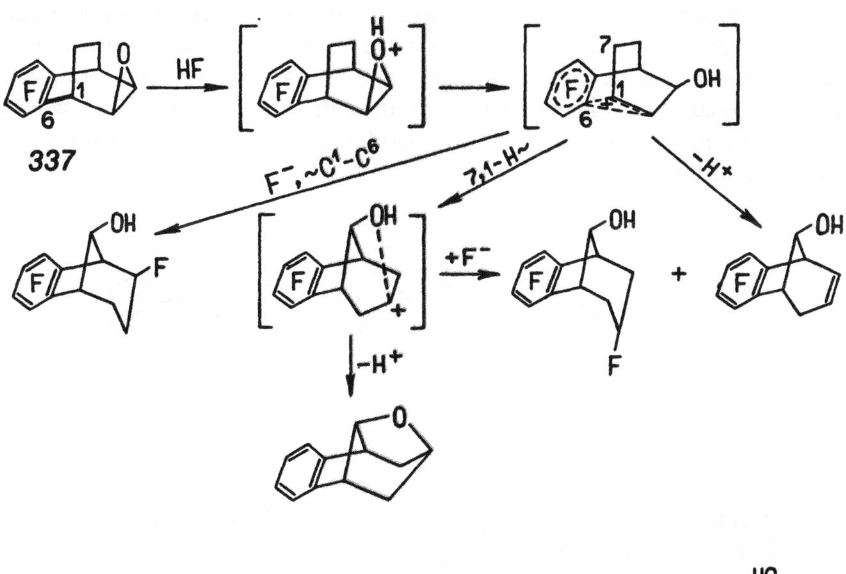

337

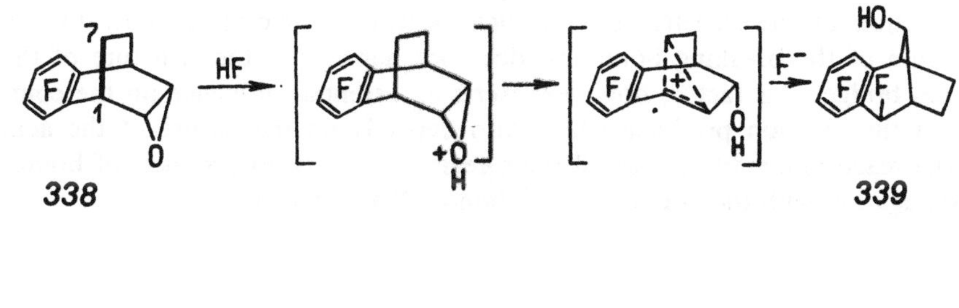

338 339

ions formed upon coordination of α-oxides with acids. The behaviour of isomeric α-oxides was studied with various configurations of the oxygen cycle with respect to the aromatic ring and the σ-bonds of the bicyclooctane system in reactions with acids [521,522].

Thus, isomeric α-oxides *337* and *338* interact with HF to form quite different reaction mixtures. The direction of the skeletal 1,2-shift in these reactions is determined by the configuration of the decomposing α-oxide: the σ-bond to be shifted is the one that can participate in the transition state of the α-oxide cycle rupture.

Assuming the σ-bond, e.g. that of $C^1—C^7$ in the α-oxide *338*, being shifted simultaneously with the splitting of the α-oxide cycle, the reaction should result in a classical benzylic cation which would further react with the fluoro-anion non-stereospecifically:

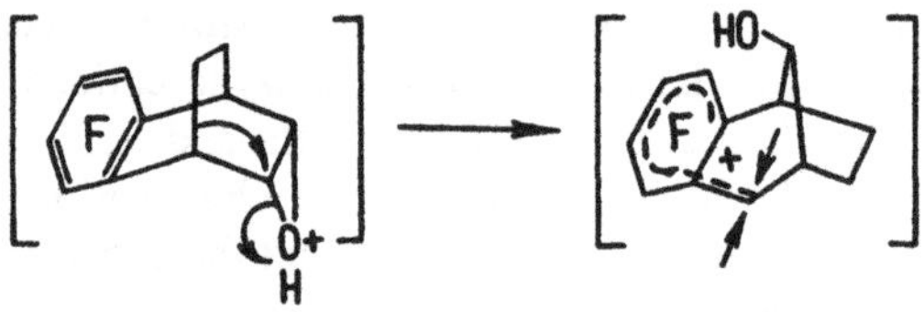

Indeed, the fluoroalcohol *339* is treated with an acid under more rigid conditions to form a mixture of isomeric alcohols differing in the fluorine configuration at C^2.

If the primary intermediate products of the ruptured α-oxide cycles in compounds *337* and *338* are classical ions it is difficult to account for the dependence between the α-oxide configuration and the skeletal shift direction. In classical ions it was reasonable to expect in both cases a shift of the $C^1—C^7$ bond with the formation of benzylic ion; besides, such ions admit a hydride shift from C^3 to C^2 which would result in the formation of a ketone from both oxides; such a ketone is not observed in the reaction mixture.

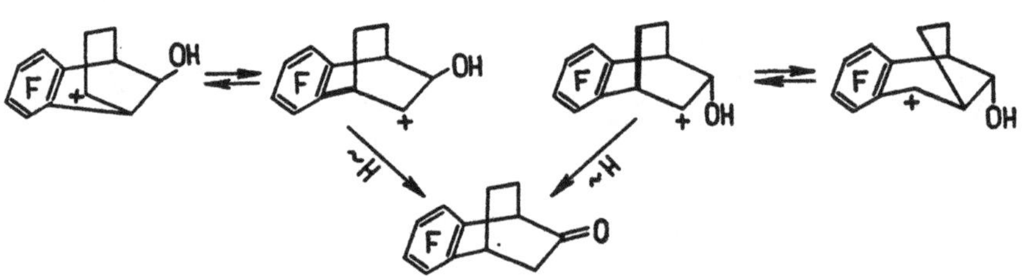

The asynchronous way of α-oxide splitting is observed only when the arising cation centre is stabilized by σ- or π-participation [523].

The direction of the skeletal shift in such systems is determined only by the configuration of the breaking bond and does not depend on the structure of the syn-located bridge; this factor, however, exerts an essential influence on the composition of the reaction products which also depends on the nature of the acid and on the reaction conditions. In all these rearrangements no products of homoallylic conjugation with the double $C^7—C^8$ bond [503] are formed.

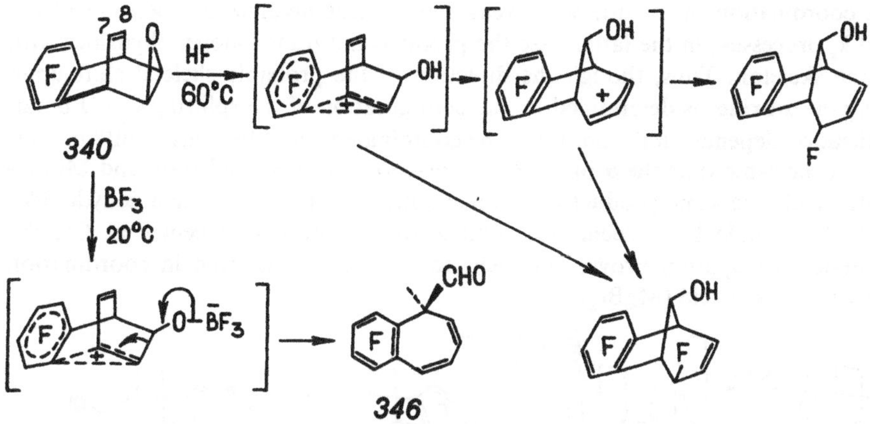

Quite differently proceeds the splitting of the α-oxide *341* which may involve π-electrons for the C—O bond rupture [524]:

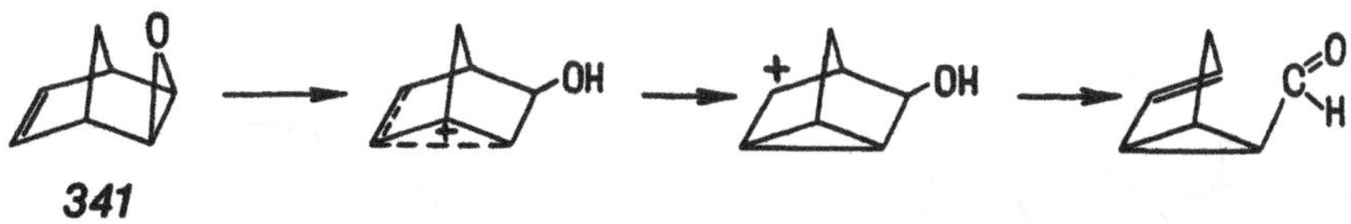

If the interaction of the α-oxide *340* with BF_3-etherate is accompanied by a deep skeletal reconstruction, then a similar reaction with the dioxide *342* results in cis-fluorohydrine *343* with the carbon skeleton retained. In the latter case the homobenzylic ion *344* does not seem to convert into a still more delocalized ion *345* (which would lead to skeleton rearrangement) since the development of the cation centre at C^1 is hindered by the negative inductive effect of the oxygen of the endo-oxide cycle. Neither does the ion *344* admit any 1,2-hydride shift with the formation of ketone which indeed was not detected among the reaction products: another test against the intermediate formation of a classical C^2-cation.

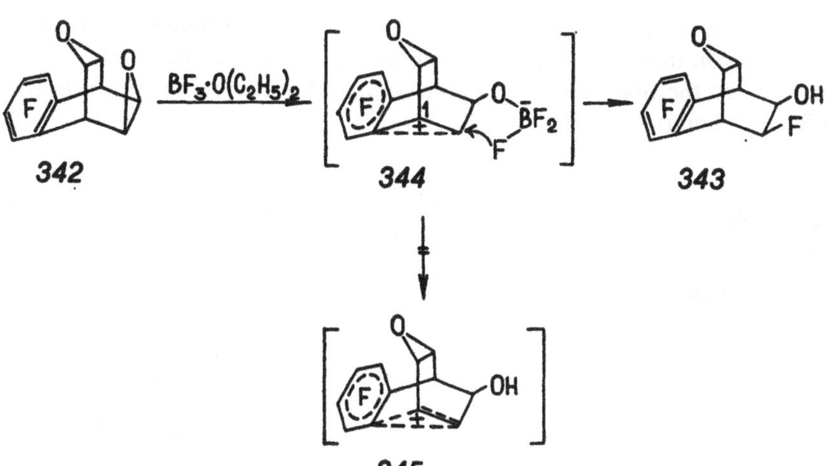

The coordination of α-oxide with weak Lewis acids involves competition of the k_Δ and k_s processes; in the latter case the products are trans-halogenohydrines with the same skeleton. Thus, though the direction of the probable skeletal shift upon opening the α-oxides is determined by the configuration of the splitting C—O bond, its realization depends on the nature of the coordinated acid (the "push-pull" mechanism). At the same time the α-oxide *340* in reactions with BF_3-etherate and $MgBr_2$-etherate yields the same product of stereospecific rearrangement the aldehyde *346*; the C^7—C^8 double bond seems to stabilize the arising cation centre at C^1; the anchimeric participation proves sufficient to split the oxide even in coordination with a weak Lewis acid ($MgBr_2$).

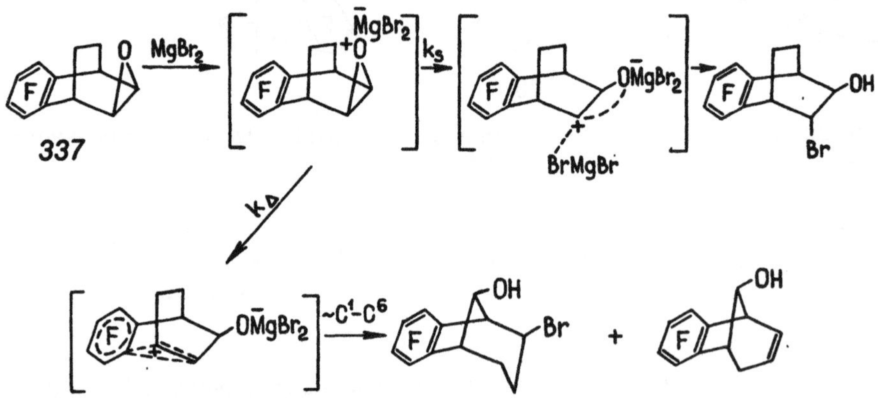

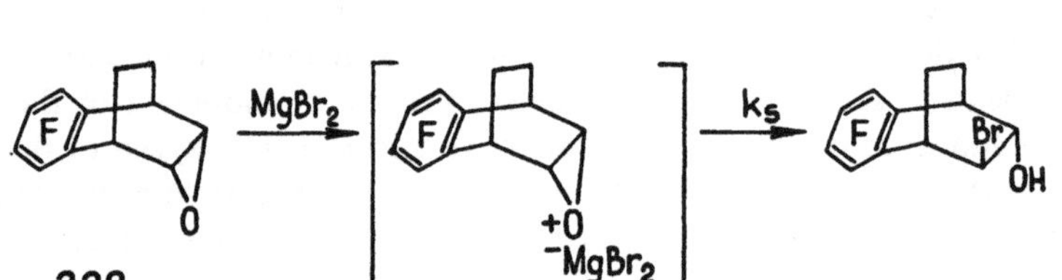

The configuration of the oxonium or halogenonium ions determines the direction of skeletal shifting on their opening; this is supported by isomeric stable cations [525] being generated for the first time from epimeric α-oxides as a result of a selective shift of different σ-bonds:

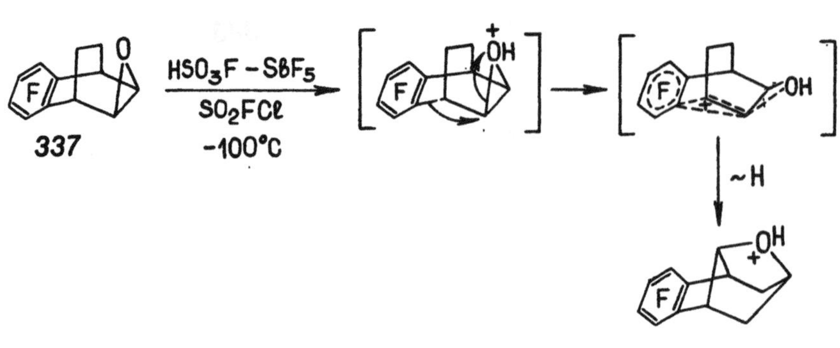

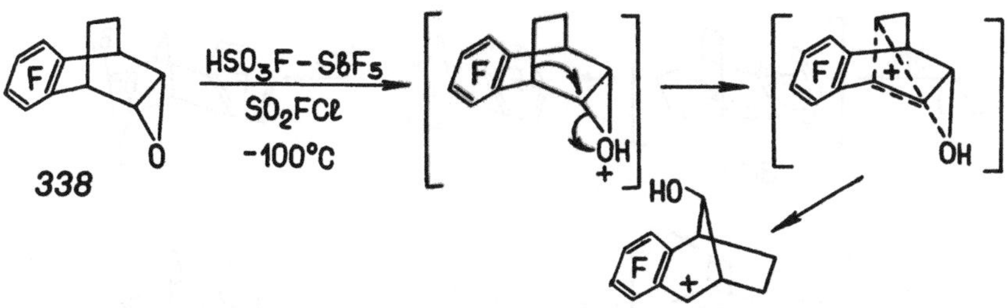

338

6.1.4 Intermediates

In the discussed halogenation reactions, as well as those of splitting the α-oxides, the "leaving group" remains in the molecules wheh the "onium" ion is opened; so the hypothetical classical carbocations, e.g., *A* and *B*, though similar, nevertheless differ in the configuration of the X-groups at the neighbouring carbon atom.

(A) (B)

For a more convincing choice between structurally different carbocations study was made of the epimeric monofunctional derivatives of polyfluorobenzocyclenes in which the splitting of the X-group must be followed by the formation of either different nonclassical ions *347* and *348* or the same classical ion *349*:

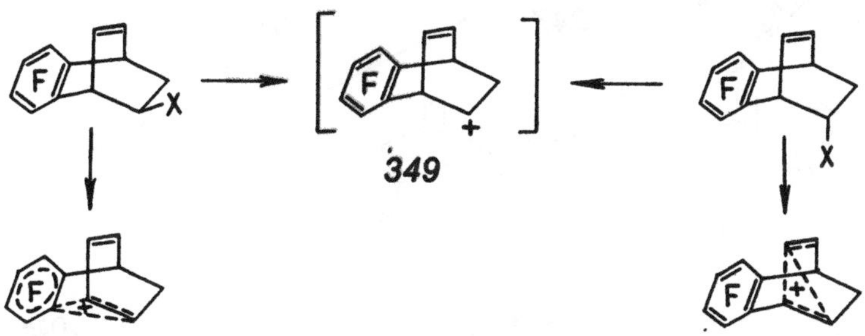

349

347 348

The interaction of the isomeric 5-iodides *350* and *351* with bromine involves the formation of different reaction mixtures which can be explained by the schemes [526] (p. 156). The assumed intermediate formation of the nonclassical ions *347* and *348* explains well both the stereospecificity in the σ- and π-bond transfer depending on the configuration of the splitting C—J bond and the stereospecificity in the addition of Br^- with the formation of monobromides. All these facts are inexplicable if bromine is taken for iodine to form the common classical ion *349*.

The iodides react with silver acetate in acetic acid somewhat differently [514]. The C—J bond in compounds *350* and *351* is ionized mainly with the participation of σ- and π-bonds; the resulting nonclassical ions *347* and *348* are stereospecifically

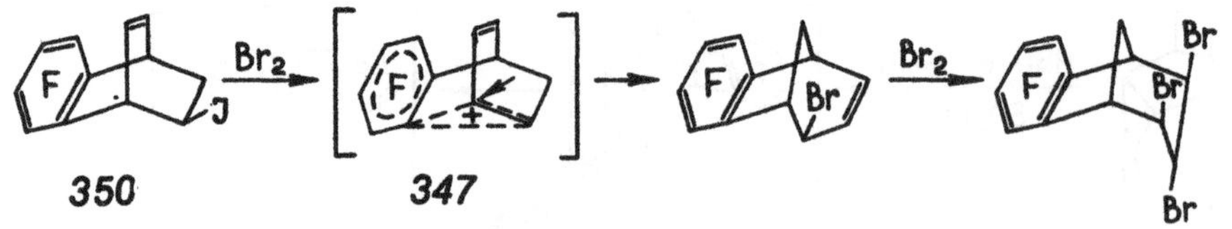

350 347

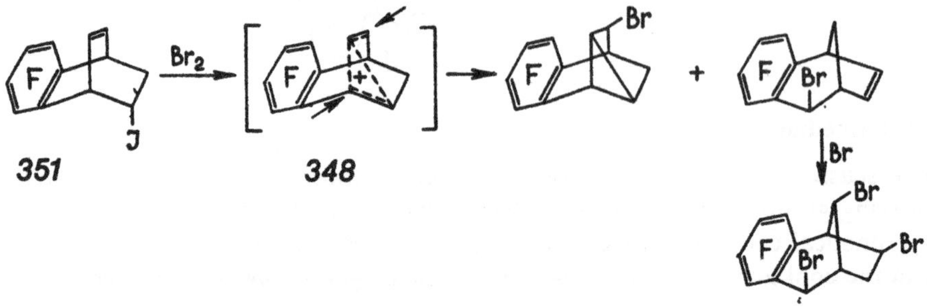

351 348

captured by the nucleophile or through a 1,2-shift the C—C bond transforms into the classical ions *318* and *352* which further yield the products of a non-stereospecific nucleophilic attack. From general considerations one can assume the allylic ion *318* to be more stable than the homobenzylic ion *347*. In the suggested scheme the attainment of the transition state through immediate conversion of *350* into the more stable ion *318* requires more energy than that on the way to ion *347*. At first

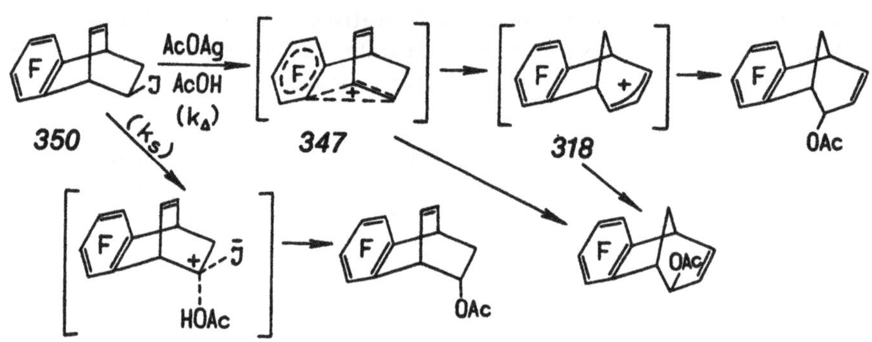

350 347 318

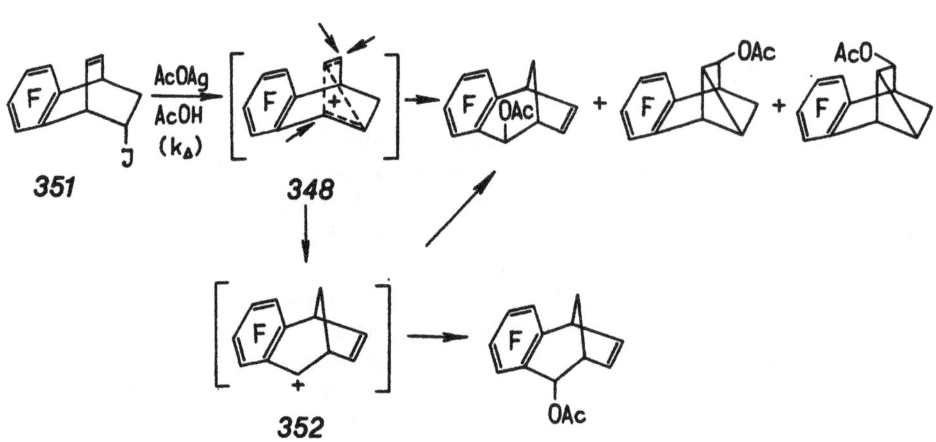

351 348 352

sight this seems to contradict the Hammond postulate; but it is precisely in such cases that this principle can be violated [527].

As noted above, the skeletal rearrangements in the S_N1 reactions and the stereospecificity in the formation of reaction products are essential criteria of the intermediate formation of nonclassical ions. Nevertheless the results obtained can be explained by assuming the C—J bond ionization to proceed with a simultaneous shift of the σ-bond $C^1—C^6$ in the isomer *350* and of the σ-bond $C^1—C^7$ in the compound *351*. The structures *347* and *348* must be transition states of ionization processes. The authors of [514], however, consider this assumption less probable because the synchronous process in question is a reaction of the S_N2 type at the bridged carbon atom and for the bridged atoms of rigid bicyclic systems such processes are known to be strongly hindered. The formation of ion *318* with bicyclo[3,2,1]octadiene skeleton consecutively from ion *347* rather than directly from compound *350* with the geometry of bicyclo[2,2,2]octadiene is also in fuller accordance with the principle of so-called "least motion" [528,529].

If the C—J bond ionization were accompanied by a simultaneous shift of, say, the σ-bond $C^1—C^6$, then the feasibility of the simultaneous rupture of the $C^2—J$ and $C^1—C^6$ bonds would chiefly depend on the nature and configuration of these bonds [530,531].

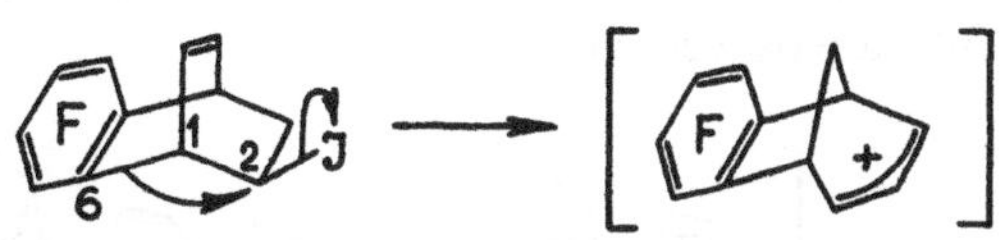

If the reagent and the substrate structure are somewhat altered while the $J—C^2—C^1—C^6$ fragment and the relative configuration of these bonds are left unchanged, then the effect of these alterations cannot be essential. But if the reaction proceeds in stages, the ratio of the (a) and (b) processes may be greatly dependent on the changes in the substrate structure and the reagent nature.

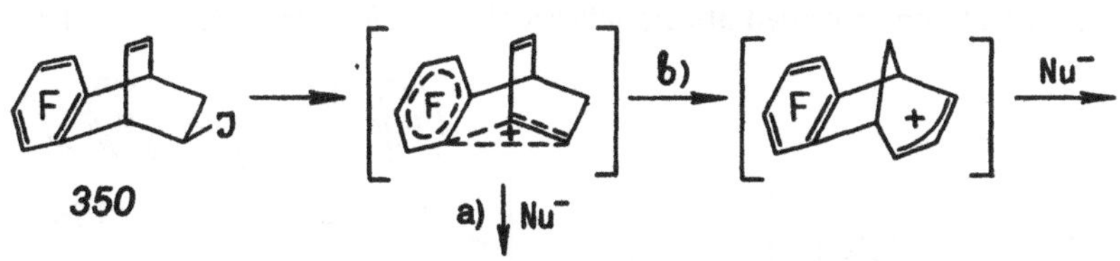

6.1.5 Acetolysis

The latter is observed in comparing the acetolysis of iodides *350* and *351* with the above reactions of the same substrates with bromine where the stereospecificity was observed not only in the shifting of bonds but in the addition of a nucleophile, i.e. ions *347* and *348* are the only product-forming ions. This difference in the rate ratio of the 1,2-bond shift and in the capture of ions *347* and *348* in reactions of iodides *350* and *351* with bromine and silver acetate can be accounted for as follows. Upon decomposition of the initially formed iodide-bromine complex the anion JBr_2^- is not only part of the ion pair, but it interacts further with the cation within the same ion pair to yield bromide. When the same iodide reacts with silver acetate the detached group moves irreversibly away as AgJ and only then is the attack by the acetate-ion possible; in the meantime there occurs a skeletal shift with the formation of an allylic ion. The stabilization of the nonclassical ion is possible not

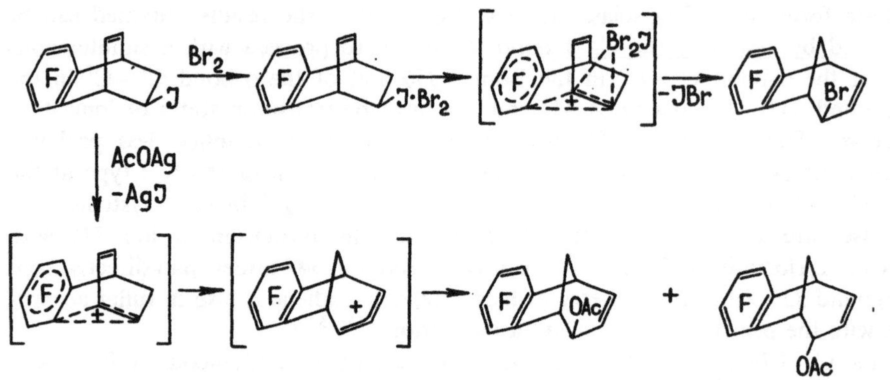

only by the counterion, but also when the α-position with respect to the initial centre of ionization is occupied by a hetereo-atom with an unshared pair of electrons, e.g. by halogen. In these reactions the nucleophilic capture also forestalls the 1,2-shift and proceeds with full stereospecificity:

It is interesting to compare the data on the interaction of compounds *350* and *353* with AcOAg under identical conditions. The latter yields equal amounts of acetates *354* and *355*. Thus, from the iodide *353*, as distinct from *350*, the nucleophilic capture of the intermediate nonclassical ion *356* forestalls the skeletal shift which

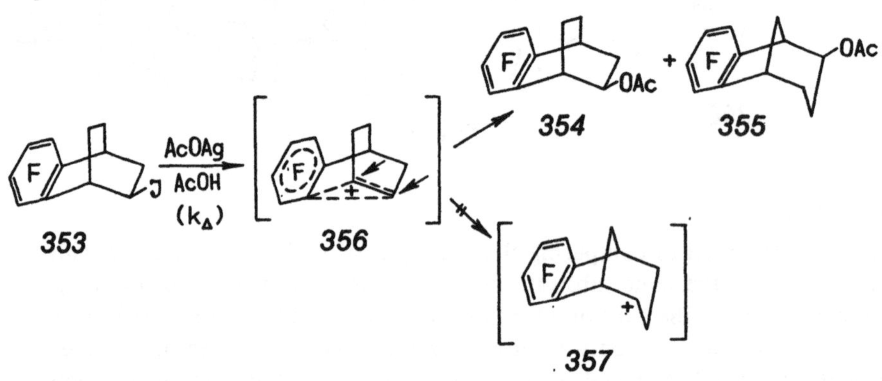

would result in a comparatively unstable ion *357*. Although the presence or absence of a double bond in the C^7—C^8 bridge does not affect the skeletal shift direction this factor influences the rate ratio of the 1,2-shift in the intermediate nonclassical ion and of its nucleophilic capture.

All the above data favour a two-stage scheme of k_{Δ} processes for the models under study. However, for nucleophilic substitution in similar substrates Japanese [502)] and American authors [496,515,532)] suggest one-stage schemes including classical carbocations as primary intermediates. Thus Cristol [496)] shows the acetolysis of cis-dichloride *358* to yield a mixture of syn-exo- *359* and syn-endo-chloroacetates *360*; the ratio of these isomers vary in time in favour of increasing amounts of the endo epimer *360*. The exo-epimer *359* was shown to isomerize into the endo epimer with a mixture $HClO_4 + CH_3COOH$ [532)]. The authors conclude the compound *358* to be acetolysed according to the scheme with the formation of the benzylic cation *361*; this is further captured mainly from the quasi-axial side, since in the transition state on attack on an ion of the *361* type results in a smaller decrease of the benzylic type p,π-overlap [532,533)].

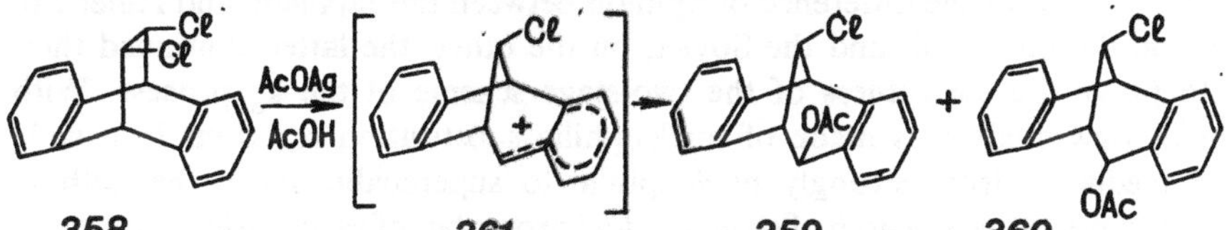

358 361 359 360

In our opinion Cristol's data agree well with the scheme of consecutive conversion of the two ions similar to the one suggested above, since the addition of Br_2 + AcOH and J_2 + AcOAg on dibenzobarrelene proceeds strictly stereospecifically.

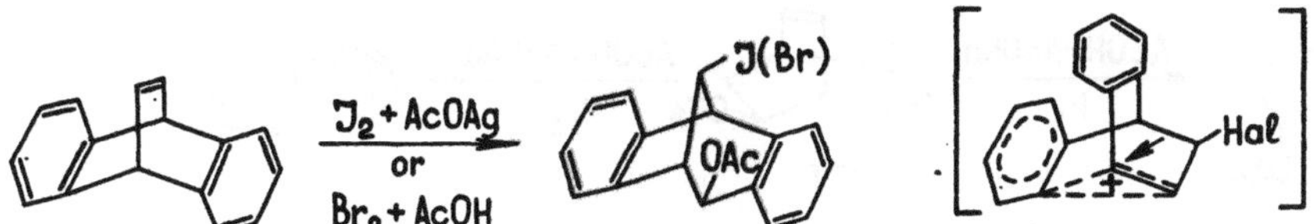

(A)

The intermediate formation of ion *A* explains well these facts, bur the above authors in this case, too, suggest the benzylic ion structure. Cristol's assertion [515)] is unclear however, that it is quite sufficient to assume "the presence of only one cationic particle"; this particle, in his opinion, cannot be ion *362* since it must rapidly and reversibly convert into the benzylic ion *363*.

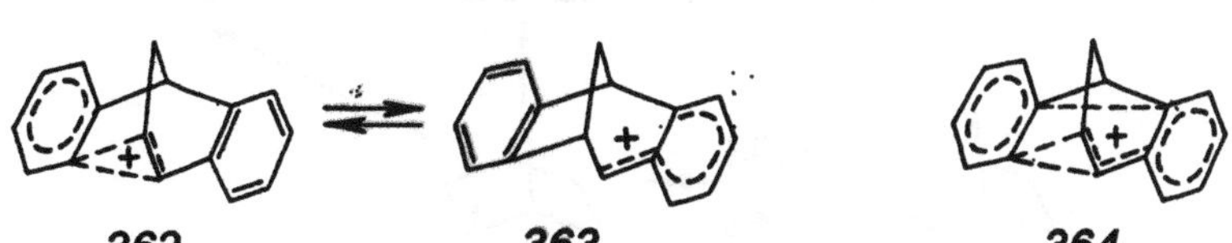

362 363 364

We believe the benzylic ion *363* will hardly convert into ion *362* since the interaction of the vacant p-orbital of C^2 in the benzylic ion *363* with the homo-located benzene ring would finally lead to the formation of an antiaromatic "homobenzylic-benzylic" ion *364*. This can be supported by Winstein [534)] showing that while the 6,7-olefine double bond in a bicyclooctadienylic anion *365* stabilizes the latter owing to the formation of a 6π-electron bishomoaromatic particle [535,536] in cation *366* the situation is quite different. Here the 6,7-double bond converts the cation into a 4π-electron antihomoaromatic bishomocyclopentadienyl system. The rate of solvolysis of p-nitrobenzoate *367* is *235* times as high as that of ester *368*. The deceleration in the latter case is stronger that one could expect with the negative inductive effect of the other olefine group considered.

365 366 367 368

With due regard for the difference of opinion between the Japanese and American chemists, on the one hand, and the Soviet, on the other, the latter continued their efforts to prove the correctness of the two-stage scheme of the k_{Δ} process. With this aim in view study was made of nucleophilic substitution reactions in a wide range of media — from strongly nucleophilic to superacidic. Also, the authors include the comparative research of structure and properties of carbocations generated from the same precursors in different media.

The results of solvolysis of epimers *369* and *370*, *371* and *372* in different media (80% acetone, AcOH, CF_3COOH) are represented by a scheme [537–539]. In the models we used the stereochemical results of reactions with the participation of σ- and π-bonds (k_{Δ}) and with the assistance of solvent (k_s) are essentially different, e.g.:

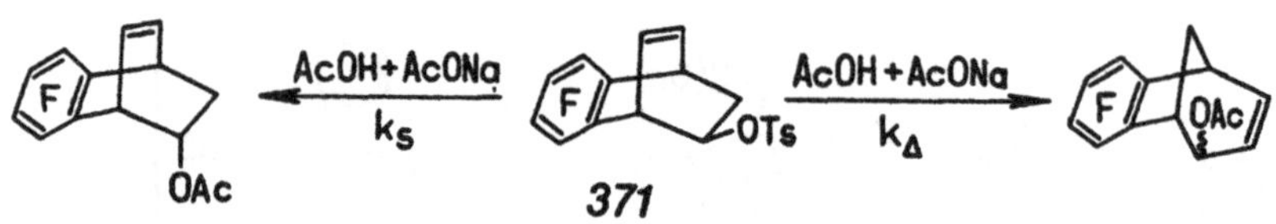

371

In using 2-unsubstituted norbornanes as model compounds the identification of the k_{Δ} and k_s processes is hindered because the latter lead to the same stereochemical result, e.g.:

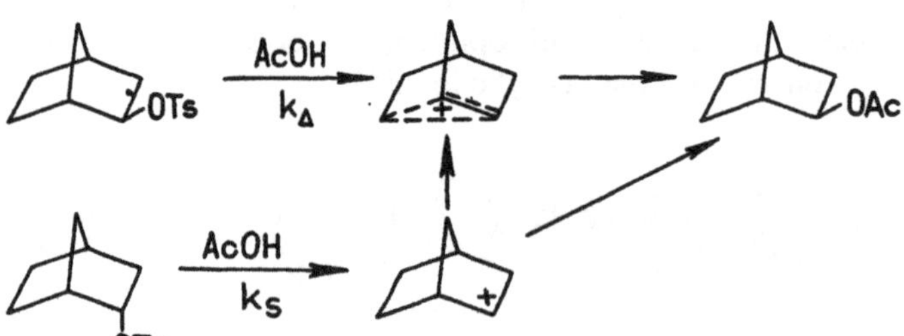

Optically active 2-norbornyl tosylates might facilitate the identification of the k_{Δ} and k_s processes. However, firstly, as noted earlier, on acetolysis of 2-endo-norbornyl tosylate only ~8% of optical activity was retained, i.e. a great part of the classical cation (k_s) converts into a nonclassical ion. No "leakage" of this kind has been observed in our models.

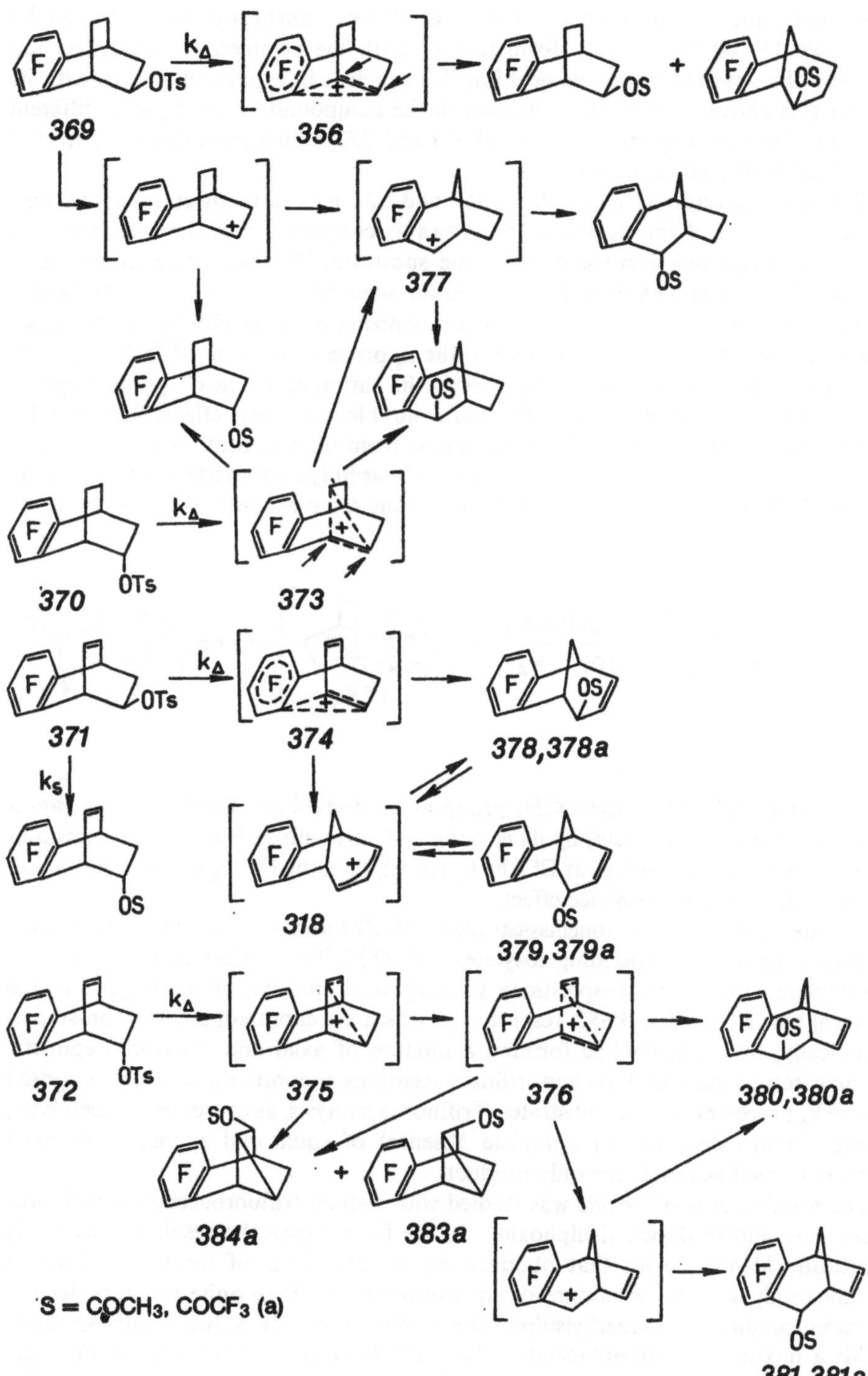

$S = COCH_3, COCF_3$ (a)

Secondly, the racemization on acetolysis of 2-exo-norbornyl brosylate can be accounted for by the primary formation of both the symmetric nonclassical ion (k_Δ) and the unsymmetrical classical ion (k_s) which is in rapid equilibrium with its epimer (see above, p. 11). The solvolysis of the compounds studied yields different reaction mixtures of epimers, — e.g., of *371* and *372*, which precludes a superrapid equilibration of classical cations.

The trifluoroacetolysis of tosylates *369* and *371* manifests an essential increase in the yield ratio of the k_Δ and k_s processes as compared with the acetolysis and solvolysis in aqueous acetone of the same substrates [538]. According to Winstein, this fact is a consequence of both the lower nucleophility of CF_3COOH (which decelerates the k_s process) and its far higher ionizing force which accelerates the k_Δ process a great deal more than it does to the k_s process ($m_\Delta > m_s$) [540]. From [538] it follows that the double bond contributes to the cation centre at C^1 (at the expense of the resonance effect, despite the unfavourable inductive effect) and thereby favours the k_Δ process. A similar effect results from the substitution in tosylate *369* of the hydrogen at C^1 by a methoxy group — the acetolysis of tosylate *369a* proceeds only by the k_Δ route, while in that of *369* under the same conditions the predominant process is k_s [541].

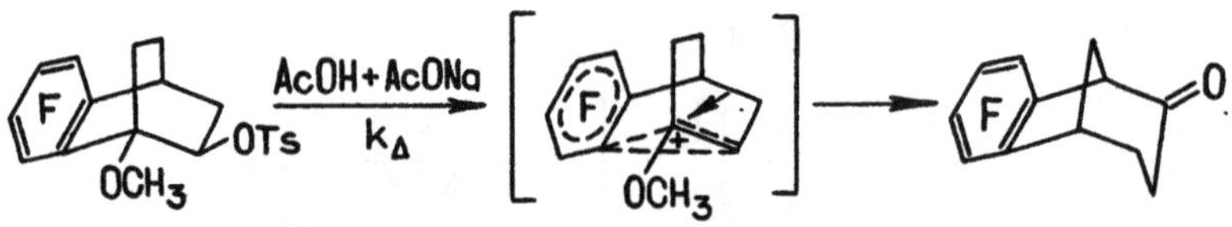

369a

These data reflect that the OCH_3 group at C^1 destabilizes the developing cation centre at C^6 (i.e. the k_s route) by its negative inductive effect. But if in the transition state a cation centre develops at C^1 (the k_Δ route), then the OCH_3 group can stabilize it due to the positive resonance effect.

For the k_Δ process the nonclassical ions *373*, *374* and *376* initially formed upon ionization of the corresponding tosylates *370–372* [538] can either be captured by a nucleophile with full stereospecificity yielding axial epimers, or rearrange into the classical ions *377*, *318* and *352*, respectively, which are captured, already non-stereospecifically, by a nucleophile forming a mixture of axial and equatorial epimers. Comparison of the acetolysis and trifluoroacetolysis supports the schemes proposed for the k_Δ processes of these substrates. Trifluoroacetolysis, as compared to acetolysis, shows a sharp decrease (or complete absence) of equatorial epimers with axial epimers formed as nearly the only products.

The reaction of tosylate *382* was studied with sodium trifluoroacetate in trifluoroacetic acid and in dimethylsulphoxide. In the former case the resulting mixture is practically identical with that obtained by the solvolysis of tosylate *372*, i.e. it contains only a slight admixture of equatorial epimer *381a* quite stable under the reaction conditions. Dimethylsulphoxide (which properly solvates carbocations) yields a mixture of trifluoroacetates *380a* and *381a* in a 1:1.8 ratio, i.e. the yield of the equatorial isomer is nearly twice as large as that of the axial one (fairly stable under the reaction conditions); in this reaction mixture the products resulting from homoallylic conjugation — compounds *383a* and *384a* — are present in negligible amounts. Apparently the latter reaction yields the classical benzylic cation *352*; the ratio of the epimers resulting from it is mainly determined by the back-side participation of the solvent and the axial-side shielding by the departing tosyloxy group. The obtained data on the trifluoroacetolysis of tosylate *382* show that the difference in the ratio of C^2 epimers upon acetolysis and trifluoroacetolysis of tosylate *372* cannot be due to that in the steric requirements of nucleophiles, but is determined above all by the different structures of intermediate carbocations.

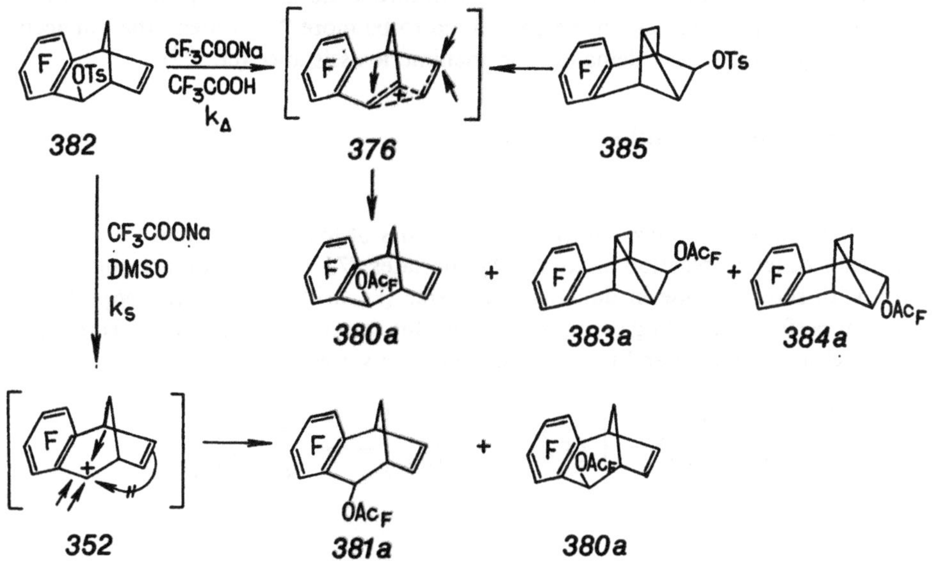

That the compounds *372*, *382* and *385* with different skeletal structures are trifluoroacetolyzed to yield practically the same product mixtures permits the assumption of the nonclassical ion *376* for all the three cases.

In trifluoroacetic acid the specific solvation of carbocations is quite insignificant [542)]; this should lead to an essential enhancement of the anchimeric assistence of σ- and π-bonds in the transition state of solvolysis and to the formation of greatly delocalized nonclassical carbocations. This effect can partly compensate that of a decrease in solvation. As for the classical cations, in passing from acetic to trifluoro acetic acid the conditions for their solvation sharply deteriorate hence their relative stability becomes much lower; this effect for classical ions with a comparatively localized charge is of far greater significance than for nonclassical ones[9].

The above conclusion [538)] can be indirectly corroborated by McManus and Peterson [544)] who have shown that with decreasing nucleophilic solvating properties of the solvent (in passing from SO_2 to SO_2FCl) the equilibrium of stable cyclic halogenonium and isomeric open carbonium ions shift towards the former. They assume that the halogenonium ions is better able to delocale the positive charge in the absence of strong solvation, than the open classical ions do.

9 This thesis is corroborated by quantum chemical calculations [543)].

The reactivity of nonclassical ions formed from the same substrates in different media may be due to the difference in the electronic structure and in the energetic characteristics of these particles, but also in the structure of the respective ion pairs and in their solvation (cf. [545]). In trifluoroacetic acid the primarily formed internal ion pairs R^+O^-Ts can be considered to be more "separated" than in acetic acid which enhances the "demand" for anchimeric assistance (cf. [546]).

6.2 Comparison with other Agents

6.2.1 Sodium Azide

While the trifluoroacetolysis of the epimeric pairs of tosylates *369* and *370*, *371* and *372* yields different reaction mixtures, the tosylates *371* and *372* interact quite differently with sodium azide in dimethylsulphoxide (i.e. with a strong nucleophile in a medium perfectly solvating carbocations); in this case only the k_s process is realized, the skeleton is retained and the configuration is inverted [538]:

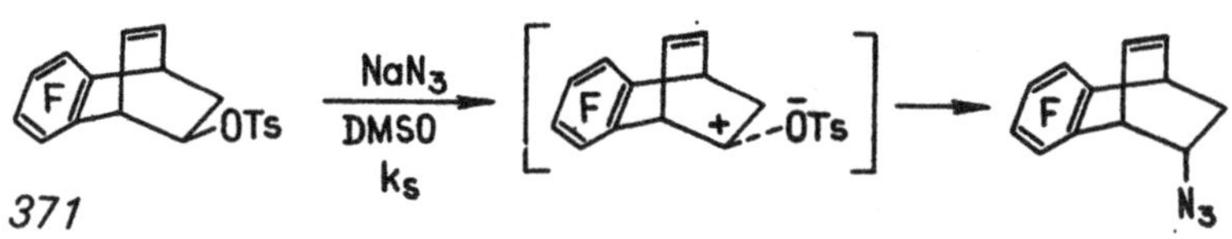

371

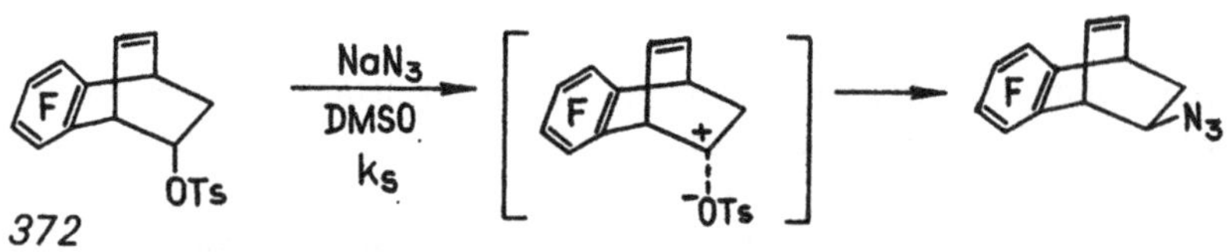

372

In 1975 Cristol, after additional research, admitted his earlier monostage scheme of solvolysis to be incorrect and suggested a two-stage scheme including the an intermediate homobenzylic ion [547]; this scheme is quite similar to the one proposed in [537].

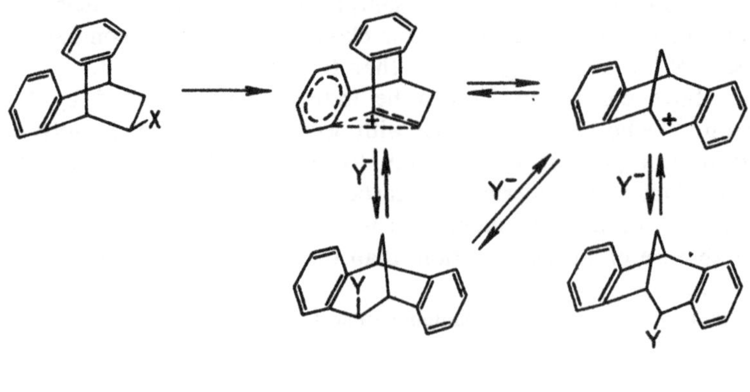

These results show the advantages of the benzobicyclo[2,2,2]octadiene and the -octene series in studying solvolysis reactions over the norbornane derivatives. When these models are used the exo and endo isomers yield different reaction mixtures which precludes the intermediate formation of a classical cation with the same skeleton. If this is formed at the stage of reaction proceeding along the k_s route it results in products with a skeleton other than that of the compounds formed in the k_Δ processes.

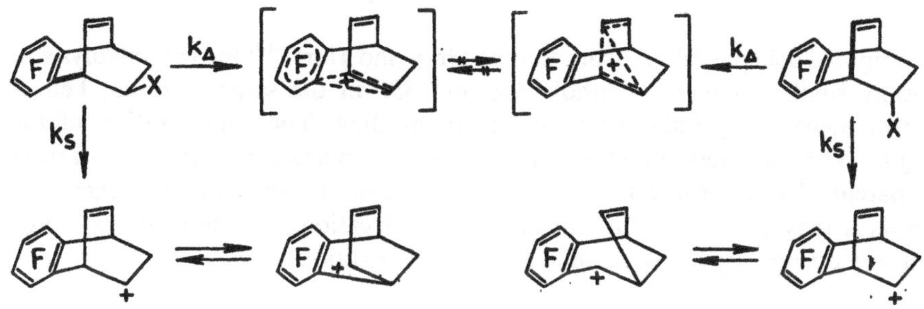

6.2.2 Acid Media

While the solvolysis of epimeric substrates take the k_Δ route formating isomeric nonclassical carbocations as primary intermediates, in the "usual" media between these carbocations there is a high energetic barrier preventing their interconversions [174]. Such interconversion, however, was observed in the solvolysis of benzobicyclooctadiene and -octene derivatives in highly acid media — mixtures of trifluoroacetic and sulphuric acids in various ratios. These investigations bridge the gap between the observation of stable carbocations and the behaviour of the same

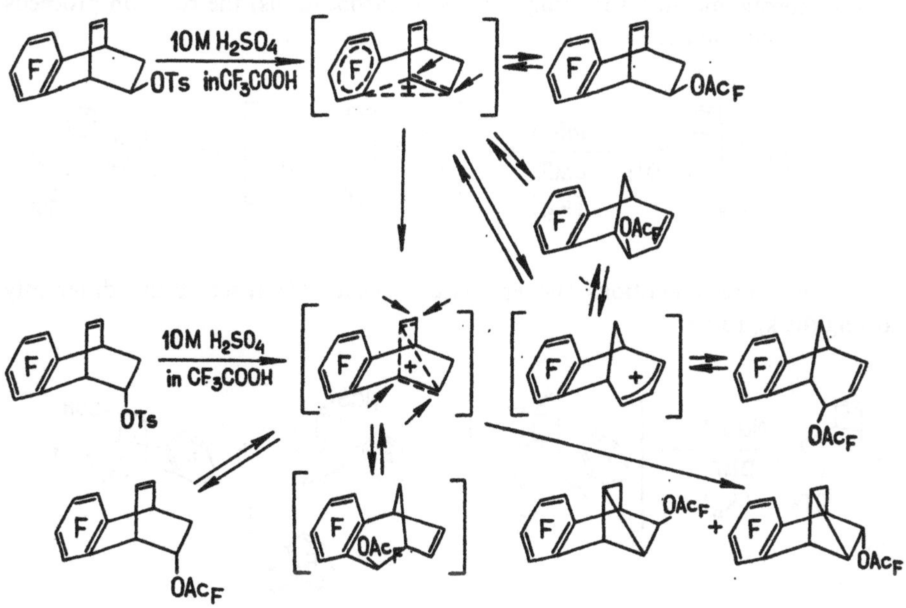

substrates in nucleophilic media. The various compositions of epimeric substrate reaction products in highly acid media under the *kinetic* control conditions correspond to isomeric nonclassical carbocations as primary intermediates. Under *thermodynamical* control epimers yield identical reaction mixtures which allow one to determine the relative stability of isomeric compounds with different skeletons. In highly acid media, when the lifetime of ions is comparatively long and their conversion into reaction products is reversible, the energetic barrier separating isomeric nonclassical ions can be overcome; this results finally in the same product mixture, e.g. [548]:

The proposed scheme differs from that of Gray and Hart [549] for the solvolysis of structurally similar epimeric alcohols *386* and *387* in the same medium, benzodihydropentalene being obtained as the main product. The configuration of the splitting C—O bond exerts no effect on the reaction course since the epimerization of the parent alcohols (or esters) occurs faster than the skeletal rearrangement. This process may go via an intermediate classical cation, i.e. along the k_s route (the authors do not suggest a mechanism of epimerization):

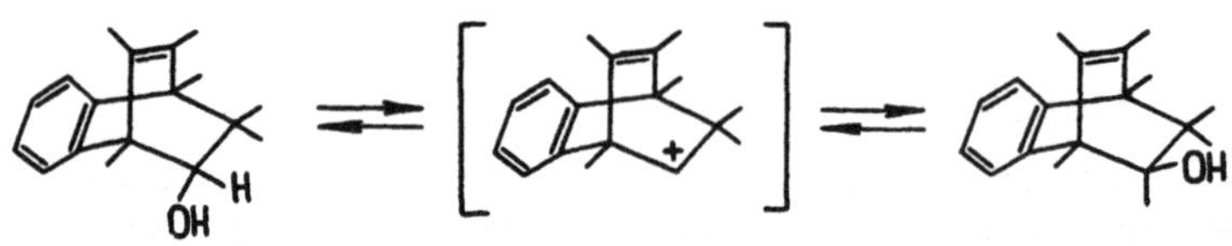

386 *387*

In [548], however, the k_s process with skeleton retention and configurarion reversal is shown to be possible only in highly nucleophilic media. Indeed, when sodium rhodanide acts on tosylate *371* in dimethylformamide (i.e. when a strong nucleophile acts in an aprotic medium solvating fairly well carbocations) the reaction proceeds along the k_s route alone:

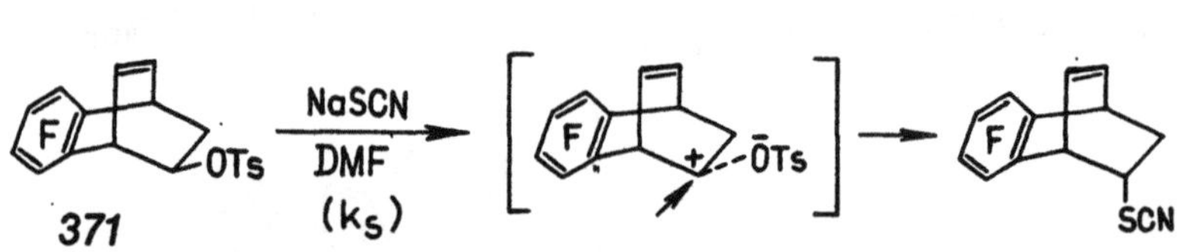

371

Under the same conditions the epimeric tosylate *372* reacts quite differently following the k_Δ route:

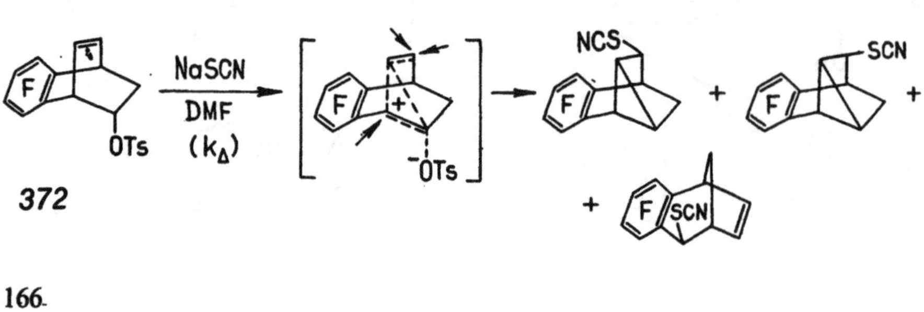

372

So sharp a difference in the conversion of epimers *371* and *372* under identical conditions is due to the far stronger π-participation of the double bond than that of the fluorinated aromatic ring in the transition state of ionization. The π-electrons of the double bond render a more effective nucleophilic assistance to the ionization of tosylate *372* than does even a rather nucleophilic thiocyanate-anion in a medium favouring the formation of localized classical cations.

From these considerations Gray's and Hart's assumption of the preliminary epimerization of the parent alcohols seems to be extremely improbable.

The isomerization of cis-2-acetoxy-3-deuterodibenzobicyclo[2,2,2]octadiene into the trans-isomer in 1 M. H_2SO_4 in AcOH is assumed [548, 550] to go via the intermediate classical 2-dibenzobicyclo[2,2,2]octadienylic cation *388*:

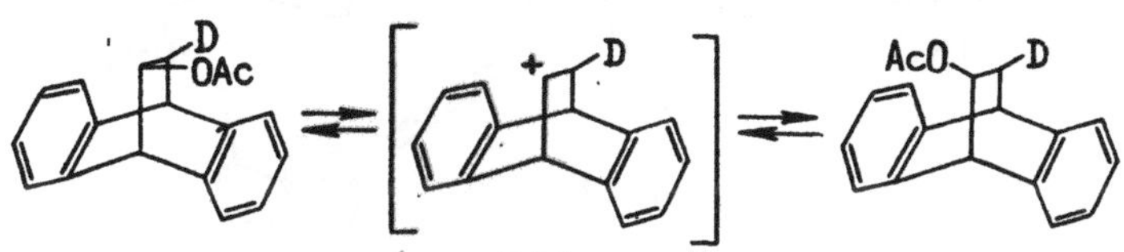

388.

The same authors showed earlier [551] that the acetolysis of cis-2-tosyloxy-3-deuterodibenzobicyclo[2,2,2]octadiene (AcONa + AcOH) manifests strict stereospecificity both in the bond transmission and the direction of nucleophilic attack; these data clearly exclude the intermediate formation the classical ion *388*. This discrepancy can be avoided if the data reported [550] on cis-trans-isomerization are accounted for by involving the interconversion of nonclassical ions:

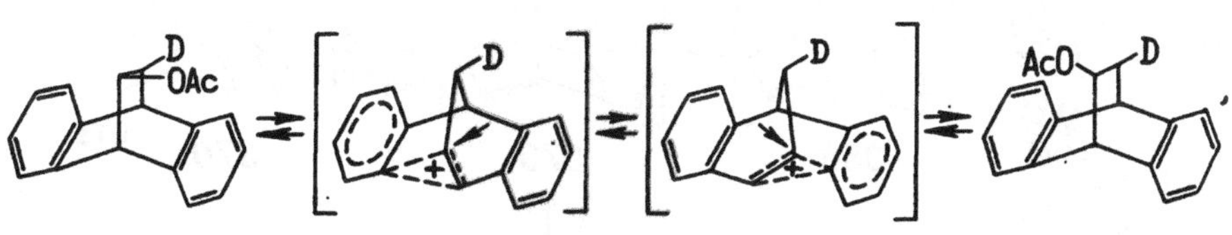

In highly acid media it has been possible to establish that in intermediate nonclassical cations there occur skeletal and hydride shifts which in less acid media (e.g., $CF_3COOH + CF_3COONa$) cannot compete with the capture of ions by nucleophiles [552]. Detailed research of solvolysis in strongly acid media greatly facilitates an understanding of ways to form stable carbocations in superacidic media from the same precursors particularly when these stable particles result from rearrangements of ions that cannot be observed. Thus, 5-substituted tetrafluorobenzobicyclo[2,2,2]octenes, 2-, 6- and 7-substituted tetrafluorobenzobicyclo[3,2,1]-octenes in $SbF_5 + SO_2FCl$ have yield the same equilibrium mixture of the ions *389* and *390* [553]. From these facts and the results of solvolysis of the same substrates in highly acid media a scheme was suggested for the formation of stable ions *389* and *390*.

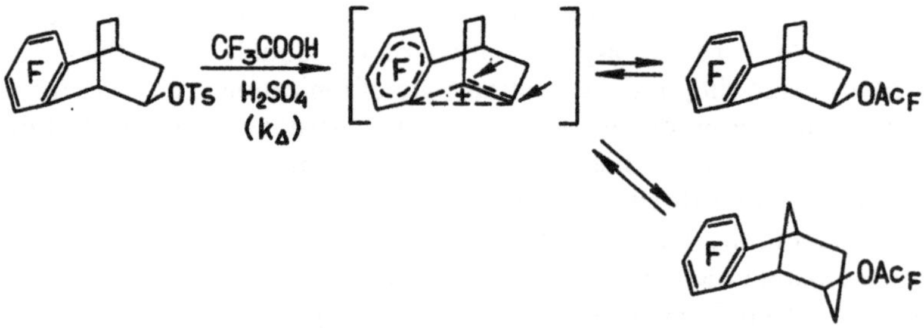
CF3COOH
H2SO4
(kΔ)
OTs
OAcF
OAcF

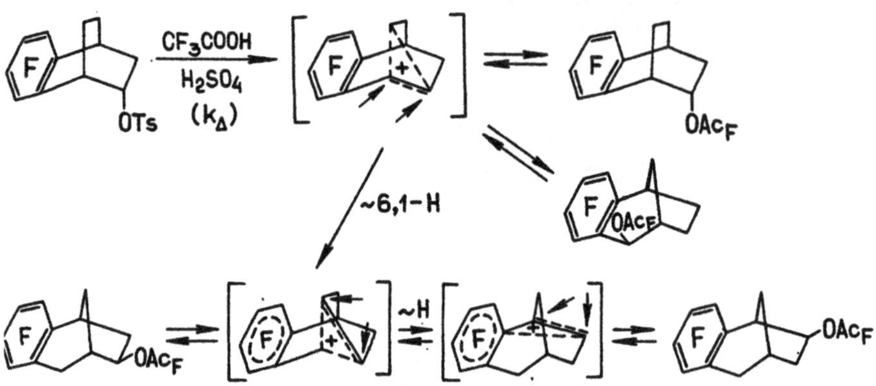
CF3COOH
H2SO4
(kΔ)
OTs
OAcF
~6,1-H
OAcF
~H
OAcF
OAcF

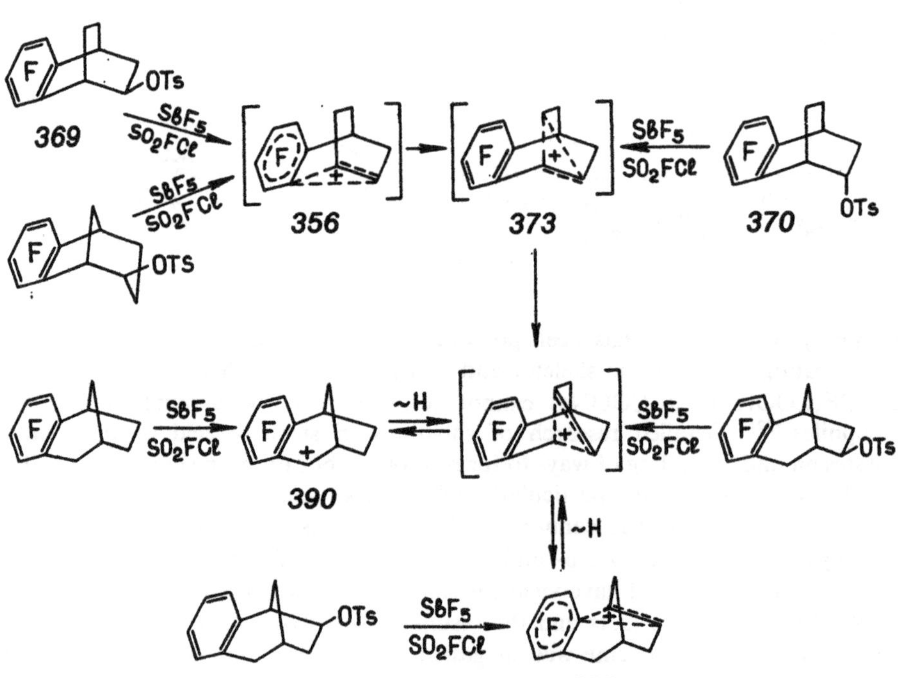
OTs
369
SbF5
SO2FCl
SbF5
SO2FCl
OTS
356
373
SbF5
SO2FCl
370
OTs
SbF5
SO2FCl
390
~H
SbF5
SO2FCl
OTs
~H
OTs
SbF5
SO2FCl
389

In [553] the epimers *369* and *370* generated identical mixtures of ions. Literary data show that until recently the same stable cations have been obtained from epimers. This is why Brown stated that for the nonclassical ion concept the solvolytic studies are far more informative than those on the generation of stable ions since only in the former epimers display different reactivity which is the most important sign of "nonclassicality" [212]. Recently, however, Kamschei et al. [541,554] have generated isomeric stable cations from epimeric substrates, e.g.:

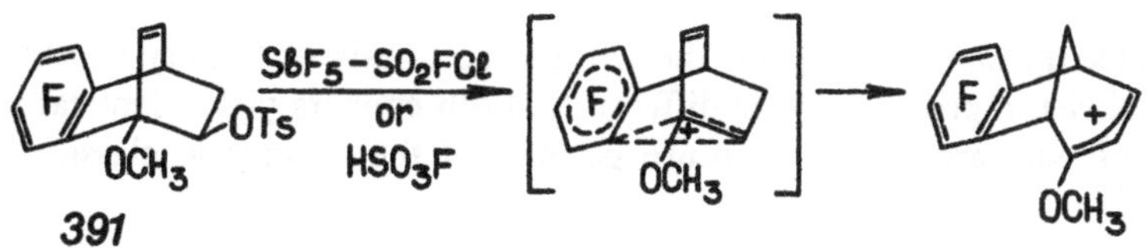

391

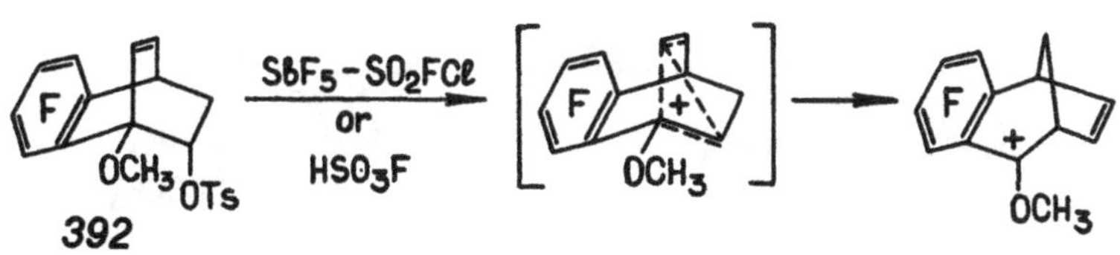

392

The epimer ionization involves a 1,2-shift to the developing cation centre of the very σ-bond which can "participate" in the transition state of the C—O bond splitting. An alternative explanation of this phenomenon may be that the ionization of epimers results in ion pairs of a classical cation with the parent skeleton, and the skeleton shift direction is determined by the position of the counterion in space so that the latter hinders the interaction of the vacant p-orbital of the classical cation with σ-bonds located on the same side as the anion. This assumption seems improbable since the ionization of tosylates *391* and *392* in HSO_3F [555] yields the same results; according to Winstein, the solvolysis of alkyl tosylates in the same medium is accompanied by the detaching of the uncharged p-toluene sulphonyl acid [556]. Also, the solvolysis rate of the endo epimer *392* was shown to exceed 700-fold that of exo epimer *391*; this agrees with the more effective π-participation of the double bond as compared to that of the aromatic ring in the ionization transition state in the same positions relative to the developing cation centre. If the ionization of the secondary tosylates *391* and *392* proceeded via a classical ion, then for steric reasons the rate ratio would be inverse (steric "hindrances" to endo epimer ionization).

The trifluoroacetolysis of the epimeric tosylates *369* and *370* follows mainly the k_Δ route with various reaction mixtures whose composition is well accounted for by the assumption of the intermediate formation of the isomeric nonclassical ions *356* and *373*:

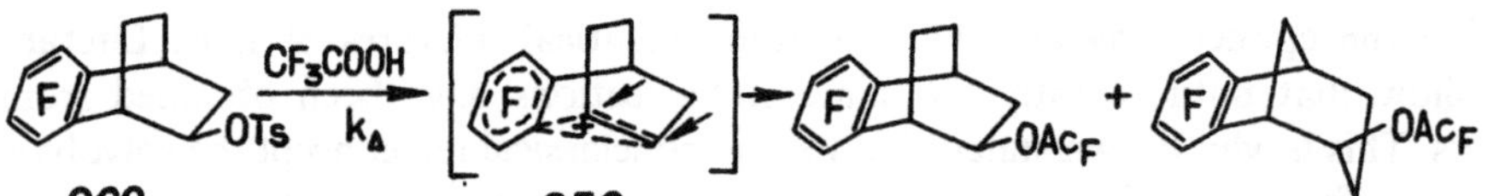

369 356

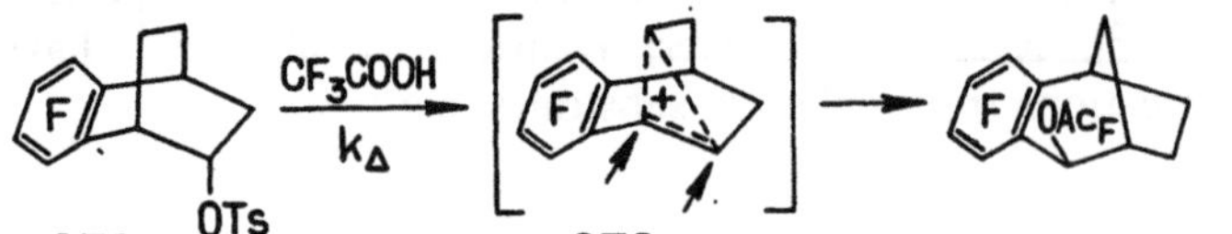

370 373

Under the same conditions the trifluoroacetolysis of tertiary substrates with the same skeleton proceeds quite differently. Both epimers yield identical product mixtures of nonstereospecific capture because of their common classical carbocation *393* [557]:

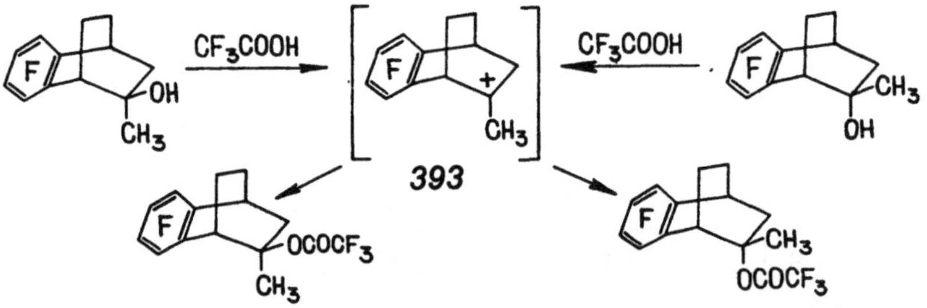

393

The classical ion *393* was generated from the same precursors in a superacid medium at —100 °C; this ion is in equilibrium with the bridged ion *394*.

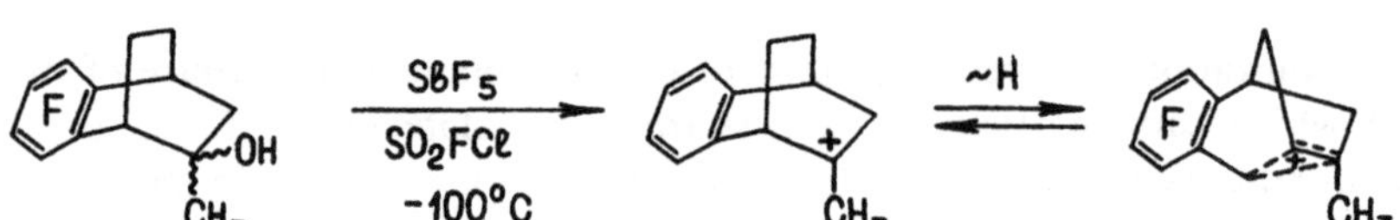

393 394

The trifluoroacetolysis of secondary tosylates *371* and *372* results in different mixtures of products formed by both the stereospecific 1,2-shift of the σ-bond and the homoallylic conjugation:

371 374

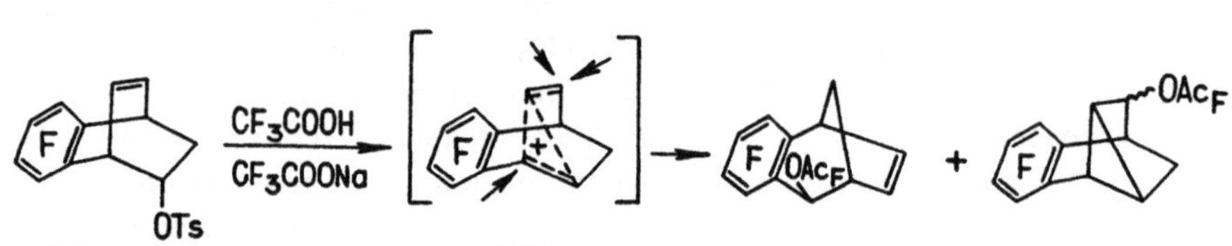

372 376

The tertiary substrates with the same skeleton under the same conditions in this case too, are solvolyzed to the same reaction mixtures. Their compositions indicate the formation of the common classical carbocation *395* which is further non-stereospecifically trapped by a nucleophile or rearranged into the homoallylic ion *396* [558]:

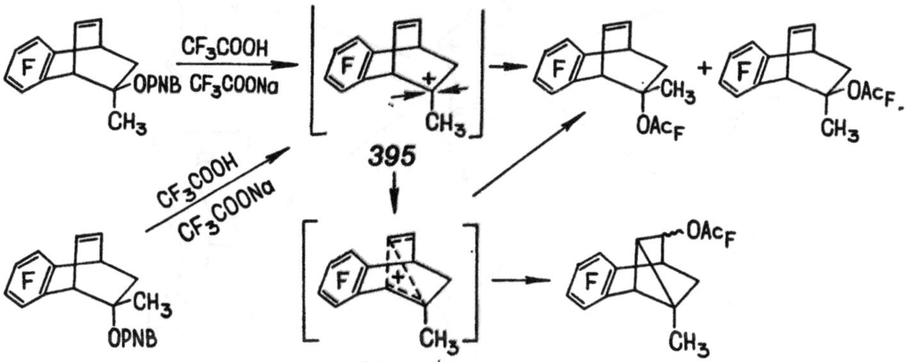

395

396

The nonclassical ion *396* was generated in a superacid medium at —100 °C; the "quenching" of this ion salt solution with trifluoroacetic acid leads to a similar product mixture [554]:

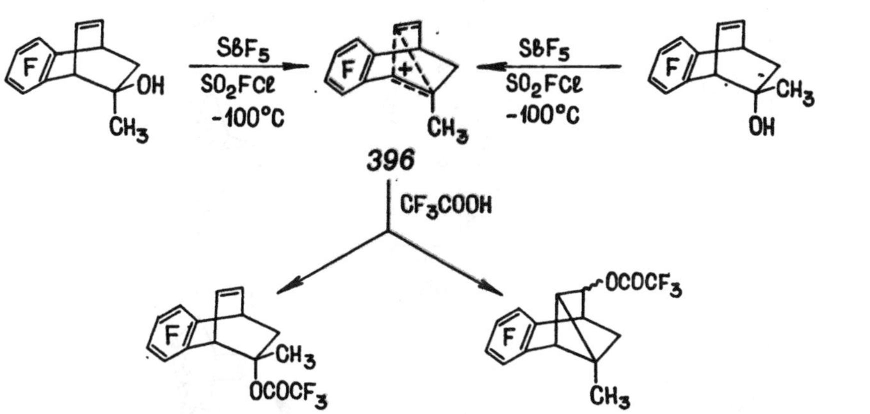

396

Tertiary and secondary cations, e.g., *374* and *395*, differ in their composition. In [559] the task was to find reactions with the secondary classical cation *349* isomeric to the nonclassical ions *374* and *376* as the primary intermediate. These reactions turn out to be the addition of compounds such as HX (CH_3COOH, CF_3COOH, HHal) to tetrafluorobenzobarrelene *296* [559,560]. Thus, for the reaction of the compound *296* with CF_3COOD the analysis of the products and the distribution of deuterium in them suggested a scheme [559] comprising the primary formation of ion pairs of the classical ion *349*, its nonstereospecific capture by a nucleophile as well as its conversion into isomeric nonclassical ions *374* and *376* (see the scheme). The fact that compounds *397* and *398* contain more deuterium in the cis-position can be accounted for by the effect of the anion position in the ion pairs of the classical cations on the direction of its predominant capture by the nucleophile.

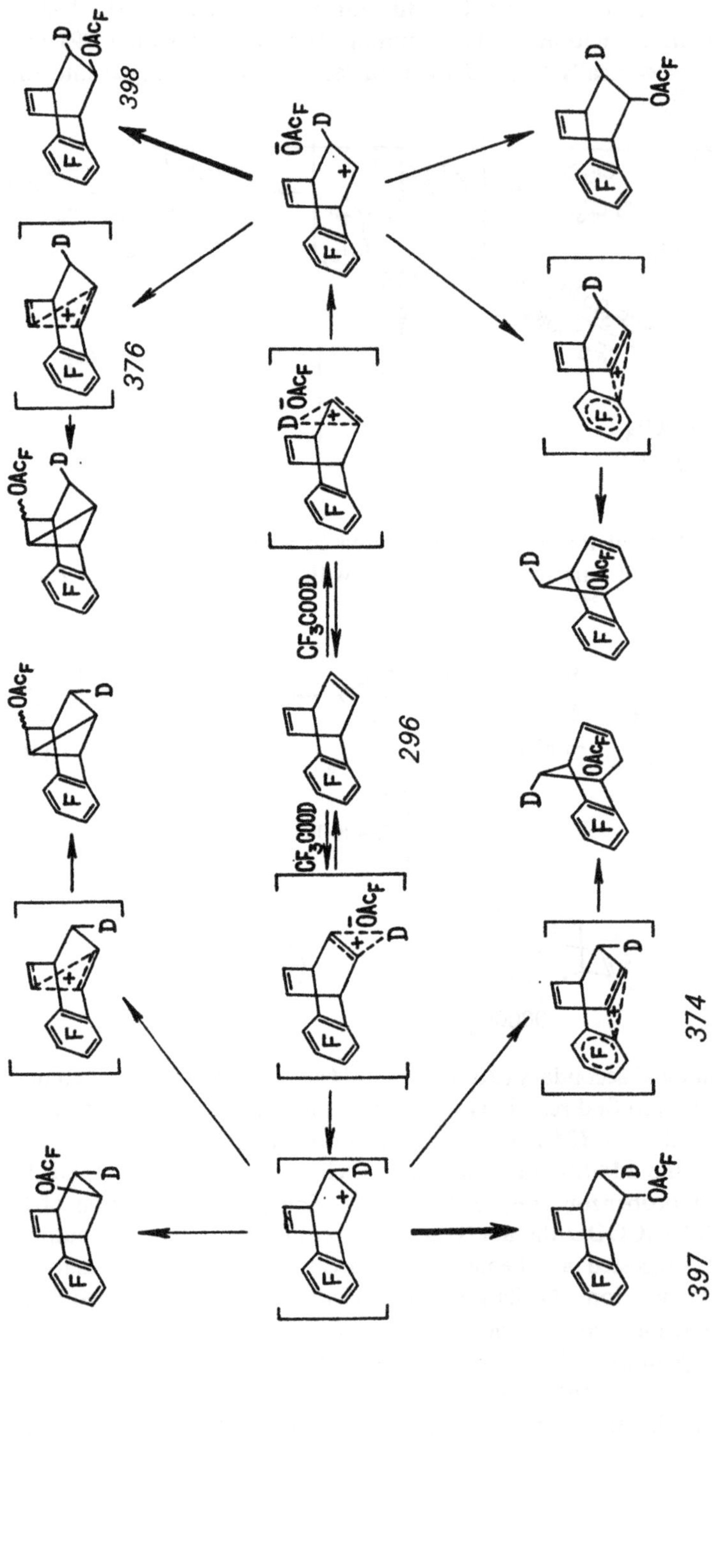
398
376
296
374
397
CF_3COOD
CF_3COOD

In good accord with this assumption [559,560] are the data by Nisnevich [525] who compared the structure of stable secondary carbocations formed on solvolysis of epimeric esters and protonation of "related" olefine:

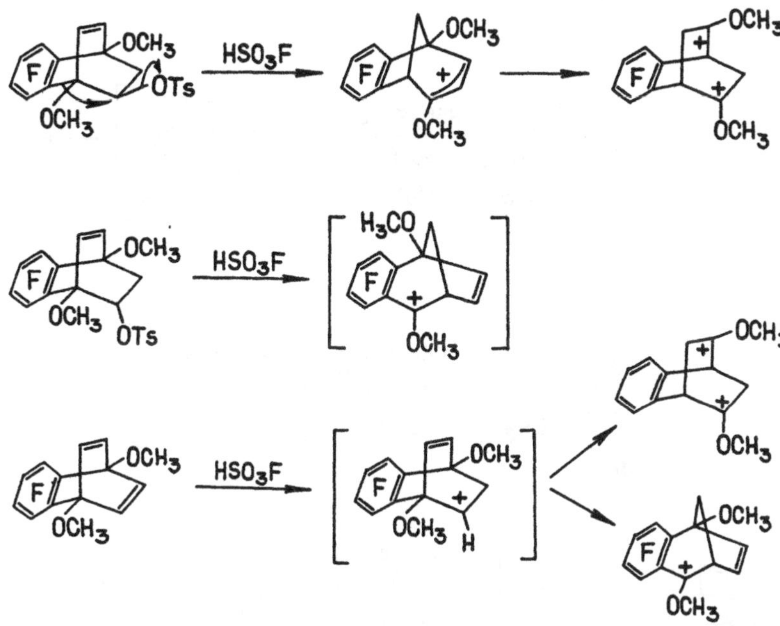

The solvolytic generation manifests a high selectivity in σ-bond shifting; thus the product structure depends on the type of the reaction resulting in a carbocation.

Some effects of the reaction type on the structure of the forming stable tertiary carbocations are analyzed [525].

1) *A comparatively weakly stabilizing* β-*substituent*; β-substituent effectively stabilizes the positive charge. Just as in secondary substrates, the epimers can yield different ions with a selective 1,2-shift of an anti-located σ-bond. The protonation of the "related" olefine yields mainly one of these ions, the stability of the forming cation playing no decisive role, e.g. (see p. 174).

As seen from the scheme, the protonation of the olefine *400* and the solvolysis of ester *399* under equal conditions have made it possible to record stable carbocations differing in their skeleton structure being one of the first examples of this kind. The olefine *400* react with $DSO_3F—SbF_5$ (—105 °C) to give ion *401* where deuterium is in the 8-exo- (~70%) and in the 8-endo-positions (~30%); consequently, the direction of the primary deuterium addition is first determined by steric factors [504]. The configuration of the deuteronium ion does not affect the skeletal shift direction, and opening it results in a classical cation. The exo ester *399* is ionized with σ- and π-participation in the transition state; this is well corroborated by comparing it with the exo alcohol *403*. Under similar conditions it is transformed into the benzylic ion *405*:

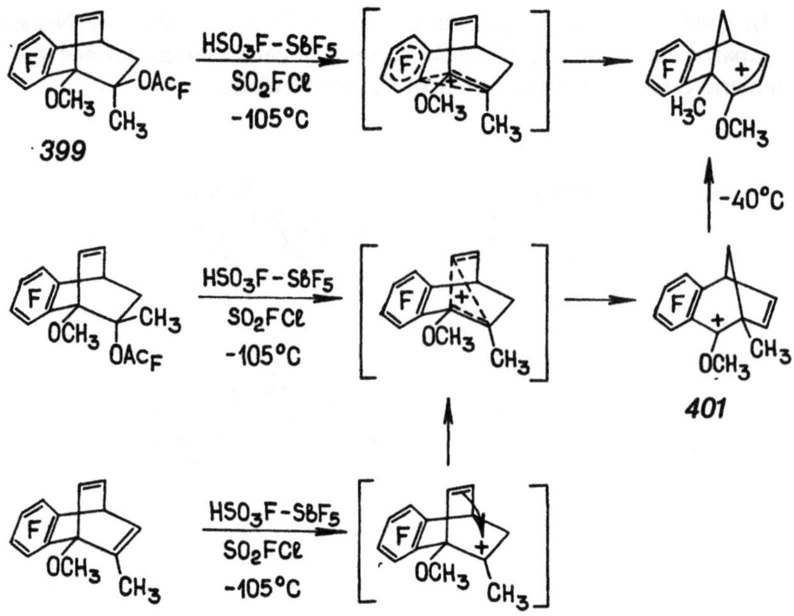

399 400 401 402

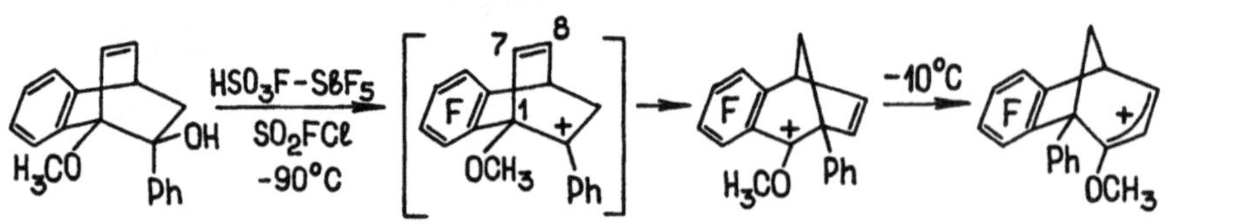

403 404 405

The fact that in ions *402* and *404* there occurs a 1,2-shift of the σ-bond 1,8 and not of the σ-bond 1,2 seems to be due to the primary homoallylic interaction of C$^+$ with the σ-bond 7,8 which makes the position of the p-orbital at C^6 more favourable for an overlap with the σ-bond 1,8 rather than 1,2 (the "sp^2-alignment factor" [561]). Note that in the case of ions *402* and *404* the 1,2-shift of the σ-bond results in a thermodynamically less stable ion.

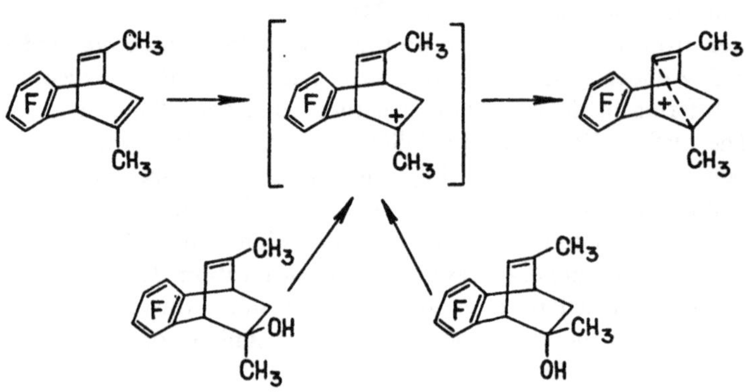

2) *A comparatively weakly stabilizing α-substituent*; the presence of a group capable of effective π-participation, the absence of a β-substituent stabilizing the positive charge — all these factors lead to the same carbocation (and, hence, the products of its "quenching") irrespective of the type of generation reaction (see p. 174).
3) The identical reaction result is also observed in the case of an *effectively stabilizing α-substituent*:

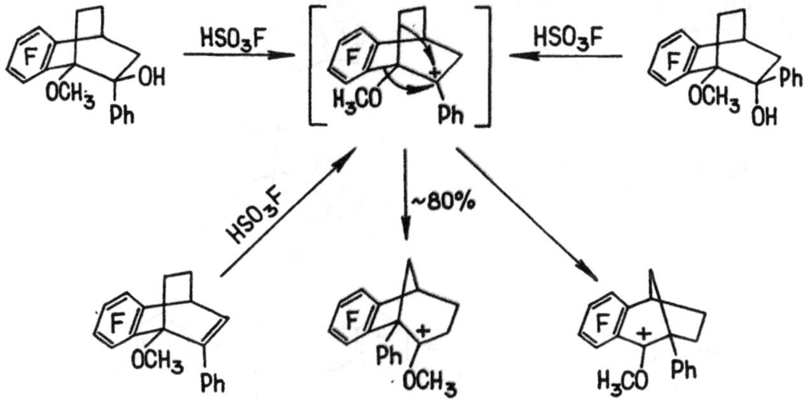

6.2.3 Is the Nonclassical Ion an Intermediate?

The solvolyse of 2,3-tetrafluorobenzobicyclo[2,2,2]octadienes and -octenes are generally accompanied by skeletal rearrangements resulting from a 1,2-shift of the σ-bond in an anti-position to the departing group, e.g.:

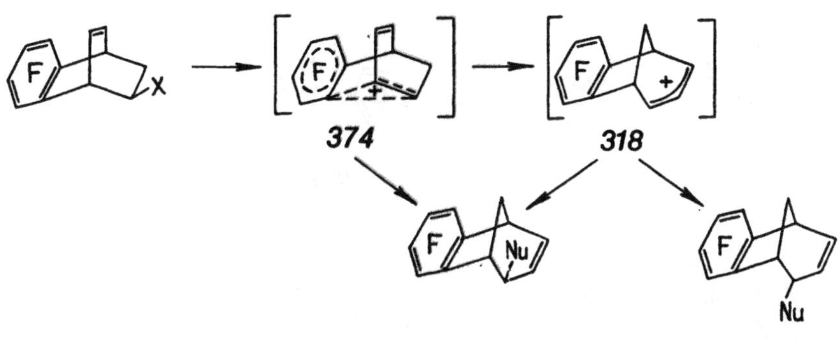

In discussing these data the question arose whether the nonclassical ion *374* is an intermediate or a transition state leading to ion *318*.

In order to simplify the choice between alternative mechanisms Slinko et al. [562] studied the solvolysis of epimeric pairs; on the one hand, the configuration of the detaching group should have a decisive effect on the composition of the reaction mixture, and, on the other, with a nonclassical ion the process of trapping it by a nucleophile should not be accompanied by skeleton rearrangement. With this aim the solvolysis of 8-substituted 3,4-tetrafluorobenzobicyclo[3,2,1]octadienes and -octenes was studied. The 8-syn-epimers are solvolyzed with the retention of the skeleton and the configuration which is well accounted for by the intermediate formation of nonclassical ions. The nucleophilic attack occurs in all cases from a sterically less accessible side; this follows from a comparison with the data on the reduction of the corresponding ketones by sodium boron hydride, e.g.:

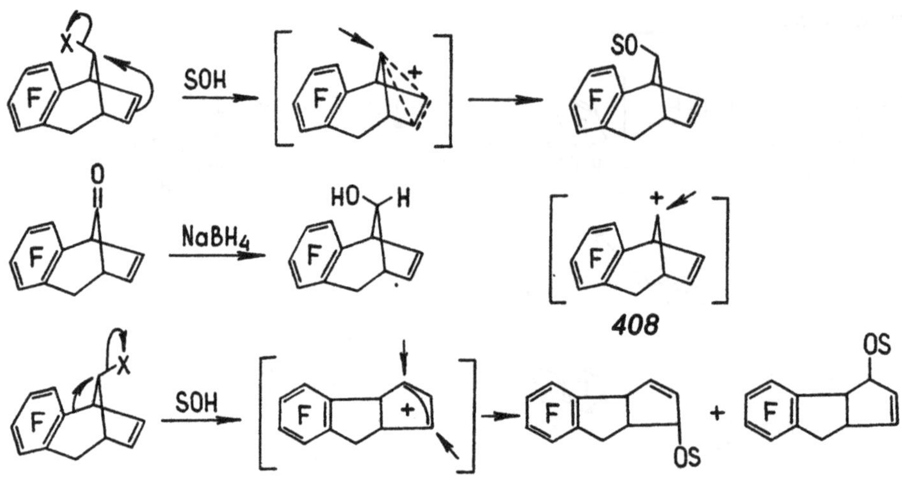

408

The suggested hypothetical schemes are fairly well corroborated by the data of Nisnevich et al. [563] who from the epimeric sulphonates *406* and *407* generated the isomeric ions *409* and *410*; the "quenching" of these ions by a nucleophile yielded product mixtures similar to those described for their solvolysis [562].

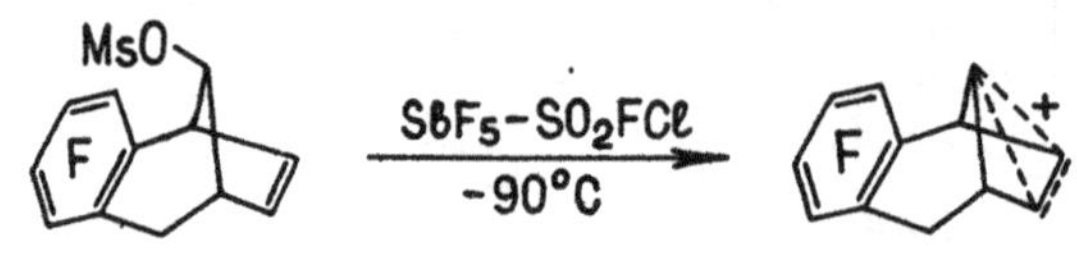

406 409

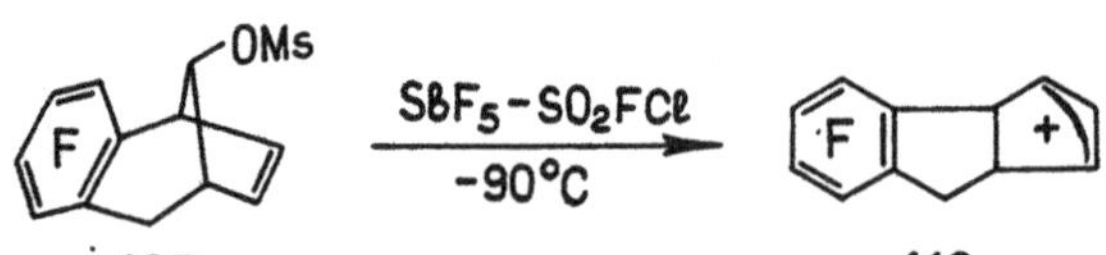

407 410

Thus, to continue earlier works [541,514] we selected an epimeric pair of secondary sulphonates from which under the "long-life" conditions we generated the isomeric stable carbocations; one of them has the structure of a nonclassical homoallylic ion with the parent skeleton, and the other is a product of the selective 1,2-shift of the C—C bond located on the anti-side relative to the detaching group.

If upon solvolysis of 8-syn-sulphonates the intermediate were the classical ion *408*, then the nucleophilic attack would be mainly expected from the anti-side. To verify this assumption the stable carbocations were generated from epimeric tertiary alcohols *411* and *412* [563]. The α-substituent at C^8 is selected since the p-anisyl group, according to Gassman and Fentiman, "levels" the π-participation of the double bond in 7-norbornenyl carbocations [391].

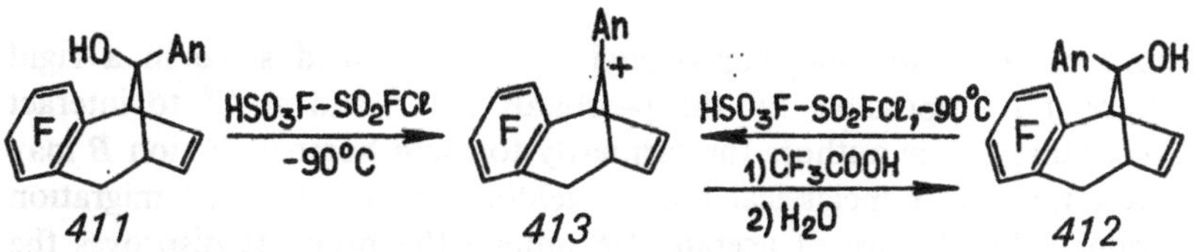

The epimeric pair of the tertiary substrates *411* and *412* was found to yield, as distinct from the secondary analogues, the same stable classical ion *413* which is attacked by a nucleophile from the anti-side, in full accord with the prediction [562].

6.3 Homoallylic Participation

The homoallylic participation was discussed in a number of publications. Thus, LeBel et al. [564,565] studied the acetolysis of 8-syn- *414* and 8-anti-tosyloxy-bicyclo-[3,2,1]octenes-2 *415*; the $k_{415}:k_{414}$ $2.6 \cdot 10^5$ at 25 °C led the authors to suggest the following scheme:

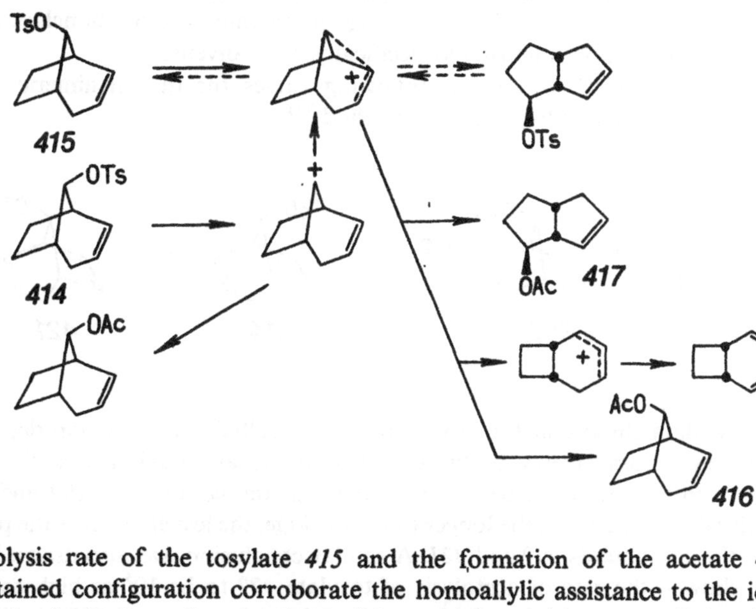

The solvolysis rate of the tosylate *415* and the formation of the acetate *416* with the retained configuration corroborate the homoallylic assistance to the ionization and the formation of a nonclassical ion which can have one of the three structures:

(A)

(B)

(C)

The structure of the bishomocyclopropenyl ion *A* is denied since in a rigid bicyclic structure it is impossible for the p-orbitals at C^3 and at C^8 to interact effectively. According to the authors the primarily forming homoallylic ion *B* may partially convert into the nonclassical ion *C* yielding acetate *417* (the migration of a vinyl group). The absence of acetate *418* among the products disproves the possibility of participation of an allylic methylene group in the ionization of tosylate *415* with the intermediate formation of ion *D*. It is interesting to note that while tosylate

(D)

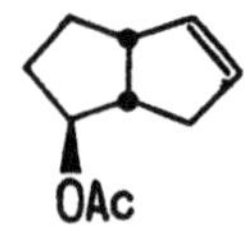

418

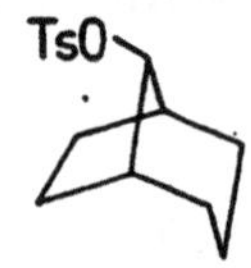

420

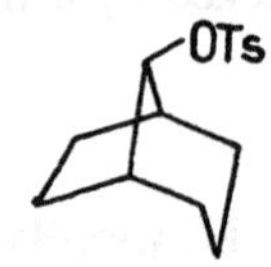

419

419 solvolyzes about 10 times as fast as tosylate *414* (the negative inductive effect of the double bond, the smaller angle 1-8-5 in compound *414*), compounds *415* and *420* react at about equal rates; this also confirms the homoallylic acceleration in the ionization of compound *415*. As seen from the scheme, the classical ion from tosylate *434* reacts along different routes: it rearranges into a nonclassical ion, into an allylic ion or reacts nonstereospecifically with solvents.

Klumpp et al. [566] report the following values for the anchimeric acceleration logarithm determined by Schleyer's method [71]:

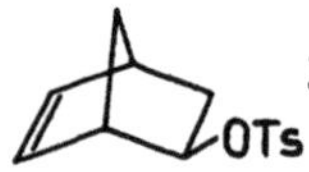

3.8
194

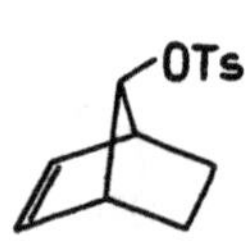

12.9
209

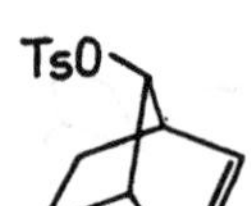

4.9
414

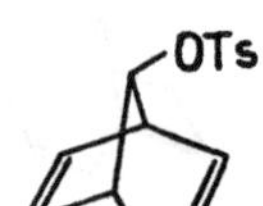

8.1
421

These data show that the efficiency of homoallylic participation depends on the symmetry or asymmetry of the double-bond location relative to the developing cation centre; the participation is stronger in the compounds *209* and *194*. Their relative distance effects: the longer the syn-bridge, the less effective is the participation as in the compounds *209* and *421*. A similar conclusion was also drawn by Hess [567] who showed the rate of acetolysis of tosylate *422* to be 10^5 as high as that of the saturated analogue *419* but far lower than that of compound *209* (the logarithm of anchimeric participation for compound *422* is 8.3).

422

419

Lambert[568] has shown that the unsymmetrical homoallylic participation in the bicyclo[2,2,2]octene series is far more intensive than in the norbornene one; this may be due to the smaller rigidity of the bicyclooctene skeleton in comparison with the bicycloheptene one[102].

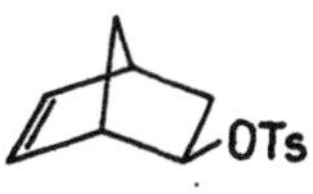

194

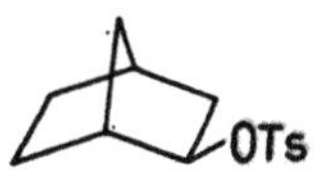

63

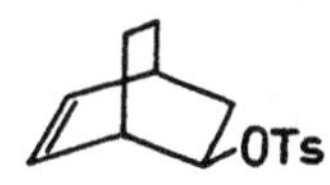

422

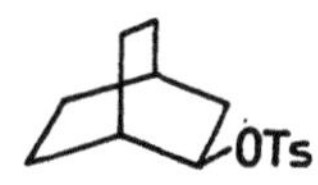

163

$k_{194}/k_{63} = 0.3$ $k_{422}/k_{163} = 63$

The effect of the π-participation becomes especially high in the series of ditosylates:

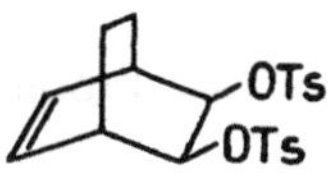

423

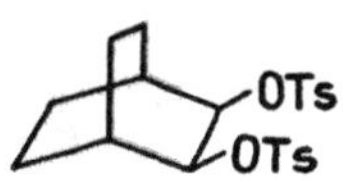

424

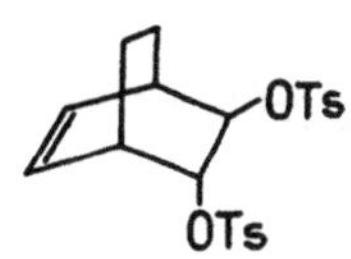

425

$k_{423}/k_{424} = 6850$

The inductive intensification of the double bond participation (6850/63 = 110) is lower than in the norbornene series (1700) due to the very weak π-participation in norbornene 2-monotosylates.

It should be noted that the trans-isomer *425* reacts 8 times as fast as the cis-isomer *423*. This fact precludes the simultaneous detaching of both tosyloxy groups and the formation of a dication. The enhanced effect of unsymmetrical homoallylic participation is attributed to an electronegative substituent (OTs) which increases the demand for the double bond participation from the developing cation centre.

The first example of the π-participation of the "twisted" double bond seems to have been discovered by Gassman[569]:

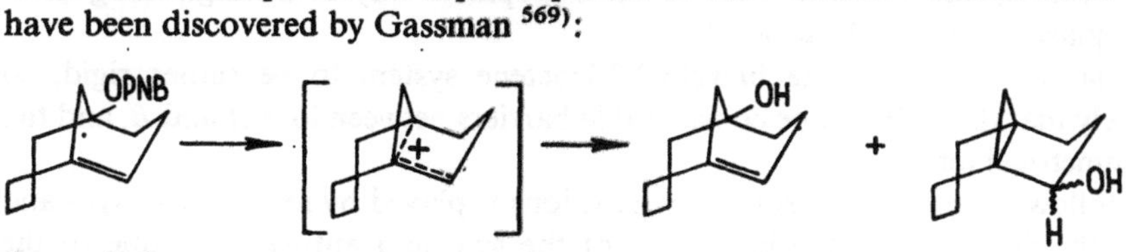

$k_{rel} = 120$

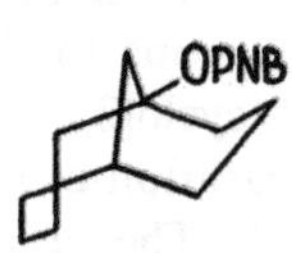

$k_{rel} = 1$

According to Berson [570], the stereochemistry of the splitting bond in multistage processes can control the skeletal rearrangement occurring even at the second stage (the "memory effect"). Thus he showed the solvolysis of 2-exo- and 2-endo-norbornylcarbinyl systems to yield quite different reaction mixtures; Berson suggested the following scheme [322–325]:

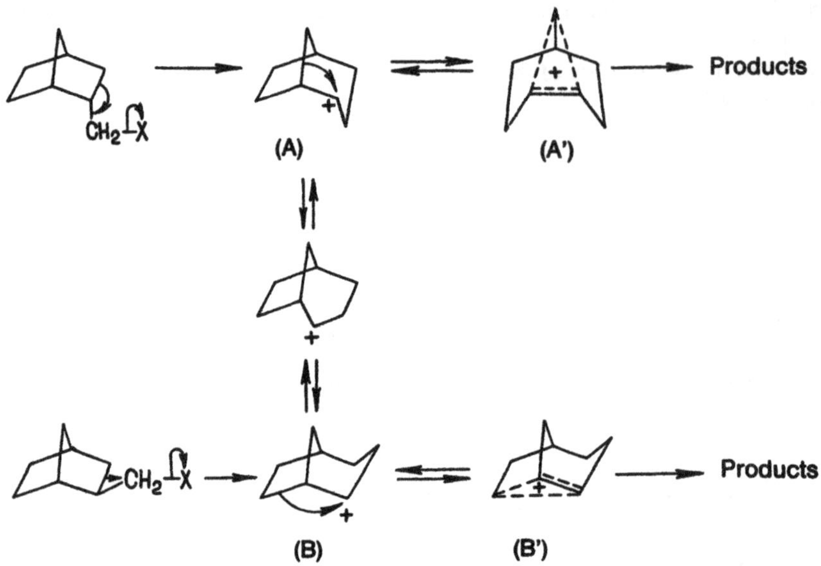

The primarily ions *A* and *B* are formed more slowly than their secondary rearrangements to the nonclassical ions *A'* and *B'*. Thus the secondary rearrangements are controlled by a group detaching in the first stage. In the case in question it is easy to assume a conformational barrier between ions *A* and *B* (the "chair" and the "bath"). If the memory effects are only due to conformational barriers then these effects must greatly decrease or even completely disappear in rigid systems, e.g., in the solvolysis of 7-syn- and 7-anti-norbornenylmethyl derivatives [571–574]. In these systems cation *D* and *E* are not converted into each other, they yield the same products as in the solvolysis of 2-endo- [319,533,575] and 2-exo-bicyclo[2,2,2]-octene-5-yl-2 tosylates [576] (cf. [577]), respectively. Cations *A* and *B* are represented in "twist"-forms, but this is no more than a graphical way of distinguishing them from the quasi-symmetrical cation *C*.

Molecular models show the bicyclo[2,2,2]octene system to be rather rigid, so it is unlikely that there should be considerable barriers between ions *A* and *B*, and the quasi-symmetrical ion *C*.

In the following scheme the role of twisted ions is played by ion pairs — syn- and anti-, i.e. the difference in the behaviour of the syn- and anti-series is due to the difference in the arrangement of counter-ions in the ion pairs which are further combined stereospecifically to form 2-bicyclo[2,2,2]octenyl derivatives. The cross-over in this system occurs due to the separation of ion pairs by the solvent with the formation of the quasi-symmetrical ion *C*. The ratio between the coupling and the separation of ion pairs depends on the nature of the solvent and the counterion.

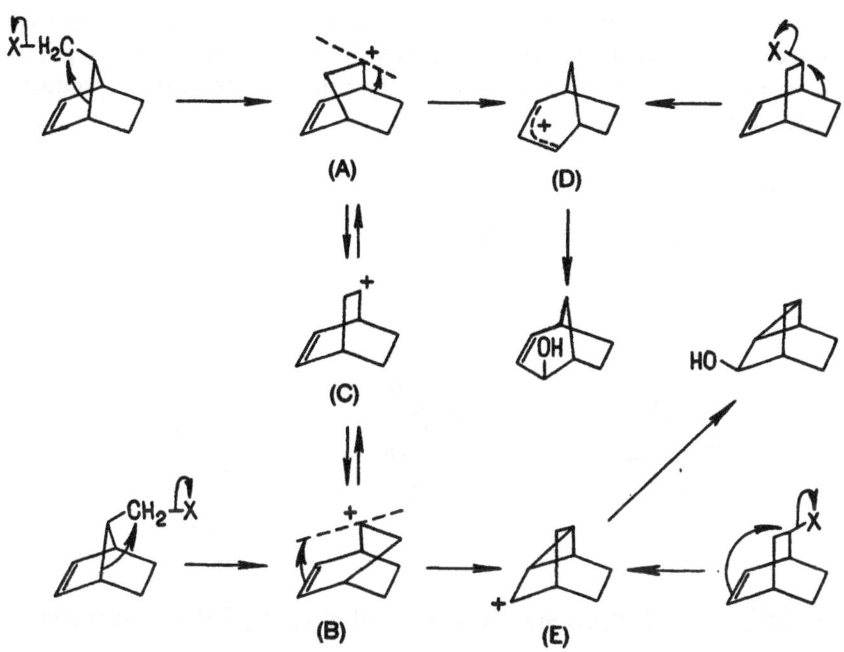

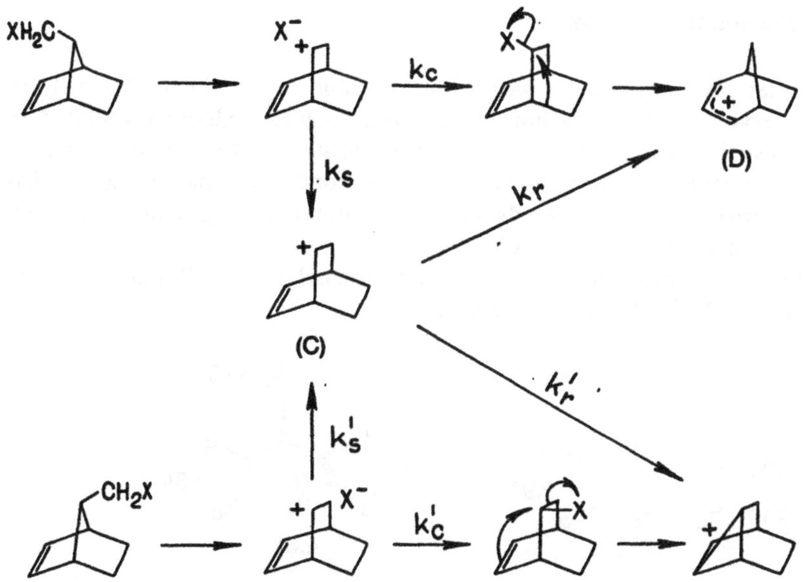

As the ionizing force of the solvent increases and its nucleophility decreases $k_c : k_s$ will fall.

Another explanation is connected with intramolecular factors. Between the "twisted" cations and the quasi-symmetrical ion *C* there are energetic barriers due to the difference in the angle strain and in the non-bonded interactions between the "twisted" ions and the transition states leading to ion *C*. Some stability is imparted to the "twisted" cations by the nonclassical bonds. If the departure of the X group is followed by the participation of σ-electrons, the primary intermediate products will be nonclassical ions.

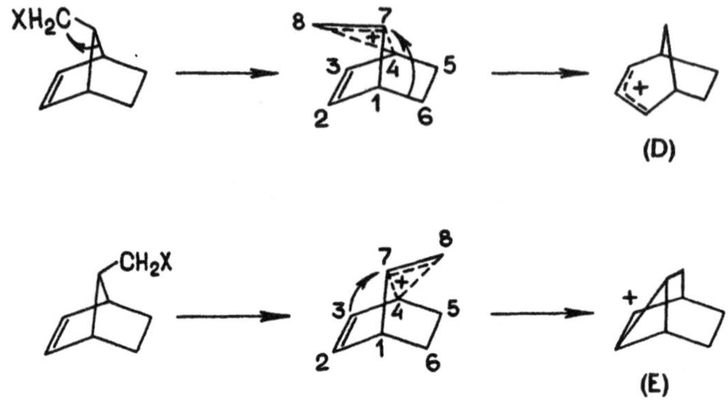

(D)

(E)

7 Homoallylic and Homobenzylic Ions under Long-Life Conditions

7.1 Bishomoaromatic System

As noted earlier, in exceptionally weakly nucleophilic media the NMR method is used to observe directly many nonclassical ions — intermediates postulated in explaining unusual rates, products and stereochemistry of the above solvolysis reactions. This enables research under stable-ion conditions may result in discovering new, earlier unknown kinds of carbocation rearrangements illustrated by the 7-norbornenyl and 7-norbornadienyl cations.

The signals in the PMR spectra of the 7-norbornenyl *19* [377,578] and the 7-norbornadienyl ions *236* [579–583] are given below.

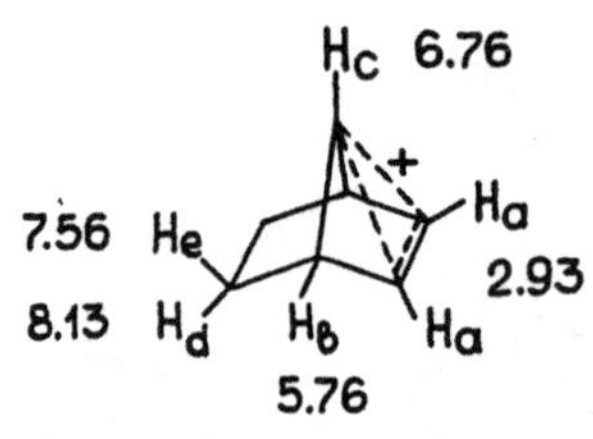

19

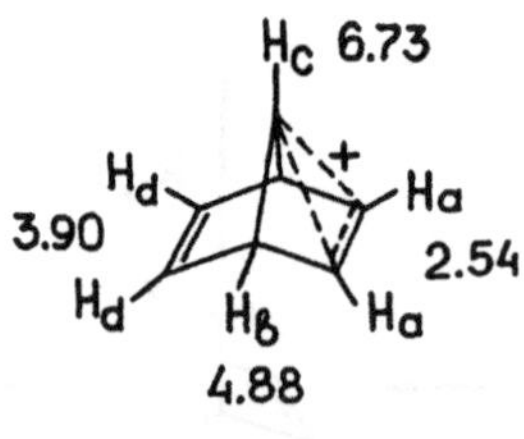

236

Judging by the *chemical shifts* the cations *19* and *236* are similar in structure and are likely to have a nonclassical symmetrical structure. Deno [48] and Brown [387] expressed an alternative viewpoint suggesting the rapid equilibrium of the tricyclic cations *219*:

219a *219b*

The symmetrical ion structure was supported by the PMR data of the 2-methyl-7-norbornenyl cation *426* [584].

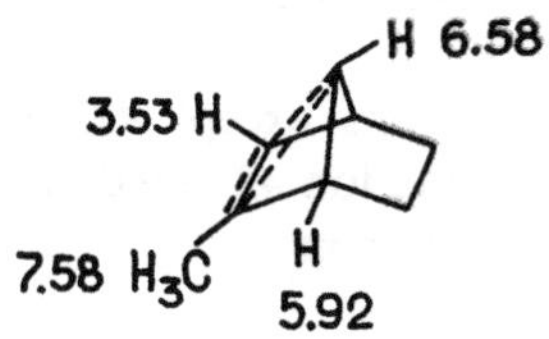

426

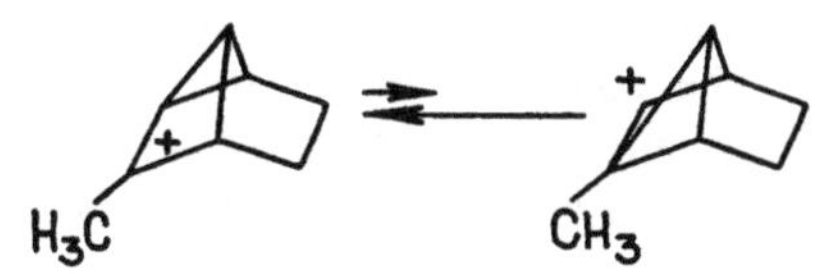

223 *223a*

The assumed equilibrium *223*$\rightleftarrows$*223a* must be very much shifted towards the tertiary ion *223*. In the spectra of α-cyclopropylcarbinyl ions the signals of α- and β-cyclopropane protons are in a comparatively high field (τ ~ 7–8) [585]. The signal of the H^3 proton (τ 3.53) of ion *426* in by 3 ppm in a lower field than the signals of H^7 (τ 6.58) and of the α-protons of cyclopropylcarbinyl ions (also by 3 ppm); on the other hand, the positions of the H^3 signals for ions *19* and *426* differ only by 0.6–0.7 ppm. These results are incompatible with the assumed equilibrium *223*$\rightleftarrows$*223a*; they indicate that the electron distributions in ions *19* and *426* are very similar, in accordance with their *bishomocyclopropenyl structure*.

7.1.1 Stabilizing Aryl Groups

The solvolysis of 7-aryl-7-norbornenyl derivatives is accompanied by a competition between the aryl and alkenyl functions in the stabilization of 7-cations. When the stabilizing action of the aryl group attains a definite level the participation of the double bond becomes insignificant [84,391]. In 1970 the 7-p-anisyl-7-norbornenyl cation *427* under the "long-life" conditions was shown to be a classical ion in which the interaction of C^7 with C^2 and C^3 is far weaker than in the 7-aryl-7-norbornenyl ions possessing a weaker stabilizing effect in comparison with the anisyl group. The classical nature of the p-anisyl cation *427* is confirmed by the following data.

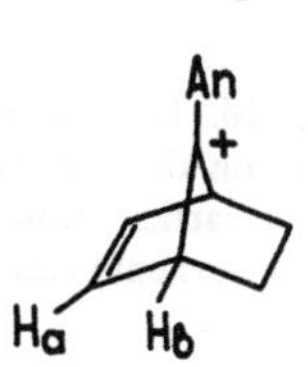

427

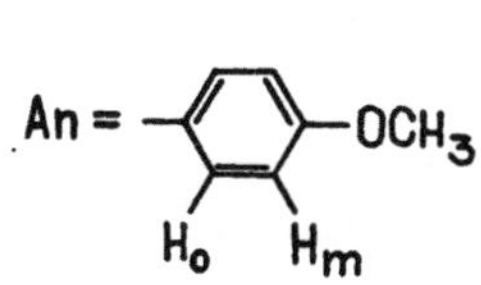

1. The signals of the aromatic protons H_0 and H_m are shifted 1.2 and 0.7 ppm to the low field as compared with their position in the precursor spectrum; this is

typical of α-p-anisyl carbocations. On the contrary, in the spectrum of the 7-phenyl-7-norbornenyl ion the signals of H_0 and H_m are shifted to a low field only by 0.4 ppm as compared with its precursor, i.e. they are in a far higher field than in the ions in which the α-phenyl group participates effectively in the delocalization of the positive charge [586].

2. In the ion *427* the signals of the H_a and H_b protons are in a far higher field than in other 7-aryl-norbornenyl ions in accordance with the smaller shift of the positive charge on C^2 and C^3.

3. The multiplicity of the signal of H_a in the ion *427* is essentially different from that in the spectra of related ions.

4. The solutions of all the 7-aryl-7-norbornenyl ions except *427* are stable in fluorosulphonic acid at low temperatures whereas the ion *427* tends to conversions connected with the addition of the acid to the double bond.

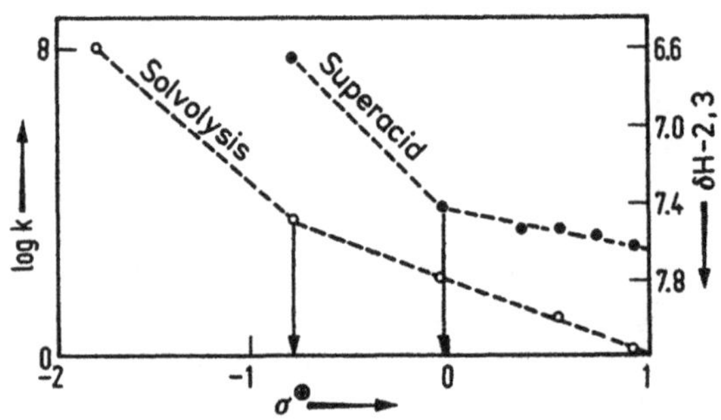

Fig. 9. Correlation of the ^1H chemical shifts at H^2 and H^3 ($\delta H^{2,3}$) in the 7-aryl-7-norbornenyl cations (superacid) and the logarithm of the solvolysis rate constants (log k) for the 7-aryl-7-norbornenyl p-nitrobenzoates (solvolysis) with σ^+ constants [243)]

In discussing the literature data on the PMR spectra of 7-arylnorbornenyl ions Farnum [243)] drew attention to the dependence of:

a) $\delta H^{2,3}$ of these ions (in superacids) and

b) the kinetic characteristics of the solvolysis of 7-aryl-7-norbornenyl-p-nitrobenzoates (log k) on σ^+ (Fig. 9).

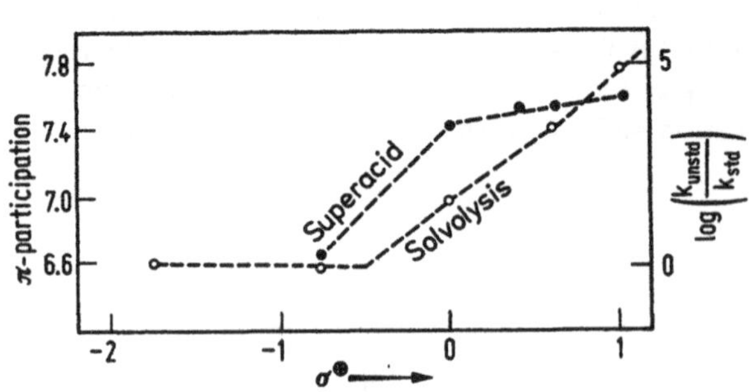

Fig. 10. The effect of increasing electron demand (increasing σ^+) on π-participation in the 7-aryl-7-norbornenyl cations [243)]

Farnum suggested a different variant of graphic dependences (Fig. 10) in which the solvolysis data are represented by the rate constant ratio of 7-aryl-7-norbornenyl-p-nitrobenzoates and the corresponding 7-aryl-7-norbornyl-p-nitrobenzoates. The solvolysis graph points to the inclusion of π-participation for the aryls more electron-seeking than anisyl. The data on the ions should be interpreted, according to Farnum, as π-participation for the aryls from anysyl to phenyl, then comes the "saturation" effect of the π-participation when the double bond participation can no longer markedly increase with increasing demand. This effect is observed for more "electron-seeking" cations, but it has not been recorded for the less electron-seeking (for the same aryls) transition states of solvolysis. According to Fig. 10 the π-participation in cations is only observed for less electron-seeking aryls than in solvolytic reactions. Farnum holds π-participation to be also possible for aryls more electron-donating than anisyl since the signal of the vinyl protons (δ 6.65) in ion *427* is considerably shifted to the low field relative to a neutral compound (δ 6.0).

By comparing the ^{13}C NMR spectra of 7-aryl-7-norbornenyl with 7-aryl-7-norbornyl ions[10] Olah [315] has recently demonstrated (Fig. 11) the interaction between the cation centre of C^7 and the π-bond, the increase in electron demand being

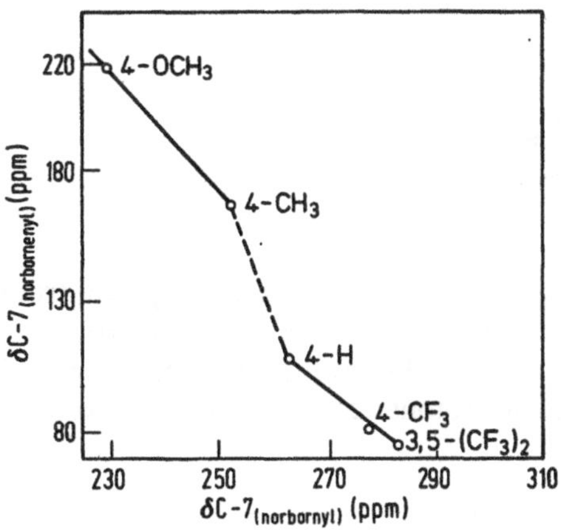

Fig. 11. Correlation of the ^{13}C NMR chemical shifts of the cationic carbons in the 7-aryl-7-norbornenyl and 7-aryl-7-norbornyl cations [315]

accompanied by the C^7 signal shifting to the high field in agreement with the intensification of π-participation. In coincidence with Farnum's data [243] Olah also notes the partial π-participation in ion *427*.

7.1.2 C^7 Cation Centre

A comparison of the ^{13}C NMR spectra between 7-norbornenyl ions and typical α-cyclopropylcarbinyl cations [587] indicates that in the former the C^7 cation centre signals are shifted 100–200 ppm to the high field. In choosing between the nonclassical structure *19* and the rapid equilibrium of α-cyclopropylcarbinyl ions one should take into account that the calculated averaged shift for the signal of $C^{2(3)}$ should be $1/2\,(250 + 80) = 165$ ppm which differs by 40 ppm from the experimental

10 There is an excellent ratio of the chemical shifts of the cation centres and those for the model classical ions 1-aryl-1-cyclopentyl ($r = 0.988$) which conforms the classical nature of the 7-aryl-7-norbornyl ions.

value. Thus the ^{13}C NMR spectra preclude the structures of both static classical ions and rapid equilibrium of α-cyclopropylcarbinyl cations for 7-norbornenyl ions. The nonclassical bishomocyclopropenyl structure of the latter agrees well with the uncommonly high values of the $^1J_{CH}$ for C^7 in ion *19* (218 cps), as well as for C^2 and C^3 in the whole series. The same conclusion is also supported by calculations according to Schleyer's additive scheme [268].

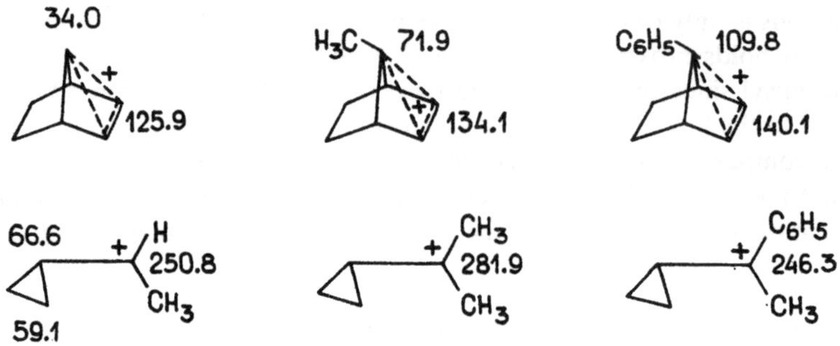

For the 7-norbornadienyl cation *236*, various formulae were suggested from chemical data [588]. However, the first PMR study of 7-norbornadienyl fluoroborate obtained from 7-chloronorbornadiene [580] showed this salt to be unsymmetrical because the PMR signals showed four types of protons (2:2:2:1). Such a spectrum corresponds either to formula *428* or to *236*.

428 *236*

A detailed PMR analysis made it possible to determine the constants $J_{7-2(3)} = 2.7$ cps which points to a bond between C^7 and the carbons of one double bond [581]. Capon [9] pointed out that the "asymmetry" of the fluoroborate in the 7-norbornadienyl ion may be due to the structural peculiarities of the ion pair. This objection, however, fell away after Richey obtained the same PMR spectrum for the 7-norbornadienyl cation in $FSO_3H—SO_2$ [589]. Hoffmann's calculations using the so-called Hückel's extended method have shown [590] that the unsymmetrical structure *236* must be 8 kcal/mole more stable than the symmetrical *429*. As the signals of the olefine protons in the PMR spectrum of the ion remained unchanged on heating up to 50 °C Story[579–581] concluded there is a considerable barrier to the *236a*⇄*236b* conversion named "bridge-flipping".

Winstein et al. tried to determine the value of this barrier[582,583,591]. Simultaneously they investigated the interesting "degenerate" rearrangements of ion *236*. Extraction of 7-norbornadienol from solution in pentane or methylene chloride with fluorosulphonic acid at −78 °C prepares a solution of the cation *236* in fluorosulphonic acid. Heating the solution up to +45 °C, does not cause an essential broadening of signals in the PMR spectrum; this, indicates of significant barrier for a degenerate rearrangement of the "bridge-flipping" type. Above +45 °C, the ion *236* is decomposed the PMR spectrum displaying several broad signals as well as a narrow signal at τ 0.76 characteristic of the tropylium ion (~25%). Despite this circumstance, the PMR spectrum of ion *236* has been successfully recorded at +77 °C.

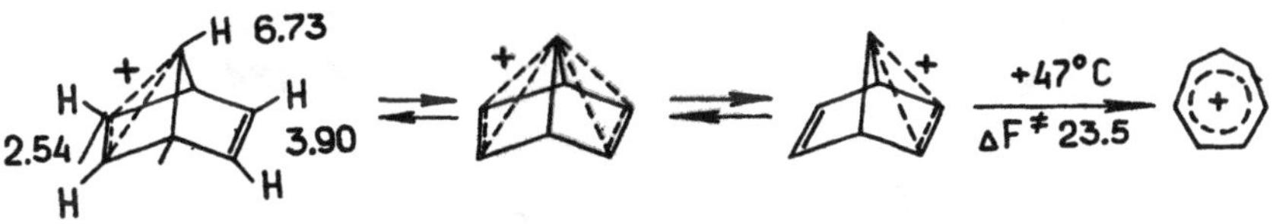

236a *429* *236b*

Even at this temperature the signal of the "bonded" olefine group protons (τ 2.54) does not markedly broaden. However, the signals τ 3.90 ("unbonded" olefine group), 4.88 (Bridge-head protons) and 6.73 (Bridged protons) broaden indicating another type of degenerate rearrangement. This rearrangement is revealed by "labelled" precursors, e.g., acetate *430*. The latter in fluorosulphonic acid at −73 °C yields ion *431* with 4 signals, those of the "unbonded" olefine fragment protons being twice as intensive as those of the "bonded" ones. On heating the cation solution to −47 °C the label is scrambled ($k = 3 \cdot 10^{-4}\ s^{-1}$). The peak intensities approached the values 2:1.6:1.6:0.8, respectively, for the protons of "bonded", "unbonded" olefine fragments, as well as for bridge-head and bridged protons.

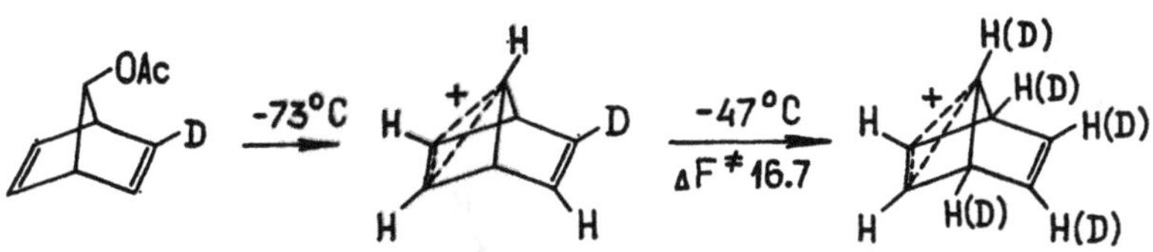

430 *431* *431a*

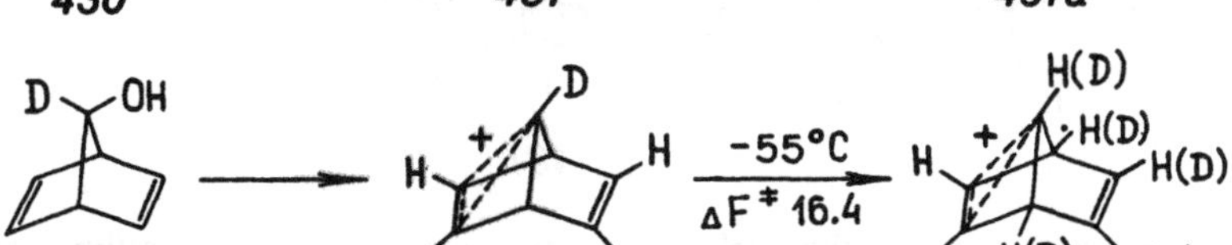

433 *432* *432a*

So, deuterium shifts to all the positions except for the two positions of the bonded olefine group. Similar results have been obtained for cation *432*. In the rearrangement in question the deuterium is incorporated sequentially at the different positions. Thus, in the ion *431* deuterium is first incorporated into the bridge-head and only then into the bridge position; in the rearrangement of ion *432* — first into a bridge-head then into the unbonded olefine fragment. Thus, isomerization is the stepwise circumambulatory motion of five carbon atoms of ion *236* with respect to the bonded vinyl group.

The simplest explanation of this degenerate rearrangement (5C-rearrangement) is the conversion of ion *236* into the bicyclo[3,2,0]heptadienyl cation *434* by 1,2-shift of the "bonded" olefine fragment from C^1 to C^4 through a transition state *A*.

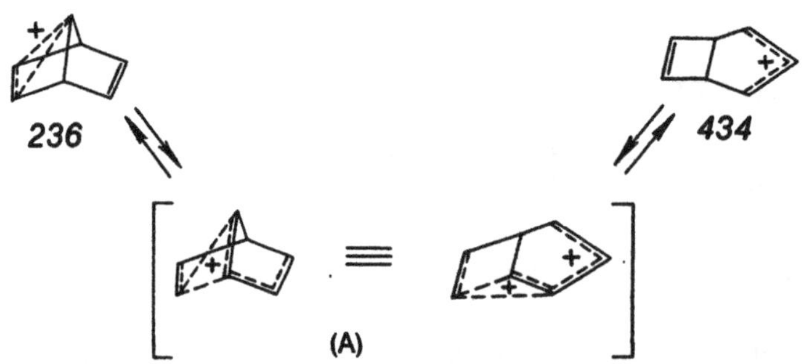

This explanation is supported by the behaviour of cis-*435* and trans-bicyclo-[3,2,0]heptadienols *436* in fluorosulphonic acid. In this case the signal intensity ratio of the protons of bonded and unbonded olefine fragments, and of the bridge-head and bridge protons of the forming 7-norbornadienyl ion is 2:1.5:1.5:1.0, i.e. in 50% of the cases deuterium is in the bridge-head and in the other 50% — in the unbonded olefine fragment.

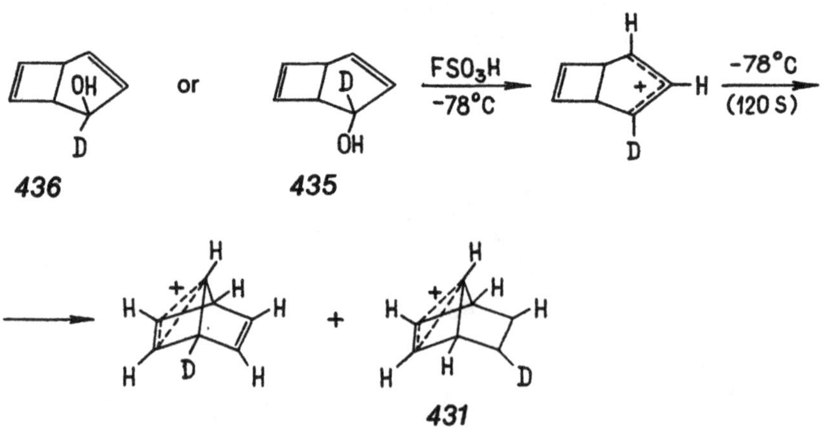

These data confirm the cycle contraction-expansion. The equilibrium of ions *236* and *434* is greatly shifted towards ion *236*; the minimum value of the equilibrium constant *434*⇄*236* at —78 °C is ~7000. It is interesting that the *236*⇄*434* transitions observed in fluorosulphonic acid under the stable ion conditions do not occur in solvolyse when the carbocation lifetime is very short.

To determine the value of the energetic barrier of the degenerate rearrangement of the "bridge-flipping" type Winstein generated the 7-norbornadienyl cation from the ether *437* [583].

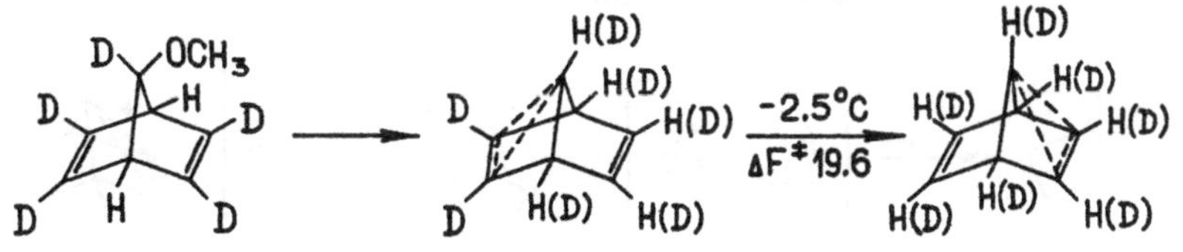

437 *438a* *438b*

The 5C-rearrangement results in cation *438a* in which either position of the bonded olefine fragment contains 1D, and each position of the unbonded fragment 2/5D. At ~0 °C protons appear in the bonded olefine fragment; the proton signal intensities of the bonded and the unbonded fragment become equalized. At —2.5 °C ΔF≠ for the latter rearrangement is 19.6 kcal/mole, 3 kcal/mole higher than ΔF≠ for the 5C rearrangement. The value 19.6 kcal/mole is the lower limit of the possible barrier value for the degenerate bridge-flipping type rearrangement.

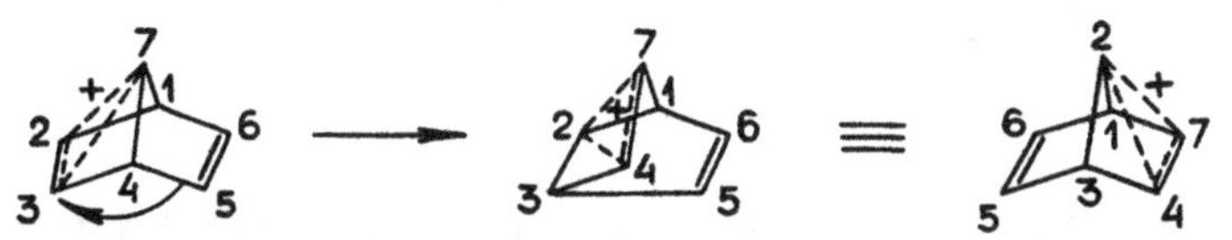

236

In 1972 Lustgarten et al. suggested another two mechanisms which, combined with 5C-rearrangement can average all the seven carbons; these mechanisms also include 1,2-bond shift in the course of which the nonbonded olefine C atoms are not scrambling [592].

At present it does not seem possible to make an unequivocal choice among all these variants.

7.1.3 Influence of the 7-Methyl Group

While in Winstein's works the bicyclo[3,2,0]heptadienyl ions were assumed to be intermediates in the rearrangements of the observed bicyclo[2,2,1]heptadienyl ions, the opposite situation was found by Hogeveen [593] in polymethylated analogues: the bicyclo[2,2,1]cations are assumed to be intermediates in the rearrangements of the observed bicyclo[3,2,0] cations.

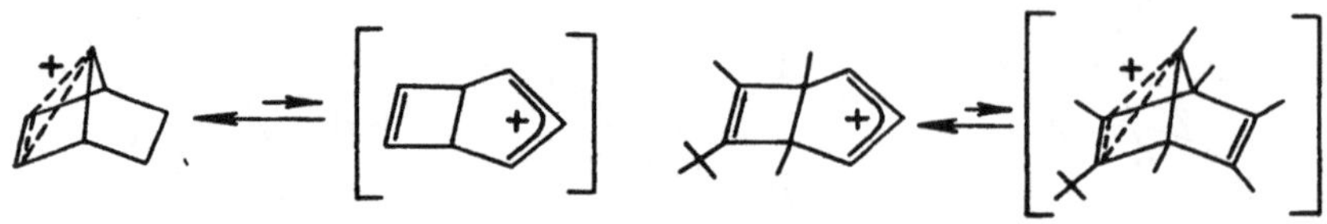

Winstein has studied the rearrangements of the 7-methylnorbornadienyl cation *439*. At —14 °C the signals of vinyl protons are averaged while those of bridge-head ones and of CH_3 groups remain narrow-band. This indicates the rearrangement *439*⇄*439a* with $\Delta F^{\neq} = 12.4$ kcal/mole. $\Delta F^{\neq}$ for this rearrangement is by 7.2 kcal/mole lower than for the rearrangement of the unsubstituted ion *236*, this is due to

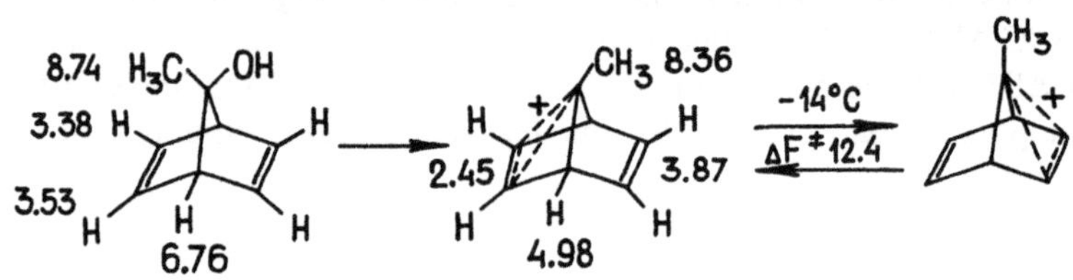

442 *439* *439a*

the stabilizing effect of the CH_3 group on the symmetrical structure *429*. The stabilizing effect of the 7-CH_3 group in the unsymmetrical ion *236* in which C^7 interacts with one of the olefine groups should be much lower. This discrepancy is due to the fact that in the symmetrical ion where the participation of olefine electrons is far less intensive for C^7 the fraction of the positive charge remains larger than in the unsymmetrical ion.

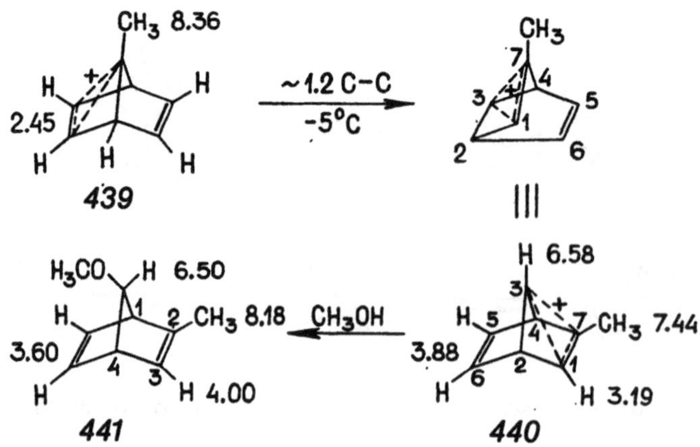

At —5 °C ion *439* is irreversibly rearranged to ion *440* with $\Delta F^{\neq} = 18.9$ kcal/mole. The simplest mechanism of the conversion *439*→*440* is the shift of the unbonded vinyl C^6 from C^1 to C^2. Another way may be: 5C-rearrangement with the subsequent bridge-flipping. The structure of ion *440* is confirmed by its PMR spectrum and conversion into compound *441* when quenching with methanol. The spectra of *439* and *440* show the greater part of their positive charge to be concentrated on C^2 and C^3 and the smaller one on C^7 (cf. the spectra of *439* and *442*, *440* and *441*). The 7-CH_3 group exerts very little influence on the hydrogen shifts in ion *439* as compared with the 7-unsubstituted ion *236* which indicates a weak interaction of the CH_3 group with the positive charge. Indeed, the chemical shifts of olefine protons in ions *439* and *236* are similar; the same applies to the shifts of CH_3 groups in ion *439* and in the precursor *442*. Thus, the 7-CH_3 group at an atom carrying a small positive charge interacts weakly with the cation electron system; at the same time the 2-CH_3 group in ion *440* interacts intensively with the positive charge, its signal is shifted 0.92 ppm to a low field in comparison with the signal of the 7-CH_3 group of ion *439*. The most stable of the monomethylnorbornadienyl cations described is, thus, ion *440*. It is reasonable to connect its PMR spectrum with the assertions of Brown [387] and Deno [48] that ion *236*, just as *19*, is in fact a pair of rapidly rearranging tricyclic ions. If this is so, then the PMR spectrum of "ion *440*" must be very close to that of ion *443*.

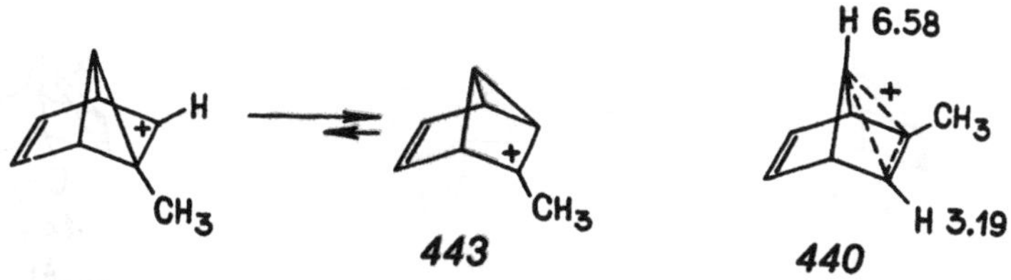

443 440

The α- and β-protons of α-cyclopropylcarbinyl ions have close chemical shifts (τ ~ 7). The signal of H^3 (τ 3.19) for ion *440* is shifted 3.4 ppm to a lower field than that of H^7 and 3 ppm to a lower field than the α-proton of cyclopropylcarbinyl ions [589]. Thus, one can add the spectral data to the above-discussed chemical and stereochemical reasons [377] against considering norbornadienyl ions as rapid equilibria of tricyclic cations.

Subsequently Winstein [594] directly observed 1-methyl-*444* and 5-methyl-substituted *445* norbornadienyl ions, to elucidate the rearrangement mechanism of ion *439* into ion *440*. At —105 °C cation *446* is rearranged into a mixture (55:45) of 1-methyl-*444* and 5-methyl-cation *445*. At —75 °C ion *444* is rearranged irreversibly into ion *445* which at —45 °C is decomposed but not converted into ion *440*. Ion *445* was obtained from another precursor — quadricyclanol *447* with intermediate rearrangement of the 7-methyl-7-quadricyclyl ion. The structure of ions *444* and *445* is established by PMR analysis and by conversion of ion *445* into syn-methyl ether *448* which is different from the anti-isomer *441* formed from ion *440*.

The isomerization of ion *444* into ion *445* results from the contraction and expansion of the cycle; here one can exclude one of the two earlier suggested mechanisms for the conversion of ion *439* into ion *440*, namely: contraction-expansion of the cycle and then rearrangement of the "bridge-flipping" type.

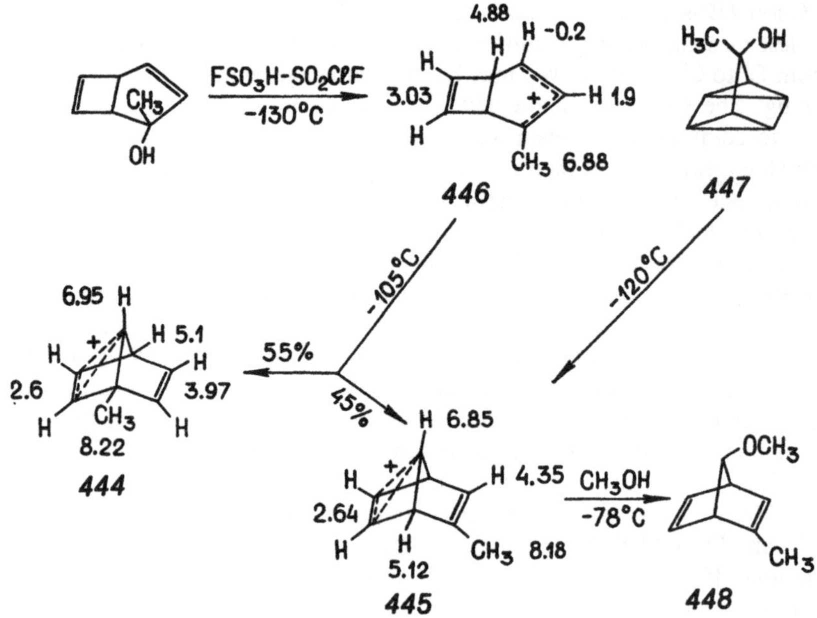

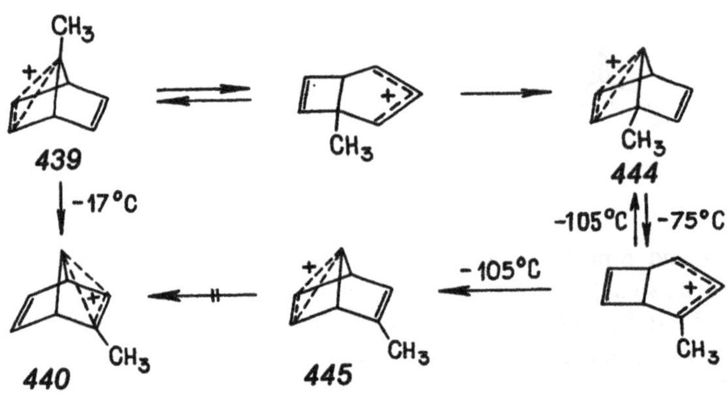

This mechanism is rejected since the ion *445* is already decomposed at —42 °C, but it is not rearranged into ion *440*. That leaves the 1,2-shift route:

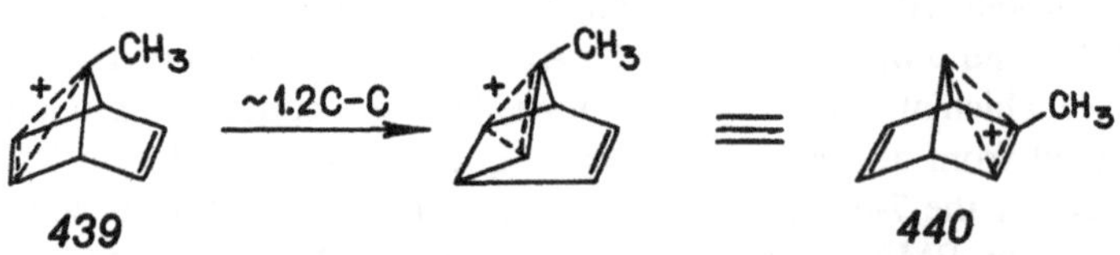

The rate of cycle contraction reaction in ion *444* is far higher than in the unsubstituted ion *236* ($\Delta F^{\neq}$ equals 13.8 and 16.6 kcal/mole, respectively), at the same time the rate of cycle expansion for ion *236* is far higher than for ion *444*. The CH_3 group seems to stabilize the allylic ion more than the nonclassical ion (the latter case does not preclude destabilization).

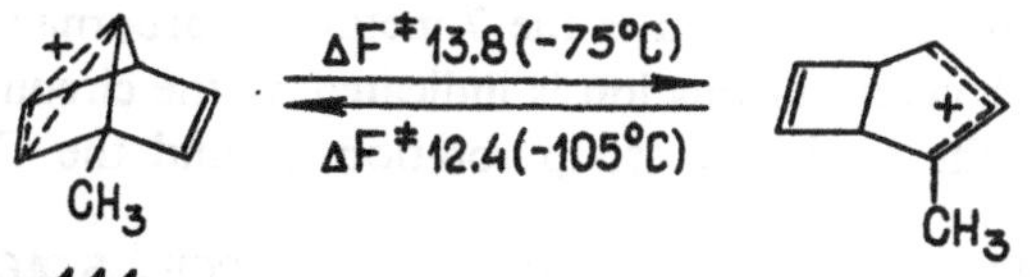

444

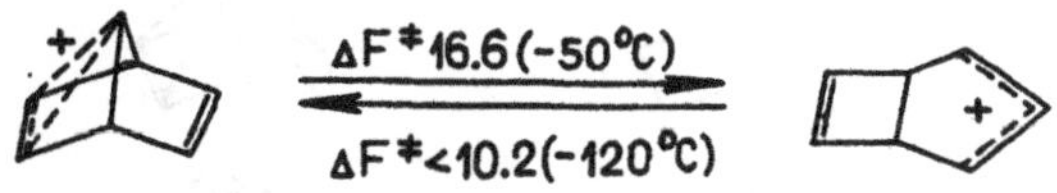

236

The most stable of all the methyl-substituted norbornadienyl cations is the 2-methyl *440*, the least stable the 1-methyl ion *444*. As for the 7- and 5-methyl ions *439* and *445*, their relative stability is still unclear.

For the 7-phenyl- *449* and the 7-methoxy-ions *450* the vinyl protons are even at —100 °C equivalent; this implies that either these ions are really symmetrical or there is a very rapid equilibrium of unsymmetrical ions.

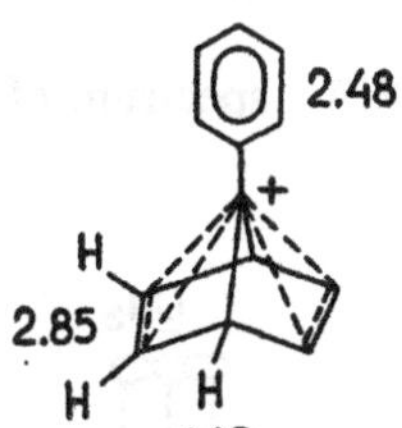

449

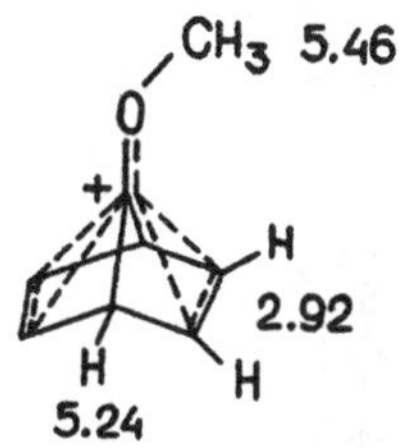

450

In the ion *449* the signal of the aromatic protons is very narrow and is located in a somewhat higher field than for the α-phenylcarbocations (τ 0.4–0.8). This indicates that the phenyl group in ion *449* participates weakly in the delocalization of the positive charge. Since in the symmetrical ion this participation should be significant one can conclude ion *449* to be unsymmetrical. Since the vinyl protons of the 7-phenyl-7-norbornadienyl ion give one signal even at —140 °C the barrier value for the bridge-flipping must amount to no more than 5.2 kcal/mole [595].

Thus, ions *236*, *439* and *449* are unsymmetrical; the barriers for the bridge-flipping are ≧19.6; 12.4 and ≦5.2 kcal/mole.

It should be noted that all the described rearrangements of 7-norbornadienyl ions — 5C-rearrangements, bridge-flipping, 1,2-bond shift — do not, as a rule, compete markedly with ion trapping by the nucleophile under the solvolytic conditions; the latter process is far faster than the above rearrangements.

7.1.4 Influence of the 7-Methoxy Group

From the above data on the dependence between an increase in the stabilizing effect of the substituent at C^7 and a decrease in the barrier to the bridge-flipping it follows that at the limit the symmetrical nonclassical ion will be more stable than the unsymmetrical one. Such an ion is the 7-methoxy-norbornadienyl cation *450* [595]. The nonclassical structure of this ion is indicated by the circumstance that the energetic barriers for the methoxyl group rotation around the C^7—O bond

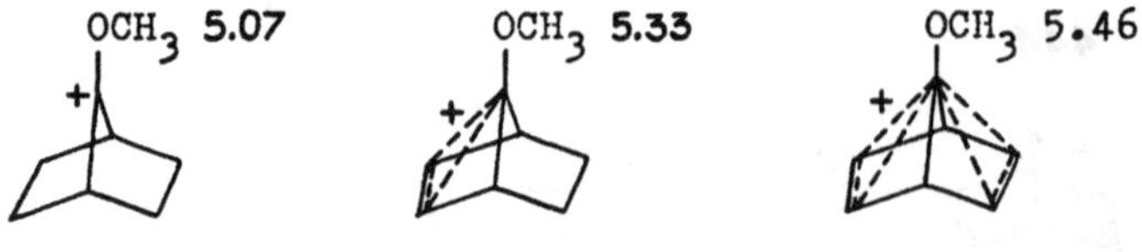

452 *451* *450*

in the ions *450* and *451* are far lower than for ion *452*. If the ion is nonclassical, it can be either symmetrical or unsymmetrical. Just as in the 7-phenyl-7-norbornadienyl ion *449* the olefine protons of ion *450* are chemically equivalent at low temperature and, hence, the spectrum corresponds either to the symmetrical nonclassical ions in rapid equilibrium ("bridge-flipping") or to one unsymmetrical ion. The following arguments favour the symmetrical structure:

1) the OCH_3 group signal of the ion *450* is shifted to a high field in comparison with that of the ion *451*;

2) the character of the olefine proton splitting in the PMR spectrum of the ion *450* is essentially different from that for ion *451*;

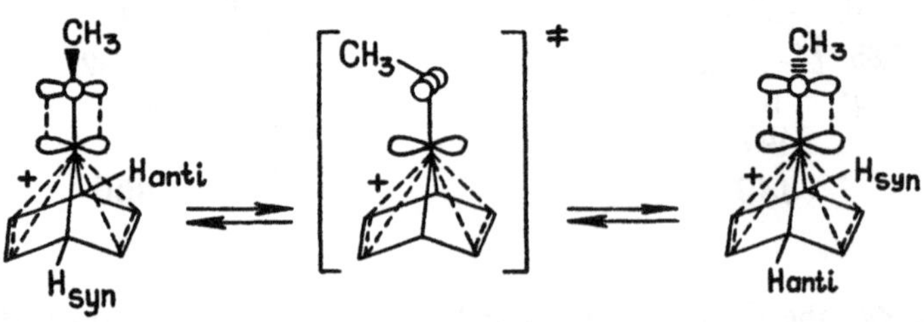

3) if the ion *450* were unsymmetrical, then the double bond would affect but slightly the barrier for syn-anti-isomerization; however, these barriers for ions *450* and *451* are rather different (14.7 and 11.7 kcal/mole);

4) the energetic barriers of conversion of 7-substituted norbornadienyl ions into corresponding tropylium ions decrease consecutively in the series 7-H (23.5 kcal/mole), 7-vinyl (17.1), 7-phenyl (15.9); at the same time for the 7-OCH_3-substituted ion the barrier is very high (24.2). The conversion into a tropylium ion seems to favour the geometry of the unsymmetrical nonclassical ion.

The above-discussed data on the 7-norbornadienyl ion clearly show it to be a "laticyclic system" [413], the homoaromatic stabilization resulting from the geometry distortion. The barrier to the bridge-flipping in the 7-norbornadienyl ion exceeds the value of 19.6 kcal/mole at 46 °C. Consequently, the stabilization of the symmetrical cation is very small. According to Grutzner, the antihomoaromaticity makes the main contribution to the ion destabilization (e.g., in the case of a hypothetical bicyclo[3,2,2]nonatrienyl cation), and the antibicycloaromatic destabilization is of small significance [414] (cf. [596]).

7.1.5 Quantum Calculations

The quantum chemical calculations for the 7-norbornadienyl ion have confirmed its unsymmetrical bishomoaromatic structure. The first calculations for this ion were made as early as 1963 [233], but this was done by using the method LCAO MO in the π-electron approximation; at the same time the nonclassical electron delocalization in the 7-norbornadienyl cation does not obviously have the π-character, but is intermediate between π- and σ-delocalizations. More reliable, therefore, are the results obtained in 1971–1972 by the CNDO/2 and the MINDO/2 method with all the valent electrons taken into account. According to these calculations, the positve charge is mainly concentrated on C^2, C^3, C^7 but not on C^5 and C^6 [597].

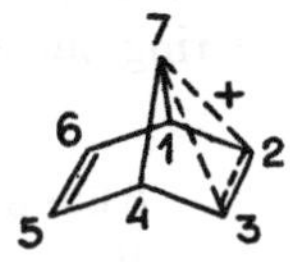

As shown by Dewar's calculations [598], in the process of bridge-flipping the symmetrical structure corresponds to the transition state and not to the intermediate, the energetic barrier amounting to 26 kcal/mole (according to Winstein's experimental data, it is $\geqq$19.6 kcal/mole).

7.2 Monohomoaromatic Systems

All the homoallylic ions so far discussed have been bishomoaromatic. A representative of the system two-electron is the pentamethylcyclobutenyl cation *453* [599], an allylic ion with unusually high 1,3-interaction. Its UV absorption maximum (245 nm) is intermediate between the allylic and cyclopropenyl cations. The calculations show the value of the 1,3-resonance integral for ion *453* to be $\beta_{13} = 0.33\ \beta_0$ [344].

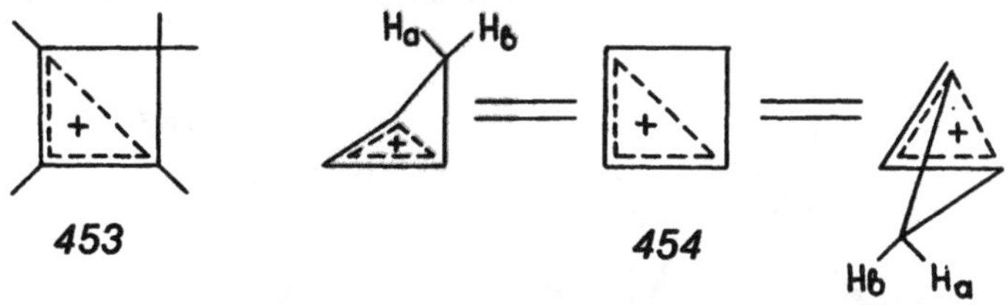

453 *454*

The convincing corroboration of Winstein's correct concept as to the monohomoaromatic structure of this type of ions are the data of the ^{13}C NMR spectra, in particular for the unsubstituted cyclobutenyl cation *454* [600,601]. In the latter case the signals of C^1 and C^3 resonate in a higher field than that of C^2 does contrary to what is observed in the spectrum of cycloalkenylic ions, but in full accord with the concept of 1,3-orbital interaction. The PMR spectra showed a reversible temperature dependence of the methyl proton signal shape which was atrributed to the nonplanar ion inversion with the intermediate formation of a planar cation. The barrier value of this inversion determined experimentally ($\Delta G^{\neq} = 8.4$ kcal/mole) agrees well with theoretical calculations [348]. The constants $^1J\,(^{13}C{-}^{1}H)$ for the allylic fragment are unusually large — 211.2 cps for $C^{1(3)}$ and 236.3 cps for C^2 which is close to the data for the cyclopropenyl ion [602]. .

7.2.1 Monohomotropylium Ions

The monohomotropylium ion *455* was the first 6-electron homoaromatic system to be described [603]. The research made has fully corroborated the delocalized nature of this ion [343,344]. Ion *455* was obtained by protonating cyclooctatetraene in conc. H_2SO_4 [604]; it can be isolated as solid salt $C_8H_9^+SbCl_6^-$ by treating the hydrocarbon with HCl and $SbCl_5$ [603].

In the PMR spectrum of ion *455* the chemical shifts show a great difference ($\Delta\delta = 5.8$ ppm) of methylene protons H_a (deshielded by the ring current) and H_b

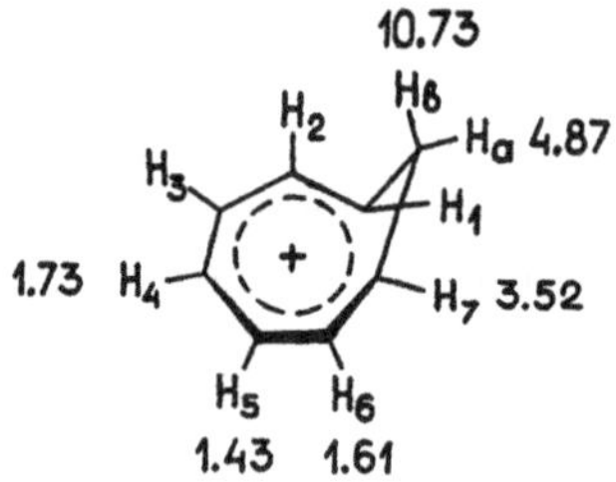

455

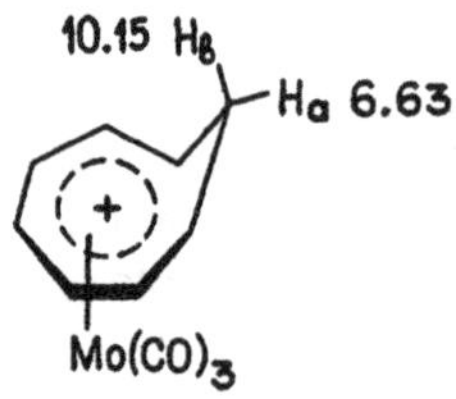

456

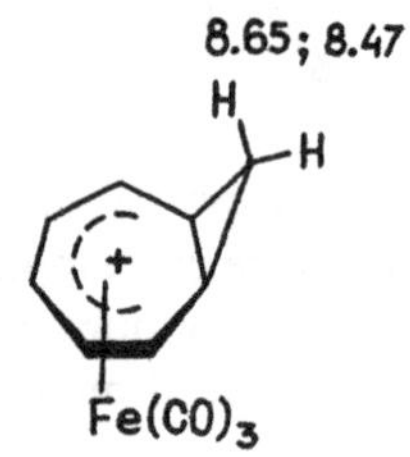

457

(shielded). The spectra of the free ion *455* and its complex with Mo $(CO)_3$ are very similar but they are different from that of the complex with $Fe(CO)_3$ *457*[11]. In the latter case the chemical shifts of the signals of H_a and H_b have practically the same value, and the ion in this complex has a classical structure *457* suggested by Wilkinson [605]. As far as the free ion *455* goes, the planar cyclooctatrienylic structure *458* must be exluded because its methylene protons are magnetically equivalent; besides, this structure, according to Winstein, is by at least 22.3 kcal/mole less stable than *455* [606].

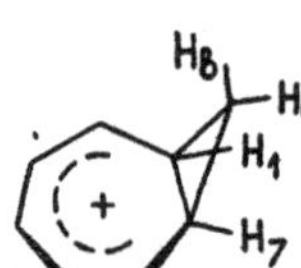

459

458

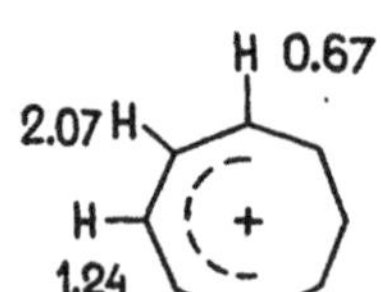

460

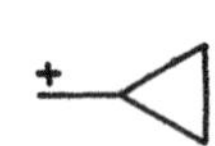

461

Deno held ion *455* to have the classical structure *459* with the "normal" cyclopropyl conjugation [48]. For such a structure one excepts the resonance of H^1 and H^7 at τ 7 or higher to be $J_{a,b} = 4.5 \div 5$ cps, $J_{a,1\,(cis)} = 8$ cps and $J_{b,1\,(trans)} = 4\text{–}5$ cps, on analogy with the known α-cyclopropylcarbinyl ions. But in reality

---

11 Molybdenum tricarbonyl is known to form complexes preferably with 6π-electron systems, ferrotricarbonyl complexes with 4π-electron ones.

the protons H^1 and H^7 of the ion are greatly deshieleded ($\tau \sim 3.6$) and $J_{b,1}$ (10 cps) is larger than the cis-constant $J_{a,1}$ (7.5 cps). On the other hand, the PMR spectrum of complex *457* fully corresponds to the structure of a type in which the cyclopropane electrons do not form part of the delocalized electron system. The charge distribution on five sp^2-hybridized atoms C^2–C^6 in ion *455* does not correspond, either, to what could be expected for the pentadienyl cation *459*; this was clearly revealed by comparing the spectrum of ion *455* with that of the cyclooctadienyl cation *460*. The chemical shifts of the H^2 and H^6 protons of ion *455* are similar to those of the tropylium ion protons. Another indication of the sharp difference between the usual α-cyclopropylcarbinyl ions from ion *455* is the comparison of the constant $J_{^{13}C-^{1}H}$ (CH_2) = 180 cps in the spectrum of ion *461* and the mean value of $J_{^{13}C-^{1}H}$ (CH_2) = 159 cps for ion *455*.

A direct proof of an aromatic ring current in ion *455* are Dauben's data on voluminal diamagnetic susceptibility. The exaltation value for ion *455* proved to be the same as that for the tropylium ion [607,608]. The presence of the ring current in ion *455* is also evidenced by the ^{13}C NMR spectral data [609].

As shown by calculation [344] the 1,7-resonance integral for ion *455* $\beta_{1,7}$ equals 0.73 β_0, while the calculated orders for the bonds 1,2-, 2,3- and 3,4-amount, respectively, to 0.69; 0.62 and 0.65 which is close to the value 0.64 for the bonds of the tropylium ion.

The essential difference between a homoallylic ion of type *455* and an α-cyclopropylcarbinyl ion of type *461* is that in the latter case the electron delocalization involves two bonds of the cyclopropyl group while in the former, only one. The available data indicate that the conversion of a bisected ion of type *461* into a homoallylic ion requires large energy expenditure and proceeds with difficulty; this is also corroborated by quantum chemical calculations. But in the ion *455* these losses are compensated by the energy gain resulting from the formation of a homoaromatic system.

The ab initio calculations made by Haddon [610] for the homotropylium *455* and the cyclooctatrienyl *458* ions confirm the homoaromatic character of the former. The nonplanarity, the shortening of the bond between the polyenyl and bridged fragments, the equal bond lengths in the polyenyl segment, the lengthening of homoconjugated bonds — all these results agree with the above conclusion.

The monohomotropylium ions are formed in quite a number of reactions:

a) protonation of the respective molecules, e.g. [611,612]:

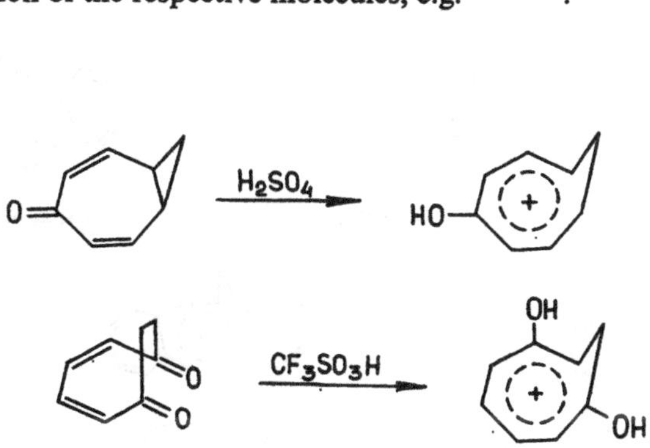

b) interaction of cyclooctatetraene with electrophiles other than a proton, e.g. [613]:

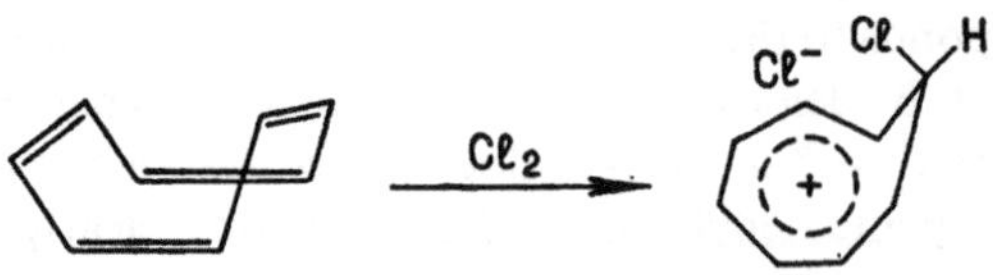

c) electrophile attack on cyclic precursors, e.g. [614]:

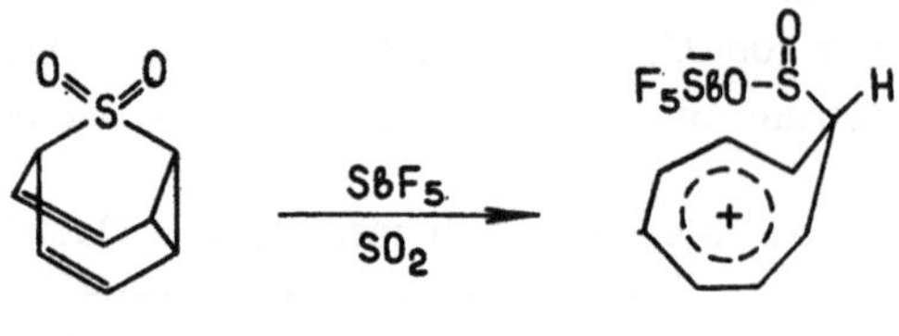

d) solvolysis of respective substrates, e.g. [615]:

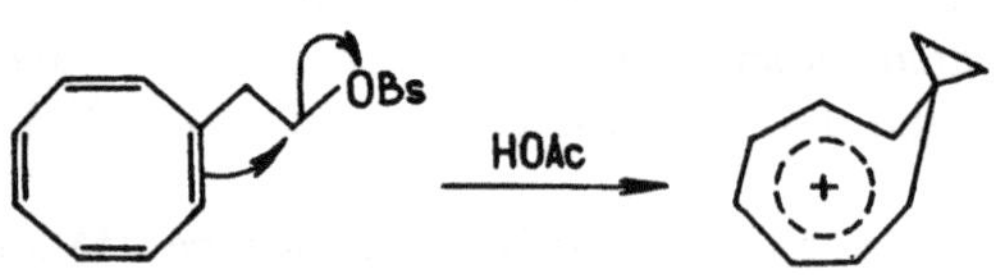

The protonation of cyclooctatetraene with D_2SO_4 at -15 °C resulted in the exo-endo-inversion of the ring, seemingly with the intermediate formation of a planar cation *462* (ΔG 22.3 kcal/mole) [606]:

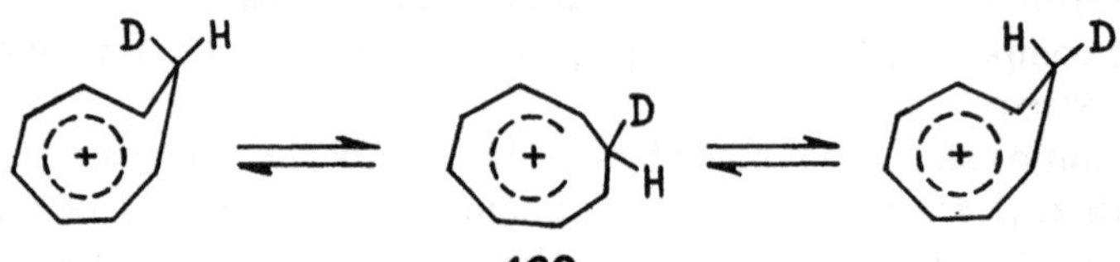

462

The generation of various substituted homotropylium ions is described [611,613, 616–623], among them that of monobenzohomotropylium [624] and dibenzohomo-

$\Delta\delta = 5.8$

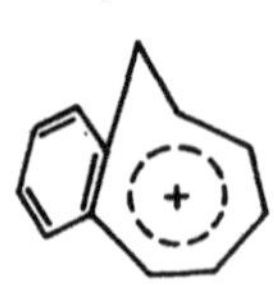

$\Delta\delta = 3.90$

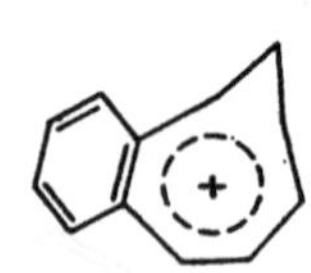

$\Delta\delta = 5.34$

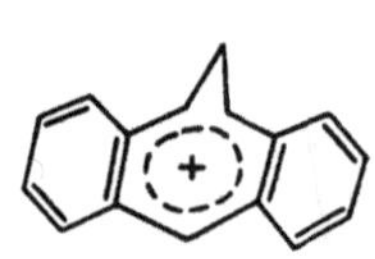

$\Delta\delta = 4.67$

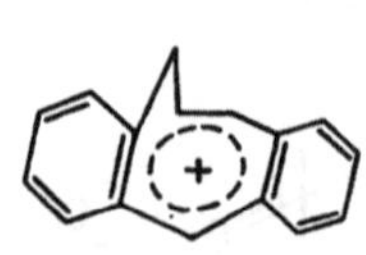

$\Delta\delta = 3.20$

tropylium [625–627]. They all display similar properties of the systems with the cyclic delocalization of 6 π-electrons. In benzo- and dibenzohomotropylium ions there is a decrease in the value of Δδ for the signals of methylene protons in the PMR spectra in comparison with ion *455* (see p. 198).
This effect is particularly significant when the benzene ring is located in the vicinity of the bridge.

Brookhart generated the 1-methoxyhomotropylium ion *463* and studied the ring inversion in this ion (ΔF$^{\neq}$ = 19.6 kcal/mole, −18 °C) proceeding obviously via a planar ion *464* and its thermal rearrangement (ΔF$^{\neq}$ = 25.7 kcal/mole, +51 °C) with the formation of a phenylmethylmethoxycarbonium ion *465* [628].

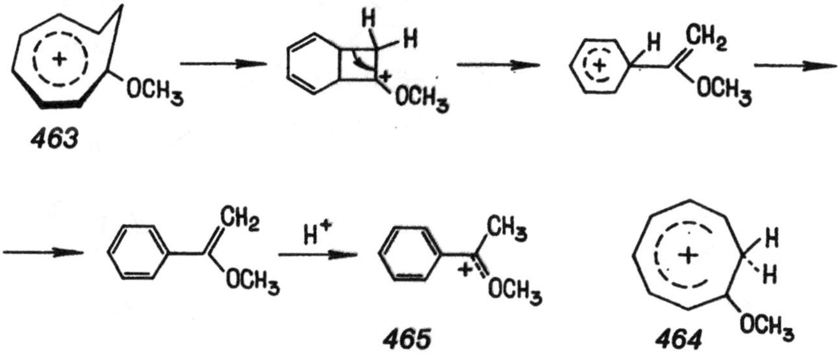

7.2.2 Bishomotropylium Ions

While only one type of monohomotropylium ion is possible, in the bishomotropylium ion, depending on the relative position of the two bridges the three types are 1,4-, 1,3- and 1,2-bishomotropylium ion [629]:

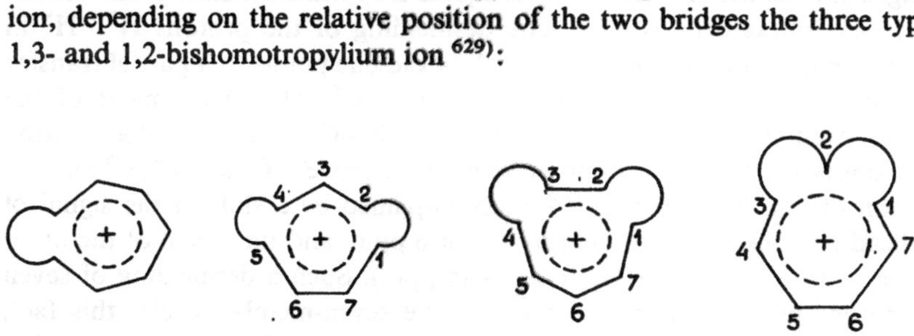

The bicyclo[4,3,0]nonatrienyl cation [466] — a representative of 1,4-bishomotropylium ions — was first observed in 1970 [622, 630]. Ion *466* was obtained from bicyclo[3,2,2]nona-3,6,8-triene-2-ol *467*, the primary product being the 9-barbaralyl-cation *468* formed at −135 °C and then, at −125 °C converted into ion *466* (ΔF$^{\neq}$ = 10.4 kcal/mole). It was also generated from alcohols *469* and *470* [629, 631]. The monomethyl analogue *466a* was obtained from alcohols *467a* and *469a* [629] as well as by the protonation of olefines *471–473* [632]:

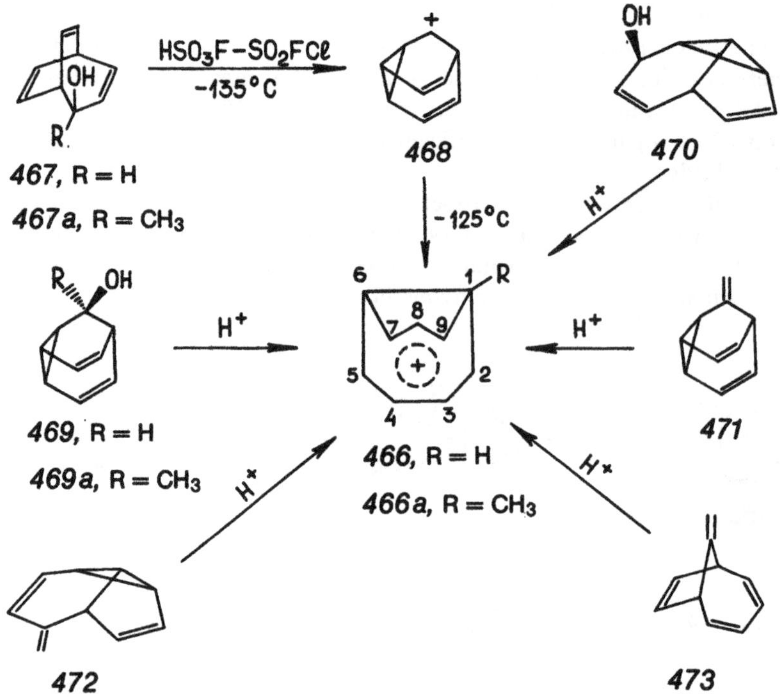

The proof of the homoaromatic character of ion *466* as well as of ion *466a* is based on its NMR spectra. In the PMR spectrum the signals of the H^7—H^9 protons are markedly shifted to a high field as compared with those of the protons H^3 (τ 1.54) and H^4 (τ 1.10) in the spectrum of cyclopentenylic cation. This indicates that the atoms C^7—C^9 are not part of an isolated allylic system.

The averaged shift of the H^2—H^5 protons equals 3.04 while the shift of the olefine protons in 1,3-cyclohexadiene is 4.22. The deshielding of the protons H^2—H^5 in ion *466* in comparison with H^1—H^4 in 1,3-cyclohexadiene points to a partial transfer of the positive charge in ion *466* to the atoms C^2—C^5. The involvement of the orbitals of the seven carbon atoms — C^7—C^9 and C^2—C^5 — in the delocalization of the π-electrons is a consequence of homointeraction between C^2 and C^9, C^5 and C^7. The conversion of ion *468* into ion *466* is accompanied by a shift of the signal ot two bridge-head protons to a high field by about 6 ppm, and the signal of the other seven protons — to a low field (totally by ~16 ppm). Such a deshielding of seven peripheral protons points to a ring current in the seven-member cycle; this fact, as well as the charge delocalization point to the homoaromatic character of cation *466*.

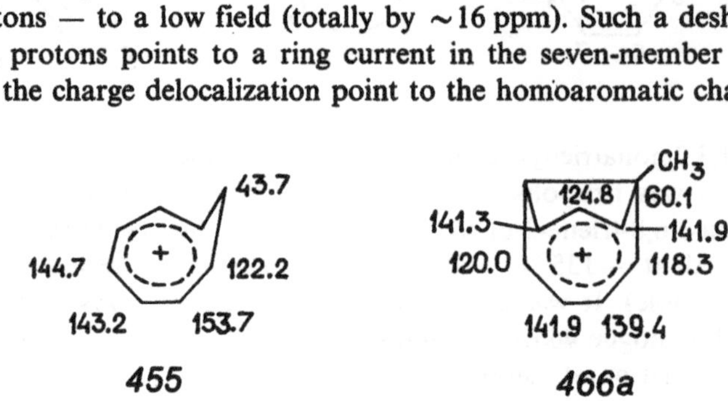

The comparison of the ^{13}C NMR spectral data of the monohomotropylium ion *455* and cation *466a* indicates a lower extent of positive charge delocalization on the seven basal carbons in the latter case (see p. 200).

So far, no one has succeeded in obtaining any kinetic data testifying to the intermediate formation of the bishomotropylium ion *466* in solvolytic media [629,631,633–636].

While ions *466a* and *466* decompose at —50 °C and —90 °C, respectively, the 1,4-bishomotropylium ion *474* in stable even at room temperature [629,637,638]:

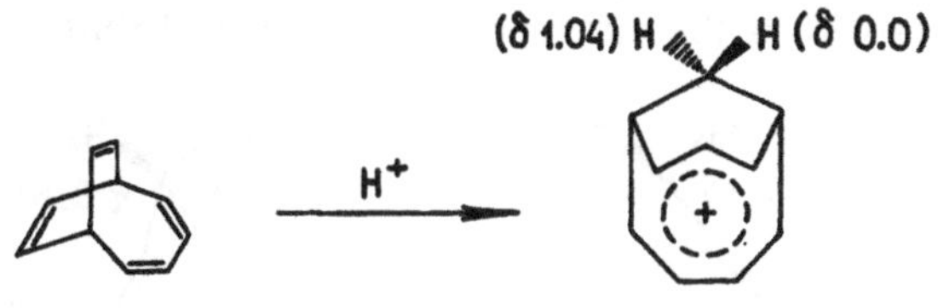

474

Its ring current manifests itself in the PMR spectrum by a shielding effect (~1 ppm) on the signals of bridge methylene protons.

The exo isomer *475* solvolyzes 3000 times as fast as compound *367*, this is a kinetic proof of the intermediate formation of ion *474* in solvolytic reactions [639]:

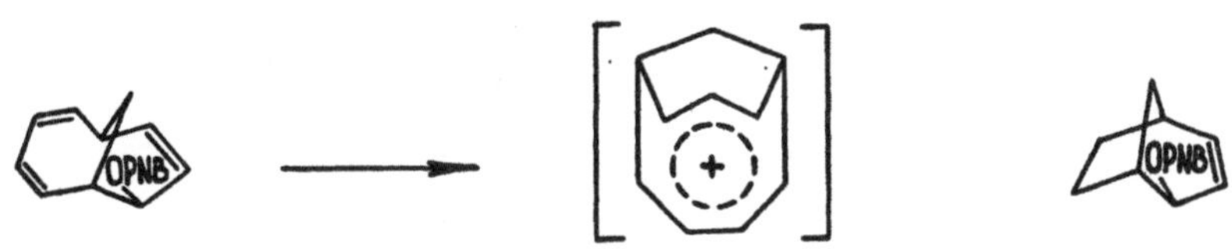

475 *474* *367*

At the same time the solvolysis rates of the endo epimers of the above compounds are sufficiently close.

The literature describes the generation other 1,4- and 1,3-bishomotropylium ions in superacids [637,638,640–643]. The 1,3-bishomotropylium ion *476* having no binding bridge is generated by the protonation of cis-bicyclo[6,1,0]nonatriene [640]:

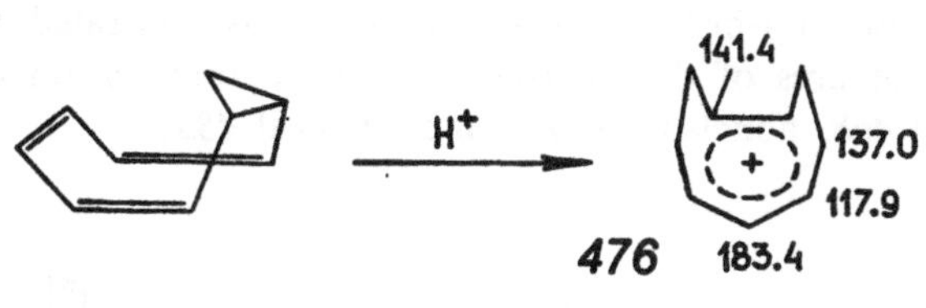

476

As seen from the ^{13}C NMR spectral data, the ring current effect in ion *476* is weaker than in monohomotropylium ions. Bishomotropylium ions exist possibly in two geometric forms — cis and trans.

In 1972 Childs[641)] investigated the protonation of cis *477* and trans-4,5-benzo-2,3:6,7-bishomotropones *478*. If the difference in the chemical shifts of methylene

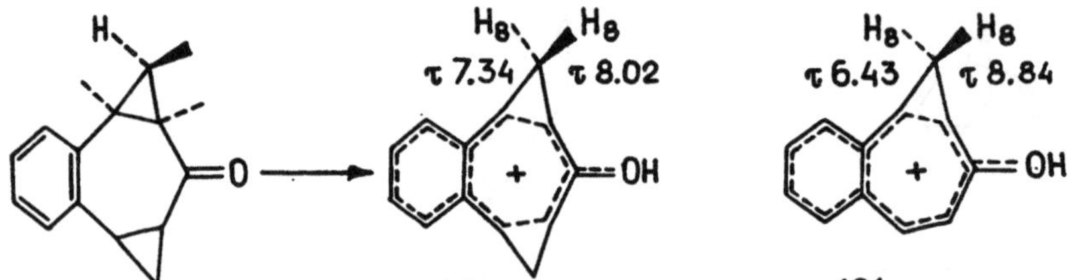

477 *479* *481*

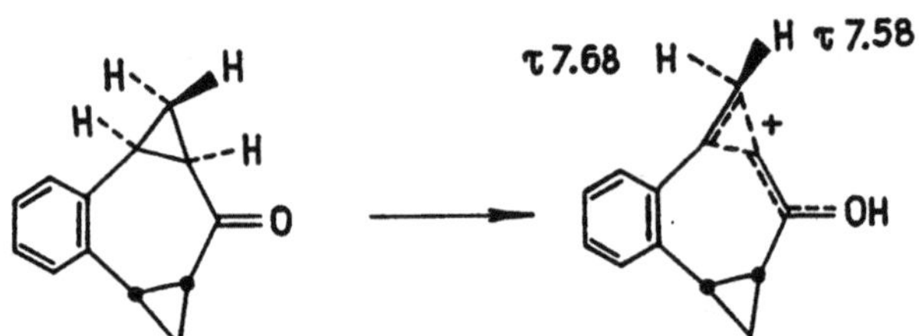

478 *480*

protons is used as a homoaromacity criterion, then ion *480* ($\Delta\delta = 0.1$ ppm) should not be considered as bishomoaromatic. In ion *481*, on the contrary, $\Delta\delta$ equals 0.7 ppm; since the close similarity of the "internal" protons leads to their Van der Waals deshielding (~1.5 ppm, see [644)]), the actual difference in the chemical shifts of the CH_2-group protons is $1.5 + 0.7 = 2.2$ ppm, quite comparable with the data for the benzomonohomotropylium ion *479*. These data show the importance of the relative arrangement of the methylene bridges and of the π-system in homoaromatic compounds.

In a number of cases 1,4-bishomotropylium ions was generated to elucidate the three possible structures of the intermediate 9-barbaralyl cation *468* [645–647)]: α-cyclopropylcarbinyl *468*, nonclassical *482* and classical *483*:

468 *482* *483*

Recently Ahlberg et al.[648] have analyzed the NMR spectra of the 9-barbaralyl cation labelled by deuterium (8 atoms) and ^{13}C (1 atom) and have chosen structure *468*.

7.3 Antihomoaromatic Systems

It is interesting to compare the properties of the above homotropylium ions, in which the breaking of the cyclopropane ring results in a homoaromatic system, with those of the bicyclo[3,1,0]hexenyl cations; in these a similar breaking would result in a 4 π-electron antihomoaromatic system similar to the cyclopentadienyl cation (cf.[649]).

As far back as 1960 Winstein[650] and L. de Vries[651] noted a high anchimeric acceleration (10^{10}) in the solvolysis of pentamethylcyclopentadienylcarbinol tosylate *484* as compared with neopentyl tosylate chosen as a model convenient for comparison. The kinetically controllable products in this reaction were olefine *485* and

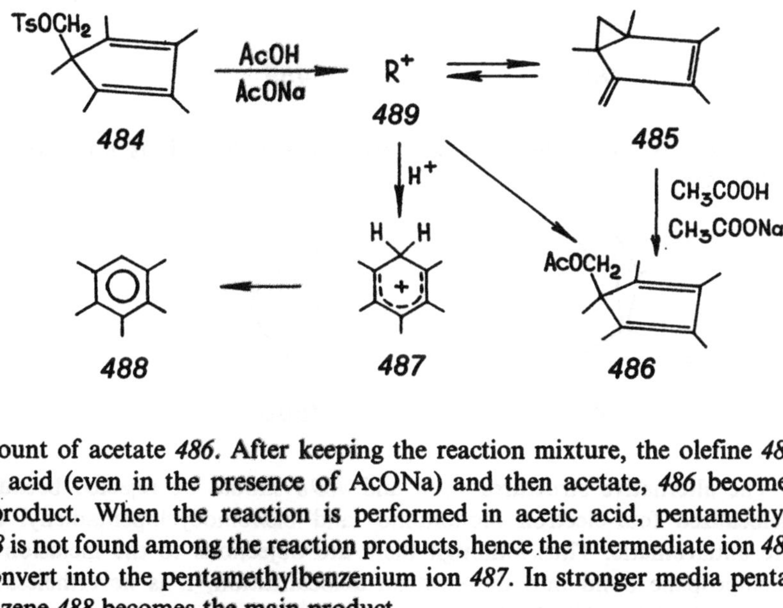

a small amount of acetate *486*. After keeping the reaction mixture, the olefine *485* adds acetic acid (even in the presence of AcONa) and then acetate, *486* becomes the main product. When the reaction is performed in acetic acid, pentamethylbenzene *488* is not found among the reaction products, hence the intermediate ion *489* does not convert into the pentamethylbenzenium ion *487*. In stronger media pentamenthylbenzene *488* becomes the main product.

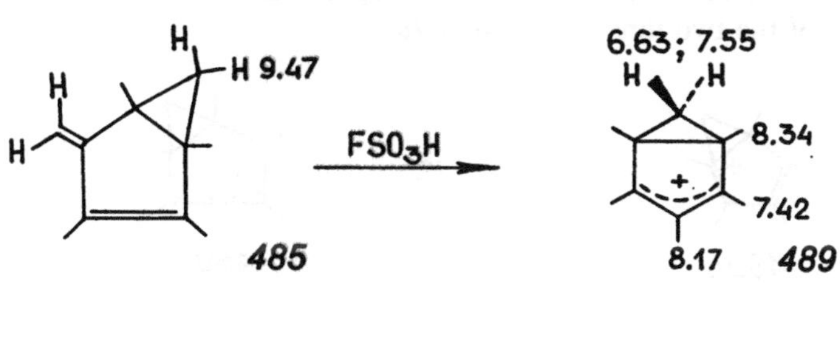

In 1968 Winstein generated the pentamethylbicyclo[3,1,0]hexenyl cation *489* from olefine *485* [652]. The cited chemical shifts (τ scale) show the methylene cyclopropane hydrogens of ion *489* to be deshielded in comparison with the respective protons of the parent olefine by 2–3 ppm. This points to essential participation of the cyclopropane fragment in positive charge delocalization.

Making use of Grutzner's approximation [653] Winstein calculated the shifts of cyclopropane methylene hydrogens to a low field as resulting from the direct field effect of the positive charge. The deshielding of the "external" proton was estimated about 1 ppm and that of the "internal" one-about 0.25 ppm. As the inductive effect through two C—C bonds is very small the remaining deshielding of ~1.5 ppm may be only due to the participation of the cyclopropane fragment in the charge delocalization.

The PMR spectrum of ion *489* can be accounted for either on the basis of the rapid equilibrium of two equivalent homoallylic (polyhomopentadienyl) ions *489* or by accepting the symmetrical structure *490*.

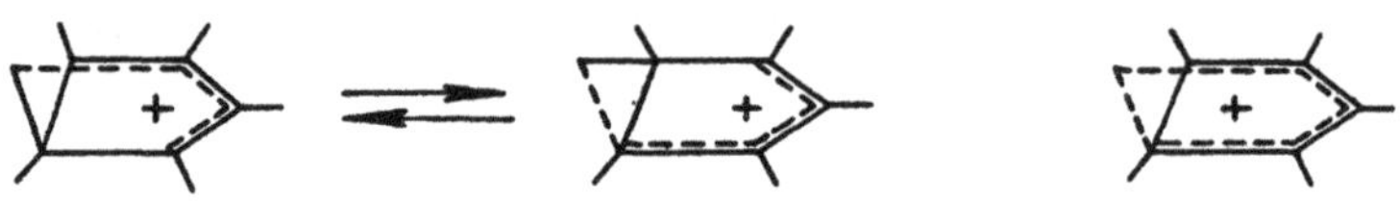

489 *489* *490*

If there is an equilibrium of equivalent structure, it is not "frozen" even at —120 °C.

489a *489b*

The alternative structures *489a* and *489b* should be rejected because they are delocalized four-electron monohomo- and bishomocyclopentadienyl cations and are antiaromatic. Thus although there are objections to the inclusion of the 1,5-cyclopropane bond into the electron delocalization, it is nevertheless possible to include the 1,6 and/or 5,6 bonds into this delocalization.

According to Story [358], Winstein's PMR spectrum of ion *489* are also met by the equilibrium of the two ions *489c* and *489d*.

489c *489d*

Described subsequently were heptamethyl-[654,655], hexamethyl-[655,656], pentamethyl-[657], tetramethyl-[657,658] and hydroxy-substituted 2-bicyclo[3,1,0]hexenylic cations[659]. In 1971 Berson[660] generated the "unsubstituted" cation *491*. In this ion the difference in the shifts of the methylene protons $\delta_{endo} - \delta_{exo}$ turned out to be +0.30 ppm which is sharply different from $\Delta\delta$ for the homotropylium ion *455*

455 *491*

(—5.8 ppm). This is due to the fact that the endo proton in ion *455* is located above the homoaromatic 6 π-electron system, and in ion *491* — over the 4 π-electron one which (to the degree to which the electron bonds $C^1—C^5$ are delocalized) is antihomoaromatic. Also, both protons of the CH_2 group in ion *491* resonate in a far lower field than in the pentamethyl analogue *489*; the methyl groups in a 5-member cycle "remove" part of the charge and thereby cause a shift of the H^6 protons to a high field. Berson also noted a sharp difference in the values of geminal constants [$J_{8,8}$ = 7.2 cps in *455*, $J_{6,6}$ = 2 cps in *491*], as well as great differences in the chemical shifts of the allylic protons 2 (4) and 3 in ion *491* (τ 0.03 and 2.51) and, vice versa, the proximity of those of protons 2 (6), 3 (5), and 4 in ion *455* (τ 1.61; 1.43 and 1.73). All this confirms the idea that the difference between ions *491* and *455* consists in precisely which bonds of the cyclopropane ring participate in the positive charge delocalization.

The bicyclo[3,1,0]hexenyl cations undergo a degenerate rearrangement "permitted" by the Woodward-Hoffmann rules — the motion of the C^6 fragment along the periphery of the cyclic polyene, — the configuration of 6-endo- and 6-exo-substituents being retained at each stage ($\Delta F^{\neq}$ = 15 kcal/mole for ion *491*). Bicyclo-[2,1,1]hexenyl cations were found not to be intermediates in this rearrangement; they would represent a so-called suprafacial 1,4-sigmatropics shift. The rearrangement proceeds according to the "least motion" principle, with a minimum change of atom coordinates[652,654–661].

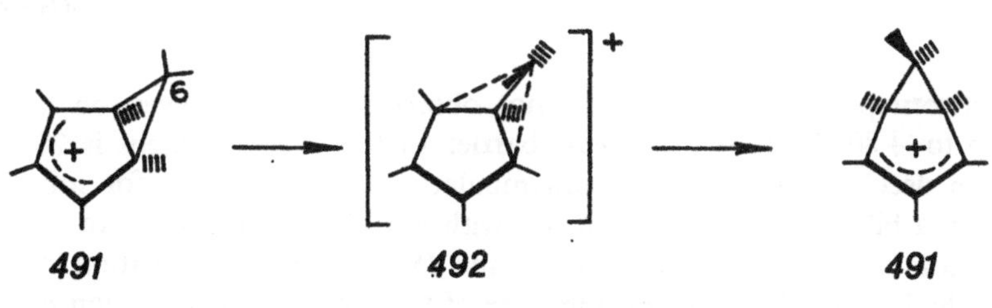

491 *492* *491*

As shown by Childs, the rate of degenerate rearrangement increases on introduction to C^6 of substituents stabilizing the positive charge. In this case the transition state can be presented as cyclopentadiene-5-methyl cation *492* in which both lobes of the formally vacant-orbital at C^6 interact with the diene. Such an interaction will be hindered by steric factors if C^6 has a voluminous substituent at the endo position; indeed, the value of $\Delta F^{\neq}$ for the degenerate rearrangement of ion *493* is 12.0 kcal/mole while for ion *494* it is 17.4 kcal/mole [655,661].

493

494

In ion *455* a similar permitted rearrangement (suprafacial 1,6-sigmatropic shift) only occurs with the configuration reversion of 8-substituents at each stage. Berson [662] attempted to detect this rearrangement by utilizing the homotropylium-4-d ion. If the sigmatropic shift did take place, the deuterium would finally be evenly distributed

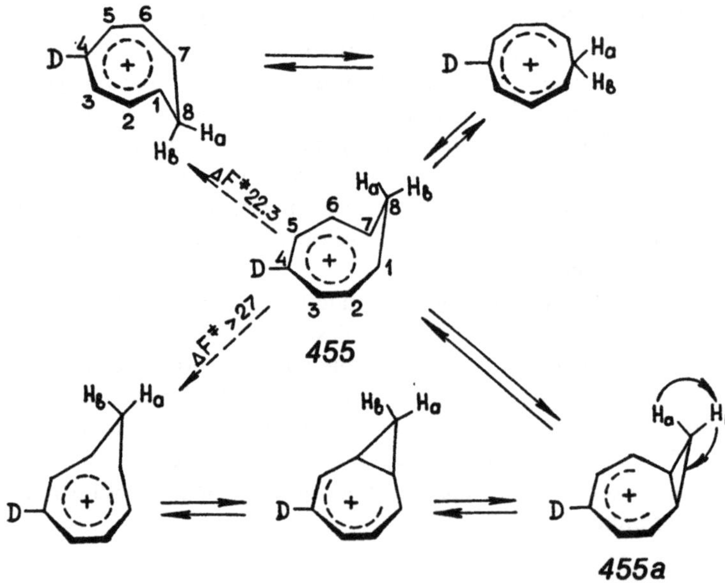

455a

between positions 1–7. This is not observed, however, even on heating the ion up to +80 °C; thus the lowest barrier to this rearrangement is 26–27 kcal/mole. The deceleration of the circumambulatory rearrangement for ion *455* by more than a billion times in comparison with ion *491* is due, firstly, to the disadvantage of homoaromaticity disturbance — the *455*→*455a* transformation.[12] Secondly, due to the fact that in the transition state of the rearrangement for ion *455a* in twisting the migrating C atom the orbital overlap cannot be effective.

12 This conclusion is supported by ab initio calculations [663].

The relative significance of these factors is not yet clear. As for the exchange of 8-exo and 8-endo hydrogens in ion *455*, it proceeds by conformational isomerization ($\Delta F^{\neq} = 22.3$ kcal/mole) [606].

Hehre [663] predicted by quantum chemical calculations that in the 8,8-dimethylhomotropylium ion, contrary to the unsubstituted ion *455*, the circumambulatory rearrangement is feasible. In 1978 Childs succeeded in generating 8,8-dimethyl-substituted ions and made the bridged atom migrate in them [664, cf. 665]:

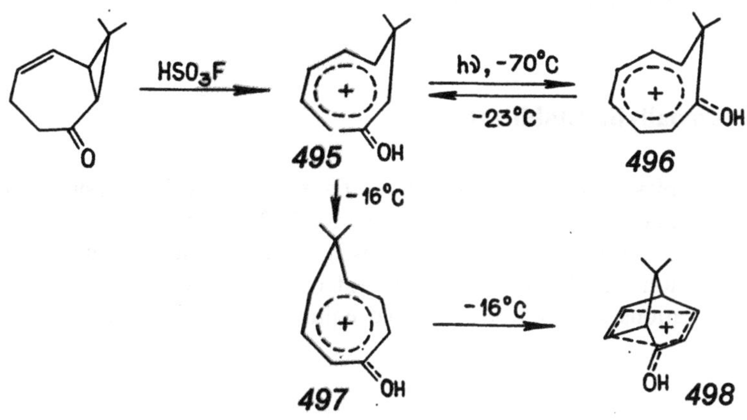

495 *496* *497* *498*

Ion *498* is stable under the reaction conditions and cannot be an intermediate in the *496*→*495* and *495*→*497* rearrangement; these rearrangements are assumed to occur through the migration of C^8 along the ion periphery. The ion *498* can be classified as "antibishomoaromatic" [534]; it proves to be more stable than the homoaromatic ions *495*–*497*.

It is interesting to compare the data on the ^{13}C NMR spectra of ions *491* and *455*. Firstly, note a great difference in the chemical shifts of methylene carbons ($\Delta\delta$ 73.5 ppm). Secondly, the "nodal" carbons C^1 and C^5 in ion *491* are far less deshielded than the respective carbons C^1 and C^7 in ion *455*.

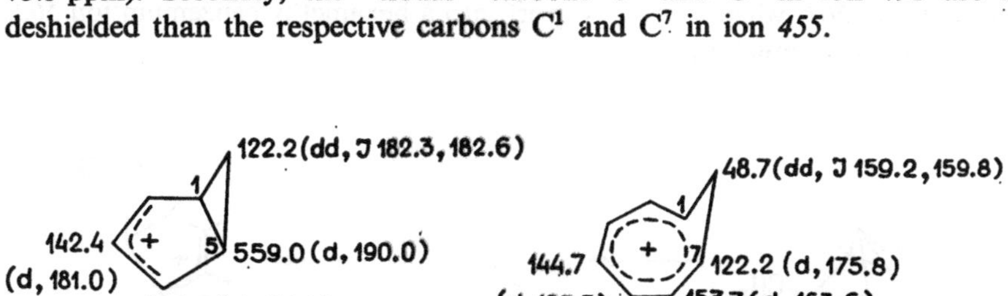

491 *455*

These data show that 1) in the ion *455* a 6 π-electron homoaromatic system is formed, and 2) the C^1—C^5 bond in ion *491* does not take any essential part in the positive charge delocalization, but a considerable part of the charge is transferred to the methylene group C^6. In the ^{13}C NMR spectrum of ion *491* the signal of the methylene carbon C^6 is located 60 ppm lower than that of the other "cyclopropane" carbons C^1 (C^5).

The opposite is valid in the ^{13}C NMR spectrum of ion *455*: the signal of the methylene carbon C^8 is 80 ppm higher than that of C^1 (C^7). The constants $^1J_{^{13}C-^1H}$ for the cyclopropane carbons of ion *491* are essentially larger than their counterparts for ion *455*. Thus, ion *491* fits better to a cyclopentenyl ion with a significant delocalization of the positive charge into the condensed cyclopropane ring with a sufficiently strong C^1—C^5 bond.

7.4 Homoallylic Stabilization

While the bicyclo[3,1,0]hexenyl cations have the potential possibility to be "antihomoaromatic" particles, the isomeric bicyclo[2,1,1]hexenyl cations related to the 7-norbornenyl ions, can be stabilized by homoallylic interaction.

Protonation of Dewar's benzene under various conditions [667,668] yields two isomeric cations *499* and *500* in rapid equilibrium. Their spectra differ from that

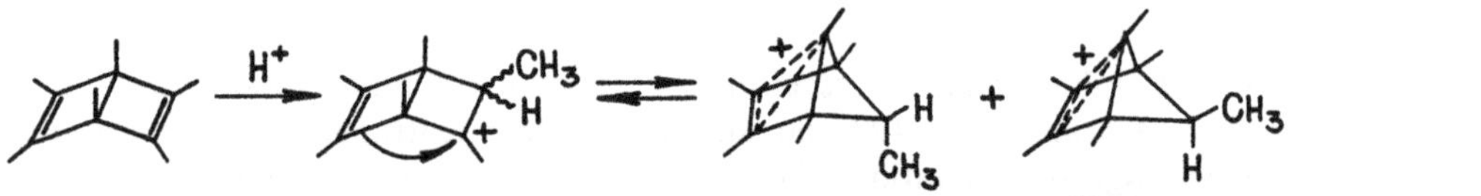

499 500

of the isomeric hexamethylbicyclo[3,1,0]hexenyl cation *501*, but agree well with the formulae of the stereoisomeric bicyclo[2,1,1]hexenyl cations *499* and *500*. Both these ions and ion *501*, when heated, convert into a hexamethylbenzenium ion *502*, but ion *501* rearranges much faster.

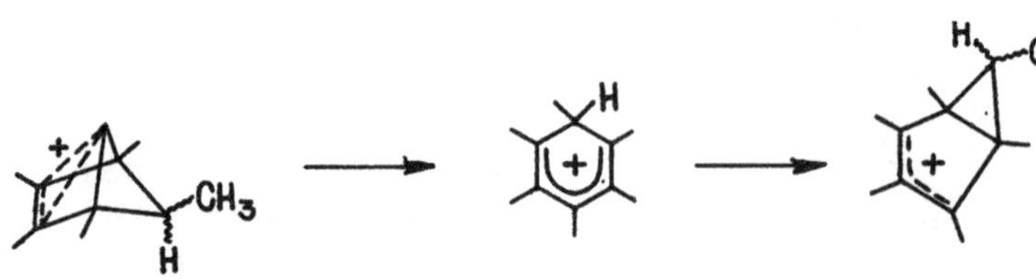

499, 500 502 501

If ions *499* and *500* are generated by treating Dewar's benzene with two equivalents of HCl at −80 °C, then on heating they are further converted according to the following scheme [669]:

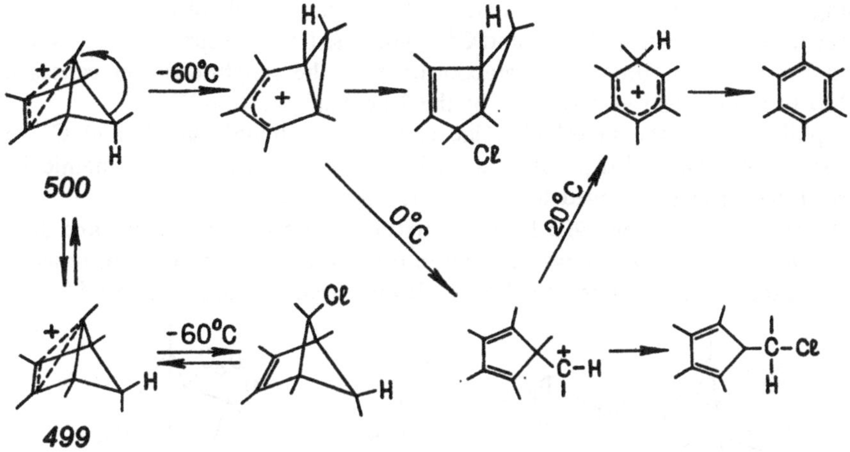

500

499

Masamune[670] has described the PMR spectrum of the unsubstituted bicyclo-[2,1,1]hexenyl cation *503*. It is compared with the PMR spectra of the 7-norbornenyl *19* and 2-norbornyl ions *5*.

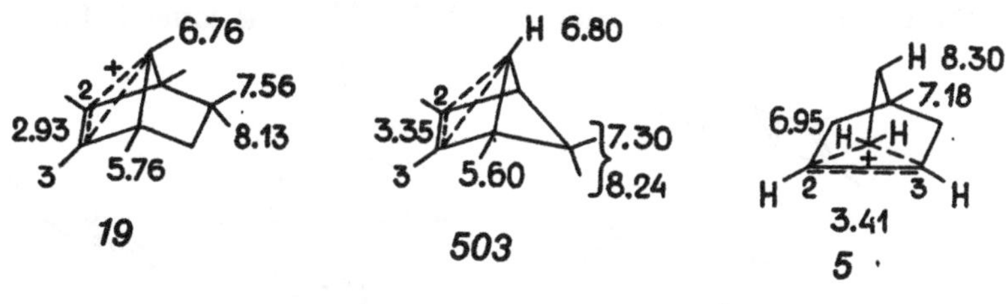

19

503

5

The chemical shifts of the "olefine" protons H^2 and H^3 and the protons at the bridged carbon are similar. Their similarity is due to the electron structure of these nonclassical bishomocyclopropenyl ions.

Hogeveen and Kwant[671] have studied the structures of 5-substituted 1,2,3,4,5,6-hexamethylbicyclo[2,1,1]hexenyl cations by PMR and ^{13}C NMR spectra. The data obtained support the nonclassical structure *B* superior to the classical structure *A* and the rapid equilibrium of the classical ions *C*.

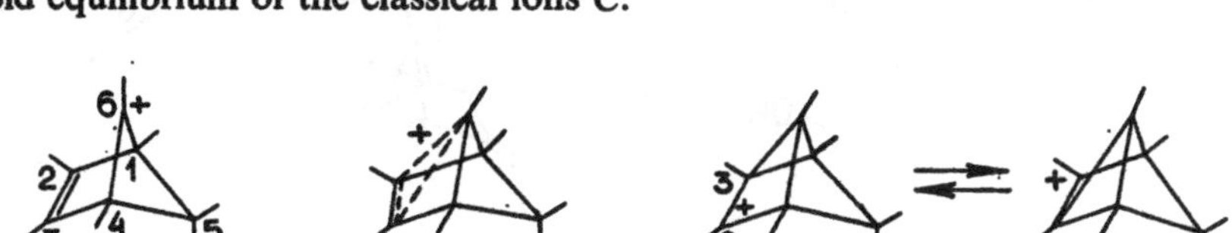

(A) (B) (C)

In the ^{13}C NMR spectra of bicyclo[2,1,1]hexenyl cations the signals of the carbon atoms C^2, C^3, C^6 are in a higher field than the similar signals in the bicyclic precursors. A similar comparison of the ^{13}C NMR spectra of the ion and its tricyclic precursor shows that the signals of the former do turn out to be in a lower field.

According to [671], the shift to a high field of the carbon signals of the nonclassical ions is due to rehybridization: the C^2 and C^3 atoms pass from the sp^2- to the sp^3-state, and then the C^6 atom becomes pentacoordinated; the rehybridization causes a signal shift to a high field as observed for the 2-norbornyl cation [184].

Nucleophiles attack the bicyclo[2,1,1]hexenyl cations mainly at C^2 and C^3 so that the kinetically controlled products are tricyclohexanes; but thermodynamically the more stable products proved to be bicyclic compounds.

With triethylamine the ion *499a* forms a mixture of tricyclohexanes *504* and *505*: their deuterium distribution shows the proton (deuterium) to detach in an unusual way — from the methyl groups added to the bridge-head carbons C^1 and C^4 [672].

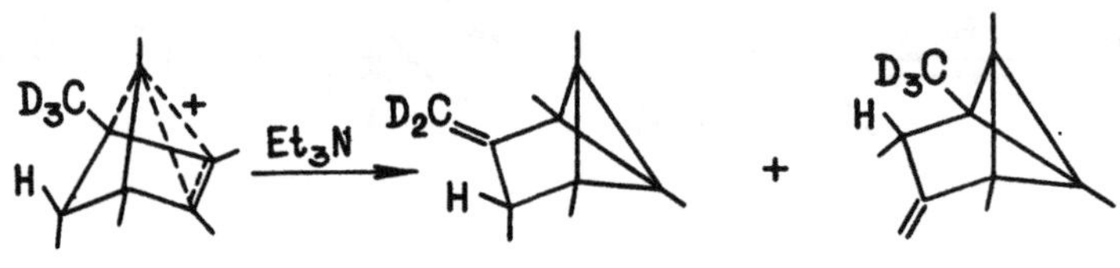

499a *504* *505*

Hogeveen and Kwant have also converted 5-substituted bicyclo[2,1,1]hexenyl cations, as well as some products of their "quenching", in a superacid medium into the unusually stable dication *506* [18, 673–675]. The latter is formally regarded

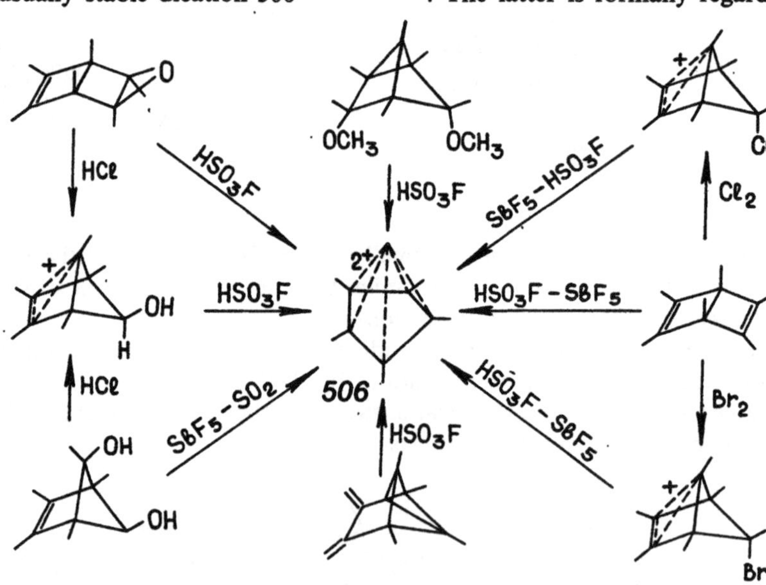

as the result of the interaction of the sp-hybridized: $\overset{+}{C}$—(CH_3) fragment and the pentamethylcyclopentadienyl cation.

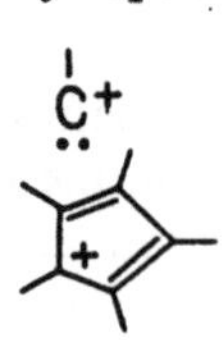

7.5 Pyramidal Mono- and Dications

The exceptional thermal stability of the dication *506* expresses itself by the PMR spectrum (in $FSO_3H—SbF_5$, 5:1 v/v) not varying between −140 to +40 °C. The comparable aromatic tetramethylcyclobutenium dication decomposes by 80% after 6 hours at −78 °C (in $FSO_3H—SbF_5—SO_2$)[676].

In all cases both the formation of dication *506* and its interaction with nucleophiles seems to proceed with the intermediate formation of monocations of type *B*. Thus, the ion *506* react with lithium aluminium hydride or triethylamine to give the following tricyclic compounds:

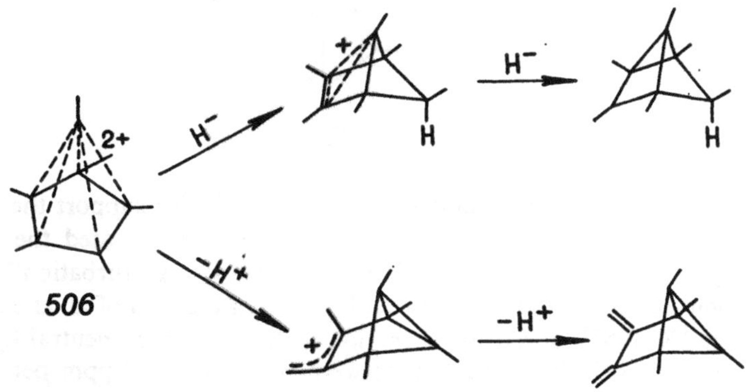

By interaction of the dication *506* with donors of hydride-ions (e.g., with isopentane) the NMR spectrum records intermediate monocations. The final product is the hexamethylbenzenium cation *502* [675]:

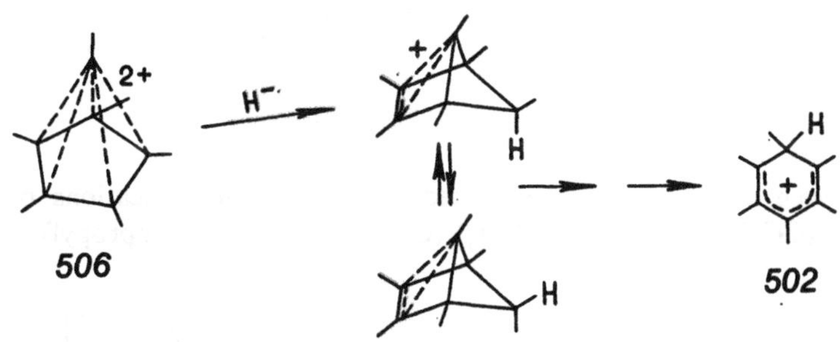

Since the hydride-ion detachment rate is greatly dependent on the system acidity, in the most strongly acid media ($HF—SbF_5$ or $HSO_3F—SbF_5$, mole ratio 1:1) ion *506* is stable even at room temperature.

The ^1H and ^{13}C NMR spectra of dication *506* (containing two signals with a 5:1 intensity ratio each), as well as the interaction with nucleophiles (obtaining tricyclic compounds)[675] show that the dication permits only two structures: one non-classical ion with a pyramidal geometry *506* or the rapid equilibration of five classical dications *506a–506e*. From the ^{13}C NMR data of ion *506* (126.3 ppm for basal and −2.0 ppm for apical carbons) after comparing them with calculated averaged shifts the authors prefer the pyramidal dication.

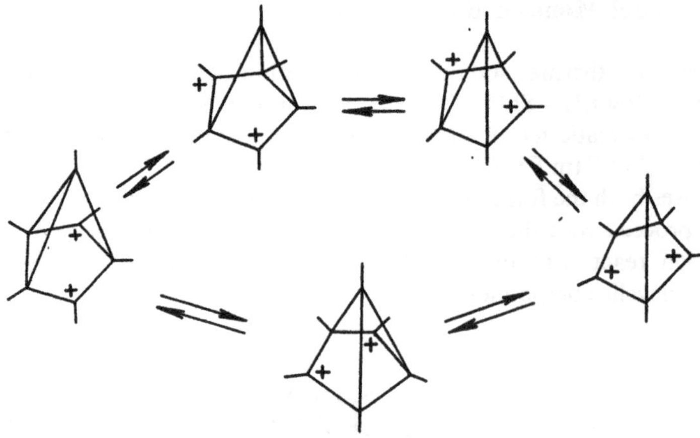

506a – e

Ab initio calculations for a model dication $(CH)_6^{2+}$ also support the nonclassical pyramidal structure [677]. Of late Hogeveen [678] has corroborated the nonclassical structure of dication *506* with both Saunders' "isotopic perturbation" method [269] and the comparison — after Schleyer et al. [268] — of the sum of all chemical carbon shifts in the ^{13}C NMR spectrum of the ion with that for the neutral hydrocarbons *507* and *508*; the calculated differences were 54.6 and 195.7 ppm per charge unit, respectively, typical for nonclassical ions.

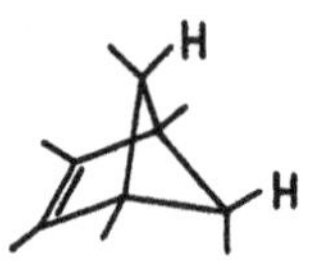

507

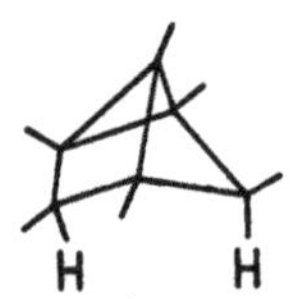

508

Hogeveen et al. [679] have generated stable pyramidal dications of the general formula $(CCH_3)_5CR^{2+}$ and $(CCH_3)_4(CR)_2^{2+}$ (R is ethyl, isopropyl). e.g.:

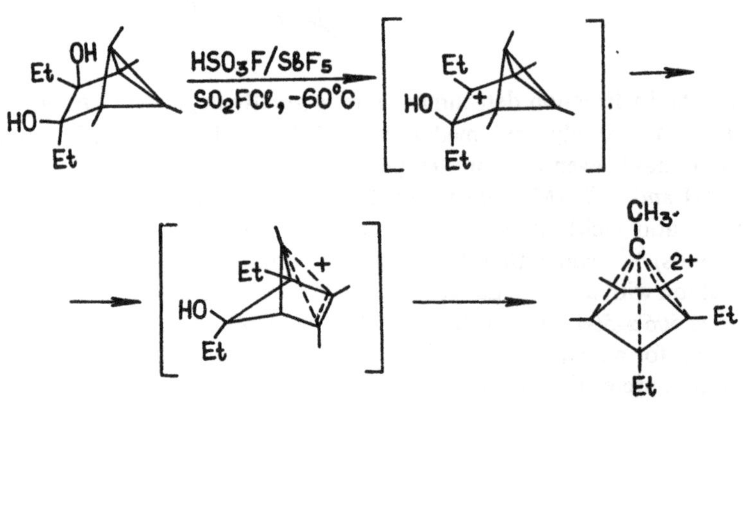

The described dications represent nonclassical ions of a special kind — pyramidal mono- and dications [18]; the latter can be regarded as a link to the elementoorganic chemistry e.g., boranes, by the peculiarities of their stereochemistry and of their bond nature. Thus, the pyramidal structure of pentaborane *509* is similar to that of monocation $(CH)_5^+$ *510* with a pentacoordinated carbon while the pyramidal dication $(CH)_6^{2+}$ *511* is similar to carborane *512*. By the peculiarities of their stereochemistry pyramidal cations are similar to metalloorganic sandwich compounds, e. g., ferrocene *513*; the dication *511* is regarded as a "semi-sandwich".

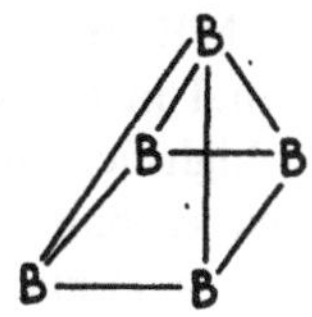

509

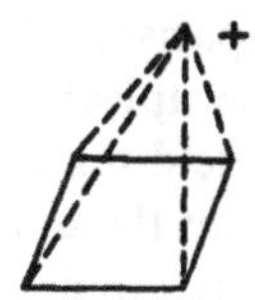

510

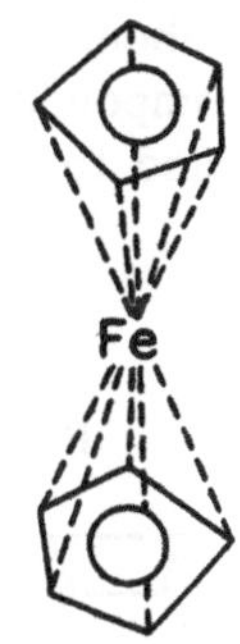

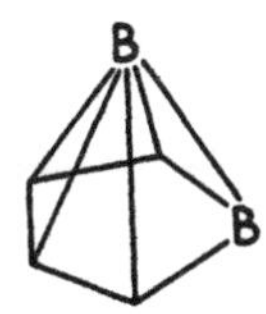

512

511

513

Originally Hoffman substantiated the possible existence of a comparatively stable cation *510* with pyramidal geometry [680]; then Masamune succeeded in generating

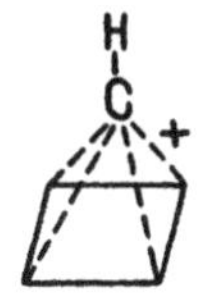

510

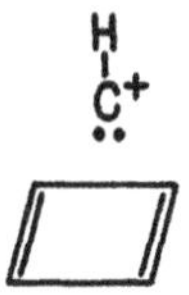

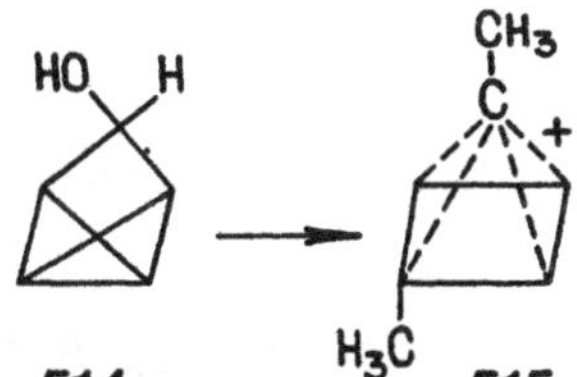

514 515

ion *515* from a homotetrahedron *514* [681]. Cation *510* can be formally regarded as resulting from the interaction of CH^+ with cyclobutadiene. The pyramidal structure of ion *515* is well corroborated by the ^1H and ^{13}C NMR spectra. An alternative variant — the equilibrium of several unsymmetrical structures *515a*⇄*515b*⇄ etc. seems less probable [681] due to the unusually high-field chemical shift (δ —23.04 ppm in a high field from TMS) of the apical C^5 atom in the ^{13}C NMR spectrum.

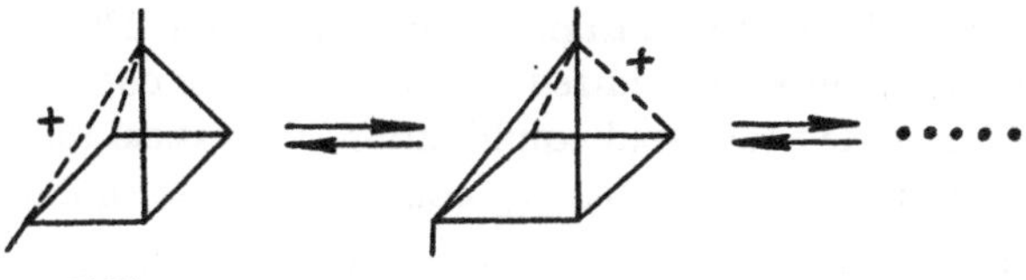

515a 515b

The quantum mechanical ab initio [682)], MINDO/3 [683)] and CNDO [684)] calculations show that ions of type *510* can really be among the most stable ones in the family of the $(CH)_5^+$ ions.

Theoretical calculations predicted for bis-homo-$(CH)_5^+$ cations the structure of a rectangular pyramid as energetically more favourable than the less symmetrical forms [685)]. Two bishomoanalogues of cation *510* were described — ions *516* [686)] and *517* [687)]. In the ^{13}C NMR spectra of either ion the signal of the apical carbon is shifted to a high field; in the ^{13}C NMR spectrum of ion *517* δC^7 is -17.5 ppm. An increase in steric strain is known to result in a shift of carbon signals in the ^{13}C NMR spectrum to a high field. Comparison of the ^{13}C NMR spectra of ion *517* and its tricyclic precursor *518* shows the apical carbon signal in the spectrum of ion *517* to be about 50 ppm in a higher field than the most high-field signals in the spectrum of compound *518*. This testifies to the unusual apical carbon hybridization in such ions and to the fact that on this atom there is but a minimum positive charge.

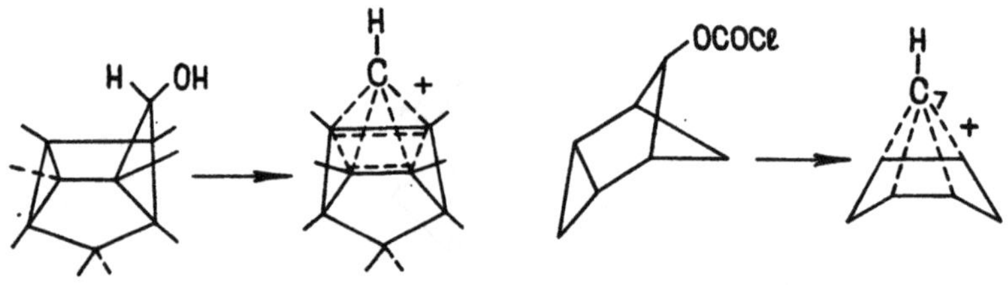

516 *518* *517*

In 1976 Hart [688)] reported new data on the ^{13}C NMR spectra of ion *516* with the $^{1}J_{^{13}C-^{1}H}$ constants (the constants in the scheme are parenthesized).

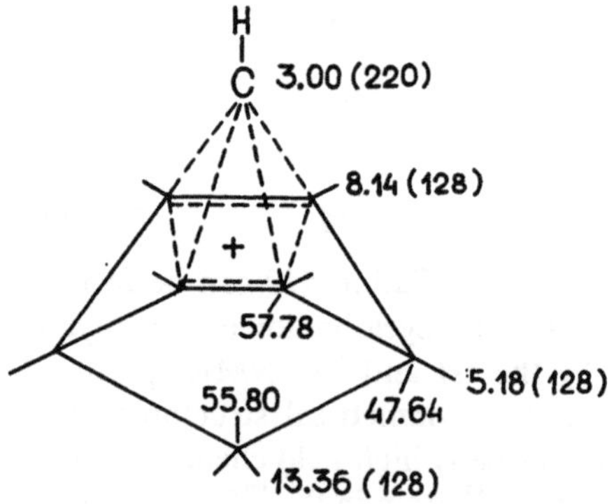

The value of the apical carbon constant is unusual: with 220 ± 5 cps it corresponds approximately to an sp-hybridized carbon. Thus the constant value is far higher than could be expected for rapid equilibration of classical ions *516a* ⇄ *516b* etc. A good model of the averaged $^{1}J_{^{13}C-^{1}H}$ constant in this equilibration is the bicyclo[2,1,0]pentane *519*. The observed constant value also exceeds the one that could be expected for the rapid equilibration of the nonclassical trishomocyclo-

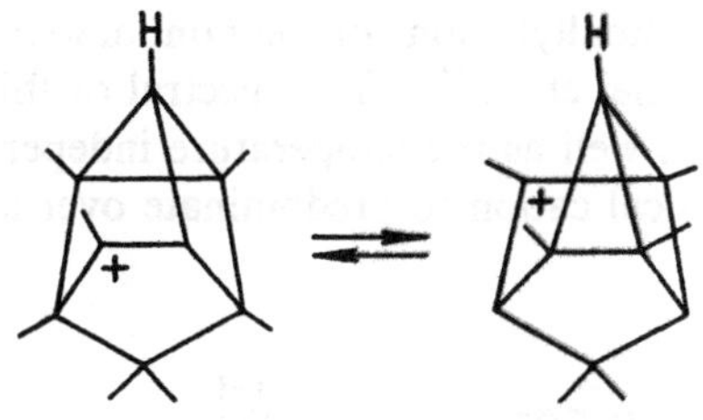

516a 516b

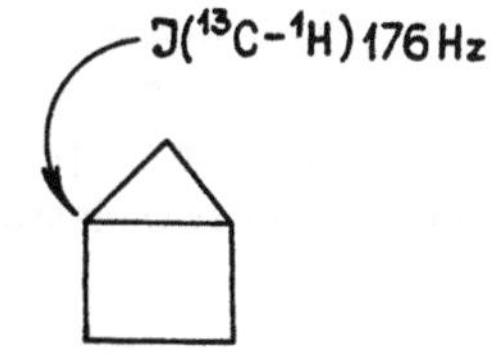

519

propenyl cations *516c*, *516d* which imitate the pyramidal structure *516*, the trishomocyclopropenyl cation *520* being used as a model [691] (see below).

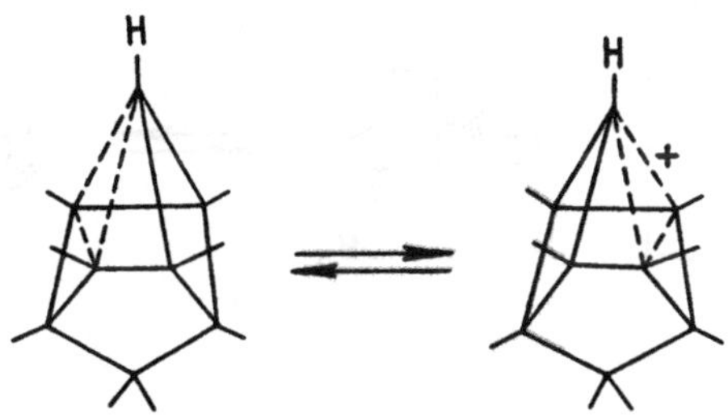

516c 516d

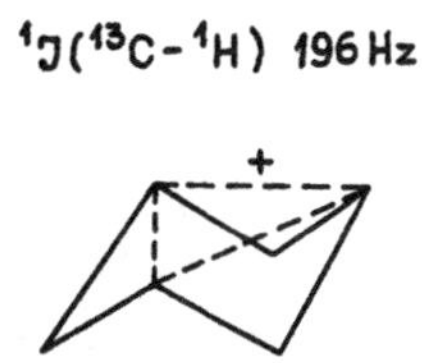

520

In ion *516* the substitution of two methyls at basal carbons by phenyls destabilized the ion; thus, ion *516* remains stable below −50 °C while its diphenyl analogue only below −100 °C [689]. An attempt to substitute the hydrogen at the apical carbon in ion *516* by methyl did not result in a pyramidal cation [690].

7.6 Nonhomoaromatic Homoallylic Ions

All the homoallylic ions discussed above belong to homoaromatic cations. The homoallylic nonhomoaromatic ions with the p-orbitals overlap the developing cation centre and only the γ-olefine carbon have been postulated as intermediates in solvolysis and electrophilic addition but until recently could not be obtained in a stable form though such attempts have been made. Thus, Hart and Kuzuya [690] have tried to obtain a stable homoallylic ion *522* from dienol *521*, however the latter interacted with HSO_3F to form the allylic ion *523*.

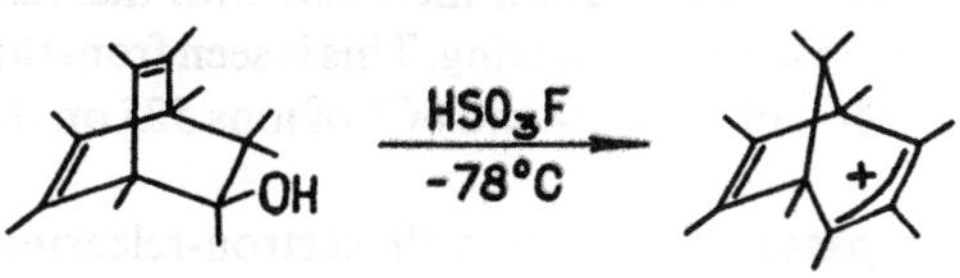

521 523

522

One of the first representatives of homoallylic ions of the nonhomoaromatic type — ion *396* — was generated by Kamschei et al. [554]. The spectral of this ion and a comparison with model compounds, as well as the temperature independent NMR spectra character support the nonclassical cation to predominate over a superrapid equilibrium of two classical ions.

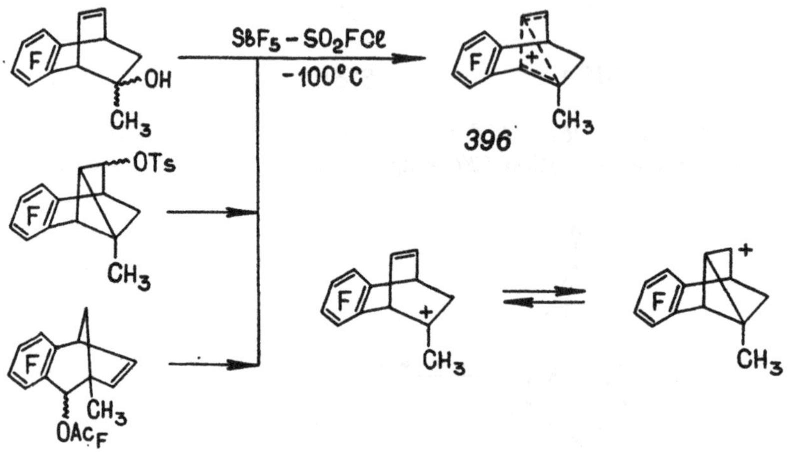

396

As already noted above, σ- or π-participation in solvolytic research has lately been established by the "method of increasing the electron demand of the developing cation centre". Kamschei [555] has applied this approach to establish the π-participation in stable 5-aryl-substituted cations of type *524* with the variable aryl nature and degree of olefine double-band electron-donating power. To separate the effects of the π-conjugation with the aryl group and of the double-band π-participation in ions *524* their spectra were compared with those of classical ions *525*.

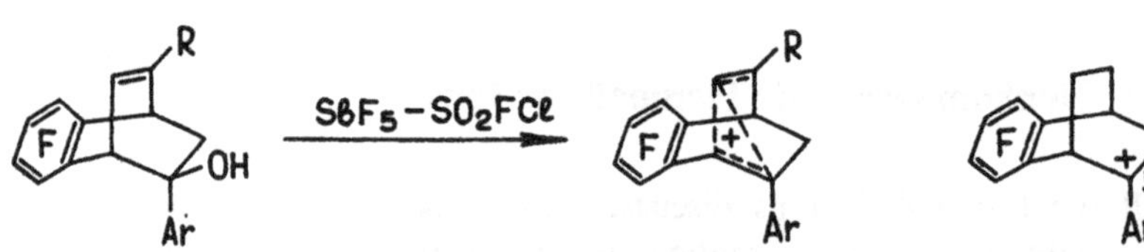

524 525

In passing from octenyl ions to octadienyl ones with the same aryls at C^5, the C^5 signals in the ^{13}C NMR spectra shift to a high field. This can be due to a transfer of part of the charge in the octadienyl cations at the expence of homoallylic interaction; the "extent" of homoallylic participation increases with increasing electron-attracting properties of substituents in the phenyl ring. This is seen from the dependence of the chemical shift values δC^5, δC^7 of ions *524* and δC^5 of ions *525* on the substituent constants σ^+ (Fig. 12).

As follows from the graph, in passing from ions with electron-releasing substituents to those with electron-seeking ones in the benzene ring the incline of the correlation line uniting the homoallylic ions *524* changes; so the value δC^5 is increased with increasing value of σ^+ (Fig. 12, b). For all the series of octenyl ions *525* the value of δC^5 is observed to increase with increasing σ^+ (Fig. 12, c).

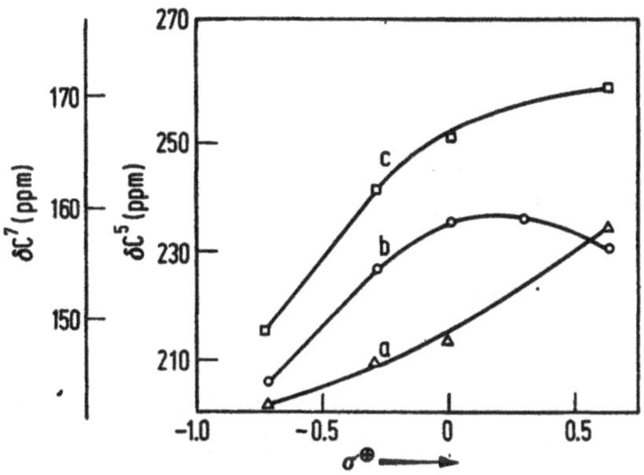

Fig. 12. Correlation of the ^{13}C chemical shifts δC^7 (a), δC^5 (b) in the ions (524) and δC^5 (c) in the ions (525) with σ^+ constants

One can assume that in ions of type *524* the positive charge delocalization proceeds not only at the expence of the π-participation of the C^7—C^8 double bond, but also by σ-participation of the C^4—C^8 bond so that the C^5, C^4 and C^8 atoms are linked together by a 2e3c bond. In model classical ions of type *525* the intensification of the electron-seeking properties of substituents is accompanied by the growth of the charge on the C^5 atom, of its chemical shift and, as expected, the proportional increase in the chemical shifts of the carbon α-atoms — both C^4 and C^6 (Fig. 13, a).

The same is observed in the ions of type *524* with donor substituents at C^5 (Fig. 13, b). As the stabilizing effect of the substituents in the latter ions decreases, the chemical shift of C^4 increases faster than that of C^6; in passing to the ions with considerable homoallylic stabilization the chemical shift of C^6 decreases with a simultaneous increase in that of C^4. This points to the growing "degree" of σ-participation in ions of type *524*. The ratio of the degree of σ-participation (which is reflected by the chemical shifts of C^4 atoms) to that of homoallylic participation (which is reflected by the differences between the chemical shifts of C^7 and C^8) is seen from the dependence in Fig. 13, c.

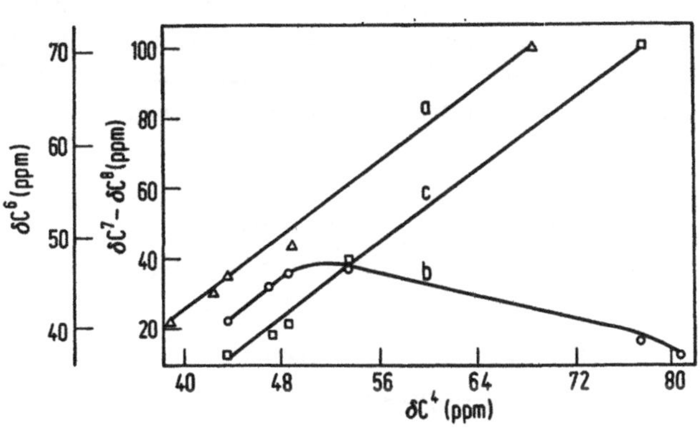

Fig. 13. Correlation of the ^{13}C chemical shifts δC^6 in the ions (524) (a) and (525) (b), also δC^7–δC^8 in the ions (524) (c) with chemical shifts δC^4

The above spectral characteristics of ions *524* agree well with the structure of homoallylic ions and contradict those of classical cations of type *526*. The choice is, however, between the nonclassical ions of type *524* and the rapid (in the time scale of NMR) equilibrium of the classical ions *526*⇄*527*. It shifts to the right as the electron-seeking properties of the aryl group intensify.

526 527

Farnum[243)] discussed a similar dilemma. The low negative value of ϱ^+ (—0.5) for the dependence of the hypothetical equilibria constant logarithms of norbornenyl and nortricyclyl ions on σ^+ argues against this hypothesis since, as a

rule, such substituents exert a stronger effect on the position of similar equilibria of cations. If one accepts the hypothesis of equilibrium of bi- and tricyclic cations, the data obtained in [555)] are explained by the low susceptibility of the position of these equilibria to the aryl group. None of the solutions of bicyclooctadienyl cations of type *524* described in this work displays any temperature dependence of the chemical shifts in the ^1H, ^{19}F or ^{13}C NMR spectra: it seems improbable for the equilibrium constants to be in all cases independent of temperature.

The results of interaction between solutions of ions *524* and nucleophiles also agree with the homoallylic structure of ions but not with the equilibrium of two cations. In reactions with one and the same nucleophile the increase in the electron-seeking properties of the aryl group at C^5 is accompanied by an increase in the tricyclic products: second, by a parallel increase of stereospecificity in the nucleophilic attack at C^5 — mainly from the endo side despite the steric accessibility of classical benzyl cations with the same skeleton from the exo side [504)], e.g.:

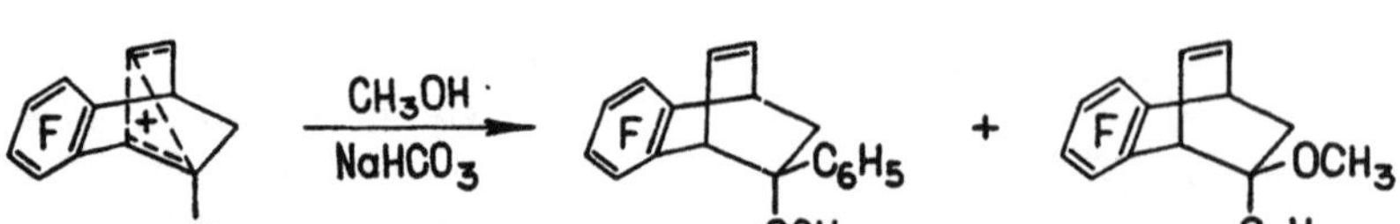

534 528 529

528 : *529* 1.3 : 1

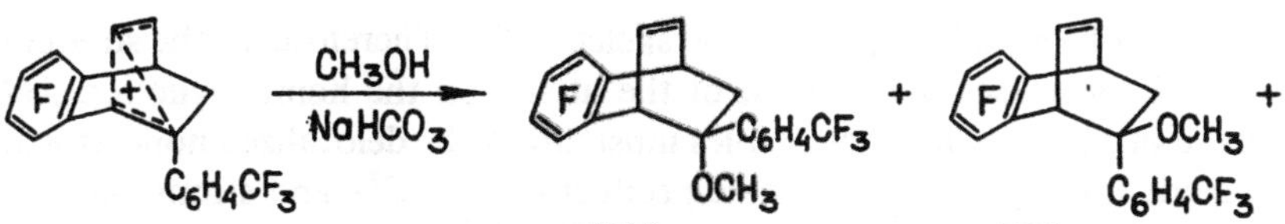

530 531

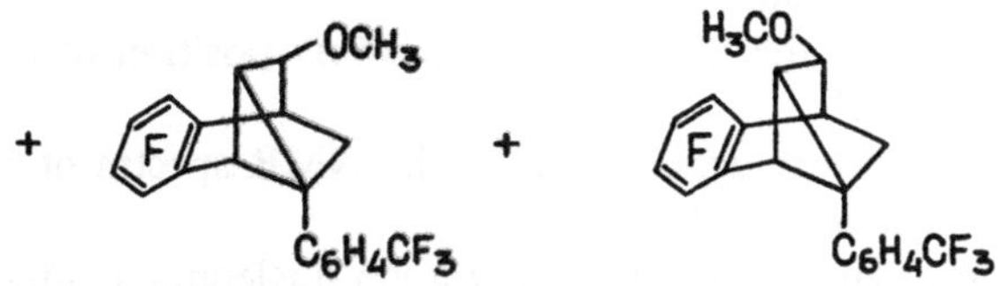

532 533

530:*531*:*532*:*533* 5:1:1.3:2.6

These facts readily account for the assumption of the homoallylic structure of the cations discussed; with increasing π-participation of the double bond the possibility of the nucleophilic "break-through" from the exo side at C^5 decreases and simultaneously an attack at C^7 becomes more probable.

Interesting data have been obtained in [555] when a cation of type *524* interacts with nucleophiles of different strength. Thus, ion *535* with a comparatively small "demand" of the cation centre reacts both with a relatively strong (CH_3OH—$NaHCO_3$), and with a weak nucleophile (AcOH—H_2SO_4—$AcOCH_3$) to form only bicyclic products of nonstereospecific attack at C^5:

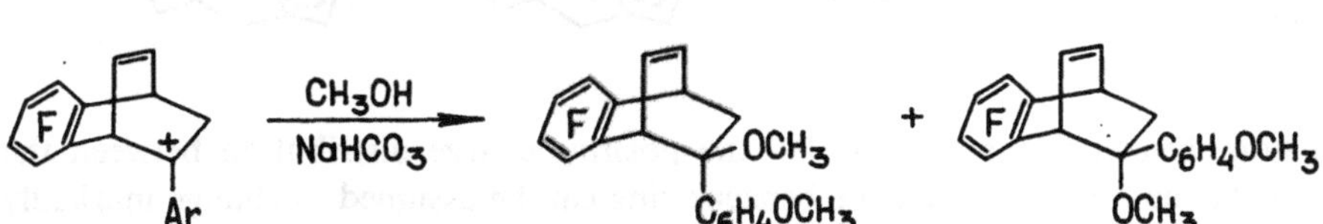

535

The cation *534* with a higher "demand" of the cationic centre reacts with CH_3OH—$NaHCO_3$ to yield the same products of a nonstereospecific attack on C^5 (see above); bur the reaction with a weaker nucleophile — AcOH—H_2SO_4—$AcOCH_3$ — results mainly in the products of a stereospecific attack on C^5 and C^7:

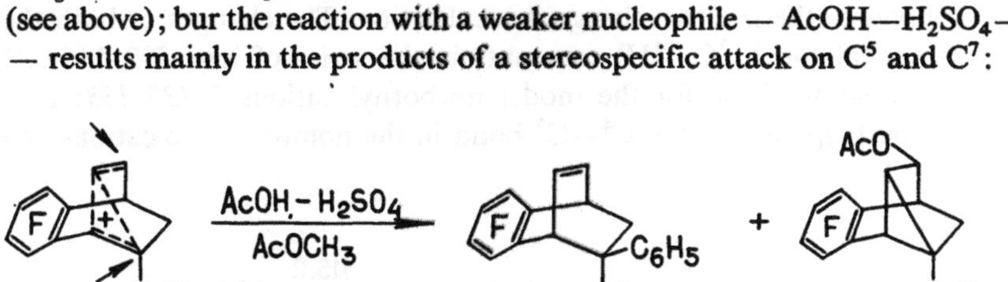

534

Thus, the "degree of manifestation" of π-participation in reactions with a nucleophile changes essentially depends on its strength. The 2,3-tetrafluorobenzobicyclo[2,2,2]octa-2,7-diene skeleton is thermodynamically less stable than the

3,4-tetrafluorobenzotricyclo[3,2,1,$0^{2,7}$]octene skeleton [548]. Therefore, as the strength of the nucleophile decreases the fraction of the attack on the homoallylic ion *534* on C^7 is likely to increase; weak nucleophiles must attack the delocalized nonclassical structure on a site leading to a more stable product (i.e. on C^7), and strong ones — on the site of the largest positive charge (i.e. on C^5) (cf. [381]); this is also the case in experiment.

All the mentioned circumstances are:
1) low susceptibility of the hypothetical equilibria position of classical ions to the nature of the aryl at C^5;
2) absence of temperature dependences in the NMR spectra of solutions of carbocation salts with octadiene skeleton;
3) intensification of endo stereospecificity upon nucleophilic attack on the C^5 atom of an ion both when the electron-seeking propertes of the aryl increase (in reacting with the same nucleophile) and when the strength of the nucleophile decreases (in reacting with the same ion). — Between the two alternatives the homoallylic ion is prefered over the rapid equilibrium of classical ions [555].

The first stable homobenzylic ions were generated in works [553,557,692,693]. A comparison between the spectral data of ion *389* and those of model ions (2-norbornyl, phenonium, polyfluoroarenonium) made it possible to reject a number of alternative structures, among them the rapid equilibria of classical or unsymmetrical homobenzylic ions, e.g. [553]:

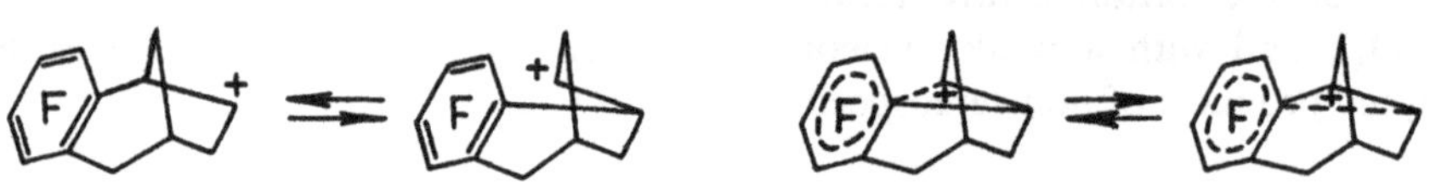

In homobenzylic ions of type *389* the positive charge distribution between the "cyclopropane" fragment and the benzene ring can be assumed to change markedly depending on the substituents and, in particular, on the changes in the "demand" of the developing cation centre at C^6. The spectral data cited below show that the decreasing "demand" of the developing cation centre at C^6 effects a decrease in the contribution of the participation of the benzene ring and of the σ-participation of the C^4—C^5 bond to the positive charge delocalization. The changes of both the chemical shifts and of the $J(^{13}C—^1H)$ constants in the series of ions *389–536–537* are practically similar to those for the model norbornyl cations *5–130–133*; they confirmed the σ-participation of the C^4—C^5 bond in the homobenzylic cations *389* and *536* [557]:

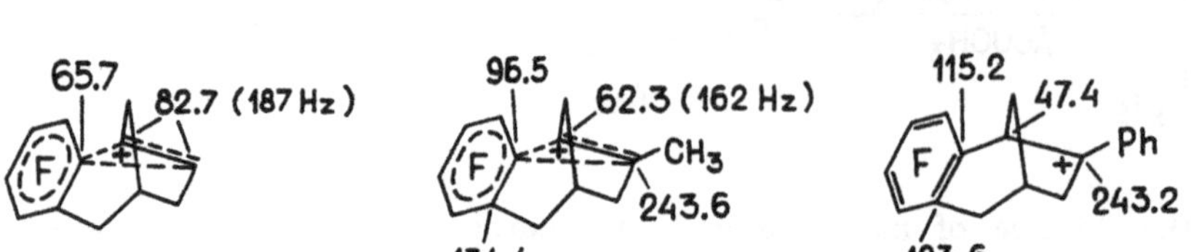

389 *536* *537*

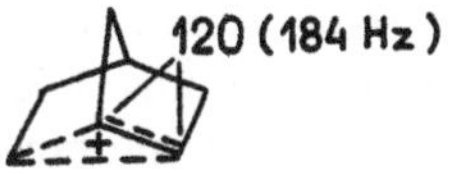

5

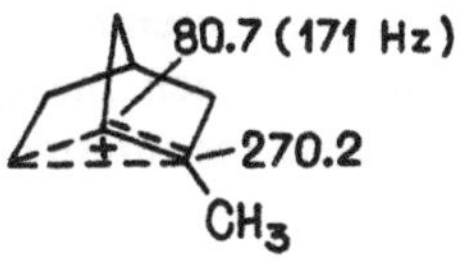

130

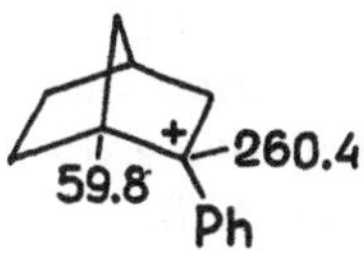

131

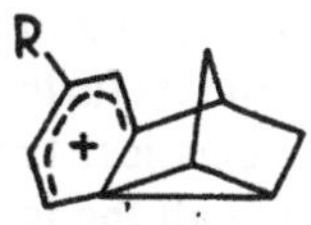

538

Olah and Liang [692] have generated ions *538* and shown that the degree of positive charge delocalization to the "cyclopropane" ring decreases as the electron-releasing properties of the substituents (R) intensify. They have also obtained tertiary 2-benzonorbornenyl cations [693]; the structure of the latter, in their opinion, can be best represented in the form of equilibrium:

The positive charge is mainly delocalized between C^2 and the benzene ring. The C^{11} atom has an intermediate hybridization between sp^2 and sp^3. The vacant p-orbital at C^2 interacts with the neighbouring σ-bond $C^1—C^{11}$.

8 β-Cyclopropylcarbinyl Cations

The interaction of the cyclopropane ring and the neighbouring vacant p-orbital in α-cyclopropylcarbinyl systems has been well investigated in both nucleophilic and superacid media [7,694–697]. Stable long-living α-cyclopropylcarbinyl ions have been generated and studied by Olah [698, cf. 699]. In recent years the research has been devoted to the interaction of the vacant p-orbital with the β-cyclopropane ring ("homocyclopropylic" interaction); since there is no survey yet, the most essential results obtained in these studies are discussed below.

8.1 Homocyclopropylic Interaction

Winstein suggested 1959 that the acetolysis of cis-bicyclo[3,1,0]hexyl tosylate *539* involves the intermediate formation of a so-called trishomocyclopropenyl cation *520* [700–702]. This equatorial isomer was found to solvolyse 50 times as fast as the axial *540*; the quantitative product of the reaction is only cis-acetate *541* with about

99% retention of configuration. The 6,6-dideuterated product yields acetate with homogeneous distribution of deuterium — each methylene group contains 1.35 + 0.03 H (calculation gives 1.33 H assuming the intermediate structure of trishomocyclopropenyl ion *520*).

The slower (50 times) solvolysis of the axial isomer *540* results in: three bicyclic acetates, two monocyclic acetates, a bicyclic olefine and two monocyclic dienes — altogether 8 main products. The acetate with the same structure is the principal product (40%) and has an inverse configuration, the label (D) being not distributed in it at all. In this case a classical cation seems to be formed, and its further rearrangements compete successfully with its conversion into a nonclassical trishomocyclopropenyl ion. At the same time for the equatorial isomer *539* all the facts observed, i.e.:

1) exclusive formation of a single product,
2) retention of configuration,
3) complete distribution of the label (D) among three positions,
4) absence of rearrangements and split-off reactions,
5) internal return in an ion pair — are fairly well supported by the assumption of the

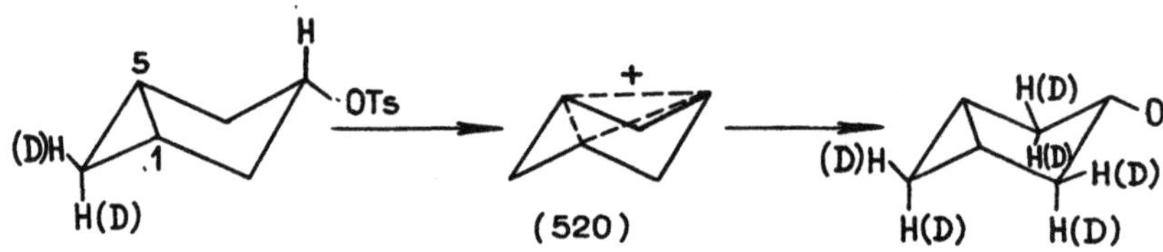

539 *520* *541*

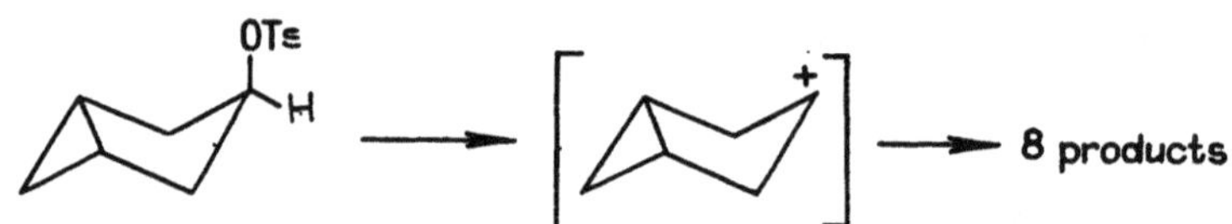

540

C^1—C^5 binding electrons participating in tosylate ionization and subsequent formation of the trishomocyclopropenyl ion [703]. Also, as shown by LeNoble [704], the activation volumes for 2-exo-norbornyl brosylate *22* and tosylate *539* practically coincide (—14 cm³/mole), but differ from those for 2-endo-norbornyl brosylate *23* and cis-tosylate *540* (—17 cm³/mole). These data confirm the transition state in the former case to be more delocalized than in the latter.

At the same time, Corey has shown the deamination of 3-deutero-cis-3-bicyclo-[3,1,0]hexylamine *542* leading to a mixture of equatorial and axial 3- and 2-bicyclohexanols without distribution of the label *D* [705]. If at C^1 and C^5 there are phenyl

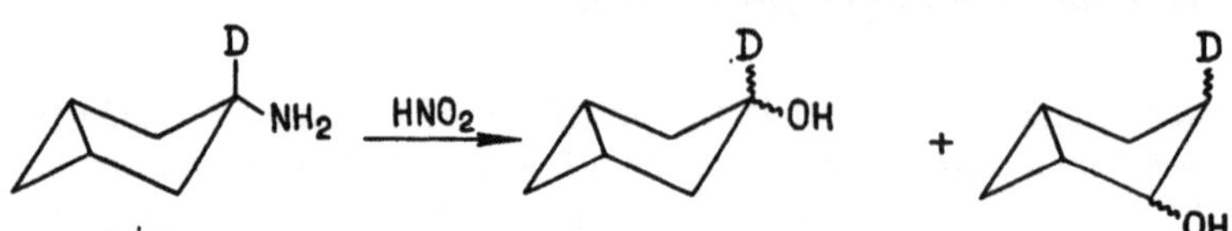

542

groups as in compound *543*, then during acetolysis no anchimeric acceleration is observed. Corey [706] assumed these reactions to yield equilibrium classical cations rather than one nonclassical one; Freeman came the same conclusion [707]. However, Winstein [343] had pointed out 1967 that compound *543*, as distinct from the unsubstituted tosylate *539* exists mainly in the bath conformation unfavourable for β-cyclopropyl interaction; therefore the solvolysis of both cis- and trans-1,5-diphenyl-substituted isomers leads to a classical cation in the bath conformation.

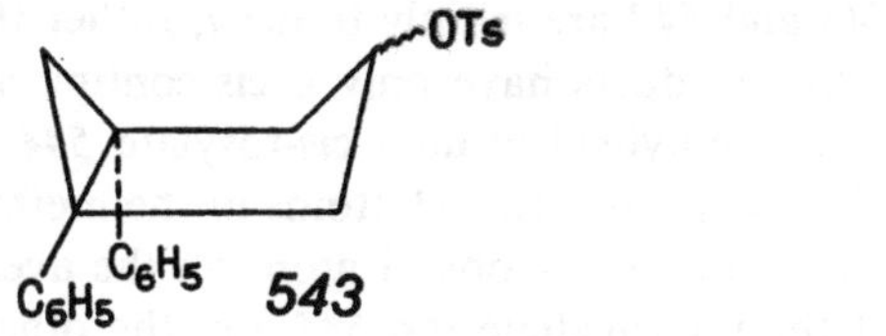

543

Table 11. Solvolysis of Some Bicyclo[3,1,0]hexyl-3-tosylates

Systems	Relative rate of cis ROTs	Rate cis/trans ratio
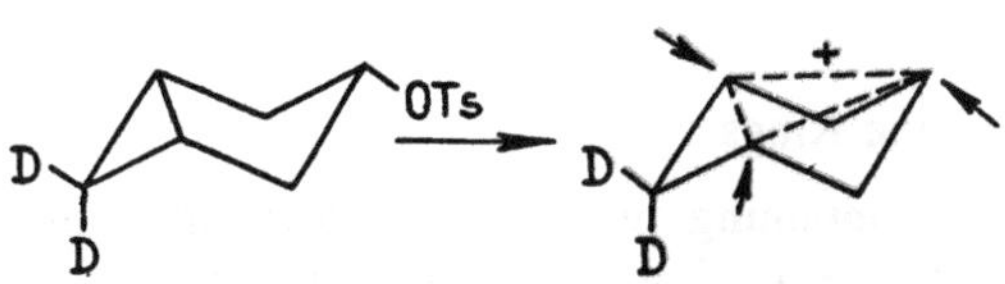 539 520 99% cis 1% trans	1.0	40
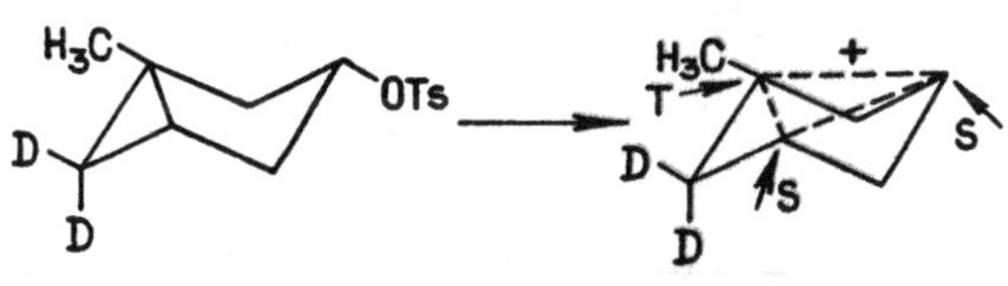 544 545 10% S; 99.7% cis 90% T; 99.9% cis	5.1	97
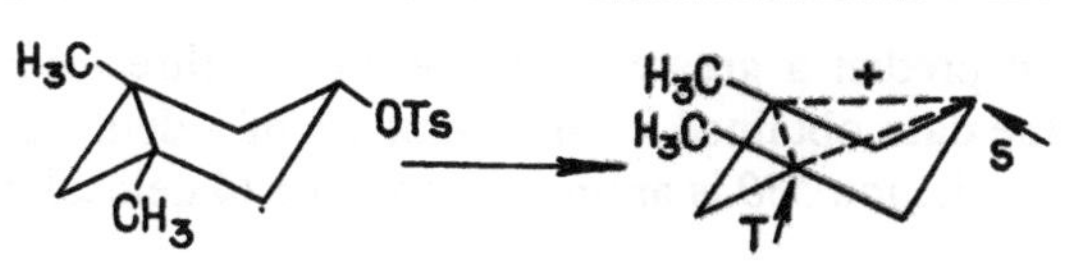 546 547 3% S 97% T; 99.9 cis	7.0	119

Interesting data have been obtained from the solvolysis of alkyl-substituted bicyclo[3,1,0]hexanes [708]. The high solvolytic activity of cis-epimers in comparison with trans-epimers, the stereochemistry of this reaction and the equilibration of deuterium in the products testify to the intermediate formation of trishomocyclopropenyl cations upon acetolysis of cis-monomethyl- and dimethylbicyclo[3,1,0]-hexyl-3-tosylates.

As seen from Table 11, the kinetically controlled products of the reaction between the methyl-substituted cations *545* and *547* are mainly tertiary, rather than secondary, acetates; but in every case the products have only a cis-configuration. Upon acetolysis of 6,6-dideuterated monomethyl-substituted cis-tosylate *544* the tertiary acetate contains, judging by the PMR spectra, two H atoms in the methylene group of the cyclopropane ring, and the secondary — one H atom on the average, in full conformity with the structure of the intermediate ion *545*. On the other hand, the secondary acetate formed on solvolysis of the trans-epimer contains two D atoms in the cyclopropane group. Thus in this case, either, no conversion is observed of a classical ion into a nonclassical one in the trans-epimer solvolysis.

Other examples of solvolysed 3-substituted cis- and trans-bicyclo[3,1,0]hexanes corroborate the intermediate formation of nonclassical trishomocyclopropenyl cations [625,704,709–711,cf.712].

8.1.1 Reactions with several Cyclopropane Rings

During solvolysis of the tosylate *548* containing two cyclopropane rings and a tosylate group in the cis-configuration one or both cyclopropane rings participate in the stabilization of the cation centre; this result in the trishomocyclopropenyl *549* or the pentahomocyclopentadienyl *550* ions [713].

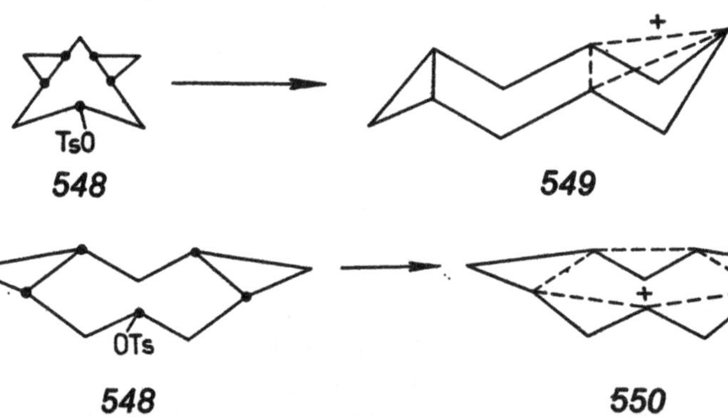

548 549

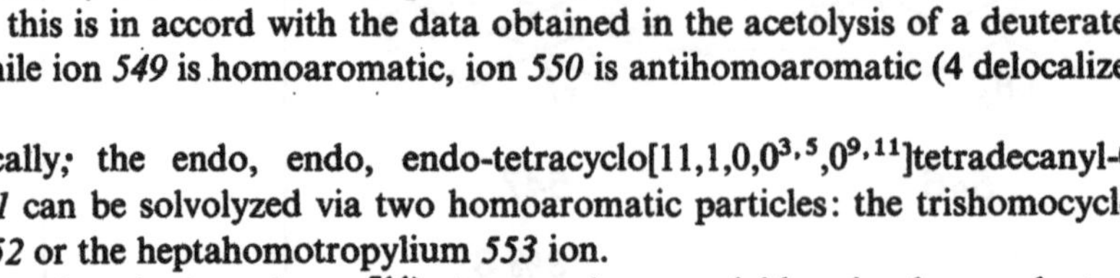

548 550

Calculations by Hückel's method predict a greater stability for the three-centre cation *549*; this is in accord with the data obtained in the acetolysis of a deuterated sample. While ion *549* is homoaromatic, ion *550* is antihomoaromatic (4 delocalized electrons).

Theoretically, the endo, endo, endo-tetracyclo[11,1,0,$0^{3,5}$,$0^{9,11}$]tetradecanyl-0-tosylate *551* can be solvolyzed via two homoaromatic particles: the trishomocyclopropenyl *552* or the heptahomotropylium *553* ion.

However, Winstein has shown [714] the reaction to yield only the product of 1,2-elimination with a 97%. Why is tosylate *551*, contrary to the similar tosylate *548*,

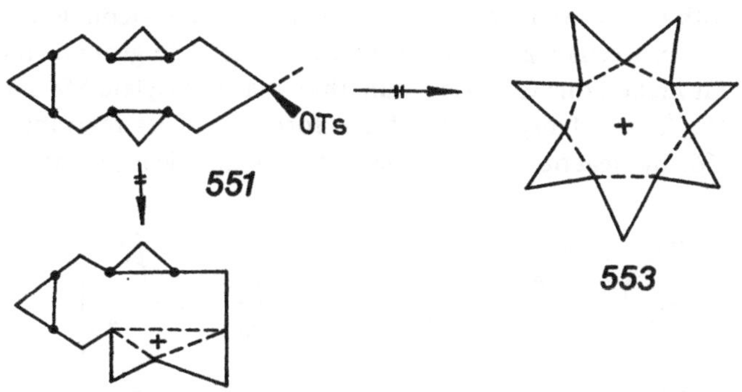

551

552

553

not converted into a trishomocyclopropenyl ion? One of the reasons may be that the geometrical conditions (distance, angle) for the interaction of the cyclopropane rings with the developing cation centre become worse as the cycle size increases. This is supported by molecular model studies, But a more important reason [715] is the ability of the middle cycles for syn-elimination under solvolytic conditions with the formation of the trans-olefine; the latter factor can actually hinder the formation of homoaromatic particles in the middle cycles.

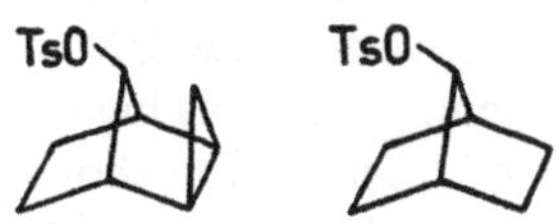

554 4

The acetolysis of the exo-8-anti-tricyclo[3,2,1,$0^{2,4}$]octyl tosylate *554* proceeds three times slower than that of the nonreactive 7-norbornyl tosylate *4* [716]. In compound *554* the p-like orbitals of C^2 and C^3 involved in the formation of the C^2—C^3 bond are remote from the 8-position where the cation centre arises; but the "face" participation of the cyclopropane fragment seem to be insignificant. At the same time the endo isomer *555* solvolyzes 10^{15} times as fast as the exo-epimer *554* [717, 718]; in this case the above-mentioned orbitals are directed to the 8-position (edge participation). The hydrolysis of this tosylate in 70% aqueous acetone results in an alcohol mixture containing 99.9% of a rearranged alcohol — endo-3-tricyclo[5,1,-0,$0^{4,8}$]octanol *556* — and 0.1% of alcohol *557* which retains the parent skeleton and configuration.

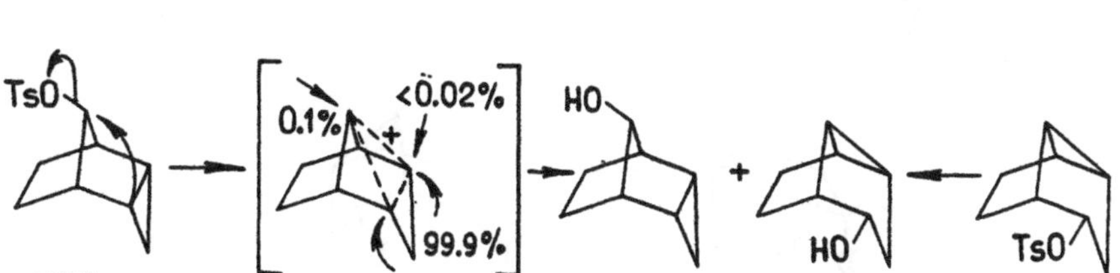

555 557 556 558

A reaction with the isomeric tosylate *558* yields the same alcohols and in the same ratio and, hence, proceeds via the same intermediates. The ground state of tosylate *555* is located far higher, due to steric strain, than that of tosylate *558* and that is what brings about the high solvolysis rate of the former and its rearrangement mainly into product *556*; the reaction is accompanied by a significant relief of steric strain.

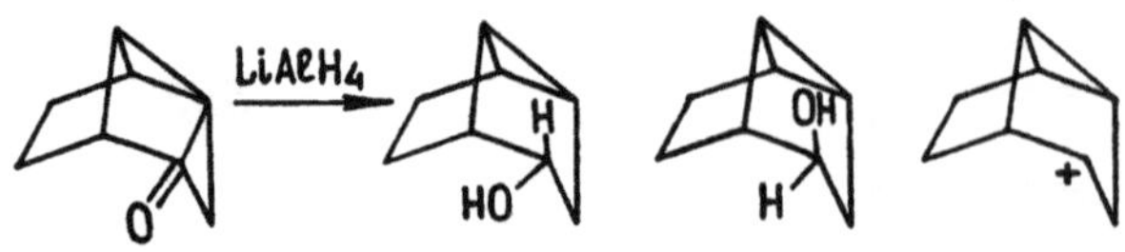

559 556 560 561

The reduction of ketone *559* by $LiAlH_4$ results mainly in the endo alcohol *556*. The thermodynamic equilibrium of exo and endo alcohols leads mainly to exo alcohol *560*. Hence, the formation of endo alcohol *556* upon solvolysis of tosylate *558* cannot be due to steric factors — the endo-side attack is sterically less favourable than the exo-side one; the resulting alcohol *556* is not a thermodynamically controllable product. Consequently, the data obtained cannot be used to assume a classical structure of the intermediate cation *561*. At the same time the participation of the C^1—C^8 bond and the intermediate non-classical ion are quite compatible with these facts. The acetolysis of the optically active tosylate *558* is accompanied by complete racemization, the rate of the latter being 3 times as high as that of acid elimination (internal return). The completeness of racemization shows the reaction to proceed via a symmetrical trishomocyclopropenyl ion or rapid equilibration of unsymmetrical cations *561* or to be accompanied by a 1,3-hydride shift from C^5 to C^3.

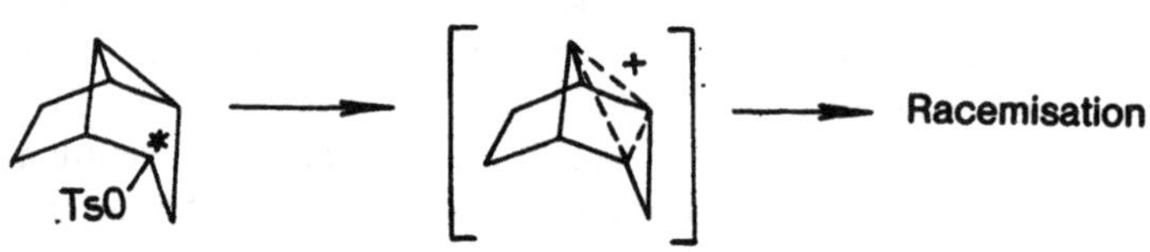

558 562

561

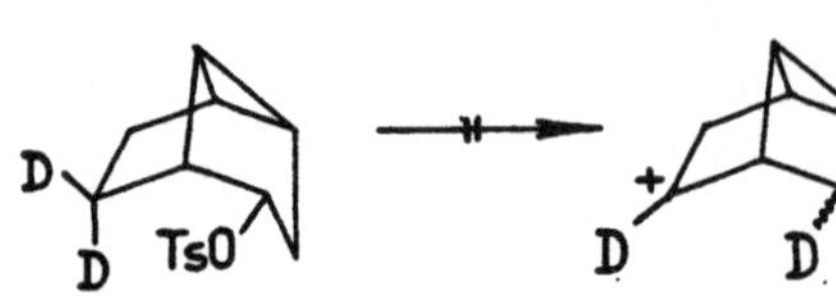

558 - d_2

The latter assumption is rejected by an experiment with a deuterated sample of tosylate *558-d_2*. The former assumption accounts for the racemization, but does not explain the exclusive formation of the thermodynamically unstable endo isomer resulting from a nucleophilic attack from a sterically unfavourable side. Thus the trishomocyclopropenyl cation *562* remains the most probable intermediate product.

8.1.2 Influence of p-Orbitals

The spatial position of the vacant p-orbital relative to the cyclopropane ring influences the efficiency of its participation in the stabilization of the cation centre [719]. In α-cyclopropylcarbinyl ions the cyclopropyl group interacts effectively with the cation centre if the vacant p-orbital is parallel to the plane of the cyclopropane ring:

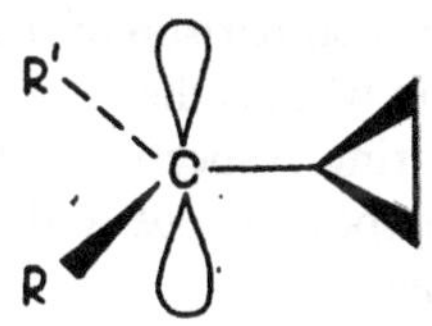

Haywood-Farmer [719] has studied the acetolysis of 4 stereoisomeric brosylates (exo-anti, exo-syn, endo-anti and endo-syn) of tricyclo[3,2,1,$0^{2,4}$]octanol-8 *563–566*.

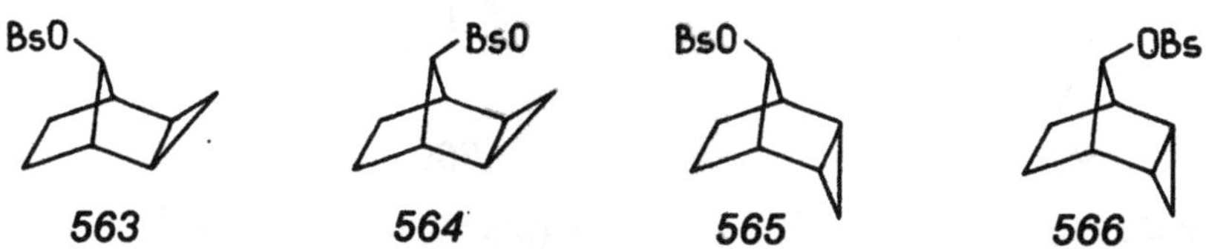

These models represent 4 variants of the cyclopropane ring position relative to the departing group at C^8. All of these positions, however, are far different from the situation in the α-cyclopropylcarbinyl cations since in all the models the position of the cyclopropane ring with respect to the vacant p-orbital at C^8 is "perpendicular" rather than parallel. For the compounds studied the authors have observed the following reactivity series: endo-anti *565* ≫ exo-syn *564* > endo-syn *566* > exo-anti *563* with a rate ratio of $10^{12}:10^4:10:1$ (100 °C).

Although in compounds *565* and *566* the steric, inductive and angular environments of the ionization centre are practically identical, nevertheless the difference in their solvolysis rates amounts to 10^{11}. The only possible reason for such a difference is the anchimeric participation of the cyclopropane ring in the ionization of *565*.

The orientation required for this participation is not like the parallel π-arrangement of orbitals in α-cyclopropylcarbinyl cations, it rather contributes to the σ-type interaction between the ionic centre and the sp^2-orbitals of cyclopropane. Here the situation is similar to the homoallylic participation of double bonds; at long distances the σ-overlap is more effective than the π-overlap [340]. The fact that compound *565* reacts 10^3 as fast as the 7-anti-norbornenyl brosylate *567* points to a more effective overlap of the vacant p-orbital with cyclopropyl orbitals than to that with double-bond orbitals.

(A)

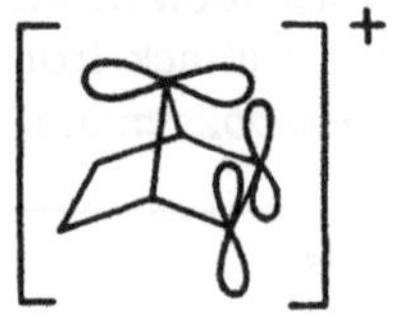

(B)

The calculations made by Hoffmann [720] have confirmed the higher efficiency of overlap *A* in comparison with *B*.

Since in the above examples the solvolysis of cyclopropane derivatives is accompanied by rearrangement to less strained structures, the comparison of the reactivities of compounds *392* and *567* cannot answer the question what is the relative efficiency of the participating double bond and cyclopropane ring respectively. It would be more reasonable to compare the acetolysis of the compounds *568* and *569* because their reactions are not accompanied by rearrangements [721].

568

k (25 °C) = $1.2 \cdot 10^{-3}$ c^{-1}

569

k (125 °C) = $7.5 \cdot 10^{-6}$ c^{-1}

Thus, the cyclopropane ring does stabilize the developing cation centre more effectively than the double bond.

Lambert et al. [722] pointed out the comparison between the participation efficiency of the cyclopropane ring and that of the double bond in different molecules being not quite correct; the difference in the inductive effects of these groups as well as the steric strain of the cyclopropane-ring interferes with the ground state of only one of the compounds to be compared. To overcome these difficulties Lambert studied the solvolysis of 3-tosyloxy-cis-bicyclo[5,1,0]octene-5 *570* whose molecule includes both the double bond and the cyclopropane ring. An analysis of the product composition has shown the double bond participation in this system to be far more effective: the π-participation in the process amounts to 80% while the participation of the three-member cycle constitutes only 20%.

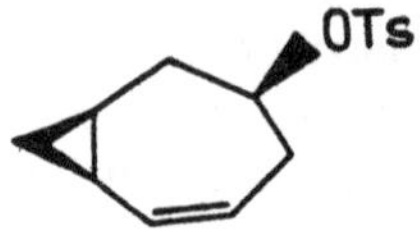

570

However, the authors rightly remark that in different models the relative efficiency of a participating group may vary for it depends on the orientation between the participating and the detaching groups, as well as on to what degree the angular strain of the cyclopropane ring can be "relieved" in the transition state of ionization. Thus, the conclusion about the relative efficiency of participating groups cannot be transferred from one system to another without preliminary critical analysis.

On the other hand, Haywood-Farmer [719] points out that the cyclopropane ring in the homoallylic position has little affect on the rate of solvolysis, e.g., in the following compounds [723–726]:

He believes it to be due to the rigid configurational requirements of homocyclopropylic participation; these conditions apply to compound *565*, namely: equal participation of both ends of the cyclopropylic bond. The high reactivity of brosylate *565* is partly due to the so-called steric strain relief during ionization.

The stereospecific formation of a small set of products, intensive skeletal rearrangements, relatively high rate of solvolysis — all this is typical only for the solvolysis of isomer *565* rather than for the compounds *563*, *564* and *566*. The acetolysis of compound *566* proceeds only 10 times as fast as that of extremely inactive 7-norbornyl brosylate and yields many products, while exo-syn-brosylate *564* is far more active, its relative rate being 10^4. In this case the ester group elimination is promoted by the spatial proximity to the methylene group of the exo cyclopropane ring. Another explanation that the ionization is accompanied by a skeletal shift of the C^1—C^7 bond resulting in a stabilized α-cyclopropylcarbinyl ion with the parallel orientation of the ring and the neighbouring vacant p-orbital. A similar

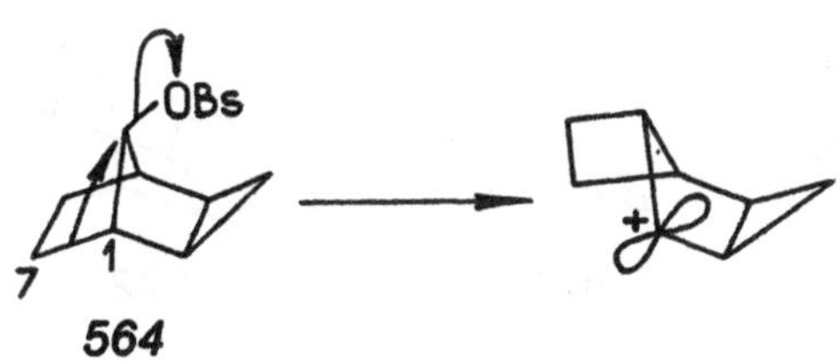

564

rearrangement of isomer *566* may also result in an α-cyclopropylcarbinyl ion, but in this case the latter will not be stabilized by the parallel arrangement of orbitals. In the least reactive of the 4 isomers compound *563* neither the cyclopropane nor the C^1—C^7 bond can participate in the ionisation stage.

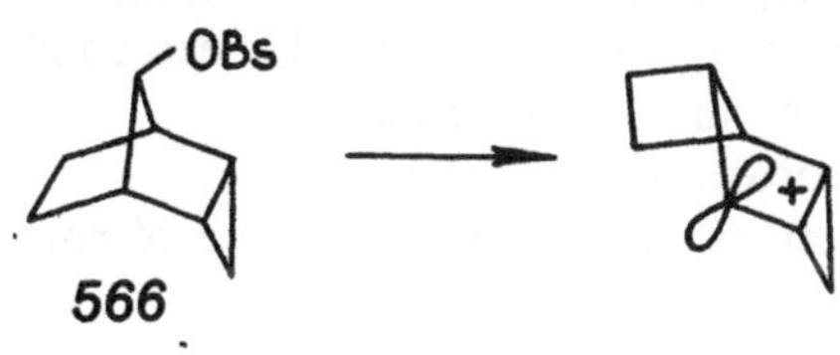

566

8.2 Cyclopropane Ring Splitting

To explain the difference in the activity of isomers *563* and *565* Tanida used the concepts of Woodward-Hoffmann [727] and Depuy [728] about the mechanism of cyclopropane ring splitting. When the cyclopropane ring is opened in both compounds the disrotatory process will proceed so that the electrons of the $C^2—C^3$ bound could approach the tail part of the developing p-orbital at C^8. In isomer *565* the disrotation will occur outward, in isomer *563* inward; in this case unfavourable steric interactions will result between the endo atoms H^2 and H^3.

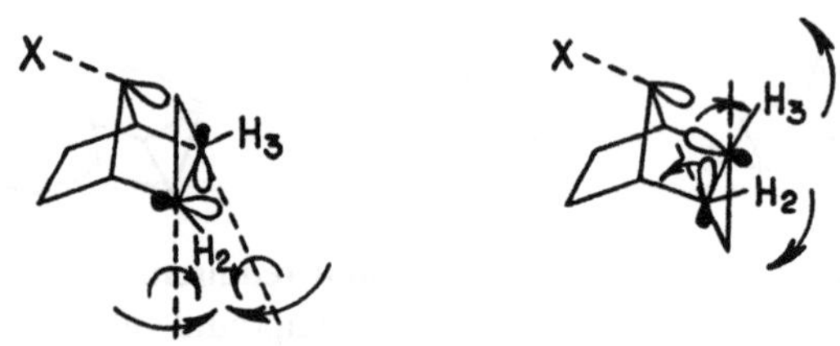

563 565

The introduction of a syn-double bond into 7-anti-norbornyl and 7-anti-benzonorbornenyl compounds, e.g., *220* and *247* increases the solvolysis rate either due to a change of steric effects in passing from the ground state to the transition one or to an additional delocalization of the charge with the participation of a syn-double bond in the transition state (homoconjugative reinforcing effect). To determine whether similar homoconjugative effects are possible with the participation of the cyclopropane ring, Battiste investigated the solvolysis of exo-syn-6-tricyclo[3,2,1,$0^{2,4}$]-octenyl-8-p-nitrobenzoate *571* and exo-anti-6,7-benzotricyclo[3,2,1,$0^{2,4}$]octenyl-8-brosylate *572* [729].

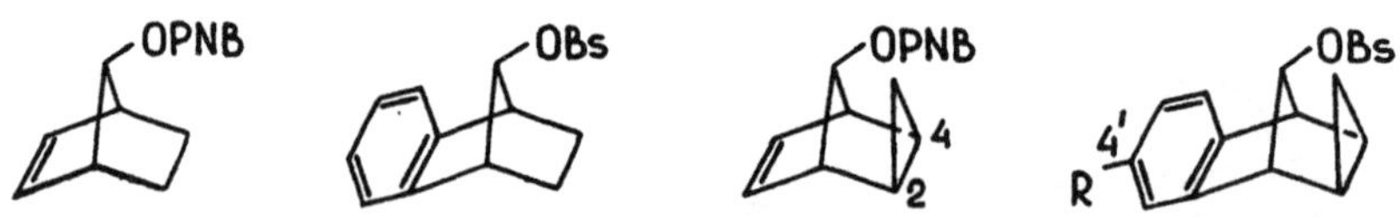

220 247 571 572

The esters *571* and *572* proved to be solvolysed at rates 10^3–10^4 times higher than the compounds *220* and *247*, respectively. The esters *571* and *572* are solvolyzed even faster than the 7-norbornadienyl analogues *573* and *574*, 5- and 45-fold, respectively.

The esters *571* and *572* are solvolysed into acetates with retained configuration in the same skeleton. The observed acceleration with the exo-condensed cyclopropane ring can be accounted for by the following reasons:

a) homoconjugation of the cyclopropane edge $C^2—C^4$;

b) weakening of unbonded interactions in the transition state of solvolysis (steric acceleration);

c) deviation of the bridged carbon atom toward the anti-double bond or the benzene ring due to unbonded interactions between the substituent X and the hydrogen of methylene group of the cyclopropane ring which should lead to an increase in the π-participation of the olefine bond or the benzene ring. The relative role of these effects is the acetolysis of exo-anti-benzotricyclo[4,2,1,$0^{2,5}$]nonyl-9-brosylate *575* proceeds at a rate intermediate between the rates for the compounds *572* and *247*.

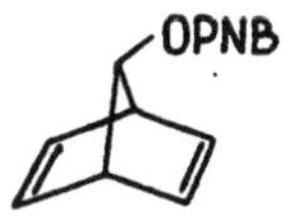

573

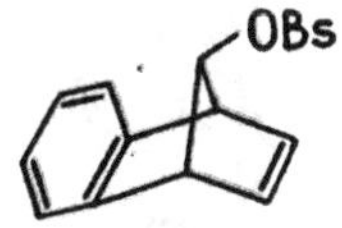

574

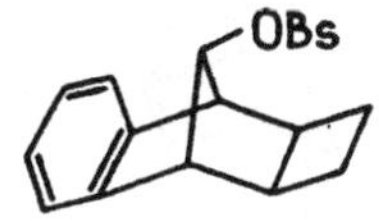

575

This result agrees with assumptions (b) and (c) because the unbonded interactions between X and the exo ring in the ground state must be less for the brosylate *575* than for the compound *572*. On the other hand, the ratio of rate constants $k_{575}:k_{247}$ ($\sim 10^2$) is comparable with the ratio $k_{572}:k_{247}$ (10^3–10^4). At the same time the difference between the direct participation efficiency of the cyclopropane edge and cyclobutane amounts to about 10^9–10^{10} [730,731]; this throws doubt upon the cyclopropane ring participation in the solvolysis of compound *572*. Then the effect of the substituent R in the 4'-position on the solvolysis rate of *572*-type compounds was studied. If the accelerating effect of the cyclopropane ring has a steric reason, then the effects of the 4'-R substituent must be similar for compounds *572* and *247*. But if it is an electronic effect, then the participation of the C^2—C^4 edge will depend on the electronic effect of the R substituent. Indeed, as Heilbronner et al. have recently shown with photoelectronic spectra [732], the exo-tricyclo[3,2,1,$0^{2,4}$]octene-6 possesses an essential homoconjugation between the p-orbitals of the double bond and the Walsh symmetrical orbitals of the cyclopropane ring, the value of this effect is similar to that for the interaction between the π-orbitals of two double bonds in norbornadiene.[13]

The 4'-R substituent turned out to exert absolutely equal effects in the solvolysis of both derivatives of *572* and *247*. All these data support the steric reason for the accelerating effect of the exo-cyclopropane ring. Accordingly the increased reactivity of exo-brosylate *564* in comparison with the endo isomer *566* (see above) is most probably caused by the steric acceleration due to the proximity of the OBs and CH_2 groups.

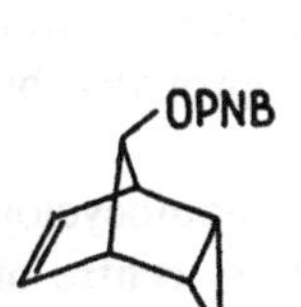

576

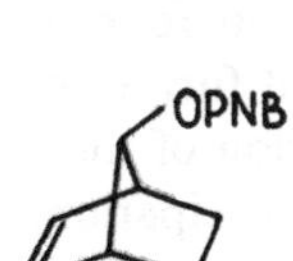

220

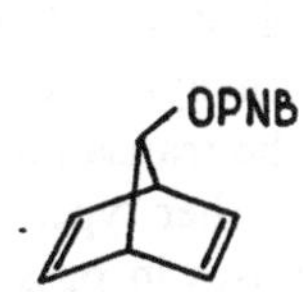

573

13 In the endo isomer the similar interaction between the double bond and the cyclopropane ring is insignificant.

As shown recently by Jovanovich and Lambert [733], syn-endo-6-tricyclo[3,2,1,$0^{2,4}$]-octenyl-8-p-nitrobenzoate *576* is solvolysed 10 times faster than compound *220* but 10 times more slowly than *573*. There are no reasons for the accelerating effect of the syn-endo ring; the solvolysis is not accompanied by skeletal rearrangements and, hence, there is no relief in the steric strain. The authors detected no bridge-flipping in the intermediate ion *A* which might lead to a more stable ion *B*.

(A) (B)

The solvolysis rate of the syn-epimer *564* is 8000 times higher than that of anti-epimer *563* [719]. Wilt et al. [734] showed compounds *577* and *578* to solvolyze

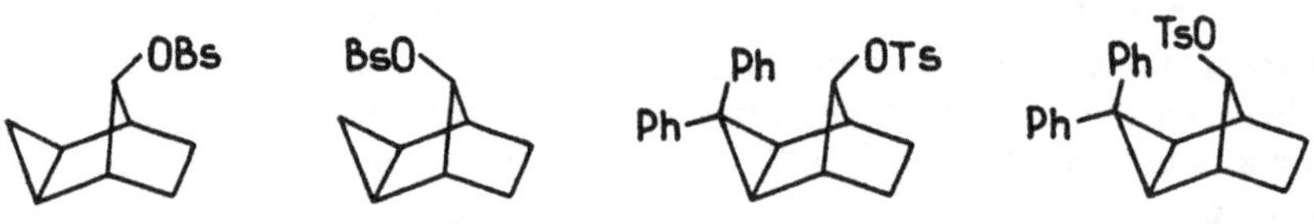

563 *564* *577* *578*

at comparable rates. According to the authors, such a sharp acceleration in the compound *577* results from the transannular participation of the phenyl group and of the subsequent disrotatory opening of the cyclopropane ring. The participation of phenyl in tosylate ionization accelerates the solvolysis more than 10^3-fold

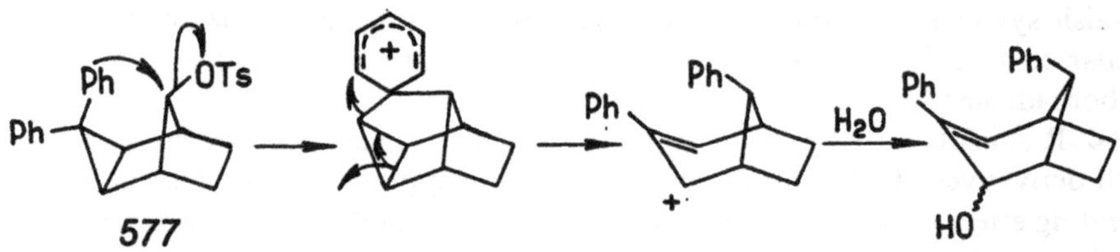

577

in comparison with that of compound *563*. In the case of syn-epimer *578* such a participation is impossible, its reactivity is somewhat lower than that of compound *564*. The solvolysis of *578* results in four products by σ-participation (as in the 7-syn-norbornenyl tosylate [395]) and further either the cyclopropylcarbinyl rearrangement or the transannular migration of the phenyl group.

Just as in other types of bond participation, the extent of homocyclopropyl interaction decreases in passing from the secondary to the tertiary centre of ionization. Thus, while the difference in the solvolysis rates of compounds *555* and *4* is 10^{14}, in those of *579* and *580* it is only 10^3 [735].

In the compound *579* stereospecificity has been observed in product formation. The compound *581*, and not *582*, is a kinetically controllable product in this rearrangement. It is also shown clearly that there is a definite upper limit for the

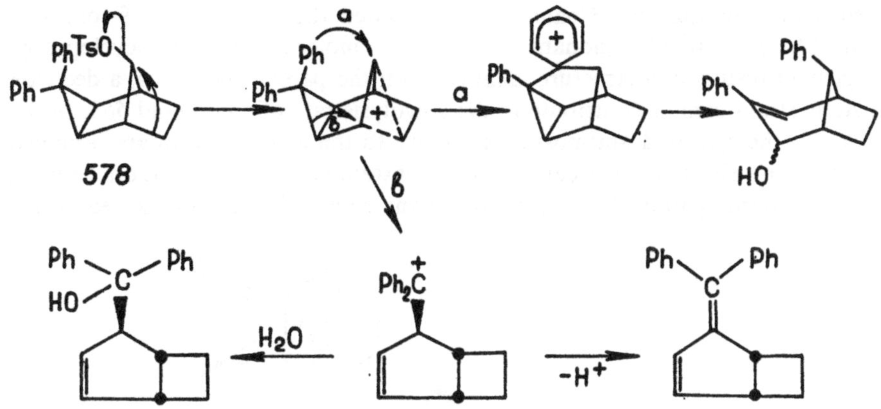

578

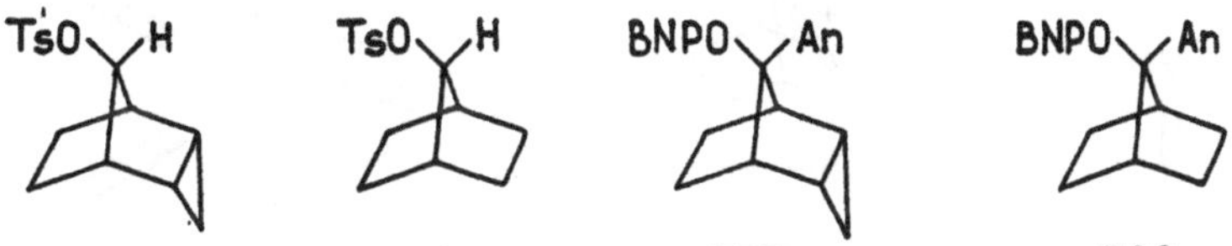

555 4 579 580

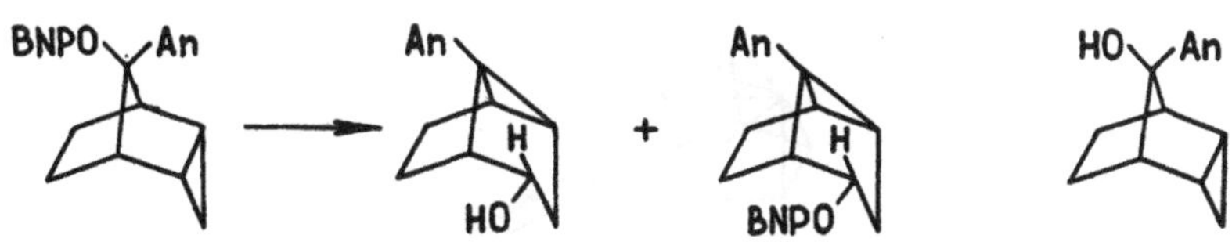

579 581 582

stabilizing effect of the aryl group on the benzyl cation centre. Thus for the anisyl group, as follows from the work under discussion, this limit amounts to $3 \cdot 10^{10}$ (levelling capacity) and this value seems to be independent of the nature of the participating group.

8.3 Contribution of Steric Strain

As the participation of the cyclopropane ring accelerates the solvolysis of compound *555* by 10^{14} in comparison with 7-norbornyl tosylate *4*, it was unclear what was the contribution by the steric strain relief because the developing tricyclic system is less strained.

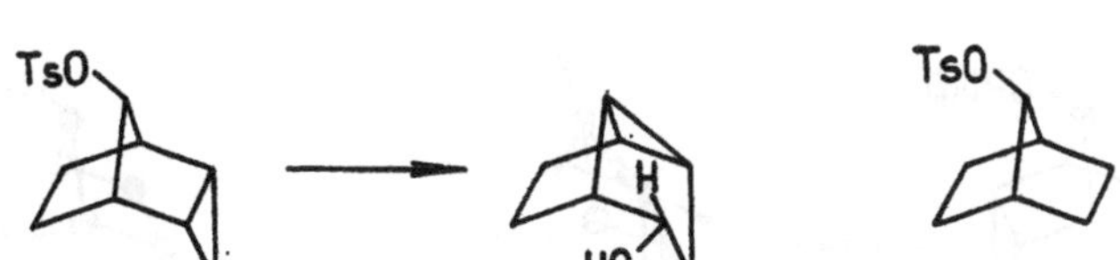

555 4

To elucidate this question Coates [736)] investigated the acetolysis of 9-tosyloxy-pentacyclo[4,3,0,$0^{2,4}$,$0^{3,8}$,$0^{5,7}$]nonane *583*. In this molecule the homocyclopropyl rearrangement results in a structure identical with the parent one. Thus, a decrease in the steric strain cannot contribute to the solvolysis rate of compound *583*. In the molecule *583* the syn- and anti-positions relative to the leaving group are occupied by cyclopropane rings, so one can study the intramolecular competition between syn- and anti-participation. If both cyclopropane rings take part in the rearrange-

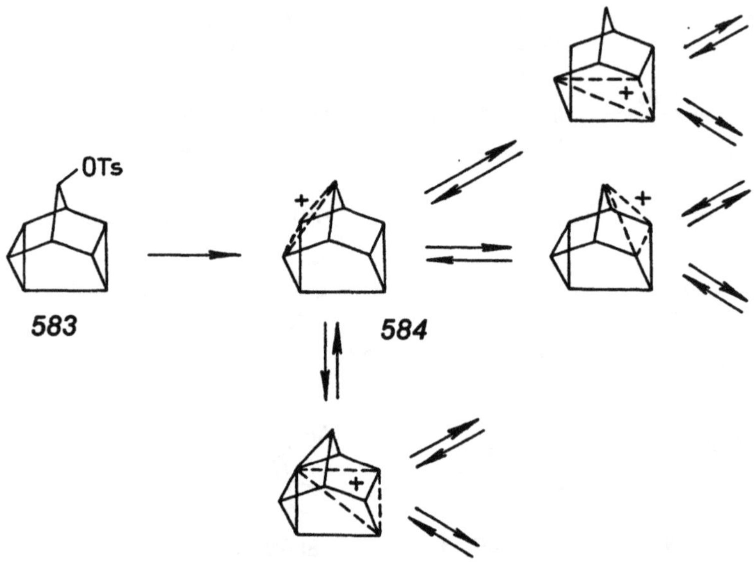

ment, all carbon atoms will be equivalent. The authors have shown the solvolysis rate of *583* to be 10^{12} times as high as that of *4*. The deuterium was distributed uniformly from compounds *583'* among the 9-, 2-, and 3-positions, and from *583''* among the 1-, 4- and 8-positions. The data obtained can be interpreted by the

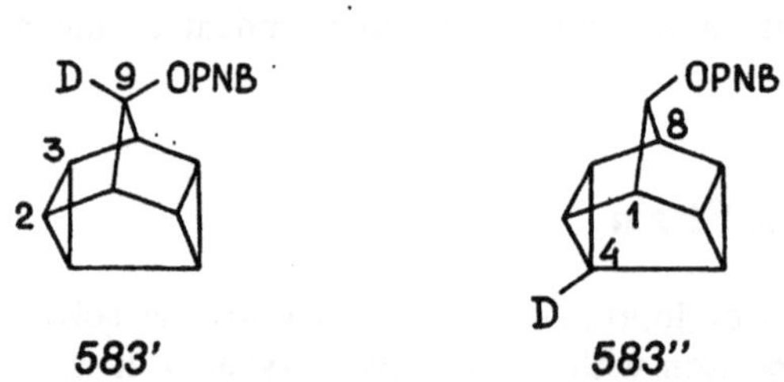

intermediate formation of a symmetrical trishomocyclopropenyl cation *584* in which only the anti-cyclopropane ring participates.

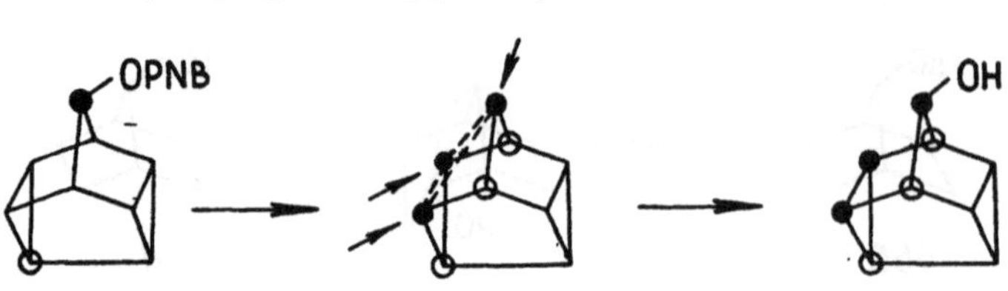

Analysis of substituent effects [737], application of the method of electron demand increase [738], direct observation of a stable ion by the methods of ^1H and ^{13}C NMR [739], theoretical analysis [740] — all these data indicate that ion *584* is really a nonclassical trishomocyclopropenyl cation.

As noted by Dauben, the reason for the high homocyclopropyl acceleration observed by Coates might be both the favourable position of the cyclopropane edge relative to the developing cation centre and the formation of a stable symmetrical (within the limit of a degenerate) cation. To verify this the authors investigated the acetolysis of 9-tosyloxypentacyclo[4,4,0,$0^{2,4}$,$0^{3,8}$,$0^{5,7}$]decane *585*; in this compound the cyclopropane ring is also located on the rear side, but the formed ion cannot take part in the degenerate rearrangement. According to Foote's correlation [320] the solvolysis rate of compound *585* could be about the same as that of cyclohexyl tosylate; in fact compound *585* solvolyzes 10^3 times faster. This acceleration should be atributed to homocyclopropyl participation. The acetolysis results in compound *586*.

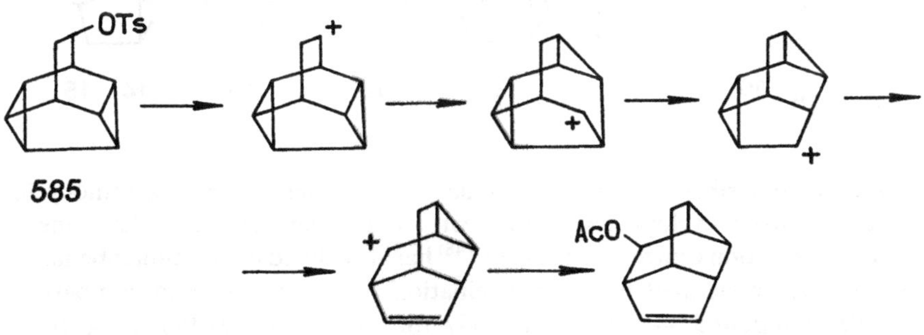

585

586

The value of the homocyclopropyl acceleration observed is lower than in the systems where the cation formed is capable of rapid degenerate rearrangement [741].

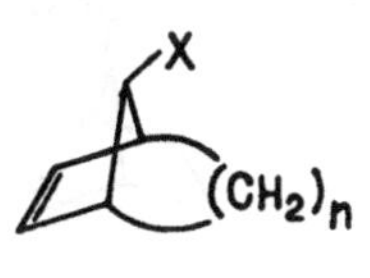

As we pointed out earlier, in bicyclic olefines of the type with decreasing number of methylene groups in the syn-located bridge the rate of solvolysis increases (in passing from n = 2 to n = 1–10^3-fold [742]). Coates [743] attempted to investigate a similar effect of shortening the syn-bridge by the homocyclopropyl participation. For this purpose the authors have shown that the acetolysis rate of 4-tosyloxy-exo-

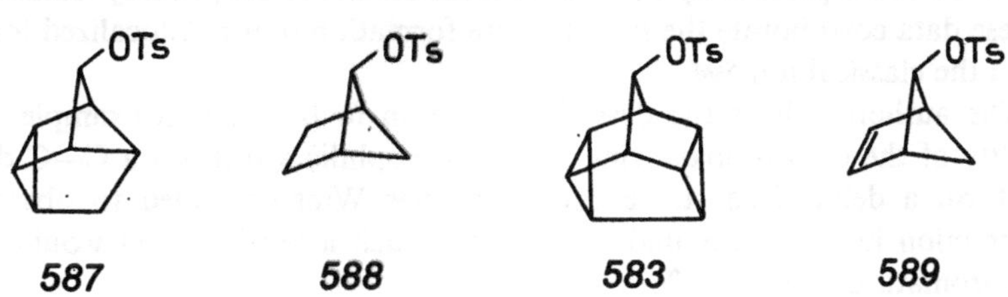

587 588 583 589

tetracyclo[3,3,0,$0^{3,6}$,$0^{2,8}$]octane *587*, although far higher than that of *588*, is nevertheless much lower than that of acetolysis of compounds *583* ($\sim 10^{-1}$–10^{-3}) and *589* ($\sim 10^{-5}$). Although in passing from compound *583* to compound *587* one should expect an essential increase in the efficiency of overlapping interacting orbitals (by analogy with homoallylic participation), but in the case of compound *587*, as the C^1 and C^2 atoms move toward the developing cation centre at C^4, the steric strain increases in the C^7 node binding the cyclopropane ring to the cyclobutane one which destabilizes the transition state and the forming trishomocyclopropenyl cation.

A comparative activity series of some homocyclopropylcarbinyl systems in acetolyses is reported by Ellen and Klump [744]:

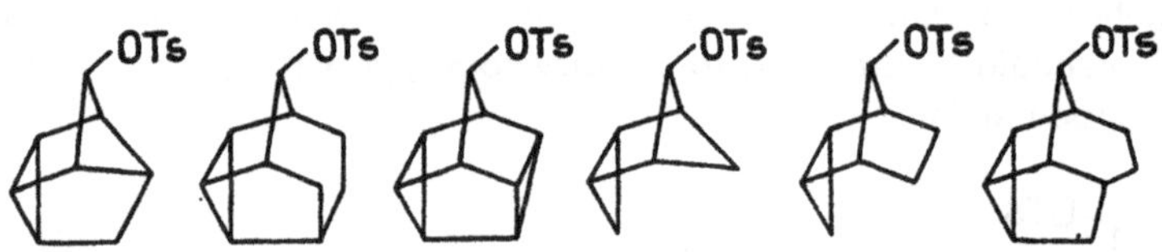

$\lg \frac{k}{k\,(7\text{-norbornyl})}$ 10 11 12 13 14 14 . 15

The literature describes several intermediate formations of the trishomocyclopropenyl cation from cyclopropane rings with a distance of more than one C atom from the ionization centre [745]. Sargent [746] has solvolysed allylic dinitrobenzoate *590* being greatly accelerated by the participation of cyclopropane in comparison with a model compound — allylic ester *591* (622-fold). The participating cyclopropane in the γ-position turned out to be more effective than that of similarly located double bond in compound *592* ($k_{590}:k_{592}$ = 23.6).

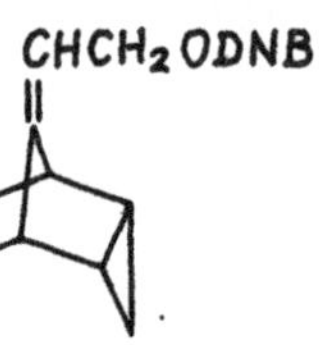

590

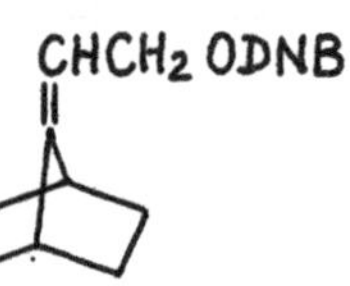

591

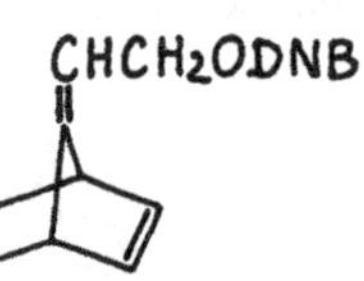

592

In the solvolysis of *590* both the solvent and the dinitrobenzoate anion attack the intermediately formed cation stereospecifically (from a sterically hindered side) at the carbon separated by 4 (!) bonds from the site of the primary ionization (>4 Å). These data corroborate the intermediate formation of σ,π-delocalized ion *593* rather than the classical ion *594*.

The authors believe the described reaction to be the first example showing the ability of the cyclopropane ring for a nucleophilic attack on a C—C double bond to form a delocalized stable cation. Earlier Winstein failed to observe such an interaction in the compound *595* because such a participation would result in an antiaromatic cation *596* [516].

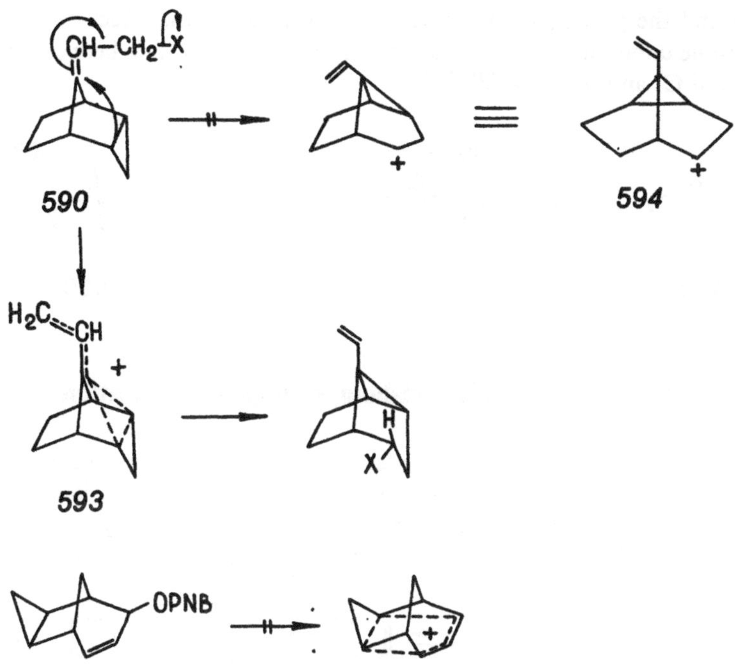

590 594 593 595 596

The extent of participation of a remote cyclopropane ring is greatly dependent on its relative orientation. Thus, in the compounds *597* and *598* the reaction centres are separated from the cyclopropane ring by 4 bonds (5 C atoms); in the first case, however, the ring participation is far more effective [747].

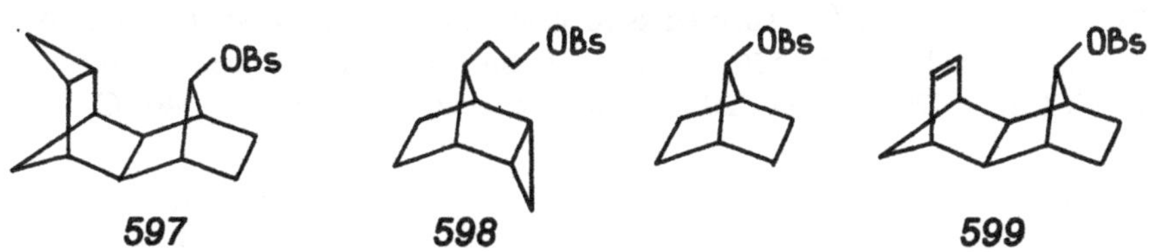

597	598		599
$k_{rel} = 10^{8.2}$	$10^{0.5}$	1	10^{10}

The less intensive participation of the three-member ring in compound *597* as compared with the double bond in the ionization of compound *599* is due to the fact that the p-orbitals in the latter are oriented right to the reaction centre while the respective cyclopropyl orbitals are directed to the region between two bridges; this is less favourable for the interaction with the developing cation centre.

The solvolysis of compound *598* in 2,2,2-trifluoroethanol proceeds 3 times as fast as that of the isomer *600*; this is accounted for by a slight effect produced by the participation of cyclopropane in the former case [748]. The authors have denoted this kind of participation as "∇_5-participation" (5 carbons between the reaction centre and the participating bond). The low value of the effect may be due to unfavourable unbonded steric interactions in the transition state between the hydrogens at C^2 and C^4 and those at C^{10} [749].

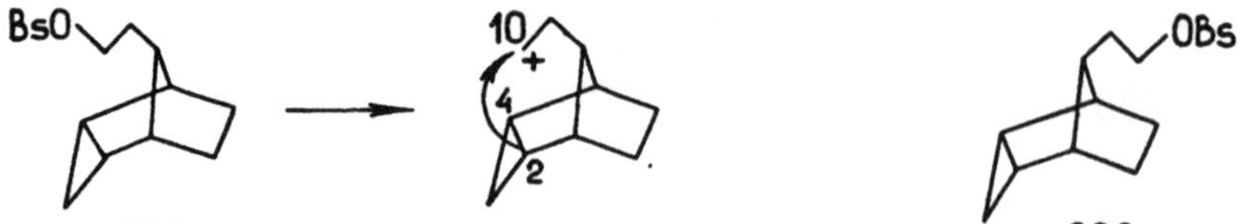

598 600

8.4 Direct Observation of a Trishomocyclopropenyl Cation

In 1972 Olah [750] attempted to observe directly a trishomocyclopropenyl cation postulated in all the works described above. However, the ionization of cis-3-bicyclo[3,1,0]hexanol in $FSO_3H—SbF_5—SO_2FCl$ resulted only in a 3-methyl-cyclopentene-2-yl cation. The primarily formed bicyclic cation seems to be very unstable,

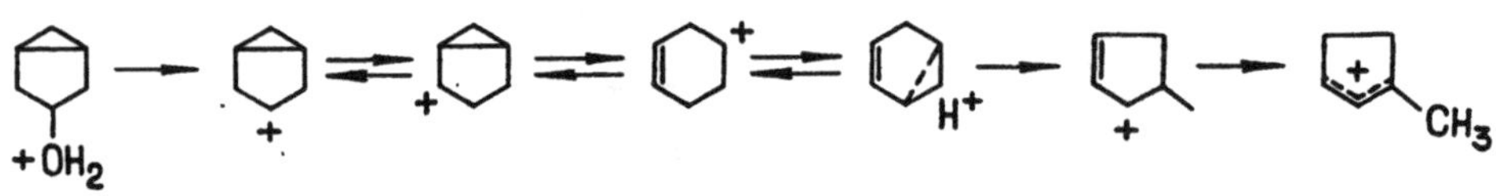

there immediately occurs a hydride shift and a cycle contraction. According to Olah, the impossibility of observing directly the trishomocyclo propenyl ion shows this ion to be unstable under the conditions where the other nonclassical ions are stable. This is reflected in a low cis: trans rate ratio of the bicyclo[3,1,0]hexanol derivatives.

In 1974 Masamune et al. [691] succeeded in generating for the first time a stable trishomocyclopropenyl cation *520* from cis-3-chlorobicyclo[3,1,0]hexane *601* in $SbF_5—SO_2FCl$ at —120 °C. Ion *520* is stable between —120 °C to —40 °C. The

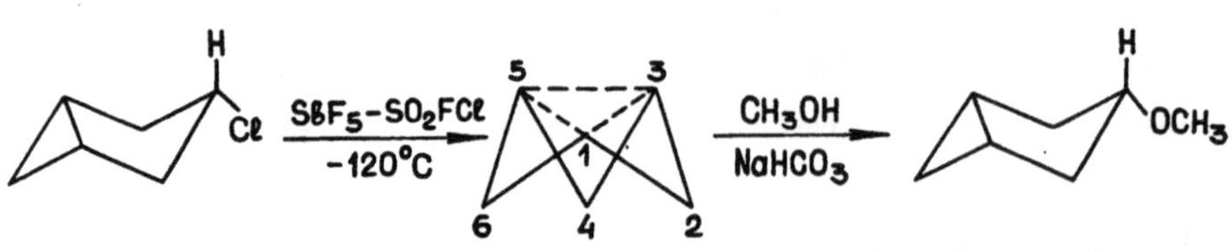

601 520 602

nonclassical structure of the cation agrees well with the character of the *D*-label distribution in the ion generated from 2,6,6-trideutero-cis-3-chlorobicyclo[3,1,0]-hexane, as well as with the results of interaction between an acid solution of a salt of ion *520* and methanol, i.e. formation of the cis-methoxyderivative *602*. Note should be made of the unusual spectral characteristics of ion *520*. Thus, the C^1, C^3 and C^5 atoms formally carrying the positive charge have an uncommonly up-field chemical shift (4.7 ppm) in the ^{13}C NMR, even more up field than the atoms C^2, C^4 and C^6 (17.3 ppm). One should also note rather high values of the $J(^{13}C—^{1}H)$ constants both for carbons contained in the 2e3c-molecular orbital (196 cps) and for methylene atoms C^2, C^4 and C^6 (168 cps). All these, at first sight unusual, characteristics of ion *520* appear to be due to the unusual hybridization of pentacoordinated atoms C^1, C^3 and C^5, as well as to the higher steric strain of the "cyclopropane" skeleton.

That Olah et al. [750)] failed to generate the stable ion *520* is attributed by Masamune et al. [691)] to the wrong choice of the precursor. According to [691)], in alcohol *603*, the primary stage is the protonation resulting in an oxonium ion *604* in which the intramolecular rearrangements leading to a methylcyclopentenyl cation proceed faster than the ionization to ion *520*.

603, X = OH
604, X = $\overset{+}{O}H_2$
605
606

Later Olah et al. [751)] showed that the trans-chloride *605* with $SbF_5—SO_2Fl$ yield a 3-cyclohexenyl cation *606*, but not the ion *520*; thus, just as under solvolytic conditions, ion *520* is formed only when σ-participation is possible in the transition state of ionisation. In the same paper it was pointed out that ion *520* could be generated from the alcohol if freshly distilled SbF_5 in SO_2FCl (—78 °C) was used; the admixtures of proton acids, the authors maintain, favour the opening of the cyclopropane ring before ionization. The ionization of 3-methyl- *607* and 3-phenyl-cis-bicyclo[3,1,0]hexanol-3 *608* in $SbF_5—SO_2FCl$ results in the 3-methyl-cyclohexenyl ion *609* and the 3-methyl-1-phenylcyclopentenyl ion *610*. That the ionization of tertiary substrates results in classical tertiary carbocations rather than trishomocyclopropenyl ions appears to be due to the hindrance in passing from the conformation of the bath to the chair which is indispensable for the formation of a delocalized ion.

607 *609* *608* *610*

Masamune et al. [752)] generated another stable trishomocyclopropenyl cation *611* with an ethane bridge; the latter is obtained from 8-chlorotricyclo[3,2,1,0^{2,4}]octane *612* or from 4-chlorotricyclo[3,3,0,0^{2,8}]octane *613* in $SbF_5—SO_2FCl$. The data of the ^{1}H and ^{13}C NMR spectra of ion *611* considering the peculiarities noted for ion *520* are used by the authors [752)] to confirm its nonclassical structure; some contribution to the distribution of the positive charge is made by the pyramidal structure *614*.

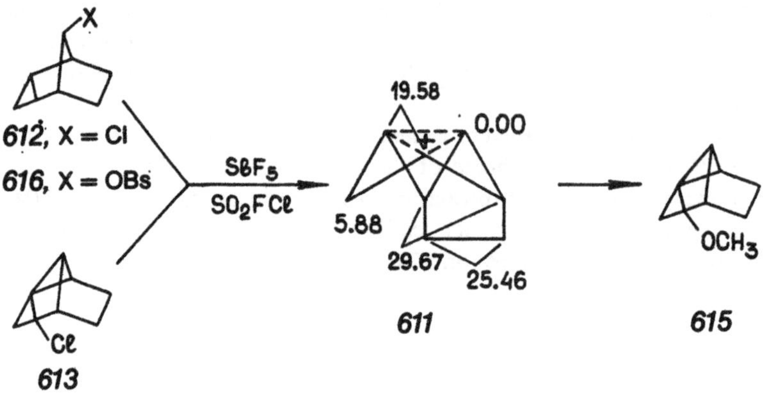

614

The acid solution of ion *611* interacts with a $CH_3OH—Na_2CO_3$ mixture to form practically the only product — ether *615*; the latter is also formed by the methanolysis of brosylate *616* [719].

Coates and Fretz [753] have noted that tosylate *539* solvolyzes 100 times more slowly than the structurally similar tosylate cis-2-hydroxytetracyclo[4,3,0,$0^{3,8}$,$0^{7,9}$]-nonane *617a*. The solvolysis of the latter, just as that of tosylate *539*, is expected to involve the intermediate formation of the trishomocyclopropenyl cation. The authors have shown, however [753], that the acetolysis of the monodeuterated derivative *617b* results in acetate *618* with the label so distributed as to suggest the intermediate formation of either a rapid equilibrium of two trishomocyclopropenyl cations or an ion of pyramidal structure.

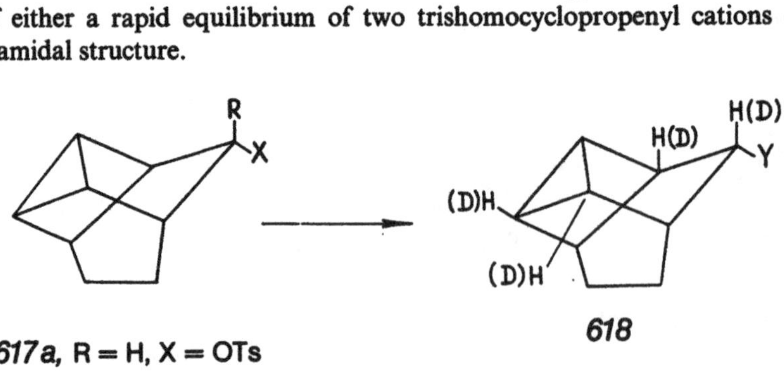

618

617a, R = H, X = OTs

617b, R = D, X = OTs

617c, R = H, X = OH

The authors [753] have generated a stable carbocation by the interaction of nonanol *617c* with SbF_5—SO_2FCl at —130 °C. Judging by the ^1H and ^{13}C NMR spectra the ion obtained should have the pyramidal structure *619*. Cited below are the chemical shifts of the signals of the hydrogens (parenthesized) and carbons of this ion.

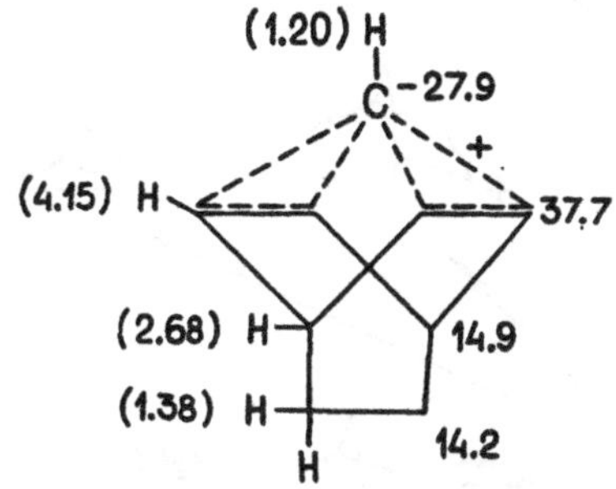

619

Note should be made of the unusual chemical shift values of protons and carbon atoms, especially of the apical carbon atom resonating in a far higher field than TMS.

8.5 Cation Centres from Unsymmetric Compounds

So far the participation of the cyclopropane ring located symmetrically to the developing cation centre has been discussed. A number of publications deal with the ring participation when the *developing cation centre* is *located unsymmetrically* to the ring edge [710]. These works show essential differences of opinion on the problem of stereochemical requirements for the interaction of the cyclopropane ring and the cation centre (cf. [754–757]). Thus, Wiberg [723] has investigated the solvolysis of three isomeric brosylates *620–622* — derivatives of tricyclo[3,2,1,$0^{2,4}$]octanols-6 and nortricyclene-3-methyl acetate *623* (cf. [758,759]). The acetolysis of all the four isomers

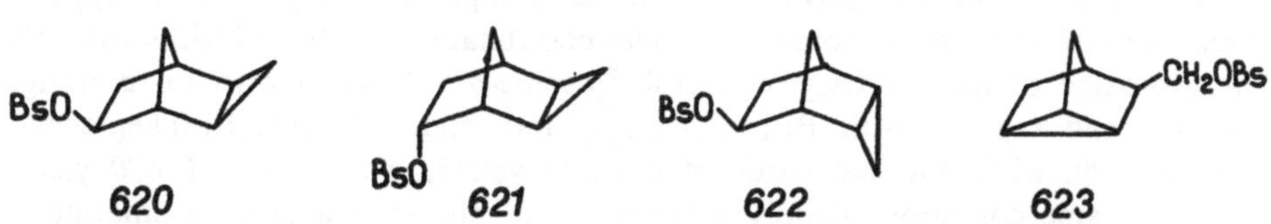

620 *621* *622* *623*

results in identical reaction mixtures; the rates of solvolysis of compounds *620* and *622* are partically equal (cf. [724]). The authors hold the participation of the cyclopropane ring inefficient; the process follows the scheme similar to that for the solvolysis of 2-norbornyl brosylates.

A change in the relative position of the cation centre and the edge of cyclopropane leads to different results. For instance, Freeman [726,760–763] has investigated the acetolysis of 8-exo *624* and 8-endo-deltacyclyl brosylates *625*; the data on

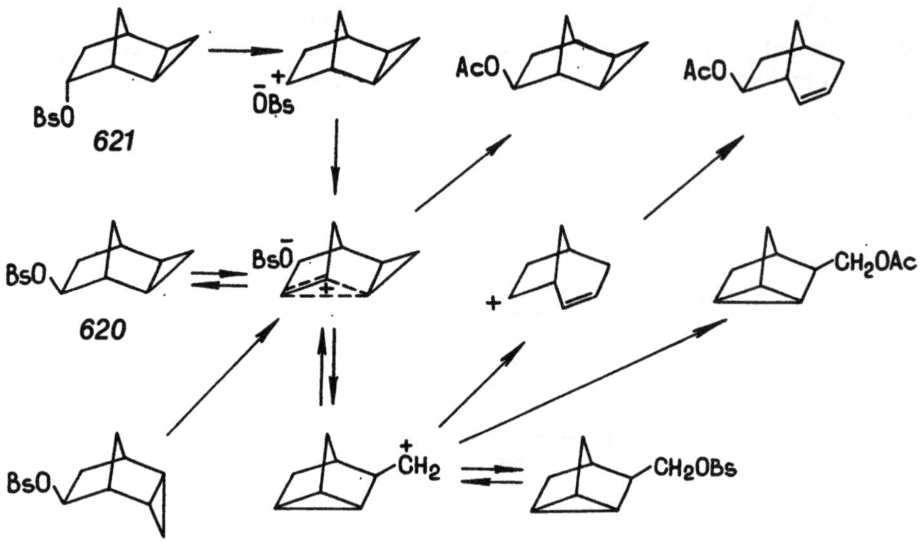

621 620 622

reaction rates, racemation of optically active substrates, distribution of deuterium in acetolysis products, and secondary isotopic effect clearly show the ionization of *624* to result directly in a nonclassical ion *626* while the ionization of *625* leads to the primary formation of a classical ion *627*.

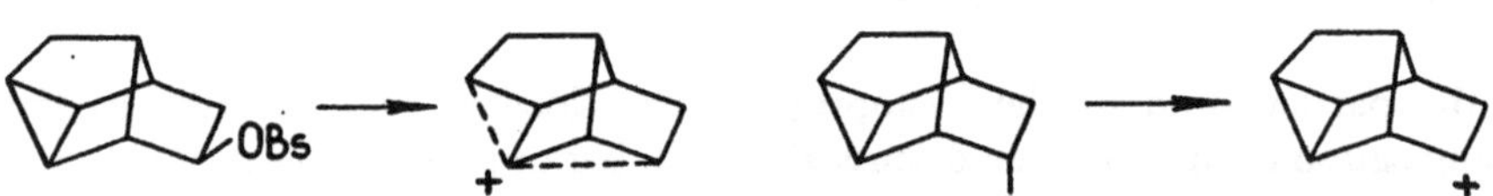

624 626 625 627

As we have noted earlier, the extent of anchimeric participation is not only greatly dependent on the relative position of the participating group and the developing p-orbital, but also on the nature of the detaching group. Thus, Sargent has discovered an effective participation of the cyclopropane ring in similar systems in reactions of opening the isomeric α-oxides coordinated with boron trifluoride [764]. The exo-oxide of the exo-tricyclo[3,2,1,$0^{2,4}$]octene-6 *628* was shown to rearrange smoothly and rapidly, with $BF_3 \cdot O(C_2H_5)_2$, into endo-7-bicyclo[4,1,0]heptene-3-aldehyde *629*, while the exo oxide of endo-tricyclo[3,2,1,$0^{2,4}$]octene-6 *630* yields under the same conditions mainly a polymer and but an insignificant amount of aldehyde *629*.

628 632 629

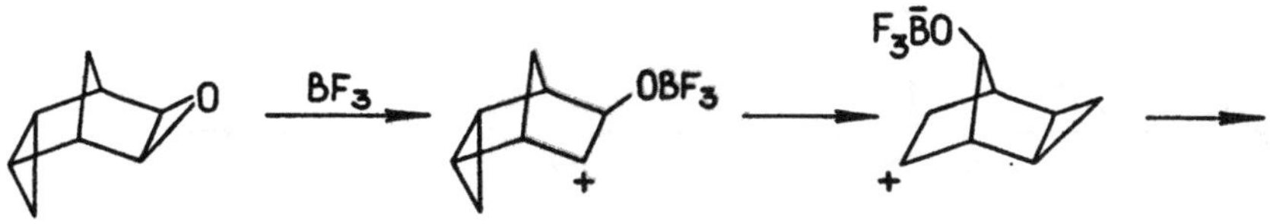

630

631 *629*

These data are due to the fact that only in the oxide *628* is it possible for cyclopropane to take part in the operning of the α-oxide cycle. In the classical ion *631* no simultaneous rearrangement into aldehyde *629* is possible (as distinct from ion *632*) since this would lead to cyclohexene with the trans-configuration of the double bond.

8.6 Participation of the Spirocyclopropane Ring

The assumption that there were sufficiently favourable conditions for the participation of the cyclopropane edge could not be verified [765], since the 7-spiro-cyclopropane ring in the 2-endo-tosyloxy-norbornane *633* [766] and the 2-endo-tosyloxy-benzonorbornene *634* fails to exert an essential effect on the rate of solvolysis.

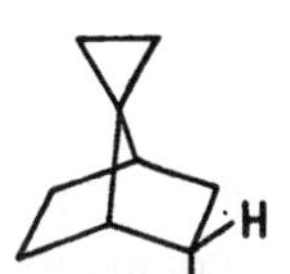

633

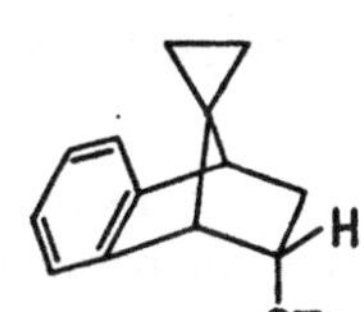

634

At the same time the acid-catalyzed decomposition of 3′-diazospiro[cyclopropane-1,7′-norbornanone-3′] *635* proceeds with the participation of C-7′-spiro-cyclopropyl group; this seems to be due to increased electron demand of the developing cation centre under the influence of the neighbouring carbonyl group [767].

Sherrod [768] and Schleyer [769] dealt with the interaction of the face of the cyclopropane ring with the cation centre. Triflate *636* was shown to solvolyze 10^4 times as fast as triflate *637*. The proximity of the cyclopropane ring "face" does not have any stabilizing effect on the developing cation centre in ion *638*. In general the approach of the electrophile to the cyclopropane edge is far more favourable than that to the cycle "face". Walsh's model [28] for the cyclopropane ring makes it possible to understand the reason for the difference in the activity of the edge and the

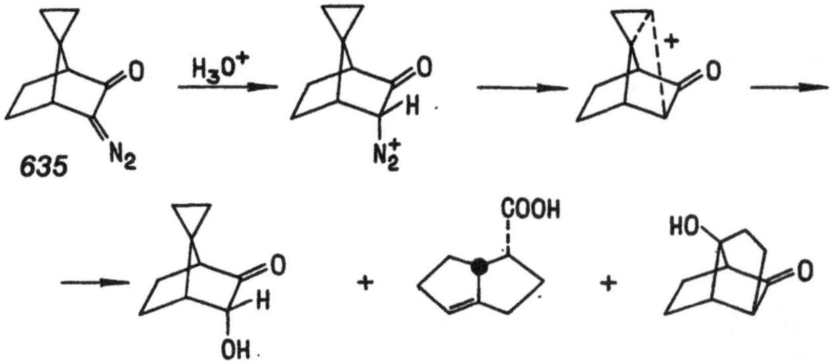

635

"face". When interacting with the "face" the vacant p-orbital mixes with three sp^2-orbitals directed to the ring centre. When the edge is attacked the vacant p-orbital mixes with peripheral p-like orbitals. The comparison of ionization potentials for the sp^2-orbitals ($\sim$14.7 ev) and p-orbitals ($\sim$11.4 ev) shows that the shift of electron density to the electrophile in the latter case proceeds more readily [770].

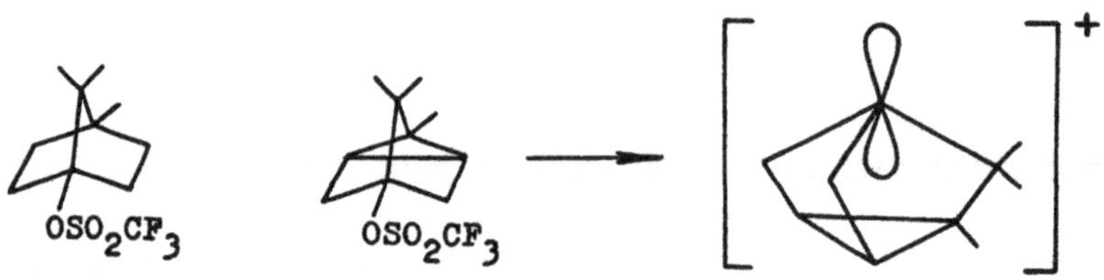

636 637 638

8.7 Comparison between Mono- and Bicyclic Systems

While in rigid bicyclic systems the participation effects of the double bond and the cyclopropane ring are comparable, in open freely rotating compounds the participation of the cyclopropane ring is far less than that of the similarly located double bond. This confirms again the significance of the rigid orientation of the ring with respect to the cation centre. Thus, Rhodes and Takino [771] showed the rates of ethanolysis, acetolysis and formolysis of 2-cyclopropylethyl brosylate *639* and isoamyl brosylate *640* to be practically equal; this points to the absence of anchimeric assistance of the ring; compound *640* was used for comparison as the best approximation to the steric requirements in the cyclopropylethyl system.

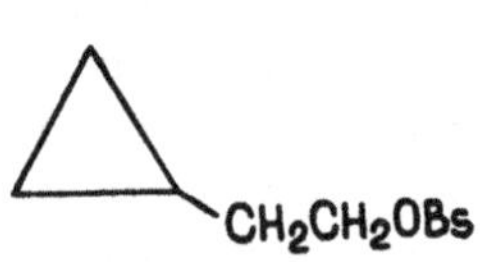

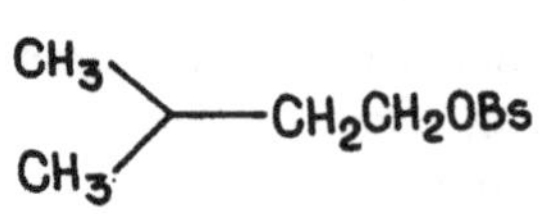

639 640

Judging by the value of m for the solvolysis of compounds *639* and *640* (0.34 and 0.33 at 75 °C), in both cases the reaction proceeds with a high degree of nucleophilic participation of the solvent (the high "N-character" [772]). Earlier Sauers and Ubersax concluded that the solvolysis of 2-cyclopropylethyl brosylate *639* proceeds with the formation of an intermediate classical cation [773]. The geometric requirements for an effective interaction of cyclopropane orbitals with the developing cation centre are far more rigid than for the double bond participation; thus, allylcarbinyl brosylate solvolyzes 17 times faster than n-butyl brosylate [774]. In Sargent's opinion [749] the absence of an effective participation of the cyclopropane ring in such systems may be due to unfavourable unbonded interactions in the transition state of the hydrogens of the cyclopropane ring with those bonded to the cation centre. This work shows the rate of solvolysis of p-nitrobenzoate *641* to be practically the same as that of cyclopentane analogue *642* but 87 times lower than that of cyclopentene one *643*.

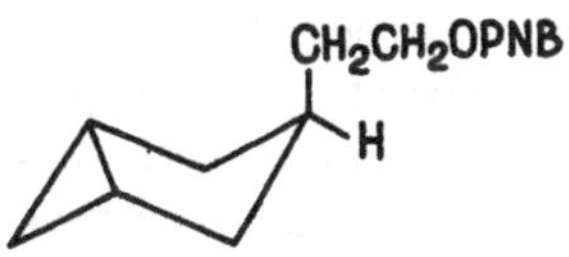

641

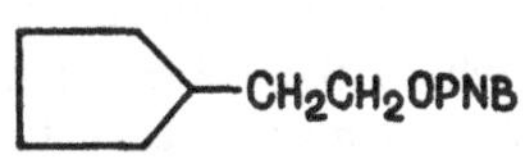

642

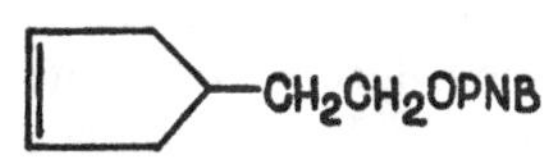

643

Dewar et al. [775] also dealt with the cyclopropyl participation in the solvolysis of 2-cyclopropylethyl brosylate *639*. They compared the solvolysis of compound *639* with its cis- *644* and trans-2'-methyl-substituted homologs *645* assuming all these compounds to react, in the absence of anchimeric ring participation, at the same

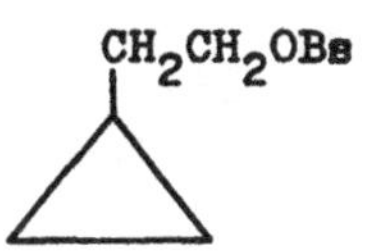

639

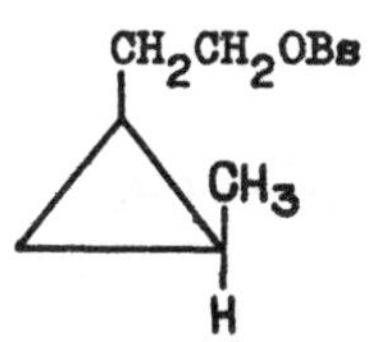

644

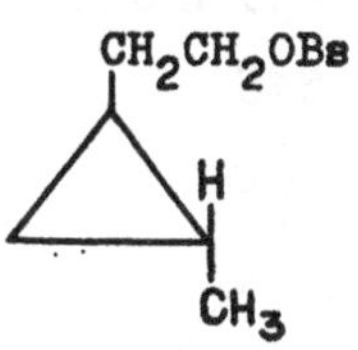

645

$k_{rel} = 1.0$ 1.33   3.24

rate. The relative rates of formolysis (75 °C) show very small differences; the authors could not draw convincing conclusions. This case, however, represents an exception, since the data peculiarities of the record an exceptionally constant character of the rates provided the environment closest to the reaction centre remains unchanged [776]. Therefore, the data obtained prove the cyclopropyl participation, at least, in the methyl-substituted *644* and *645*. This assumption is favoured by the rearrangement products; as the primary carbocations are unstable, the solvolysis without anchimeric participation must proceed with an S_N2-attack by the solvent resulting exclusively in unrearranged products. According to Dewar the structure of the intermediate nonclassical ion can be represented by the π-complexes *646* or *647* depending on which bond of the ring exerts an anchimeric effect; in any case, however, the presence of a CH_3 group in the trans- and not in the cis-position is the most favourable for participation.

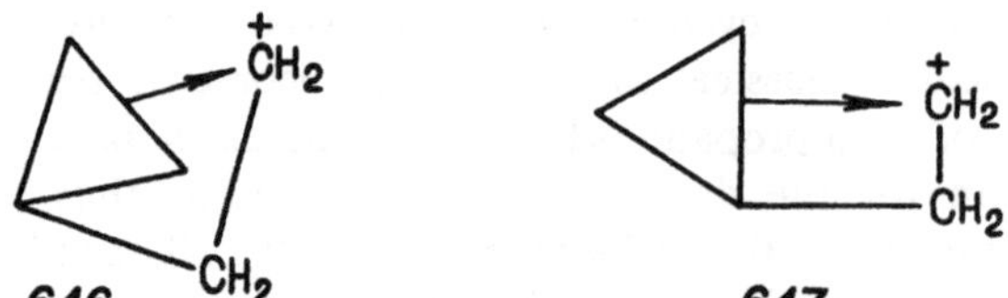

646 647

Later on Rhodes et al.[777] showed that in reactions of trifluoroethanolysis of 2-cyclopropylethyl tosylate and other conformationally mobile systems the cyclopropane ring is an efficient β-participating group; at the same time in less ionizing, more nucleophilic media the participation of the cyclopropane ring in ionization processes for simple acyclic systems cannot effectively compete with that of the solvent. In order to "relieve" the latter effect eclipsing the potential participation of the cyclopropane ring, Rhodes and Takino investigated the solvolysis of 2-cyclopropyl-2-methylpropyl brosylate *648* where the solvent participation is practically suppressed[778]. The compound *648* was acetolyzed 8 times as fast as 2,2,3-trimethylbutyl brosylate *649*; the latter was chosen as a steric model to establish the effect of the steric strain relief at the quaternary C atom.

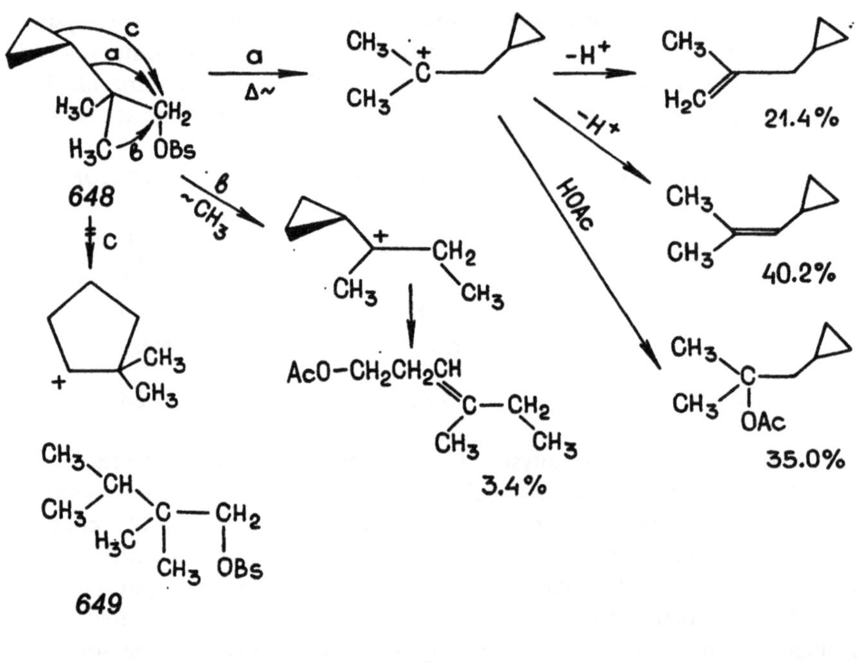

Judging by the reaction products the solvolysis involves the predominant migration of a cyclopropane radical rather than that of methyl. These data indicate that in the migration of the cyclopropyl fragment the transition state is located lower than in that of the methyl or isopropyl group although in the other systems the migrating ability of the secondary alkyl groups is higher than that of the cyclopropyl one[779,780].

The relative stability of the formed tertiary carbocations upon migration of the cyclopropane (a) or the methyl group (b) is not of great account since route (a) leads to a less stable cation.

The stabilization of the transition state should result from the delocalization of the ring σ-bonds which can be represented as an interaction of the homocyclopropenyl type *A*[713] similar to the protonation of cyclopropane[781]. This picture of interaction is, however, contradicted by the absence of cyclopentyl products.

The second type of interaction *B*, does not lead to the products observed. The simplest mechanism — the usual Wagner-Meerwein rearrangement — accounts neither for the higher rate of solvolysis nor for the predominant migration of cyclopropane. The C-type transition state comprising the delocalization of two cyclopropane bonds accounts for the observed rates and products (cf.[782]); however the authors admit that further research is necessry for a deeper understanding of the structure of the transition state and of the intermediates.

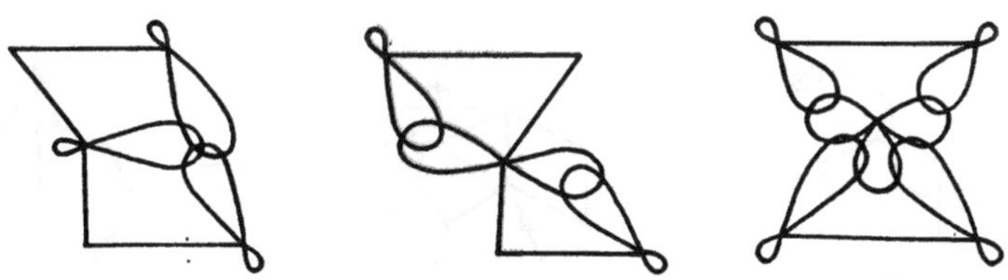

(A) (B) (C)

In other works it has also been shown that in β-cyclopropylethyl cations the migrating ability of the cyclopropyl group is much higher than that of the methyl group (e.g.[783]; cf.[784]); the stereochemistry of the migrating cyclopropyl group is wholly retained and the methyl-for-hydrogen substitution in the ring diminishes the migrating ability to make it 3 times lower[785].

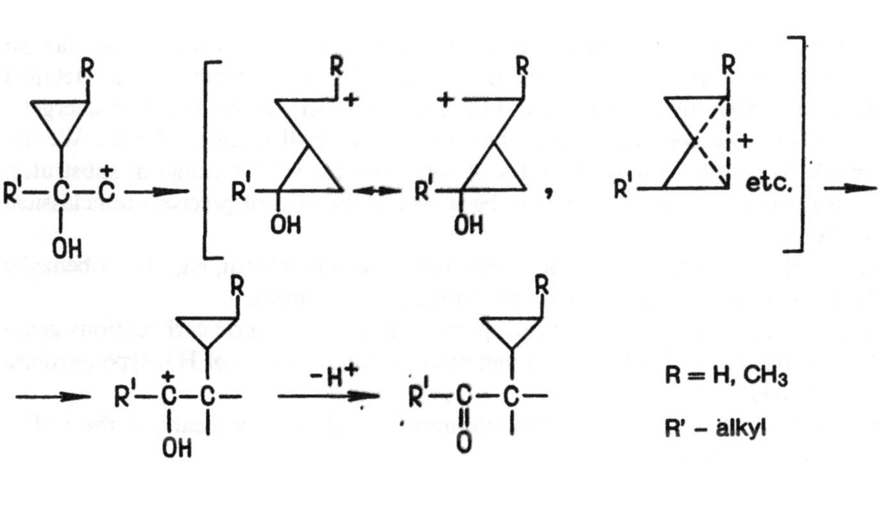

Theoretical calculations show the edge participation to be possible not only for cyclopropane but also for cyclobutane [786,787]. The first experimental works [730,731] reported the following relative rates of acetolysis (25 °C):

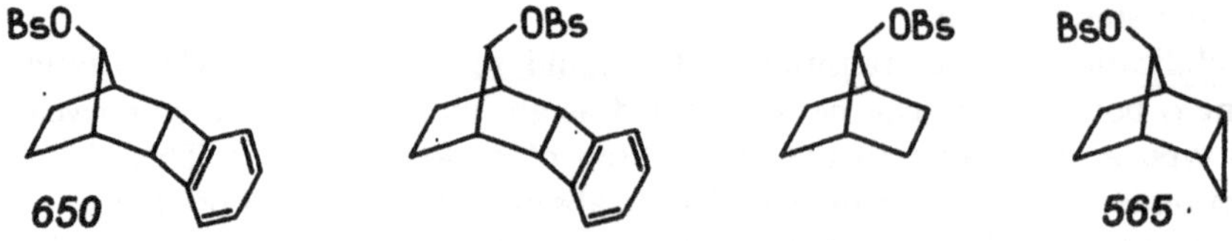

650 | 565

650			565
72200	1.1	1	10^5

The acceleration caused by the participation of the cyclobutane ring ($10^{4.9}$) upon solvolysis of compound *650* amounts to about a third of the anchimeric effect of the cyclopropane ring in the compound *565*. The calculated extent of bond bending for cyclobutane is also about a third of bond bending for cyclopropane. Similar data have also been obtained in the acetolysis of compound *651*.

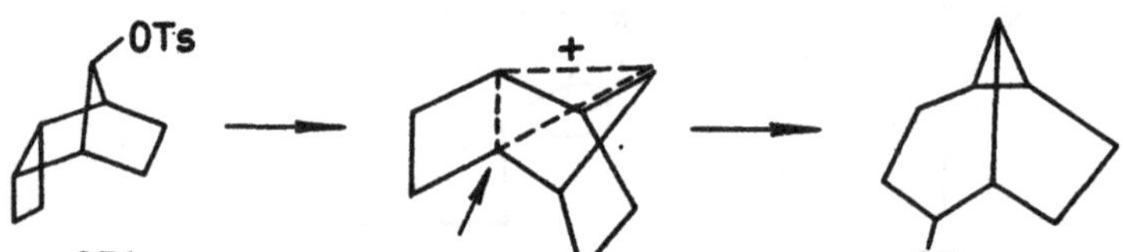

651

For the anchimeric assistance of the four-member cycle on solvolysis of α-cyclobutylcarbinyl compounds see [788,789], and for the inclusion of the cyclobutane ring in homoaromatic conjugation see [790].

9 Conclusion

In the chemistry of carbocations there are many unsolved problems. So far no general experimental technique has been worked out to establish the detailed structure of carbocations in the gas phase or in superacid media. The energetic characteristics have been determined but for a very small number of ions. We are still unable to predict exactly the rates of solvolysis for a wide range of substrates. There also remain specific problems to be solved in the field of precisely nonclassical carbocations:

1) generation of new types of nonclassical ions in a stable form, e.g., homobenzylic ions with a condensed benzene and "cyclopropane" fragment;

2) comparison of the structure and properties of stable isomeric carbocations generated in the reactions of substitution and electrophilic addition of HX-type particles to double bonds;

3) investigation of the structure of stable nonclassical ions by means of the ESCA spectra and X-ray analysis;

4) generation from epimers of stable isomeric nonclassical ions;
5) investigation of the chemical properties of stable nonclassical ions — their interactions with nucleophiles of various force, intramolecular rearrangements;
6) comparison of the structure and properties of carbocations formed from the same precursors in a wide range of media — from strongly nucleophilic to super-non-nucleophilic;
7) connection of the problem of nonclassical ions with that of "cold" and "hot" cations (solvolysis and deamination);
8) quantum chemical calculations of the spectral characteristics of isomeric classical and nonclassical carbocations;
9) application of data on nonclassical carbocations to the planning of works on synthesis (prediction of the stereochemistry of products);
10) investigation of nonclassical effects by using natural objects, first of all terpenes, where the substrate structure itself predetermines the possibility of long-range interactions of the developing cation centre with σ- and π-bonds;
11) investigation of quantitative regularities of processes with the formation of intermediate nonclassical ions applying the methods of molecular mechanics.

What is to-day's state of "wrestling" between the adherents and opponents to the concept of nonclassical carbocations? But way of answer recall the words of the well-known expert in the field of carbocation chemistry P. Schleyer [13]:
"... This marks the beginning of the end of this controversy, but the need for careful scientific work, soundly based interpretations and constructive criticism will always remain".

10 References

1. Bartlett, P. D.: Nonclassical Ions, New York, 1965
2. Sargent, G. D.: Quart. Revs., *20*, 301 (1966)
3. Brown, H. C.: Chem. Eng. News, *45*, 87 (1967)
4. Brown, H. C.: Chem. Brit., *2*, 199 (19666)
5. Gream, G. E.: Rev. Pure Appl. Chem., *16*, 25 (1966)
6. Berson, J. A.: In: Molecular rearrangements, vol. 1, New York, 1963, p. 111–231
7. Hanack, M., Schneider, H. J.: Angew. Chem. Int. Ed., *6*, 666 (1967); Fortsch. Chem. Fortsch., *8*, 554 (1967)
8. Dewar, M. J. S., Marchand, A. P.: Ann. Rev. Phys. Chem., *16*, 321 (1965)
9. Capon, B.: Quart. Revs., *18*, 45 (1964)
10. Huckel, W.: J. Prakt. Chem., *28*, 27 (1965)
11. Bethel, D., Gold, V.: Carbonium ions, New York. Acad. Press, 1967
12. Carbonium Ions, ed. Olah, G., Schleyer, P. v. R., vol. 3, New York, 1972
13. Brown, H. C.: The Nonclassical Ion Problem, New York, Plenum Press, 1977
14. Olah, G. A.: Chem. Brit., *8*, 281 (1972)
15. Olah, G. A.: Acc. Chem. Res., *9*, 41 (1976)
16. Olah, G. A.: Angew. Chem. Int. Ed., *12*, 173 (1973)
17. Hogeveen, H.: Adv. Phys. Org. Chem., *10*, 29 (1973)
18. Hogeveen, H., Kwant, P.: Acc. Chem. Res., *8*, 413 (1975)
19. Paquette, L.: Angew. Chem. Int. Ed., *17*, 106 (1978)
20. Brown, H. C.: Acc. Chem. Res., *6*, 377 (1973)
21. Brown, H. C.: Tetrahedron, *32*, 179 (1976)
22. Topics in Current chemistry, Vol. 80, 1979

23. Nevell, T. P., de Salas, E., Wilson, C. L.: J. Chem. Soc., *1939*, 1188
24. Winstein, S., Trifan, D.: J. Am. Chem. Soc., *71*, 2953 (1949)
25. Winstein, S., Trifan, D.: J. Am. Chem. Soc., *74*, 1147, 1154 (1952)
26. Tanida, H., Tsushima, T.: J. Am. Chem. Soc., *92*, 3397 (1970)
27. Dewar, M. J. S.: J. Chem. Soc., *1946*, 406
28. Walsh, A. D.: J. Chem. Soc., *1947*, 89
29. Dewar, M. J. S.: The Molecular orbital theory of organic chemistry, New York, McGraw-Hill, 1969
30. Brown, F., Hughes, E., Ingold, C., Smith, J.: Nature, *168*, 65 (1951)
31. Streitwieser, A., Jr.: Molecular orbital theory, New York, Wiley, 1961
32. Trehanovsky, W. S.: J. Org. Chem., *30*, 1666 (1965)
33. Longuet-Higgins, H. C.: J. Chem. Phys., *46*, 268 (1949)
34. Olah, G. A.: J. Am. Chem. Soc., *94*, 808 (1972)
35. Olah, G., Schlosberg, R.: J. Am. Chem. Soc., *90*, 2726 (1968); Alah, G., Helpern, Y., Shen, J., Mo, Y.: J. Am. Chem. Soc., *93*, 1251 (1971); Olah, G. A., Olah, J. A.: J. Am. Chem. Soc., *93*, 1256 (1971); Olah, G. A., Lin, H. C.: J. Am. Chem. Soc., *93*, 1259 (1971)
36. Sargent, G. D.: In: Carbonium Ions, ed. Olah, G. A., Schleyer, P. v. R., vol. 3, New York, 1972, p. 1099
37. Winstein, S., Morse, B., Grundwald, E., Jones, H., Corse, J., Marshall, H.: J. Am. Chem. Soc., *74*, 1127 (1952)
38. Winstein, S., Clippinger, E., Howe, R., Vogelfanger, E.: J. Am. Chem. Soc., *87*, 376 (1965)
39. Lin, Y., Nickon, A.: J. Am. Chem. Soc., *92*, 3496 (1970)
40. Schleyer, P. v. R., Donaldson, M. M., Watts, W. E.: J. Am. Chem. Soc., *87*, 375 (1965)
41. Jones, L. B., Jones, V. K.: J. Org. Chem., *36*, 1017 (1971)
42. Howe, R. K., Winstein, S.: J. Org. Chem., *38*, 2797 (1973)
43. Baker, R., Mason, T.: J. Chem. Soc., *1971B*, 1144
44. Baker, R., Hudec, J.: Chem. Commun., *1967*, 929
45. Corey, E. J., Glass, R. S.: J. Am. Chem. Soc., *89*, 2600 (1967)
46. Roberts, J. D., Lee, C. C.: J. Am. Chem. Soc., *73*, 5009 (1951)
47. Roberts, J. D., Lee, C. C., Saunders, W. A.: J. Am. Chem. Soc., *76*, 4501 (1954)
48. Deno, N. C.: Progress in physical organic chemistry, New York, Wiley, 1963
49. Bartlett, P. D., Bank, S., Crawford, R., Schmid, G.: J. Am. Chem. Soc., *87*, 1288 (1965)
50. Lawton, R. G.: J. Am. Chem. Soc., *83*, 2399 (1961)
51. Bartlett, P. D., Sargent, G. D.: J. Am. Chem. Soc., *87*, 1297 (1965)
52. Spurlock, L. A., Cox, W. G.: J. Am. Chem. Soc., *91*, 2961 (1969)
53. Spurlock, L. A., Parks, T. E.: J. Am. Chem. Soc., *92*, 1279 (1970)
54. Bartlett, P. D., Bank, S.: J. Am. Chem. Soc., *83*, 2591 (1966)
55. Bartlett, P. D.: Annalen, *653*, 45 (1962)
56. Lee, C. C., Hahn, B. S.: J. Am. Chem. Soc., *91*, 6420 (1969)
57. Collins, C. J., Lietzke, M. N.: Am. Chem. Soc., *89*, 6565 (1967)
58. Lee, C. C., Lam, L. K. M.: J. Am. Chem. Soc., *88*, 2834 (1966)
59. Brown, H. C.: In: Transition state. Chem. Soc. Special Publ. N° 16, p. 140–158, 174–178 (1962)
60. Colter, A., Friedrich, E., Holness, N., Winstein, S.: J. Am. Chem. Soc., *87*, 378 (1965)
61. Winstein, S., Holness, N.: J. Am. Chem. Soc., *77*, 5562 (1955)
62. Brown, H. C., Chloupek, F. J.: J. Am. Chem. Soc., *85*, 2322 (1963)
63. Brown, H. C., Rei, M.-H.: J. Am. Chem. Soc., *86*, 5004, 5009 (1964)
64. Brown, H. C., Bell, H. M.: J. Am. Chem. Soc., *86*, 5003, 5006, 5007 (1964)
65. Brown, H. C., Chloupek, F. J., Rei, M.-H.: J. Am. Chem. Soc., *86*, 1247 (1964)
66. Brown, H. C., Chloupek, F. J., Rei, M.-H.: J. Am. Chem. Soc., *86*, 1248 (1964)
67. Zbinden, R., Hall, H.: J. Am. Chem. Soc., *82*, 1215 (1960)
68. Halford, J. O.: J. Chem. Phys., *24*, 830 (1956)
69. Schleyer, P. v. R., Hicholas, R. D.: J. Am. Chem. Soc., *83*, 182 (1961)
70. Angell, C. L., Krueger, P. J., Lauzon, R., Leitch, L. C., Noack, K., Smith, R. J., Jones, R. N.: Spectrochim, Acta, *15*, 926 (1959)
71. Schleyer, P. v. R.: J. Am. Chem. Soc., *86*, 1854, 1856 (1964)
72. Brown, H. C., Takeuchi, K.: J. Am. Chem. Soc., *90*, 2691, 2693 (1968)

73. Brown, H. C., Takeuchi, K.: J. Am. Chem. Soc., *88*, 5336 (1966)
74. Takeuchi, K., Brown, H. C.: J. Am. Chem. Soc., *90*, 2693 (1968)
75. Winstein, S., Morse, B., Grunwald, E., Schreiber, K., Corse, J.: J. Am. Chem. Soc., *74*, 1113 (1952)
76. Howe, R., Carter, P., Winstein, S.: J. Org. Chem., *37*, 1433 (1972)
77. Dewar, M. J. S.: Ann. Reports, *48*, 118 (1951)
78. Streitwieser, A.: Solvolytic Displacement Reactions, New York, 1962, p. 62
79. Brown, H. C., Ikegami, S.: J. Am. Chem. Soc., *90*, 7122 (1968)
80. Ikegami, S., Jagt, D., Brown, H. C.: J. Am. Chem. Soc., *90*, 7124 (1968)
81. Brown, H. C., Ravindranathan, M., Takeuchi, K., Peters, E. N.: J. Am. Chem. Soc., *97*, 2899 (1975)
82. Brown, H. C., Takeuchi, K.: J. Am. Chem. Soc., *99*, 2679 (1977)
83. Gassman, P. G., Zeller, J., Lumb, J.: Chem. Commun., *1968*, 69
84. Gassman, P. G., Fentiman, A. F.: J. Am. Chem. Soc., *92*, 2549 (1970)
85. Brown, H. C., Rao, C. G.: J. Org. Chem., *44*, 133 (1979)
86. Brown, H. C., Ravindranathan, M.: J. Am. Chem. Soc., *97*, 2895 (1975)
87. Peters, E. N.: J. Org. Chem., *42*, 3015 (1977)
88. Battiste, M. A., Fiato, R. A.: Tetrahedron Lett., *1975*, 1255
89. Brown, H. C., Takeuchi, K., Ravindranathan, M.: J. Am. Chem. Soc., *99*, 2684 (1977)
90. Brown, H. C., Ikegami, S.: J. Am. Chem. Soc., *98*, 253 (1976)
91. Brown, H. C., Peters, E. N.: J. Am. Chem. Soc., *97*, 1927 (1975)
92. Brown, H. C., Takeuchi, K.: J. Am. Chem. Soc., *90*, 5268 (1968)
93. McManus, S. P., Harris, J. M.: J. Org. Chem., *42*, 1422 (1977)
94. Jorgensen, W. L.: J. Am. Chem. Soc., *99*, 3840 (1977)
95. Peters, E. N.: J. Am. Chem. Soc., *98*, 5627 (1976)
96. Wilcox, C. F., Jesaitis, R. G.: Tetrahedron Lett., *1967*, 2567
97. Wilcox, C. F., Jesaitis, R. G.: Chem. Commun., *1967*, 1046
98. Brown, H. C., Rao, G.: J. Org. Chem., *44*, 3536 (1979)
99. Lenoir, D., Röll, W., Ipaktschi, J.: Tetrahedron Lett., *1976*, 3073
100. Lambert, J. B., Mark, H. W.: Tetrahedron Lett., *1976*, 1765
101. Lambert, J. B., Mark, H. W.: J. Am. Chem. Soc., *100*, 2501 (1978)
102. Lambert, J. B., Mark, H. W., Holcomb, A. G., Maguar, E. S.: Acc. Chem. Res., *12*, 317 (1979)
103. Kleinfelter, D. C., Miller, J. M.: J. Org. Chem. *38*, 4142 (1973)
104. Schleyer, P. v. R., Kleinfelter, D. C.: 138th Meeting of the American Chemical Society, New York, *1960*, Abstr., p. 43P
105. Gassman, P. G., Marshall, J. L.: J. Am. Chem. Soc., *88*, 2822 (1966)
106. Gassman, P. G., Marshall, J. L.: Tetrahedron Lett., *1968*, 2429, 2433
107. Gassman, P. G., Marshall, J. L.: J. Am. Chem. Soc., *91*, 5527 (1969)
108. Gassman, P. G., Hornback, J. M.: J. Am. Chem. Soc., *91*, 4280 (1969)
109. Gassman, P. G., Hornback, J. M.: J. Am. Chem. Soc., *94*, 7010 (1972)
110. Muneyuki, R., Yano, T.: J. Am. Chem. Soc., *92*, 746 (1970)
111. Hartman, G. D., Traylor, T. G.: J. Am. Chem. Soc., *97*, 6147 (1975)
112. Wilcox, C. F., Loew, L. M., Jesaitis, R. G., Belin, S., Hsu, J. N.: J. Am. Chem. Soc., *96*, 4061 (1974)
113. Wu, J. W.-T., Thornton, E. R.: J. Org. Chem., *40*, 1041 (1975)
114. Dewar, M. J. S., Dougherty, R. C.: The PMO theory of organic Chemistry, Plenum Press, New York, 1975
115. Brown, H. C., Jagt, D. L., Schleyer, P. v. R., Fort, R. C., Watts, W. E.: J. Am. Chem. Soc., *91*, 6848 (1969)
116. Brown, H. C., Rao, C. G., Jagt, D. L.: J. Am. Chem. Soc., *101*, 1780 (1979)
117. Harris, J. M., Mount, D. L., Raber, D. J.: J. Am. Chem. Soc., *100*, 3139 (1978)
118. Brown, H. C., Ravindranathan, M., Chloupek, F. J., Rothberg, I.: J. Am. Chem. Soc., *100*, 3143 (1978)
119. Brown, H. C., Rao, C. G.: J. Org. Chem., *45*, 2113 (1980)
120. Slutsky, J., Bingham, R. C., Schleyer, P. v. R., Dickason, W. C., Brown, H. C.: J. Am. Chem. Soc., *96*, 1969 (1974)

121. Nordlander, J. E., Gruetzmacher, R. G., Kelly, W. J., Jindal, S. P.: J. Am. Chem. Soc., *96*, 181 (1974)
122. Jensen, F. R., Smart, B. E.: J. Am. Chem. Soc., *91*, 5686, 5688 (1969)
123. Brown, H. C., Gnedin, B. G., Takeuchi, K., Peters, E. N.: J. Am. Chem. Soc., *97*, 610 (1975)
124. Dewar, M. J. S.: Hyperconjugation, New York, 1962
125. Hanstein, W., Berwin, H., Traylor, T.: J. Am. Chem. Soc., *92*, 829 (1970)
126. Clinton, N. A., Brown, R. S., Traylor, T. G.: J. Am. Chem. Soc., *92*, 5228 (1970)
127. Traylor, T. G., Hanstein, W., Berwin, H. J., Clinton, N. A., Brown, R. S.: J. Am. Chem. Soc., *93*, 5715 (1971)
128. Grob, C. A., Ostermayer, F.: Helv. Chim. Acta, *45*, 1119 (1962)
128a. Nugent, W. A., Wu, M., Fehler, T. P., Kochi, J. K.: Chem. Commun., *1976*, 456
129. Radom, L., Pople, J., Buss, V., Schleyer, P. v. R.: J. Am. Chem. Soc., *93*, 1813 (1971); *94*, 311 (1972)
130. Hoffmann, R., Radom, L., Pople, J., Schleyer, P. v. R., Hehre, W., Salem, L.: J. Am. Chem. Soc., *94*, 6221 (1972)
131. Olah, G. A., Liang, G.: J. Am. Chem. Soc., *95*, 3792 (1973)
132. Scheppele, S. E.: Chem. Rev., *72*, 511 (1972)
133. Murr, B. L., Conkling, J. A.: J. Am. Chem. Soc., *92*, 3462, 3464 (1970)
134. Lee, C. C., Wong, E. W.: J. Am. Chem. Soc., *86*, 2752 (1964)
135. Lee, C. C., Wong, E. W.: Canad. J. Chem., *43*, 2254 (1965)
136. Lee, C. C., Wong, E. W.: Tetrahedron, *21*, 532 ((1965)
137. Collins, C., Bowman, F.: Isotope effects in chemical reactions, New York, 1970, p. 201
138. Jerkunica, J. M., Borcic, S., Sunko, D. E.: Chem. Commun., *1967*, 1302
139. Jerkunica, J. M., Borcic, S., Sunko, D. E.: Chem. Commun., *1968*, 1489
140. Schaefer, J. P., Dagani, M. J., Weinberg, D. S.: J. Am. Chem. Soc., *89*, 6938 (1967)
141. Schaefer, J. P., Foster, J. P., Dagani, M. J., Honig, L. M.: J. Am. Chem. Soc., *90*, 4497 (1968)
142. Murr, B. L., Nickon, A., Swartz, T. D., Werstiuk, N. H.: J. Am. Chem. Soc., *89*, 1730 (1967); Jerkunica, J. M., Borcic, S., Sunko, D. E.: J. Am. Chem. Soc., *89*, 1732 (1967)
143. Bunton, C. A., O'Connor, C.: Chem. Ind., *1965*, 1182
144. Howe, R., Friedrich, E. C., Winstein, S.: J. Am. Chem. Soc., *87*, 379 (1965)
145. Beckmann, S., Mezgen, R.: Chem. Ber., *98*, 2738 (1956)
146. Noyce, D. S., Denny, D. B.: J. Am. Chem. Soc., *72*, 5743 (1950)
147. Brown, H. C., Lin, K.-T.: J. Am. Chem. Soc., *89*, 3898, 3900 (1967)
148. Peters, E. N., Brown, H. C.: J. Am. Chem. Soc., *96*, 265 (1974)
149. Schleyer, P. v. R., Kleinfelter, D. C., Richey, H. G.: J. Am. Chem. Soc., *85*, 479 (1963)
150. Corey, E. J., Casanova, J., Vatakencherry, P., Winter, R.: J. Am. Chem. Soc., *85*, 169 (1963)
151. Skell, P., Maxwell, R.: J. Am. Chem. Soc., *84*, 3963 (1962)
152. Fong, F. K.: J. Am. Chem. Soc., *96*, 7638 (1974)
153. Woods, W. G., Carboni, R. A., Roberts, J. D.: J. Am. Chem. Soc., *78*, 5653 (1966)
154. Goering, H., Degani, M.: J. Am. Chem. Soc., *91*, 4506 (1969)
155. Beckmann, S., Geger, H., Shaber-Kiechle, M.: Chem. Ber., *92*, 2419 (1959)
156. Berson, J. A., Grubb, P. W.: J. Am. Chem. Soc., *87*, 4016 (1965)
157. Collins, C. J., Benjamin, B. M.: J. Am. Chem. Soc., *89*, 1652 (1967)
158. Benjamin, B. M., Collins, C. J.: J. Am. Chem. Soc., *88*, 1556, 1558 (1966)
159. Benjamin, B. M., Collins, C. J.: Tetrahedron Lett., *1966*, 5477
160. Collins, C. J., Cheema, L. K., Werth, R. G., Benjamin, B. M.: J. Am. Chem. Soc., *86*, 4913 (1964)
161. Berson, J. A., Hammons, J. H., McRowe, A. W., Bergman, R. G., Remanick, A., Houston, D.: J. Am. Chem. Soc., *89*, 2561, 2590 (1967)
162. Hirsjarvi, P., Heinonen, K., Pirilä, L.: Suom. Kem., *B37*, 77 (1964)
163. Voughan, W. R., Doetschel, C. T., Goodrow, M. H., Warren, C. I.: J. Am. Chem. Soc., *85*, 2282 (1963)
164. Finch, A. M., Vaughan, W. R.: J. Am. Chem. Soc., *87*, 5520 (1965)
165. Gream, G. E., Wege, D.: Tetrahedron, *22*, 2583 (1966)
166. Kleinfelter, D. C., Schleyer, P. v. R.: J. Am. Chem. Soc., *83*, 1556 (1961)

167. Kleinfelter, D. C., Dye, T. E.: J. Am. Chem. Soc., *88*, 3174 (1966)
168. Berson, J. A.: J. Am. Chem. Soc., *87*, 3248 (1965)
169. Schleyer, P. v. R.: J. Am. Chem. Soc., *89*, 699, 701 (1967)
170. Carey, A., Tremfer, H.: J. Org. Chem., *34*, 4 (1969)
171. Buchell, A. W., Wilder, P.: J. Am. Chem. Soc. *89*, 5721 (1967)
172. Saimon, J. R., Whittaker, D.: Chem. Commun., *1967*, 491
173. Breslow, D. S., Edwards, E. I., Leone, R., Schleyer, P. von R.: J. Am. Chem. Soc., *90*, 7097 (1968)
174. Cristol, S. J., Sequeira, R. M., Mayo, G. O.: J. Am. Chem. Soc., *90*, 5564 (1968)
175. Bunton, C. A., O'Connor, C., Whittaker, D.: J. Org. Chem., *32*, 2812 (1967)
176. Baker, R., Mason, T.: Chem. Commun., *1969*, 120
177. Bartlett, P., Fickes, G. N., Haupt, F. C., Helgeson, R.: Acc. Chem. Res., *3*, 177 (1970)
178. Davies, A. G., Tudor, R.: J. Chem. Soc., *1970B*, 1815
179. Brown, H. C., Rei, M. H.: J. Am. Chem. Soc., *90*, 6116 (1968)
180. Werstiuk, N. H., Taillefer, R.: Canad. J. Chem., *48*, 3966 (1970)
181. Jindal, S. P., Tidwell, T. T.: Tetrahedron Lett., *1971*, 783
182. Collins, C., Harding, C.: J. Am. Chem. Soc., *91*, 7194 (1969)
183. Saunders, M., Schleyer, P. von R., Olah, G. A.: J. Am. Chem. Soc., *86*, 5680 (1964)
184. Olah, G. A., White, A. M., DeMember, J., Commeyras, A., Lui, C. J.: J. Am. Chem. Soc., *92*, 4627 (1970)
185. Berson, J. A., Bergman, R. G., Hammons, J. H., McRowe, A. W.: J. Am. Chem. Soc., *89*, 2581 (1967)
186. Lee, C. C., Hahn, B. S., Lam, L. K., Woodcock, D. J.: Canad. J. Chem., *48*, 3831 (1970)
186a. Oxer, G. W., Wege, D.: Tetrahedron Lett., *1971*, 457
187. Fischer, W., Grob, C. A., Sprecher, G. V., Waldner, A.: Tetrahedron Lett., *1979*, 1905
188. Brown, H. C., Peters, E. N.: Proc. Nat. Acad. Sci., *71*, 132 (1974)
189. Berson, J. A., Remanick, A.: J. Am. Chem. Soc., *86*, 1749 (1964)
190. Streitwieser, A.: J. Org. Chem., *22*, 861 (1957)
191. Dilling, W. L., Alford, J. A.: Tetrahedron Lett., *1971*, 761
192. Streitwieser, A., Schaeffer, W. D.: J. Am. Chem. Soc., *79*, 2888 (1957)
193. Collins, C. J.: Acc. Chem. Res., *4*, 315 (1971); Collins, C. J.: Quart. rev., *29*, 251 (1975)
194. Semenow, D., Shih, C., Young, W.: J. Am. Chem. Soc., *80*, 5472 (1958)
195. Raaen, F., Benjamin, B. M., Collins, C. J.: Tetrahedron Lett., *1971*, 2613
196. Wilder, P., Hsieh, W.: J. Org. Chem., *36*, 2552 (1971)
197. Rengaraju, S., Berlin, K.: Tetrahedron, *27*, 2399 (1971)
198. Renk, E., Roberts, J.: J. Am. Chem. Soc., *83*, 878 (1961)
199. Kirmse, W.: Ang. Chem. Int. Ed., *15*, 251 (1976)
200. Collins, C., Raaen, V., Eckart, M.: J. Am. Chem. Soc., *92*, 1787 (1970)
201. Collins, C., Benjamin, B., Raaen, V., Glover, T., Eckart, M.: Ann. Chem., *739*, 7 (1970)
202. Collins, C., Benjamin, B.: J. Am. Chem. Soc., *92*, 3182, 3183 (1970)
203. Collins, C. J., Benjamin, B. M.: J. Org. Chem., *37*, 4358 (1972)
204. Collins, C. J., Glover, I. T., Eckart, M. D., Raaen, V. F., Benjamin, B. M., Benjaminov, B. S.: J. Am. Chem. Soc., *94*, 899 (1972)
205. White, E. H., Ryan, T. J., Field, K. W.: J. Am. Chem. Soc., *94*, 1360 (1972)
206. Berson, J. A., Foley, J. W., McKenna, J. M., Junge, H., Donald, D., Luibrand, R., Kundu, N., Libbey, W., Poonian, M., Gajewski, T., Allen, J.: J. Am. Chem. Soc., *93*, 1299 (1971)
207. Blattel, R. A., Yates, P.: Tetrahedron Lett., *1972*, 1069, 1073
208. Kirmse, W., Arend, G., Siegfried, R.: Ang. Chem., Int. Ed., *9*, 165 (1970)
209. Kirmse, W., Arend, G.: Chem. Ber., *105*, 2738 (1972)
210. Kirmse, W.: Ang. Chem., *84*, 833 (1972)
211. Goering, H. L., Clevenger, J. V.: J. Am. Chem. Soc., *94*, 1010 (1972)
212. Brown, H. C., Ravindranathan, M., Rao, C. G., Chloupek, F. J., Rei, M.-H.: J. Org. Chem., *43*, 3667 (1978)
213. Schleyer, P. von R., Watts, W. E., Fort, R. C., Comisarow, M. B., Olah, G. A.: J. Am. Chem. Soc., *86*, 5679 (1964)
214. Olah, G. A., Commeyras, A., Lui, C. Y.: J. Am. Chem. Soc., *90*, 3882 (1968)
215. Olah, G. A., White, A. M.: J. Am. Chem. Soc., *91*, 3954 (1969)
216. Olah, G. A., White, A. M.: J. Am. Chem. Soc., *91*, 3956 (1969)

217. Olah, G. A., DeMember, J. R., Lui, C. Y., White, A. M.: J. Am. Chem. Soc., *91*, 3958 (1969)
218. Olah, G. A., Jeuell, C. L., White, A. M.: J. Am. Chem. Soc., *91*, 3961 (1969)
219. Olah, G. A., White, A. M.: J. Am. Chem. Soc., *91*, 5801 (1969)
220. Olah, G. A., White, A. M.: J. Am. Chem. Soc., *91*, 6883 (1969)
221. Olah, G. A., Liang, G., Mateescu, G. D., Riemenschneider, J. L.: J. Am. Chem. Soc., *95*, 8698 (1973)
222. Saunders, M., Vogel, P., Hagen, E., Rosenfeld, J.: Acc. Chem. Res., *6*, 53 (1973)
223. Lee, C. C., Cessna, A. J., Ko, E. C., Vassie, S.: J. Am. Chem. Soc., *95*, 5688 (1973)
224. Klopman, G.: J. Am. Chem. Soc., *91*, 89 (1969)
225. Schleyer, P. von R., Williams, J. E., Blanchard, K. R.: J. Am. Chem. Soc., *92*, 2377 (1970)
226. Bodor, N., Dewar, M. J. S.: J. Am. Chem. Soc., *93*, 6685 (1971)
227. Dewar, M. J. S., Haddon, R. C., Komornicki, A., Rzepa, H.: J. Am. Chem. Soc., *99*, 377 (1977)
228. Bingham, R. C., Dewar, M. J. S., Lo, D. H.: J. Am. Chem. Soc., *97*, 1294 (1975)
229. Köhler, H.-J., Lischka, H.: J. Am. Chem. Soc., *101*, 3479 (1979)
230. Goetz, D. M., Schlegel, H. B., Allen, L. C.: J. Am. Chem. Soc., *99*, 8118 (1977)
231. Wenke, G., Lenoir, D.: Tetrahedron, *35*, 489 (1979)
232. Winstein, S.: J. Am. Chem. Soc., *87*, 381 (1965)
233. Piccolini, R. J., Winstein, S.: Tetrahedron, *19*, 423 (1963)
234. Schleyer, P. von R., Stang, P. J., Raber, D. J.: J. Am. Chem. Soc., *92*, 4725 (1970)
235. Brouwer, D. M., Mackor, E. L., MacLean, C.: Carbonium Ions, *2*, 837 (1980)
236. Haseltine, R., Wong, N., Sorensen, T., Jones, A.: Can. J. Chem. *53*, 1891 (1975)
237. Olah, G. A., Mateescu, G. D., Riemenschneider, J. L.: J. Am. Chem. Soc., *94*, 2529 (1972)
238. Olah, G. A., Clifford, P. R., Jeuell, C. L.: J. Am. Chem. Soc., *92*, 553 (1970)
239. Farnum, D. G., Mehta, G.: J. Am. Chem. Soc., *91*, 3256 (1969)
240. Farnum, D. G., Wolf, A. D.: J. Am. Chem. Soc., *96*, 5166 (1974)
241. Olah, G. A., Prakash, G. K., Liang, G.: J. Am. Chem. Soc., *99*, 5683 (1977)
242. Kelly, D. P., Spear, R. J.: Aust. J. Chem., *31*, 1209 (1978)
243. Farnum, D. G., Botto, R. E., Chambers, W. T., Lam, B.: J. Am. Chem. Soc., *100*, 3847 (1978)
244. Coxon, J. M., Steel, P. J., Coddington, J. M., Rae, I. D., Jones, A. J.: Aust. J. Chem., *31*, 1223 (1978)
245. Nickon, A., Lin, Y.: J. Am. Chem. Soc., *91*, 6861 (1969)
246. Olah, G. A., DeMember, J. R., Lui, C. J., Porter, R. D.: J. Am. Chem. Soc., *93*, 1442 (1971)
247. Sorensen, T. S., Ranganayakulu, K.: Tetrahedron Lett., *1972*, 2447
248. Haseltine, R. P., Sorensen, T. S.: Can. J. Chem., *53*, 1067 (1975)
349. Jones, A. J., Huang, E., Haseltine, R., Sorensen, T. S.: J. Am. Chem. Soc., *97*, 1133 (1975)
250. Olah, G. A., Liang, G.: J. Am. Chem. Soc., *96*, 189 (1974)
251. Kramer, G. M.: Adv. Phys. Org. Chem., *11*, 179 (1975)
252. Clark, D. T., Cromarty, B. J., Colling, L.: J. Am. Chem. Soc., *99*, 8120 (1977)
253. Eaton, D. F., Traylor, T. G.: J. Am. Chem. Soc., *96*, 1226 (1974)
254. Brown, H. C., Peters, E. N.: J. Am. Chem. Soc., *95*, 2400 (1973)
255. Olah, G. A., Westerman, P. W.: J. Am. Chem. Soc., *95*, 7530 (1973)
256. Brown, H. C., Ravindranathan, M., Peters, E. N.: J. Org. Chem., *42*, 1073 (1977)
257. Brown, H. C., Peters, E. N.: J. Am. Chem. Soc., *99*, 1712 (1977)
258. Olah, G. A., Spear, R. J.: J. Am. Chem. Soc., *97*, 1539 (1975)
259. Olah, G. A., Westerman, P. W., Forsyth, D. A.: J. Am. Chem. Soc., *97*, 3419 (1975)
260. Volz, H., Shin, J., Streicher, H.: Tetrahedron Lett., *1975*, 1297
261. Farnum, D. G.: Adv. Phys. Org. Chem., *11*, 123 (1975)
262. Olah, G. A., Forsyth, D. A.: J. Am. Chem. Soc., *97*, 3137 (1975)
263. Ditchfield, R., Miller, D.: J. Am. Chem. Soc., *93*, 5287 (1971)
264. Spielvogel, B. F., Nutt, W. R., Izydore, R. A.: J. Am. Chem. Soc., *97*, 1609 (1975)
265. Williams, R. E.: Third Int. Meeting on Boron Chemistry, Ettol, Germany, 1976, abstr. N 31
266. Olah, G. A., Prakash, G. K.S., Liang, G., Henold, K. L., Haigh, G. B.: Proc. Nat. Acad. Sci. USA, *74*, 5217 (1977)

267. Cheremisin, A. A., Schastnev, P. V.: OMR, *14*, 327 (1980); Cheremisin, A. A., Schastnev, P. V.: Izv. SO Acad. Nauk USSR, 1980, 105
268. Schleyer, P. v. R., Lenoir, D., Mison, P., Liang, G., Prakash, G. K. S., Olah, G. A.: J. Am. Chem. Soc., *102*, 683 (1980)
269. Saunders, M., Telkowski, L., Kates, M.: J. Am. Chem. Soc., *99*, 8070 (1977)
270. Saunders, M., Kates, M., Wiberg, K., Pratt, W.: J. Am. Chem. Soc., *99*, 8072 (1977)
271. Saunders, M., Kates, M. R.: J. Am. Chem. Soc., *99*, 8071 (1977)
272. Saunders, M., Kates, M. R.: J. Am. Chem. Soc., *102*, 6868 (1980)
273. Servis, K. L., Shue, F.-F.: J. Am. Chem. Soc., *102*, 7233 (1980)
274. Hogeveen, H., Baardman, F., Roobeek, C. F.: Rec. trav., *89*, 227 (1970)
275. Hogeveen, H., Roobeek, C.: Tetrahedron Lett., *1969*, 4941
276. Hogeveen, H., Gaasbeek, C.: Rec. trav., *88*, 719 (1969)
277. Hogeveen, H.: Rec. trav., *89*, 74 (1970)
278. Sorensen, T. S.: Acc. Chem. Res., *9*, 257 (1976)
279. Solomon, J. J., Field, F. A.: J. Am. Chem. Soc., *98*, 1567 (1976)
280. Arnett, E. M., Petro, C.: J. Am. Chem. Soc., *100*, 2563 (1978)
281. Arnett, E. M., Petro, C.: J. Am. Chem. Soc., *100*, 5408 (1978)
282. Arnett, E. M., Petro, C.: J. Am. Chem. Soc., *100*, 5402 (1978)
283. Arnett, E. M., Pienta, N., Petro, C.: J. Am. Chem. Soc., *102*, 398 (1980)
284. Fărcaşiu, D.: J. Org. Chem., *46*, 223 (1981)
285. Haseltine, R., Ranganayakulu, K., Wong, N., Sorensen, T.: Can. J. Chem., *53*, 1901 (1975)
286. Schleyer, P. v. R., Chandrasckhar, J.: J. Org. Chem. *46*, 225 (1981)
287. Dewar, M. J. S.: Chem. Brit., *11*, 97 (1975)
288. Meinwald, J., Gassman, P.: J. Am. Chem. Soc., *85*, 57 (1963)
289. Meinwald, J., Crandall, J.: J. Am. Chem. Soc., *88*, 1292 (1966)
290. Seybold, G., Vogel, P., Saunders, M., Wiberg, K.: J. Am. Chem. Soc., *95*, 2045 (1973)
291. Olah, G., DeMember, J., Commeyras, A., Bribes, J.: J. Am. Chem. Soc., *93*, 459 (1971)
292. Brouwer, D. M., v. Doorn, J. A.: Rec. trav., *89*, 88 (1970)
293. Hogeveen, H., Gaasbeek, C.: Rec. trav., *88*, 1305 (1969)
294. Pople, J. A., Segal, G. A.: J. Chem. Phys., *43*, S136 (1965); *44*, 3289 (1966)
295. Olah, G. A., Liang, G., Jindall, S. P.: J. Am. Chem. Soc., *98*, 2508 (1976)
296. Brown, H. C., Ravindranathan, M., Rao, C. G.: J. Am. Chem. Soc., *99*, 2359 (1977)
297. Schmitz, L. R., Sorensen, T. S.: J. Am. Chem. Soc., *102*, 1645 (1980)
298. Huang, E., Ranganayakulu, K., Sorensen, T. S.: J. Am. Chem. Soc., *94*, 1780 (1972)
299. Brouwer, D. M., Hogeveen, H.: Prog. Phys. Org. Chem., *9*, 204 (1972)
300. Kelly, D. P., Underwood, G. R., Barron, P. F.: J. Am. Chem. Soc., *98*, 3106 (1976)
301. Kirchen, R. P., Sorensen, T. S.: J. Am. Chem. Soc., *99*, 6687 (1977)
302. Olah, G. A., Donovan, D. J.: J. Am. Chem. Soc., *99*, 5026 (1977)
303. Olah, G. A., Berrier, A. L., Arvanaghi, M., Prakash, G. K. S.: J. Am. Chem. Soc., *103*, 1122 (1981)
304. Gassman, P. G., Hornback, J. M.: J. Am. Chem. Soc., *89*, 2487 (1967)
305. Gassman, P. G., Hornback, J. M.: J. Am. Chem. Soc., *90*, 6238 (1968)
306. Miles, F. B.: J. Am. Chem. Soc., *89*, 2488 (1967)
307. Miles, F. B.: J. Am. Chem. Soc., *90*, 1265 (1968)
308. Gassman, P. G., Hornback, T. M., Pascone, T. M.: Tetrahedron Lett., *1971*, 1425
309. Dilling, W. L., Alford, J. A.: Tetrahedron Lett., *1971*, 761
310. Dilling, W. L., Reineke, C. E., Plepys, R. E.: J. Org. Chem. *34*, 2605 (1969)
311. Dilling, W. L., Plepys, R. A., Kroenig, R. D.: J. Am. Chem. Soc., *91*, 3404 (1969)
312. Paquette, L. A., Meehan, G. V., Wise, L. D.: J. Am. Chem. Soc., *91*, 3231 (1969)
313. Funke, B., Winstein, S.: Tetrahedron Lett., *1971*, 1477
314. Svensson, T., Winstein, S.: J. Am. Chem. Soc., *94*, 2336 (1972)
315. Olah, G. A., Berrier, A. L., Arvanaghi, M., Prakash, G. K. S.: J. Am. Chem. Soc., *103*, 1122 (1981)
316. Walborsky, H. M., Baum, M. E., Youssef, A. A.: J. Am. Chem. Soc., *83*, 988 (1961)
317. Youseff, A. A., Baum, M. E., Walborsky, H. M.: J. Am. Chem. Soc., *81*, 4709 (1959)
318. Goering, H. L., Sloan, M. F.: J. Am. Chem. Soc., *83*, 1397 (1961)
319. Goering, H. L., Sloan, M. F.: J. Am. Chem. Soc., *83*, 1992 (1961)

320. Foote, C. S.: J. Am. Chem. Soc., *86*, 1853 (1964)
321. Walborsky, H. M., Webb, J., Pitt, C. G.: J. Org. Chem. *28*, 3214 (1963)
322. Berson, J. A., Willner, D.: J. Am. Chem. Soc., *86*, 609 (1964)
323. Berson, J. A., Willner, D.: J. Am. Chem. Soc., *86*, 675 (1964)
324. Berson, J. A., Reynolds-Warnhoff, P.: J. Am. Chem. Soc., *84*, 683 (1962)
325. Berson, J. A., Reynolds-Warnhoff, P.: J. Am. Chem. Soc., *86*, 595 (1964)
326. Goering, H. L., Fickes, G. N.: J. Am. Chem. Soc., *90*, 2848 (1968)
327. Goering, H. L., Fickes, G. N.: J. Am. Chem. Soc., *90*, 2856 (1968)
328. Goering, H. L., Fickes, G. N.: J. Am. Chem. Soc., *90*, 2862 (1968)
329. Kaplan, F.: J. Am. Chem. Soc., *92*, 1445 (1970)
330. Kwart, H., Irvine, J.: J. Am. Chem. Soc., *91*, 5541 (1969)
331. Kirchen, R. P., Sorensen, T. S., Wagstaff, K.: J. Am. Chem. Soc., *100*, 6761 (1978)
332. Kirchen, R. P., Sorensen, T. S.: Chem. Commun., *1978*, 769
333. Kirchen, R. P., Ranganayakulu, K., Rauk, A., Singh, B. P., Sorensen, T. S.: J. Am. Chem. Soc., *103*, 588 (1981)
334. Kirchen, R. P., Okazawa, N., Ranganayakulu, K., Rauk, A., Sorensen, T.: J. Am. Chem. Soc., *103*, 597 (1981)
335. Kirchen, R. P., Sorensen, T. S.: J. Am. Chem. Soc., *101*, 3240 (1979)
336. Prelog, V., Traynham, J. G.: "Molecular rearrangements", *1*, 593 (1963); Cope, A. C., Martin, M. M., McKervey, M. A.: Quart. rev. Chem. Soc., *20*, 119 (1966)
337. Winstein, S., Adams, R.: J. Am. Chem. Soc., *70*, 838 (1948)
338. Winstein, S., Schlesinger, A.: J. Am. Chem. Soc., *70*, 3528 (1948)
339. Winstein, S., Walborsky, H., Schreiber, K.: J. Am. Chem. Soc., *72*, 5795 (1950)
340. Simonetta, M., Winstein, S.: J. Am. Chem. Soc., *76*, 18 (1954)
341. Winstein, S.: J. Am. Chem. Soc., *81*, 6524 (1959)
342. Winstein, S., Sonnenberg, J.: J. Am. Chem. Soc., *83*, 3244 (1961)
343. Winstein, S.: Chem. Soc. Special Publ., N 21, 5 (1967)
344. Winstein, S.: Quart. revs., *23*, 141 (1969)
345. Hehre, W. J.: J. Am. Chem. Soc., *94*, 8908 (1972); *95*, 5807 (1973); *96*, 5207 (1974)
346. Haddon, R. C.: J. Am. Chem. Soc., *97*, 3608 (1975); Tetrahedron Lett., *1974*, 2797
347. Haddon, R. C.: Tetrahedron Lett., *1976*, 863
348. Jorgensen, W. L.: J. Am. Chem. Soc., *97*, 3082 (1975); *98*, 6784 (1976)
349. Roberts, J. D., Bennett, W., Armstrong, R.: J. Am. Chem. Soc., *72*, 3329 (1950)
350. Roberts, J. D., Bennett, W.: J. Am. Chem. Soc., *76*, 4623 (1954)
351. Winstein, S., Kosower, E.: J. Am. Chem. Soc., *81*, 4399 (1959)
352. Winstein, S., Shatavsky, M., Norton, C., Woodward, R.: J. Am. Chem. Soc., *77*, 4183 (1965)
353. Schleyer, P. von R.: J. Am. Chem. Soc., *80*, 1700 (1958)
354. Roberts, J. D., Lee, C. C., Saunders, W. H.: J. Am. Chem. Soc. *77*, 3034 (1955)
355. Cristol, S. J., Morrill, T. C., Sanchez, R. A.: J. Am. Chem. Soc., *88*, 3087 (1966)
356. Lee, C. C., Hahn, B. S.: J. Am. Chem. Soc., *92*, 2583 (1970)
357. Cristol, S. J., Beimborn, D. A.: J. Am. Chem. Soc., *95*, 3651 (1973)
358. Story, P. R., Clark, B. C.: Carbonium Ions *3*, 1018 (1972)
359. Vogel, P., Delseth, R., Quarroz, D.: Helv. Chim. Acta, *58*, 508 (1975)
360. Sonney, J.-M., Vogel, P., Burger, U.: Tetrahedron Lett., *1978*, 825
361. Burger, U., Sonney, J.-M., Vogel, P.: Tetrahedron Lett., *1978*, 829
362. Spurlock, L., Cox, W.: J. Am. Chem. Soc., *93*, 146 (1971)
363. Bly, R. S., Swindell, R. T.: J. Org. Chem., *30*, 10 (1965)
364. Wilt, J., Parsons, C.: J. Org. Chem., *33*, 694 (1968)
365. Lambert, J. B., Holcomb, A. G.: J. Am. Chem. Soc., *93*, 2994 (1971)
366. Olah, G. A., Mateescu, G.: J. Am. Chem. Soc., *92*, 1430 (1970)
367. DePuy, C. H., Ogawa, I. A., McDonald, J. C.: J. Am. Chem. Soc., *83*, 1668 (1961)
368. Baker, R., Salter, J.: J. Chem. Soc., *1971B*, 757
369. Peters, E. N., Brown, H. C.: J. Am. Chem. Soc., *94*, 5899 (1972)
370. Peters, E. N., Brown, H. C.: J. Am. Chem. Soc., *94*, 7920 (1972)
371. Peters, E. N., Brown, H. C.: J. Am. Chem. Soc., *95*, 2397; 2398 (1973)
372. Brown, H. C., Peters, E. N.: J. Am. Chem. Soc., *94*, 7442 (1975)

373. Brown, H. C., Peters, E. N., Ravindranathan, M.: J. Am. Chem. Soc., *97*, 7449 (1975)
374. Hehre, W. J., Hiberty, P. C.: J. Am. Chem. Soc., *94*, 5917 (1972)
375. Winstein, S., Shatavsky, M.: J. Am. Chem. Soc., *78*, 592 (1956)
376. Winstein, S., Lewin, A., Pande, K.: J. Am. Chem. Soc., *85*, 2324 (1963)
377. Brookhart, M., Diaz, A., Winstein, S.: J. Am. Chem. Soc., *88*, 3135 (1966)
378. Tufariello, J., Mich, T., Lorence, R.: Chem. Commun., *1967*, 1202
379. Tufariello, J. J., Lorence, R. J.: J. Am. Chem. Soc., *91*, 1547 (1969)
380. Lhomme, J., Diaz, A., Winstein, S.: J. Am. Chem. Soc. *91*, 1549 (1969)
381. Diaz, A., Brookhart, M., Winstein, S.: J. Am. Chem. Soc., *88*, 3133 (1966)
382. Tanida, H., Hata, Y.: J. Org. Chem., *30*, 977 (1965)
383. Tanida, H., Tsuji, T., Iries, T.: J. Am. Chem. Soc., *88*, 864 (1966)
384. Spurlock, L. A., Mikuriya, Y.: J. Org. Chem., *36*, 1549 (1971)
385. Tufariello, J., Rowe, D.: Chem. Commun., *1972*, 1066
386. Cooke, B. J. A., Story, P. R.: Tetrahedron Lett., *1974*, 1705
387. Brown, H. C., Bell, H. M.: J. Am. Chem. Soc., *85*, 2324 (1963)
388. Gassman, P. G., Patton, D. S.: J. Am. Chem. Soc., *91*, 2160 (1969)
389. Olah, G. A., Kelly, D. P., Jeuell, C. L., Porter, R. D.: J. Am. Chem. Soc., *92*, 2544 (1970)
390. Gassman, P. G., Pascone, J. M.: J. Am. Chem. Soc., *95*, 7801 (1973)
391. Gassman, P. G., Fentiman, A. F.: J. Am. Chem. Soc., *91*, 1545 (1969)
392. Lustgarten, R. K., Lhomme, J., Winstein, S.: J. Org. Chem. *37*, 1075 (1972)
393. Bly, R. K., Bly, R. S.: J. Org. Chem., *28*, 3165 (1963)
394. Clark, F. R. S., Warkentin, J.: Can. J. Chem., *51*, 4090 (1973)
395. Winstein, S., Stafford, E.: J. Am. Chem. Soc., *79*, 505 (1957)
396. Tanida, H., Tsuji, T., Irie, T.: J. Org. Chem., *31*, 3941 (1966)
397. Huisgen, R., Rüchardt, C.: Annalen, *601*, 1, 21 (1956)
398. Zollinger, H.: Azo and Diazo Chemistry, New York, 1961, p. 93–101, 123–136
399. Ohorodnyk, H. O., Santry, D. P.: J. Am. Chem. Soc., *91*, 4711 (1969)
400. Coulson, C. A., Dewar, M. J. S.: Dis. Faraday Soc., *2*, 54 (1947)
401. Hoffmann, R.: Acc. Chem. Res., *4*, 1 (1971)
402. Ohorodnyk, H. O., Santry, D. P.: Chem. Commun., *1969*, 510
403. Tsuji, T., Ishitobi, H., Tanida, H.: Tetrahedron Lett., *1972*, 3083
404. Bartlett, P. D., Rice, M. R.: J. Org. Chem., *28*, 3351 (1963)
405. Lambert, J. B., Finzel, R. B., Belec, C. A.: J. Am. Chem. Soc., *102*, 3281 (1980)
406. Humski, K., Sendijarevic, V., Shiner, V.: J. Am. Chem. Soc., *95*, 7722 (1973)
407. Mallory, T. B., Bauman, L. E., Carreira, L. A.: Top. Stereochem., *11*, 142 (1979)
408. Lambert, J. B., Featherman, S. I.: J. Am. Chem. Soc., *99*, 1542 (1977)
409. Peters, E. N., Brown, H. C.: J. Am. Chem. Soc., *97*, 7454 (1975)
410. Winstein, S., Ordronneau, C.: J. Am. Chem. Soc., *82*, 2084 (1960)
411. Story, P. R.: J. Am. Chem. Soc., *83*, 3347 (1961)
412. Goldstein, M. J.: J. Am. Chem. Soc., *89*, 6357 (1967)
413. Goldstein, M. J., Hoffmann, R.: J. Am. Chem. Soc., *93*, 6193 (1971)
414. Grutzner, J. B., Winstein, S.: J. Am. Chem. Soc., *94*, 2200 (1972)
415. Allred, E. L., Hinshaw, J. C.: Tetrahedron Lett., *1968*, 1293
416. Paquette, L. A., Dunkin, I. R.: J. Am. Chem. Soc., *96*, 1220 (1974); Paquette, L. A., Dunkin, I. R.: J. Am. Chem. Soc., *97*, 2243 (1975)
417. Allred, E. L., Lyon, G. D., Stroebel, G.: J. Am. Chem. Soc., *101*, 3415 (1979)
418. Bartlett, P. D., Giddings, W. P.: J. Am. Chem. Soc., *82*, 1240 (1960)
419. Tanida, H.: J. Am. Chem. Soc., *85*, 1703 (1963)
420. Tanida, H., Tsuji, T., Ishitobi, H.: J. Am. Chem. Soc., *86*, 4904 (1964)
421. Tanida, H., Tsuji, T., Ikegami, S., Ishitobi, H.: J. Am. Chem. Soc., *89*, 2928 (1967)
422. Tanida, H., Ishitobi, H.: J. Am. Chem. Soc., *88*, 3663 (1966)
423. Brown, H. C., Okamoto, Y.: J. Am. Chem. Soc., *80*, 4979 (1958)
424. Tanida, H., Tsushima, T., Irie, T.: Tetrahedron Lett., *1970*, 4331
425. Muneyuki, R., Tanida, H.: J. Am. Chem. Soc., *90*, 656 (1968)
426. Winstein, S., Grunwald, E., Ingraham, L.: J. Am. Chem. Soc., *70*, 821 (1948)
427. Winstein, S., Grunwald, E.: J. Am. Chem. Soc., *70*, 828 (1948)
428. Bunton, C. A.: Nucleophilic Substitution at a Saturated Carbon atom, New York, 1963, p. 62

429. Rei, M. N., Brown, H. C.: J. Am. Chem. Soc., *88*, 5335 (1966)
430. Braddon, D. V., Wiley, G. A., Dirlam, J., Winstein, S.: J. Am. Chem. Soc., *90*, 1901 (1968)
431. Tanida, H., Ishitobi, H., Irie, T.: J. Am. Chem. Soc., *90*, 2688 (1968)
432. Brown, H. C., Trittle, G. L.: J. Am. Chem. Soc., *90*, 2689 (1968)
433. Brown, H. C., Trittle, G. L.: J. Am. Chem. Soc., *88*, 1320 (1968)
434. Brown, H. C., Liu, K.-T.: J. Am. Chem. Soc., *91*, 5909 (1969)
435. Brown, H. C., Ikegami, S., Liu, K.-T.: J. Am. Chem. Soc., *91*, 5911 (1969)
436. Giddings, W. P., Dirlam, J.: J. Am. Chem. Soc., *85*, 3900 (1963)
437. Lancelot, C., Cram, D., Schleyer, P.: Carbonium Ions, vol. 3, New York, 1972, p. 1426
438. Dirlam, J. P., Diaz, A., Winstein, S., Giddings, W. P., Hanson, G. C.: Tetrahedron Lett., *1969*, 3133
439. Tanida, H., Ishitobi, H., Irie, T., Tsushima, T.: J. Am. Chem. Soc., *91*, 4512 (1969)
440. Tanida, H., Tsuji, T., Teratake, S.: J. Org. Chem., *32*, 4121 (1967)
441. Tanida, H., Irie, T., Tsushima, T.: J. Am. Chem. Soc., *92*, 3404 (1970)
442. Kleinfelter, D.: Diss. Abstr., *22*, 428 (1961)
443. Tanida, H., Tsushima, T.: J. Am. Chem. Soc., *93*, 3011 (1971)
444. Bartell, L. S.: J. Am. Chem. Soc., *83*, 3567 (1961)
445. Streitwieser, A., Jagew, R., Fahey, R.: J. Am. Chem. Soc., *80*, 2326 (1958)
446. Lenoir, D., Ipaktschi, J., Schleyer, P. v. R.: Annalen, *750* 28 (1971)
447. Paquette, L. A., Dunkin, I. R.: J. Am. Chem. Soc., 3067 (1973)
448. Dirlam, J. P., Winstein, S.: J. Am. Chem. Soc., *91*, 5905 (1969)
449. Dirlam, J. P., Winstein, S.: J. Am. Chem. Soc., *91*, 5907 (1969)
450. Brown, H. C., Ikegami, S., Liu, K.-T., Tritle, G. L.: J. Am. Chem. Soc., *98*, 2531 (1976)
451. Goering, H. L., Chang, C.-S., Clevenger, J. V.: J. Am. Chem. Soc., *96*, 7602 (1974)
452. Goering, H. L., Chang, C.-S., Masilamani, D.: J. Am. Chem. Soc., *100*, 2506 (1978)
453. Hata, Y., Tanida, H.: J. Am. Chem. Soc., *91*, 1170 (1969)
454. Tanida, H., Irie, T.: J. Org. Chem., *36*, 2777 (1971)
455. Traylor, T. G.: Acc. Chem. Res., *2*, 152 (1969)
456. Fahey, R. C.: Topics in Stereochemistry, *3*, 253 (1969)
457. Cristol, S. J., Caple, R.: J. Org. Chem., *31*, 2741 (1966)
458. Kwart, H., Nyce, J.: J. Am. Chem. Soc., *86*, 2601 (1964)
459. Stille, J. K., Sonnenberg, F. M., Kinstle, T. H.: J. Am. Chem. Soc., *88*, 4922 (1966)
460. Schleyer, P. v. R.: J. Am. Chem. Soc., *89*, 3901 (1967)
461. Caple, R., Tan, H., Hsu, F.: J. Org. Chem., *33*, 1542 (1968)
462. Dewar, M. J. S., Fahey, R. C.: Angew. Chem. Int. Ed., *3*, 245 (1964)
463. Jefford, C. W., Wojnarowski, W.: Tetrahedron Lett., *1968*, 193
464. Cristol, S. J., Seifert, W. K., Johnson, D. W., Jurale, J. B.: J. Am. Chem. Soc. *84*, 3918 (1962)
465. Cristol, S. J., Morrill, T. C., Sanchez, R. A.: J. Org. Chem., *31*, 2719 (1966)
466. Bach, R. D., Heinneike, H. F.: J. Am. Chem. Soc., *92*, 5589 (1970)
467. Inagaki, S., Fujimoto, H., Fukuri, K.: J. Am. Chem. Soc., *98*, 4054 (1976)
468. Tidwell, T. T., Traylor, T. G.: J. Org. Chem., *33*, 2614 (1968)
469. Brown, H. C., Kawakami, J. H.: J. Am. Chem. Soc., *92*, 1990 (1970)
470. Brown, H. C., Liu, K.-T.: J. Am. Chem. Soc., *93*, 7335 (1971)
471. Brown, H. C.: J. Am. Chem. Soc., *89*, 3898 (1967)
472. Brown, H. C., Kawakami, J. H., Liu, K.-T.: J. Am. Chem. Soc., *92*, 5536 (1970)
473. Brown, H. C., Kawakami, J. H., Liu, K.-T.: J. Am. Chem. Soc., *92*, 3816 (1970)
474. Brown, H. C., Liu, K.-T.: J. Am. Chem. Soc., *97*, 600 (1975)
475. Brown, H. C., Liu, K.-T.: J. Am. Chem. Soc., *97*, 2469 (1975)
476. Brown, H. C., Kawakami, J. H.: J. Am. Chem. Soc., *97*, 5521 (1975)
477. Cristol, S. J., Morrill, T. C., Sanchez, R. A.: J. Org. Chem., *31*, 2719, 2726, 2733, 2738 (1966)
478. Cristol, S. J., Sallivan, J. M.: J. Am. Chem. Soc., *93*, 1967 (1971)
479. Poutsma, M.: J. Am. Chem. Soc., *87*, 4293 (1965)
480. Roberts, J. D., Johnson, F. O., Carboni, R. A.: J. Am. Chem. Soc., *76*, 5692 (1954)

481. Roberts, J. D., Trumbull, E. R., Bennett, J. W., Armstrong, R.: J. Am. Chem. Soc., *72*, 3116 (1950)
482. Kaplan, L., Kwart, H., Schleyer, P.: J. Am. Chem. Soc., *82*, 2341 (1960)
483. Marshall, D. R., Reynolds-Warnhoff, P., Warnhoff, E. W., Robinson, J. R.: Canad. J. Chem., *49*, 885 (1971)
484. Winstein, S.: J. Am. Chem. Soc., *83*, 1516 (1961)
485. Wilt, J., Gutman, G., Ranus, W., Zigman, A.: J. Org. Chem., *32*, 893 (1967)
486. Cristol, S. J., Nachtigall, G. W.: J. Org. Chem., *32*, 3727 (1967)
487. Meinwald, J., Wiley, G.: J. Am. Chem. Soc., *80*, 3667 (1958)
488. Caple, R., Hsu, F., Ilenda, C.: J. Org. Chem., *33*, 4111 (1968)
489. Berson, J. A., Swidler, R.: J. Am. Chem. Soc., *76*, 4060 (1953)
490. Mautecon, J., Cortes, L., Payo, E., Salazar, A.: J. Org. Chem., *32*, 3796 (1967)
491. Caple, R., Chen, G., Nelson, J.: J. Org. Chem., *36*, 2870 (1971)
492. Martin, M., Koster, R.: J. Org. Chem., *33*, 3428 (1968)
493. Limasova, T. I., Rumjanzeva, A. G., Kollegova, M. I., Chernjachovskaya, S. G.. Berus, E. I., Barkhash, V. A.: J. Org. Chem. (USSR), *7*, 751 (1971)
494. Grob, C., Gagneux, A.: Helv. Chim. Acta, *1959*, 1753
495. Cristol, S., Arganbright, R., Tanner, D.: J. Org. Chem., *28*, 1374 (1963)
496. Cristol, S., Parungo, F. P., Plorde, D.: J. Am. Chem. Soc., *87*, 2870 (1965)
497. Miller, R., Stiles, M.: J. Am. Chem. Soc., *85*, 1798 (1963)
498. Brewer, J., Eckard, I., Heaney, H., Marples, B.: J. Chem. Soc., *1968C*, 664
499. Voroztsov, I. N., Ivanova, N. G., Barkhash, V. A.: Izv. Akad. Nauk SSSR, Ser. Khim., *1967*, 1514
500. Lobanova, T. P., Berus, E. I., Barkhash, V. A.: J. Org. Chem. (USSR), *39*, 2332 (1969)
501. Lobanova, T. P., Slinko, N. M., Derendjaev, B. G., Barkhash, V. A.: J. Org. Chem. (USSR), *6*, 2473 (1970)
502. Tanida, H., Tori, K., Kitahonoki, K.: J. Am. Chem. Soc., *89*, 3212 (1967)
503. Povolotskaya, N. N., Scheinman, A. U., Derendjaev, B. G., Kollegova, M. I., Barkhash, V. A.: J. Org. Chem. (USSR), *7*, 760 (1971)
504. Mironova, M. K., Povolotskaya, N. N., Barkhash, V. A.: Izv. Akad. Nauk SSSR, Ser. Khim., *1976*, 679
505. Voroztsov, I. N., Berus, E. I., Derendjaev, B. G., Barkhash, V. A.: J. Org. Chem. (USSR), *39*, 2326 (1969)
506. Voroztsov, I. N.: Diss., Novosibirsk, 1968
507. Lobanova, T. P.: Diss., Novosibirsk, 1973
508. Barkhash, V. A.: Diss., Novosibirsk, 1977
509. Povolotskaya, N. N., Limasova, T. I., Berus, E. I., Eksner, O., Barkhash, V. A.: J. Org. Chem. (USSR), *6*, 1603 (1970)
510. Povolotskaya, N. N., Derendjaev, B. G., Barkhash, V. A.: J. Org. Chem. (USSR), *9*, 1878 (1973)
511. Povolotskaya, N. N.: Diss., Novosibirsk, 1972
512. Slinko, N. M., Derendjaev, B. G., Kollegova, M. I., Barkhash, V. A.: J. Org. Chem. (USSR), *9*, 1901 (1973)
513. Povolotskaya, N. N., Kollegova, M. I., Berus, E. I., Barkhash, V. A.: J. Org. Chem. (USSR), *6*, 2331 (1970)
514. Lobanova, T. P., Barkhash, V. A.: J. Org. Chem. (USSR), *9*, 2281 (1973)
515. Cristol, S., Parungo, F., Plorde, D., Schwarzenbach, K.: J. Am. Chem. Soc., *87*, 2879 (1965)
516. Diaz, A., Harris, D., Sakai, M., Winstein, S.: Tetrahedron Lett., *1971*, 303
517. Slinko, N. M., Derendjaev, B. G., Kollegova, M. I., Barkhash, V. A.: J. Org. Chem. (USSR), *9*, 2058 (1973)
518. Slinko, N. M.: Diss., Novosibirsk, 1975
519. Slinko, N. M., Mamatjuk, V. I., Barkhash, V. A.: J. Org. Chem. (USSR), *13*, 69 (1976)
520. Lobanova, T. P., Derendjaev, B. G., Kollegova, M. I., Barkhash, V. A.: J. Org. Chem. (USSR), *9*, 1883 (1973)
521. Povolotskaya, N. N., Slinko, G. D., Kollegova, M. I., Derendjaev, B. G., Petrov, A. K., Barkhash, V. A.: J. Org. Chem. (USSR), *9*, 1869 (1973)

522. Povolotskaya, N. N., Spivak, A. Yu., Derendjaev, B. G., Barkhash, V. A.: J. Org. Chem. (USSR), *9*, 1873 (1973)
523. Bly, R. S., Konizer, G. B.: J. Org. Chem., *34*, 2346 (1969)
524. Meinwald, J., Labana, S., Chadha, M.: J. Am. Chem. Soc., *85*, 582 (1963)
525. Nisnevich, G. A.: Diss., Novosibirsk, 1982
526. Lobanova, T. P., Derendjaev, B. G., Kollegova, M. I., Barkhash, V. A.: J. Org. Chem. (USSR), *9*, 2273 (1973)
527. Streitwieser, A.: Chem. Revs., *56*, 571 (1956)
528. Rice, F. O., Teller, E.: J. Chem. Phys., *6*, 489 (1938)
529. Hine, J.: J. Am. Chem. Soc., *93*, 3701 (1971)
530. Grob, C. A.: Ang. Chem., Int. Ed., *8*, 535 (1969)
531. Henbest, H., Millward, B.: J. Chem. Soc., *1960*, 3575
532. Cristol, S. J., Tanner, D.: J. Am. Chem. Soc., *86*, 3122 (1964)
533. Goering, H. L., Towns, D. L.: J. Am. Chem. Soc., *85*, 2295 (1963)
534. Diaz, A., Sakai, M., Winstein, S.: J. Am. Chem. Soc., *92*, 7477 (1970)
535. Brown, J. M.: Chem. Commun., *1967*, 639
536. Winstein, S., Ogliaruso, M., Sakai, M., Nicholson, J.: J. Am. Chem. Soc., *89*, 3656 (1967)
537. Spivak, A. Yu., Chertok, V. S., Derendjaev, B. G., Barkharsh, V. A.: J. Org. Chem. (USSR), *9*, 2288 (1973)
538. Spivak, A. Yu., Lobanova, T. P., Chertok, V. S., Podgornaya, M. I., Barkhash, V. A.: J. Org. Chem. (USSR), *12*, 1210 (1976)
539. Spivak, A. Yu.: Diss., Novosibirsk, 1977
540. Winstein, S., Fainberg, A., Grunwald, E.: J. Am. Chem. Soc., *79*, 4146 (1957)
541. Kamschii, G. T., Spivak, A. Yu., Mamatjuk, V. I., Barkhash, V. A.: J. Org. Chem. (USSR), *14*, 2523 (1978)
542. Nordlander, J. E., Kelly, W. J.: J. Am. Chem. Soc., *91*, 996 (1969)
543. Jorgensen, W., Munroe, J. E.: Tetrahedron Lett., *1977*, 581
544. McManus, S. P., Peterson, P. E.: Tetrahedron Lett., *1975*, 2753
545. Olah, G. A., Lukas, J.: J. Am. Chem. Soc., *89*, 4739 (1967)
546. Lambert, J. B., Putz, G. J.: J. Am. Chem. Soc., *95*, 6313 (1973)
547. Cristol, S. J., Kochansky, M. C.: J. Org. Chem., *40*, 2171 (1975)
548. Spivak, A. Yu., Makarova, Z. S., Barkhash, V. A.: J. Org. Chem. (USSR), *12*, 1416 (1976)
549. Gray, A. C., Hart, H.: J. Am. Chem. Soc., *90*, 2569 (1968)
550. Cristol, S. J., Bopp, R. J.: J. Org. Chem., *39*, 1336 (1974)
551. Cristol, S. J., Bopp, R. J., Johnson, A. E.: J. Org. Chem., *34*, 3574 (1969)
552. Spivak, A. Yu., Makarova, Z. S., Barkhash, V. A.: J. Org. Chem. (USSR), *12*, 1424 (1976)
553. Kamschii, G. T., Mamatjuk, V. I., Slinko, N. M., Barkhash, V. A.: J. Org. Chem. (USSR), *12*, 2546 (1976)
554. Kamschii, G. T., Mamatjuk, V. I., Barkhash, V. A.: J. Org. Chem. (USSR), *16*, 85 (1980)
555. Kamschii, G. T.: Diss., Novosibirsk, 1980
556. Reich,I. L., Diaz, A., Winstein, S.: J. Am. Chem. Soc., *91*, 5637 (1969)
557. Kamschii, G. T., Mamatjuk, V. I., Spivak, A. Yu., Barkhash, V. A.: J. Org. Chem., (USSR), *15*, 1221 (1979)
558. Spivak, A. Yu., Barkhash, V. A.: J. Org. Chem. (USSR), *16*, 74 (1980)
559. Spivak, A. Yu., Barkhash, V. A.: J. Org. Chem. (USSR), *12*, 1223 (1976)
560. Slinko, N. M., Barkhash, V. A.: J. Org. Chem. (USSR), *14*, 2286 (1978)
561. Nickon, J. A., Weglein, R. C.: J. Am. Chem. Soc., *97*, 1271 (1975)
562. Slinko, N. M., Mironova, M. K., Barkhash, V. A.: J. Org. Chem. (USSR), *12*, 1907 (1976)
563. Nisnevich, G. A., Mamatjuk, V. I., Barkhash, V. A.: J. Org. Chem. (USSR), (1982)
564. LeBel, N. A., Spurlock, L. A.: Tetrahedron, *20*, 215 (1964)
565. LeBel, N. A., Huber, J. E., Zalkow, L. H.: J. Am. Chem. Soc., *84*, 2226 (1962)
566. Klumpp, G., Ellen, G., Bickelhaupt, F.: Rec. trav., *88*, 474 (1969)
567. Hess, B. A.: J. Am. Chem. Soc., *93*, 1000 (1971)
568. Lambert, J. B., Holcomb, A. G.: J. Am. Chem. Soc., *93*, 3952 (1971)
569. Gassman, P. G., Lein, G. M., Yamaguchi, R.: Tetrahedron Lett., *1976*, 3113
570. Berson, J. A.: Ang. Chem., Int. Ed., *7*, 779 (1968)

571. Berson, J. A., Gajewski, J. J.: J. Am. Chem. Soc., *86*, 5020 (1964)
572. Berson, J. A., Gajewski, J. J., Donald, D. S.: J. Am. Chem. Soc., *91*, 5550 (1969)
573. Berson, J. A., Donald, D. S., Libbey, W. J.: J. Am. Chem. Soc., *91*, 5580 (1969)
574. Berson, J. A., Poonian, M. S., Libbey, W. J.: J. Am. Chem. Soc., *91*, 5567 (1969)
575. Goering, H. L., Greiner, R. W., Sloan, M. F.: J. Am. Chem. Soc., *83*, 1391 (1961)
576. LeBel, N. A., Huber, J. E.: J. Am. Chem. Soc., *85*, 3193 (1963)
577. Fraser, R. R., Farrell, S. O.: Tetrahedron Lett., *1962*, 1143
578. Richey, H. G., Lustgarten, R. K.: J. Am. Chem. Soc., *88*, 3136 (1966)
579. Story, P. R., Saunders, M.: J. Am. Chem. Soc., *84*, 4876 (1962)
580. Story, P. R., Saunders, M.: J. Am. Chem. Soc., *82*, 6199 (1960)
581. Story, P. R., Snyder, L. C., Douglas, D. C., Anderson, E. W., Koornegav, R. L.: J. Am. Chem. Soc., *85*, 3630 (1963)
582. Lustgarten, R. K., Brookhart, M., Winstein, S.: J. Am. Chem. Soc., *89*, 6350 (1967)
583. Brookhart, M., Lustgarten, R. K., Winstein, S.: J. Am. Chem. Soc., *89*, 6352 (1967)
584. Lustgarten, R. K., Gassman, P. G., Patton, D. S., Brookhart, M., Winstein, S., Richey, H. G., Nichols, J. D.: Tetrahedron Lett., *1970*, 1699
585. Pittman, C. U., Olah, G. A.: J. Am. Chem. Soc., *87*, 5123 (1965)
586. Farnum, D. G.: J. Am. Chem. Soc., *89*, 2970 (1967)
587. Olah, G. A., Liang, G.: J. Am. Chem. Soc., *97*, 6803 (1975)
588. Winstein, S., Ordronneau, C.: J. Am. Chem. Soc., *82*, 2084 (1960)
589. Richey, H. G., Lustgarten, R. K.: J. Am. Chem. Soc., *88*, 864 (1966)
590. Hoffmann, R.: J. Am. Chem. Soc., *86*, 1259 (1964)
591. Brookhart, M., Lustgarten, R. K., Winstein, S.: J. Am. Chem. Soc., *89*, 6354 (1967)
592. Lustgarten, R. K., Brookhart, M., Winstein, S.: J. Am. Chem. Soc., *94*, 2347 (1972)
593. Hogeveen, H., Kruchten, E. M.: Topics in Current Chem., *80*, 90 (1979)
594. Lustgarten, R. K., Brookhart, M.: J. Am. Chem. Soc., *90*, 7364 (1968)
595. Lustgarten, R. K., Brookhart, M., Winstein, S.: Tetrahedron Lett., *1971*, 141
596. Moncur, M. V., Grutzner, J. B.: J. Am. Chem. Soc., *95*, 6449 (1973)
597. Yoneda, S., Yoshida, Z., Winstein, S.: Tetrahedron, *28*, 2395 (1972)
598. Dewar, M. J. S., Schoeller, W. W.: Tetrahedron, *27*, 4401 (1971)
599. Katz, T. J., Gold, E. H.: J. Am. Chem. Soc., *86*, 1600 (1964)
600. Olah, G. A., Staral, J. S., Liang, G.: J. Am. Chem. Soc., *96*, 6233 (1974)
601. Olah, G. A., Staral, J. S., Spear, R. J., Liang, G.: J. Am. Chem. Soc., *97*, 5489 (1975)
602. Breslow, R., Groves, J. T., Ryan, G.: J. Am. Chem. Soc., *89*, 5048 (1967)
603. Rosenberg, J. L., Mahler, J. E., Pettit, R.: J. Am. Chem. Soc., *84*, 2842 (1962)
604. Winstein, S., Kaesz, H. D., Kreiter, C. G., Friedrich, E. C.: J. Am. Chem. Soc., *87*, 3267 (1965)
605. Davison, A., McFarlane, W., Pratt, L., Wilkinson, G.: J. Chem. Soc., *1962*, 4821
606. Winstein, S., Kreiter, C. G., Braumann, J. I.: J. Am. Chem. Soc., *88*, 2047 (1966)
607. Dauben, H. J., Wilson, J. D., Laity, J. L.: J. Am. Chem. Soc., *90*, 811 (1968)
608. Dauben, H. J., Wilson, J. D., Laity, J. L.: J. Am. Chem. Soc., *91*, 1991 (1969)
609. Paquette, L. A., Broadhurst, M. J., Warner, P., Olah, G. A., Liang, G.: J. Am. Chem. Soc., *95*, 3386 (1973)
610. Haddon, R. C.: J. Org. Chem., *44*, 3608 (1979)
611. Chapman, O. L., Fugiel, R. A.: J. Am. Chem. Soc., *91*, 215 (1969)
612. Oda, M., Layama, Y., Miyasaki, H., Kitahara, Y.: Ange. Chem., *87*, 414 (1975)
613. Boche, G., Hechtl, W., Huber, H., Huisgen, R.: J. Am. Chem. Soc., *89*, 3344 (1967); Huisgen, R., Boche, G., Huber, H.: J. Am. Chem. Soc., *89*, 3345 (1967)
614. Paquette, L. A., Jacobsson, U., Ley, S. V.: J. Am. Chem. Soc., *98*, 152 (1976); Gasteiger, J., Huisgen, R.: J. Am. Chem. Soc., *94*, 6541 (1972)
615. Kitching, W., Henzel, K. A., Paquette, L. A.: J. Am. Chem. Soc., *97*, 4643 (1975)
616. Brookhart, M., Ogliaruso, M., Winstein, S.: J. Am. Chem. Soc., *89*, 1965 (1967)
617. Holmes, J. D., Pettit, R.: J. Am. Chem. Soc., *85*, 2531 (1963)
618. Keller, C. E., Pettit, R.: J. Am. Chem. Soc., *88*, 604 (1966)
619. Keller, C. E., Pettit, R.: J. Am. Chem. Soc., *88*, 606 (1966)
620. Mateescu, G. D., Nenitzescu, C. D., Olah, G. A.: J. Am. Chem. Soc., *90*, 6235 (1968)
621. Paquette, L., Marpass, J., Barton, T.: J. Am. Chem. Soc., *91*, 4714 (1969)

622. Ahlberg, P., Harris, D., Winstein, S.: J. Am. Chem. Soc., *92*, 2146 (1970)
623. Sugimura, Y., Soma, N.: Tetrahedron Lett., *1970*, 1721
624. Merk, W., Pettit, R.: J. Am. Chem. Soc., *90*, 814 (1968)
625. Childs, R. F., Winstein, S.: J. Am. Chem. Soc., *89*, 6348 (1967)
626. Childs, R. F., Brown, M. A., Anet, F. A. L., Winstein, S.: J. Am. Chem. Soc., *94*, 2175 (1972)
627. Willner, I., Minsky, A., Rabinovitz, M.: J. Org. Chem., *44*, 4440 (1979)
628. Brookhart, M. S., Atwater, M. A.: Tetrahedron Lett., *1972*, 4399
629. Ahlberg, P., Harris, D. L., Roberts, M., Warner, P., Seidl, P., Sakai, M., Cook, D., Diaz, A., Dirlam, J., Hamaberger, H., Winstein, S.: J. Am. Chem. Soc., *94*, 7063 (1972)
630. Ahlberg, P., Harris, D., Winstein, S.: J. Am. Chem. Soc., *92*, 4454 (1970)
631. Cook, D., Diaz, A., Dirlam, J., Harris, D., Sakai, M., Winstein, S., Barborak, J., Schleyer, P. v. R.: Tetrahedron Lett., *1971*, 1405
632. Pquette, L. A., Oku, M., Farnham, W. B., Olah, G. A., Liang, G.: J. Org. Chem., *40*, 700 (1975)
633. Kende, A. S., Bogard, T. L.: Tetrahedron Lett., *1967*, 3383
634. Barborak, J. C., Daub, J., Follweiler, D. M., Schleyer, P. v. R.: J. Am. Chem. Soc., *91*, 7760 (1969)
635. Barborak, J. C., Schleyer, P. v. R.: J. Am. Chem. Soc., *92*, 3184 (1970)
636. Grutzner, J. B., Winstein, S.: J. Am. Chem. Soc., *92*, 3186 (1970)
637. Roberts, M., Hamberger, H., Winstein, S.: J. Am. Chem. Soc., *92*, 6346 (1970)
638. Schroeder, G., Prange, U., Bowman, N. S., Oth, J. F. M.: Tetrahedron Lett., *1970*, 3251
639. Seidl, P., Roberts, M., Winstein, S.: J. Am. Chem. Soc., *93*, 4089 (1971)
640. Warner, P., Winstein, S.: J. Am. Chem. Soc., *93*, 1284 (1971)
641. Corver, H. A., Childs, R. F.: J. Am. Chem. Soc., *94*, 6201 (1972)
642. Goldstein, M. J., Kline, S. A.: J. Am. Chem. Soc., *95*, 935 (1973)
643. Groves, J. T., Ma, K. W.: J. Am. Chem. Soc., *97*, 4434 (1975)
644. Winstein, S., Anet, F. A. L., Carter, P., Bouru, A. J. R.: J. Am. Chem. Soc., *87*, 5247, 5249 (1965)
645. Ahlberg, P.: Chem. Scr., *2*, 231 (1972)
646. Engdahl, C., Ahlberg, P.: J. Am. Chem. Soc., *101*, 3940 (1979)
647. Engdahl, C., Jonsäll, G., Ahlberg, P.: J. Chem. Soc. Chem. Commun., *1979*, 626
648. Ahlberg, P., Engdahl, C., Jonsäll, G.: J. Am. Chem. Soc., *103*, 1583 (1981)
649. Breslow, R., Hill, R., Wasserman, E.: J. Am. Chem. Soc., *86*, 5349 (1964)
650. Winstein, S., Battiste, M.: J. Am. Chem. Soc., *82*, 5244 (1960)
651. deVries, L.: J. Am. Chem. Soc., *82*, 5242 (1960)
652. Childs, R. F., Sakai, M., Winstein, S.: J. Am. Chem. Soc., *90*, 7144 (1968)
653. Grutzner, J. B.: Ph. D. Thesis, Melbourne, 1967
654. Childs, R. F., Winstein, S.: J. Am. Chem. Soc., *90*, 7146 (1968)
655. Childs, R. F., Winstein, S.: J. Am. Chem. Soc., *96*, 6409 (1974)
656. Koptyug, V. A., Kusubova, L. I., Isaev, I. S., Mamatyuk, V. I.: Chem. Commun., *1969*, 389
657. Childs, R. F., Sakai, M., Parrington, B. D., Winstein, S.: J. Am. Chem. Soc., *96*, 6403 (1974)
658. Childs, R. F., Parrington, B.: Chem. Commun., *1970*, 1540
659. Swatton, D., Hart, H.: J. Am. Chem. Soc., *89*, 5075 (1967)
660. Vogel, P., Saunders, M., Hasty, N. M., Berson, J. A.: J. Am. Chem. Soc., *93*, 1551 (1971)
661. Childs, R. F., Zeya, M.: J. Am. Chem. Soc., *96*, 6418 (1974)
662. Berson, J. A., Jenkins, J. A.: J. Am. Chem. Soc., *94*, 8907 (1972)
663. Hehre, W. J.: J. Am. Chem. Soc., *94*, 8908 (1972); Hehre, W. J.: J. Am. Chem. Soc., *96*, 5207 (1974)
664. Childs, R. F., Rogerson, C. V.: J. Am. Chem. Soc., *100*, 649 (1978)
665. Scott, L. T., Brunsvold, W. R.: J. Am. Chem. Soc., *100*, 6535 (1978)
666. Olah, G. A., Liang, G., Jindal, S. P.: J. Org. Chem., *40*, 3259 (1975)
667. Hogeveen, H., Volger, H.: Rec. trav., *87*, 385 (1968)
668. Raquette, L. A., Krow, G. R., Bollinger, J. M., Olah, G. A.: J. Am. Chem. Soc., *90*, 7147 (1968)
669. Hogeveen, H., Kwant, P. W., Schudde, E. P., Wadl, P. A.: J. Am. Chem. Soc., *96*, 7518 (1974)
670. Masamune, S.: Chem. Commun., *1969*, 243

671. Hogeveen, H., Kwant, P. W.: J. Am. Chem. Soc., *95*, 7315 (1973)
672. Heldeweg, R. F., Hogeveen, H., Zwart, L.: Tetrahedron Lett., *1977*, 2535
673. Hogeveen, H., Kwant, P.: Tetrahedron Lett., *1973*, 1665
674. Hogeveen, H., Kwant, P., Postma, J., van Duynen, P. Th.: Tetrahedron Lett., *1974*, 4351
675. Hogeveen, H., Kwant, P.: J. Am. Chem. Soc., *96*, 2208 (1974)
676. Olah, G. A., Bollinger, J. M., White, A. M.: J. Am. Chem. Soc., *91*, 3667 (1969)
677. Jonkman, H. T., Nieuwpoort, W. C.: Tetrahedron Lett., *1973*, 1671
678. Hogeveen, H., van Kruchten, E. M. G. A.: J. Org. Chem., *46*, 1350 (1981)
679. Giordano, C., Heldeweg, R. F., Hogeveen, H.: J. Am. Chem. Soc., *99*, 5181 (1977)
680. Stohrer, W.-D., Hoffman, R.: J. Am. Chem. Soc., *94*, 1661 (1972); Yoneda, S., Yoshida, L.: Chem. Lett., *1972*, 607
681. Masamune, S., Sakai, M., Ona, H.: J. Am. Chem. Soc., *94*, 8955 (1972); Masamune, S., Sakai, M., Ona, H., Jones, A. J.: J. Am. Chem. Soc., *94*, 8956 (1972)
682. Hehre, W. J., Schleyer, P. von R.: J. Am. Chem. Soc., *95*, 5837 (1973)
683. Dewar, M. J., Haddon, R. C.: J. Am. Chem. Soc., *95*, 5836 (1973)
684. Kollmar, H., Smith, H., Schleyer, P. von R.: J. Am. Chem. Soc., *95*, 5834 (1973)
685. Morio, K., Maşamune, S.: Chem. Letters, *1974*, 1107
686. Hart, H., Kuzya, M.: J. Am. Chem. Soc., *96*, 6436 (1974)
687. Masamune, S., Sakai, M., Kemp-Jones, A. V., Ona, H., Venot, A., Nakashima, T.: Angew. Chem., Int. Ed., *12*, 769 (1973)
688. Hart, H., Willer, R.: Tetrahedron Lett., *1978*, 4189
689. Hart, H., Kuzuya, M.: J. Am. Chem. Soc., *97*, 2450 (1975)
690. Hart, H., Kuzuya, M.: J. Am. Chem. Soc., *97*, 2459 (1975)
691. Masamune, S., Sakai, M., Kemp-Jones, A., Nakashima, T.: Canad. J. Chem., *52*, 855 (1974)
692. Olah, G., Liang, G.: J. Am. Chem. Soc., *97*, 2236 (1975)
693. Olah, G., Liang, G.: J. Am. Chem. Soc., *98*, 6304 (1976)
694. Breslow, R.: "Molecular rearrangements", *1*, New York, 259 (1963)
695. Richey, H. J.: Carbonium Ions, *3*, New York, 1972
696. Wiberg, K. B., Ashe, A. J.: Carbonium Ions, *3*, New York, 1972
697. Poulter, C. D., Winstein, S.: J. Am. Chem. Soc., *94*, 2297 (1972)
698. Olah, G. A., Jeuell, C. L., Kelly, D. P., Porter, R. D.: J. Am. Chem. Soc., *94*, 146 (1972)
699. Staral, J. S., Yavari, I., Roberts, J. D., Prakash, G. K. S., Donovan, D. J., Olah, G. A.: J. Am. Chem. Soc., *100*, 8016 (1978); Staral, J. S., Roberts, J. D.: J. Am. Chem. Soc., *100*, 8018 (1978)
700. Winstein, S., Sonnenberg, J., de Vries, L.: J. Am. Chem. Soc., *81*, 6523 (1959)
701. Winstein, S., Sonnenberg, J.: J. Am. Chem. Soc., *83*, 3235 (1961)
702. Winstein, S., Friedrich, E., Baker, R., Lin, Y.: Tetrahedron, Suppl. *8*, p. 2, 6211 (1966)
703. Winstein, S., Sonnenberg, J.: J. Am. Chem. Soc., *83*, 3244 (1961)
704. Le Noble, W. J., Yates, B. L., Scaplehorn, A. W.: J. Am. Chem. Soc., *89*, 3751 (1967)
705. Corey, E. J., Dawson, R. L.: J. Am. Chem. Soc., *85*, 1782 (1963)
706. Corey, E. J., Uda, H.: J. Am. Chem. Soc., *85*, 1788 (1963)
707. Freeman, M. F., Grostic, M. F., Raymond, F. A.: J. Org. Chem., *30*, 771 (1965)
708. Lin, Y.: Thesis UCLA, 1967
709. Norin, T.: Tetrahedron Lett., *1964*, 37
710. Kropp, R. J.: J. Am. Chem. Soc., *88*, 4926 (1966)
711. Gassman, P. G., Zalar, F. V.: J. Am. Chem. Soc., *88*, 2252 (1966)
712. Friedrich, E. C., Saleh, M. A.: J. Am. Chem. Soc., *95*, 2617 (1973); Friedrich, E. C., Saleh, M. A., Winstein, S.: J. Org. Chem., *38*, 860 (1973)
713. Winstein, S., Bruck, P., Radlick, P., Baker, R.: J. Am. Chem. Soc., *86*, 1867 (1964)
714. Thies, R. W., Sakai, M., Whalen, D., Winstein, S.: J. Am. Chem. Soc., *94*, 2270 (1972)
715. Prelog, V.: Pure Appl. Chem., *6*, 545 (1963)
716. Haywood-Farmer, J., Pincock, R. E., Wells, J. I.: Tetrahedron, *22*, 2007 (1966)
717. Tanida, H., Tsuji, T., Irie, T.: J. Am. Chem. Soc., *89*, 1953 (1967)
718. Battiste, M. A., Deyrup, C. L., Pincock, R. E., Haywood-Farmer, J.: J. Am. Chem. Soc., *89*, 1954 (1967)
719. Haywood-Farmer, J. S., Pincock, R. E.: J. Am. Chem. Soc., *91*, 3020 (1969)
720. Hoffmann, R.: Tetrahedron Lett., *1965*, 3819

721. Ellen, G., Klump, G.: Tetrahedron Lett., *1974*, 3637
722. Lambert, J. B.,, Iovanovich, A. P., Hamersma, J. W., Koeng, F. R., Oliver, S. S.: J. Am. Chem. Soc., *95*, 1570 (1973)
723. Wiberg, K. B., Wenzinger, G. R.: J. Org. Chem., *30*, 2278 (1965)
724. Colter, A. K., Musso, R. C.: J. Org. Chem., *30*, 2462 (1965)
725. Wilcox, C. F., Jesaitis, R. G.: Tetrahedron Lett., *1967*, 2567
726. Freeman, P. K., Balls, D. M.: Tetrahedron Lett., *1967*, 437
727. Woodward, R. B., Hoffmann, R.: J. Am. Chem. Soc., *87*, 395 (1965)
728. DePuy, C. H., Schnack, L. G., Hausser, J. W.: J. Am. Chem. Soc., *88*, 3343 (1966)
729. Battiste, M. A., Ranken, P. F., Edelman, F.: J. Am. Chem. Soc., *93*, 6276 (1971)
730. Battiste, M. A., Nebzydoski, J. W.: J. Am. Chem. Soc., *92*, 4450 (1970)
731. Sakai, M., Diaz, A., Winstein, S.: J. Am. Chem. Soc., *92*, 4452 (1970)
732. Bischof, P., Heilbronner, E., Prinzbach, H., Martin, H.: Helv. Chim. Acta, *54*, 1072 (1971)
733. Javanovich, A. P., Lambert, J. B.: J. Chem. Soc., *1971B*, 1129
734. Wilt, J. W., Malloy, T. P.: J. Am. Chem. Soc., *92*, 4747 (1970)
735. Gassman, P. G., Fentiman, A. F.: J. Am. Chem. Soc., *92*, 2251 (1970)
736. Coates, R. M., Kirkpatrick, J. L.: J. Am. Chem. Soc., *92*, 4883 (1970)
737. Coates, R. M., Fretz, E. R.: J. Am. Chem. Soc., *99*, 297 (1977)
738. Brown, H. C., Ravindranathan, M.: J. Am. Chem. Soc., *99*, 299 (1977)
739. Coates, R. M., Fretz, E. R.: J. Am. Chem. Soc., *97*, 2538 (1975)
740. Jorgensen, W. L.: Tetrahedron Lett., *1976*, 3033
741. Dauben, W. G., Schallhern, C. H.: J. Am. Chem. Soc., *93*, 2254 (1971)
742. Masamune, S., Takada, S., Nakatsuka, N., Vukov, R., Cain, E.: J. Am. Chem. Soc., *91*, 4322 (1969)
743. Coates, R. M., Yano, K.: Tetrahedron Lett., *1972*, 2289
744. Ellen, G., Klump, G.: Tetrahedron Lett., *1974*, 2995
745. Eakin, M. A., Martin, J., Parker, W.: Chem. Commun., *1967*, 955
746. Sargent, G. D., Herkenham, M. A.: J. Am. Chem. Soc., *94*, 2893 (1972)
747. Battiste, M., Haywood-Farmer, J., Malkus, H., Seidl, P., Winstein, S.: J. Am. Chem. Soc., *92*, 2144 (1970)
748. Muneyuki, R., Yano, T., Tanida, H.: J. Am. Chem. Soc., *91*, 2408 (1969)
749. Sargent, G. D., Taylor, R. L., Demish, W. A.: Tetrahedron Lett., *1968*, 2275
750. Olah, G. A., Liang, G., Mo, Y.: J. Am. Chem. Soc., *94*, 3550 (1972)
751. Olah, G. A., Prakash, G. K. S., Rawdah, T. N., Whittaker, D., Rees, J. C.: J. Am. Chem. Soc., *101*, 3935 (1979)
752. Masamune, S., Sakai, M., Kemp-Jones, A. V.: Canad. J. Chem., *52*, 858 (1974)
753. Coates, R. M., Fretz, E. R.: Tetrahedron Lett., *1977*, 1955
754. Goodman, A. L., Eastman, R. H.: J. Am. Chem. Soc., *86*, 908 (1964)
755. Baird, R., Winstein, S.: J. Am. Chem. Soc., *85*, 567 (1963)
756. Kosower, E., Ito, M.: Proc. Chem. Soc., *1962*, 25
757. Cromwell, N. H., Schumaker, F. H., Adelfand, J. L.: J. Am. Chem. Soc., *83*, 974 (1961)
758. Sauers, R. R., Beisler, J. A.: Tetrahedron Lett., *1964*, 2181
759. Sauers, R. R., Beisler, J. A., Freilich, H.: J. Org. Chem., *32*, 569 (1967)
760. Freeman, P. K., Balls, D. M.: J. Org. Chem., *32*, 2354 (1967)
761. Freeman, P. K., Blazevich, J. N.: Chem. Commun., *1969*, 1357
762. Freeman, P. K., Balls, D., Blazevich, J.: J. Am. Chem. Soc., *92*, 2051 (1970)
763. Freeman, P. K., Stevenson, B. K.: J. Am. Chem. Soc., *95*, 2890 (1973)
764. Sargent, G. D., Harrison, M. J., Khoury, G.: J. Am. Chem. Soc., *91*, 4937 (1969)
765. Lenoir, D., Schleyer, P. v. R., Ipaktschi, J.: Lieb. Ann., *750*, 28 (1971)
766. Wilcox, C. F., Jesaitis, R. G.: Tetrahedron Lett., *1967*, 2567
767. Yates, P., Fenwick, J.: J. Am. Chem. Soc., *93*, 4618 (1971)
768. Sherrod, S. A., Bergman, R. G., Gleicher, G. J., Morris, D. G.: J. Am. Chem. Soc., *92*, 3469 (1970)
769. Bingham, R. C., Sliwinski, W. F., Schleyer, P. v. R.: J. Am. Chem. Soc., *92*, 3471 (1970)
770. Sargent, G. D., Harrison, M. J., Khoury, G.: J. Am. Chem. Soc., *91*, 4937 (1969)
771. Rhodes, Y. E., Takino, T.: J. Am. Chem. Soc., *90*, 4469 (1968)
772. Winstein, S., Grunwald, E., Jones, H.: J. Am. Chem. Soc., *73*, 2700 (1951)

773. Sauers, R. R., Ubersax, R. W.: J. Org. Chem., *31*, 495 (1966)
774. Servis, K. L., Roberts, J. D.: J. Am. Chem. Soc., *86*, 3773 (1964)
775. Dewar, M. J. S., Harris, J. M.: J. Am. Chem. Soc., *92*, 6557 (1970)
776. Streitwieser, A.: Chem. Rev., *56*, 571 (1956)
777. Takakis, I., Rhodes, Y.: Tetrahedron Lett., *1978*, 2475; Rhodes, Y. E., Takakis, I. M., Schueler, P. E., Weiss, R. A.: Tetrahedron Lett., *1978*, 2479
778. Rhodes, Y. E., Takino, T.: J. Am. Chem. Soc., *92*, 5270 (1970)
779. Berson, J. A., Wege, D., Clarke, D. M., Bergman, R. G.: J. Am. Chem. Soc., *90*, 3240 (1968)
780. Berson, J. A., Bergman, R. G., Clarke, G. M., Wege, D.: J. Am. Chem. Soc., *91*, 5601 (1969)
781. Deno, N. C., LaVietes, D., Mockus, J., Sholl, P. C.: J. Am. Chem. Soc., *90*, 6457 (1968)
782. Sanders, D. A., Oliver, J. P.: J. Am. Chem. SoC., *90*, 5910 (1968)
783. Schono, T., Fujita, K., Kumai, S., Watanabe, T., Nishiguchi, T.: Tetrahedron Lett., *1972*, 3249
784. Hanack, M., Ensslin, H.: Lieb. Ann., *713*, 49 (1968)
782. Sanders, D. A., Oliver, J. P.: J. Am. Chem. Soc., *90*, 5910 (1968)
786. Coulson, C. A., Goodman, T. H.: J. Chem. Soc., *1962*, 2851
787. Coulson, C. A., Goodman, T. H.: J. Chem. Soc., *1963*, 3161
788. Yano, K., Isobe, M., Yoshida, K.: J. Am. Chem. Soc., *100*, 6166 (1978)
789. Roberts, D. D.: J. Org. Chem. *39*, 1267 (1974)
790. Haddon, R. C.: Tetrahedron Lett., *1974*, 4303

Rearrangements of Carbocations by 1,2-Shifts

Vyacheslav Gennadievich Shubin

Institute of Organic Chemistry, Siberian Division of the USSR Academy of Sciences,
630090 Novosibirsk-90, USSR

Table of Contents

Main Denotations and Abbreviations

R^m = Migrating Atom or Group (Migrant)
C^o = Carbon Atom from which a Migrant is Shifting
C^t = Carbon Atom (Carbonium Centre) toward which a Migrant is Shifting
MT = Migration Tendency
CMA = Competitive Migratory Aptitude
MCR = Migrant-Carrying Remnant
PA = Proton Affinity

1 Introduction

Rearrangements of carbocations are widespread in organic chemistry. The tendency of carbonium ions to rearrangements due to their "internal instability" is one of most characteristic features of these electron-deficient structures. Carbonium ion rearrangements have been dealt with in numerous publications both original ones and reviews (see, e.g., [1-13]).

The key stage of carbocation rearrangements is generally the *1,2-shift* of a migrating atom or group (migrant, R^m) to an electron-deficient carbon, the carbonium centre. Various atoms and groups, such as hydrogen, halogens, alkyls, aryls, different electron-withdrawing groups [14-16] and even such unusual groups as $AlCl_3^-$ [17] can act as migrants.

One of the most fruitful concepts of organic chemistry is that of a substituent possessing a set of intrinsic individual properties. By analogy, the migrants would be assigned a specific property, "intrinsic migratory aptitude" which by virtue of being "intrinsic" should be transferable quite unchanged from one rearranging carbocation to another. It is necessary, however, to note the relative character of the migrant concept and, hence, that of *migratory aptitude*. It is evident that the universality of migratory aptitude of a given migrant is in principle impossible since by dividing mentally the rearranging carbocation into two parts — the migrant and the migrant-carrying remnant — we "break the unbreakable", although the approximate constancy of migratory aptitude of some migrants may be maintained within rather a wide range.

Carbocation rearrangements have been studied for a long time with the result that many of their features have been elucidated. However, these processes, even at a qualitative level, are far from being fully enough described as yet. As noted in one of the monographs [18], "the weighting of the relative importance of the various factors in rearrangements of carbonium ions remains a difficult problem". Bearing in mind the complexity of these reactions one can assume that the best approach to the problem consists in *studying simple ("elementary") structural models* of carbonium ions to reveal the main reactivity-governing factors and to establish the structure-reactivity relationships. A complete picture of the rearrangement (even at the qualitative level) does not seem to be obtainable without establishing such "elementary" relationships.

Until recently the situation in this field has been unfavourable because the kinetic characteristics of the key step of rearrangement — 1,2-shift of the migrant — have practically remained inaccessible. Due to the low availability of these data the usual approach to the quantitative description of rearrangements has consisted in determining the migratory aptitude of various atoms and groups by the ratio of products resulting from the competing 1,2-shifts (*competitive migratory aptitude* — CMA). The CMA concept was subjected to criticism [19], cf. Ref. [4,15,20,21]; it was, therefore, proposed to use the relative rates of the 1,2-shifts of different migrants bonded to the same MCR as a more adequate characteristic of their ability to migrate [19] (*migration tendency*, MT[1] $= k^{R_i^m}/k^{R_o^m}$).

1 Unfortunately, these two notions are frequently confused in the literature.

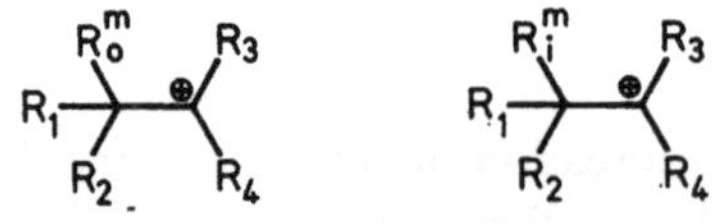

The problem of the quantitative description of the 1,2-shift process thus consists in the determination of the MT and the quantitative description of environmental factors — the structural effects of the *migrant-carrying remnant* (MCR, "short-range" environment) and those of the medium ("long-range" environment).

The development of the chemistry of the so-called "long-lived" carbonium ions combined with that of the NMR method has opened new opportunities for investigating the rearrangements of carbonium ions: it has become possible to obtain the kinetic characteristics of the very 1,2-shift processes — the key step of numerous rearrangements.

In this book the available data are considered on the rates of 1,2-shifts in terms of the above approach. The rearrangements of "long-lived" carbonium ions as well as those assumed to involve the formation of short-lived carbonium ions or the like structures will be considered here.

Within the period that has elapsed since our first review in Russian [13] was published, a great number of works have appeared that are connected with the problem under discussion. We do not want give here an exhaustive review of the new data, but we hope to reflect the main works dealing with the matter in question. No consideration is given to such questions as the rearrangements of vinyl cations [22]; see also Ref. [23]. The numerous studies of carbocation rearrangements by quantum chemical methods have also largely remained beyond the framework of this book.

2 Migration Tendencies

2.1 General Discussion

The numerous atoms and atom groups capable of a 1,2-shift in carbocation rearrangement can be divided into σ-, π- and n-migrants (cf. Ref. [24]). Table 1 gives the data on the migration tendency (MT) of typical representatives of these migrants: H, CH_3, C_2H_5, C_6H_5, Cl in degenerate[2] and nondegenerate rearrangements of long-lived carbocations, as well as in some reactions accompanied by 1,2-shifts. Table 2 has listed the corresponding data for alkyl groups.

As seen from Tables 1 and 2, the values of MT of different migrants vary depending on the structure of the rearranging cation in rather a wide range, but for $R^m = C_2H_5$, C_6H_5, Cl its value (R^m: CH_3) seems to be constant in the qualitative sense ($R^m > CH_3$), cf. also the data for $R^m = CH_2Cl$ [83], Br, Cl [4,24]. The higher value of MT of ethyl relative to that of methyl may be due to the more effective stabilization of the transi-

A

2 Since rearrangements usually occur in achiral media the cases of 1,2-shifts from the chiral carbon atom accompanied by enanthiomerization can also be referred to the category of degenerate processes.

tion state A by the fragment R of the migrant CH_2R in the case of $R = CH_3$ ($R^m = C_2H_5$) relative to $R = H$ ($R^m = CH_3$), cf. Ref. [64,84]. Note in this connection that on the basis of the large value of the methyl migrational isotope effect in dienone-phenol rearrangement in Ref. [85] the migrant group is concluded to bear a good deal of positive charge (cf. Ref. [86,87]). The lower MT of such groups as CH_2Cl [83,88] and $CH_2CH_2N^+(CH_3)_3$ [34] (relative to CH_3) in degenerate rearrangements of arenonium ions, as well as the lower MT of a benzyl group and of substituted benzyl groups in solvolytic rearrangements of neopentyl-type tosylates (Table 3) seem to indicate an essential role of the inductive effect. This is in accord with the similarity of MT of such groups as C_2H_5, n-C_3H_7, n-C_4H_9, i-C_4H_9 and with the linear relationship between the MT logarithms of para-substituted benzyl groups and the respective σ^0-constants [94].

The qualitative constancy of MT is also observed for $R^m = H$ ($H > CH_3$) (in degenerate rearrangements) which may point to the qualitative constancy of differences in the "intrinsic barriers" [93,95–97] of hydrogen and methyl shifts (cf. Ref. [16]). But in the non-degenerate rearrangement of α-hydroxycarbonium ions (Table 1, item 25) there occurs an inversion of the MT $H:CH_3$ which may be due to the disturbing thermodynamic factor reflecting the difference in the relative stability of the parent and the rearranged structures depending in the migrant [10]. The inversion of the MT $H:CH_3$ is also characteristic of the non-degenerate *exo*-3,2-shifts in the rearrangements of norbornyl cations [52] and for the solvolytic rearrangement of tosylates (Table 1, item 29).

The inversion of the MT of some hydrocarbon migrants (n-C_3H_7, *sec*-C_4H_9, $CH_2C_6H_5$) in passing from the rearrangements of arenonium ions to those accompanying the solvolysis and dehydration (Table 2, 3) seems to be due to the inversion of the relative contributions from the resonance and inductive components of the bulk electron effect produced by the fragments of the migrant surrounding the latter's central atom (cf. Ref. [87,98]). This reflects the differences between the transition states of the 1,2-shifts in the carbocations with a "fully developed" carbonium centre and in the species with a "developing" one. This assumption is supported by a decrease in the MT of the ethyl group in passing from the rearrangements of arenonium ions to the reactions of solvolysis and dehydration, as well as by a linear relationship between the logarithms of the 1,2-shift rate constants of the meta- and para-substituted benzyl groups in the degenerate rearrangement of 1-X-benzyl-1,2,3,4,5,6-hexamethylbenzenonium ions and the σ_x^+ constants: [89]

$$\lg k^{x}_{-50\,^\circ C} = (1.81 \pm 0.03) - (4.45 \pm 0.09)\,\sigma_x^+$$

$$r = -0.999\,, \quad s = 0.07\,, \quad n = 5\,.$$

Let us consider one more example of MT inversion. According to the estimation reported in Ref. [99] the free-energy barrier of the 1,2-shift of the $COOC_2H_5$ group in the exothermal rearrangement by the following scheme is about 6 kcal/mole,

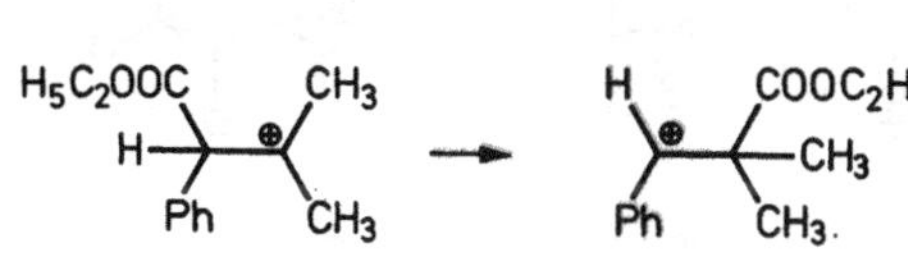

Table 1. Migration Tendencies of the Hydrogen and Chlorine Atoms and of Ethyl and Phenyl Groups[a]

	Rearranging compound	Solvent	t °C	H	Cl	C_2H_5	C_6H_5	Ref.
1	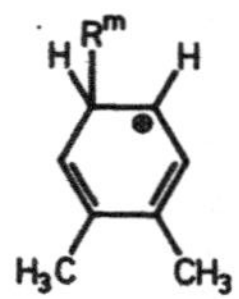	$HF—SbF_5—SO_2ClF$[b]	25	$\approx 10^8–10^9$				27)
2	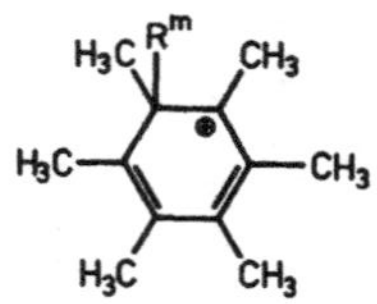	$HSO_3F—SbF_5—SO_2ClF$[c]	25	$2.0 \cdot 10^5$				29), cf. 30)
		HSO_3F[c]	25		$2.8 \cdot 10^2$			31)
		CF_3COOH	25			31[d]		
		$CF_3SO_3H—SO_2ClF$[c]	25				4.3	35)
3	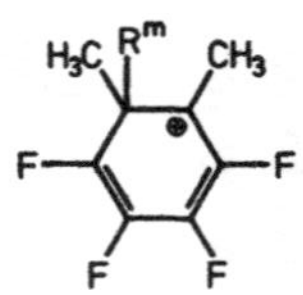	$HF—SbF_5—SO_2ClF$	− 42		$1.7 \cdot 10^{-2}$[e]			37)
4	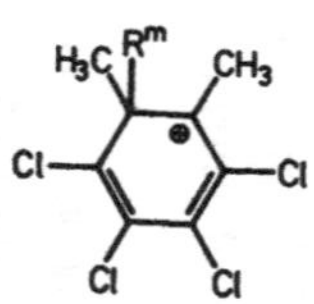	$HF—SbF_5—SO_2ClF$	25		0.2[e]			37)
5	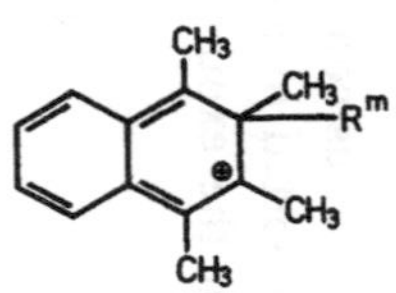	$CF_3SO_3H—SbF_5$	62	$2.9 \cdot 10^6$[f]				39)

	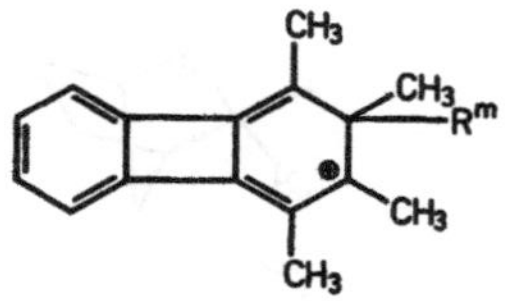	$CF_3SO_3H—SO_2ClF$[g]	25	$3.5 \cdot 10^4$				[41]
7	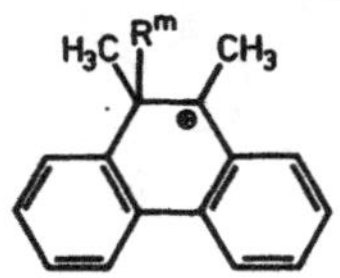	$HSO_3F—SbF_5—SO_2ClF—SO_2F_2$[h]	25	$1.1 \cdot 10^2$	$\approx 1 \cdot 10^7$			[29]
		$CF_3SO_3H—SO_2ClF—SO_2F_2$[h]	−110					[31]
		$HCl—Al_2Cl_6—CH_2Cl_2$	25			13		[43,44]
		$HCl—Al_2Cl_6—CH_2Cl_2$	25				18	[42,44,45]
		$HSO_3F—SO_2ClF$	25				17	[42,46]
8	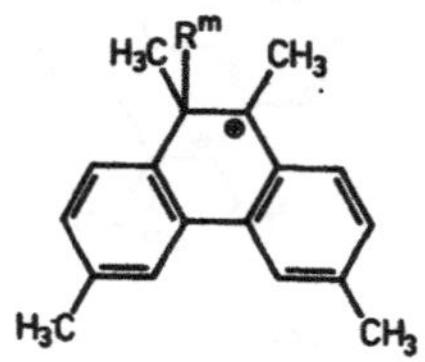	$HSO_3F—SO_2ClF—SO_2F_2$[h]	25	$4.4 \cdot 10^2$				[29]
		$CF_3SO_3H—SO_2ClF$[h]	−110		$7.1 \cdot 10^7$			[31]
		$HSO_3F—SO_2ClF$	25				12	[42,46]
9	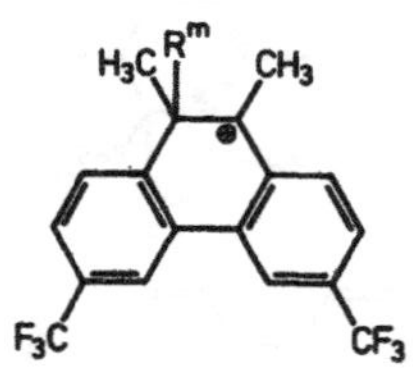	$HSO_3F—SbF_5—SO_2ClF—SO_2F_2$[h]	25	$2.0 \cdot 10^2$				[29]
		$HSO_3F—SO_2ClF—SO_2F_2$[h]	25				$2.0 \cdot 10^2$	[46]
10	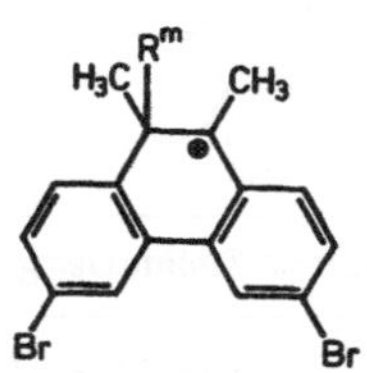	$HSO_3F—SbF_5—SO_2ClF—SO_2F_2$[h]	25	$1.1 \cdot 10^2$				[29]
		$HSO_3F—SO_2ClF$	25				20	[42,46]

Table 1. (continued)

	Rearranging compound	Solvent	t °C	H	Cl	C_2H_5	C_6H_5	Ref.
11	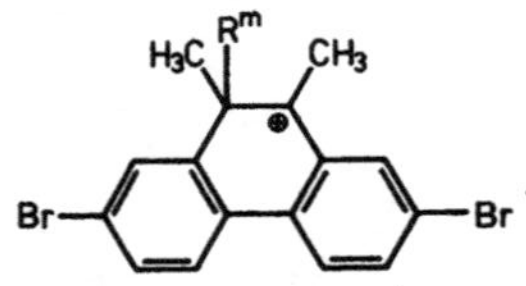	$HSO_3F—SO_2ClF—SO_2F_2$[h)]	25				43	[46)]
12	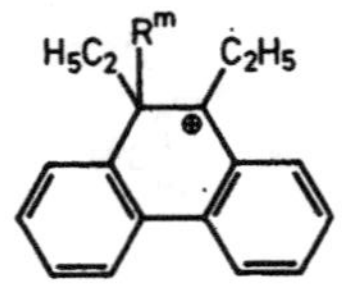	$HCl—Al_2Cl_6—CH_2Cl_2$	25			≈10		[47,48)]
13	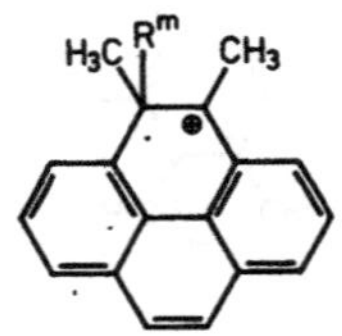	$HSO_3F—SO_2ClF$	25				20	[42,46)]
14	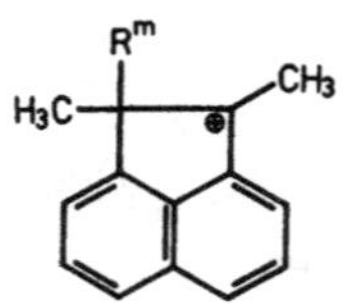	$HSO_3F—SbF_5—SO_2ClF$[i)]	25	$7.2 \cdot 10^3$				[50)]
		$HSO_3F—SO_2ClF$[i)]	25		$2.9 \cdot 10^6$			[31)]
		HSO_3F[i)]				[j)]		
		CF_3SO_3H	25				$1.1 \cdot 10^4$	[49,51)]
15	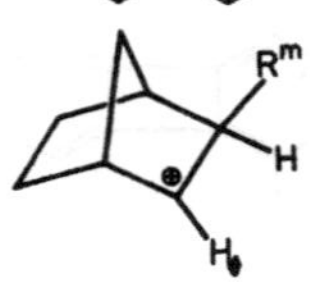	$SbF_5—SO_2—SO_2F_2$[k)]	25	$1.6 \cdot 10^2$				[53)]

16	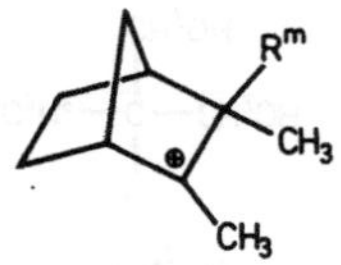	HSO_3F—SO_2ClF	25	6				54,55)
17	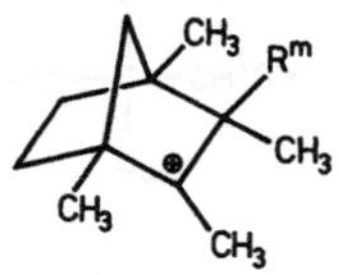	HSO_3F—SO_2ClF	− 94	$3.3 \cdot 10^{2}$ [l]				56, 57, cf. 54)
18	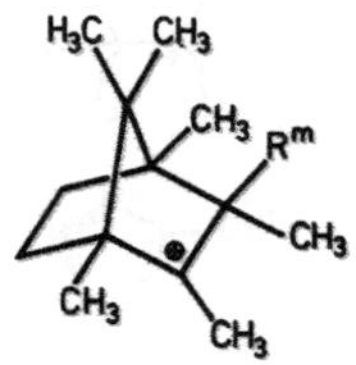	HSO_3F—SO_2ClF	− 50	$\approx 2.6 \cdot 10^{5}$ [l]				57)
19	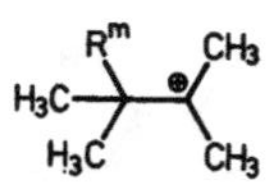	SO_2ClF—SO_2F_2 [n] SbF_5—SO_2ClF	−136	12 [o]	p)	r)	q)	59)
20	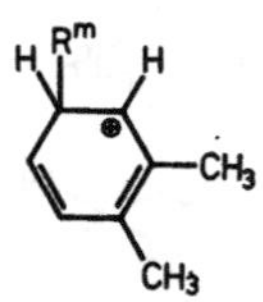	HF—SbF_5—SO_2ClF [b]	25	$1.3 \cdot 10^{7}$				27)
21	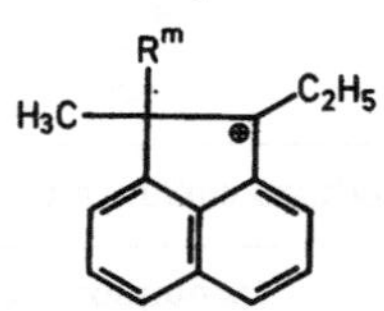	HSO_3F—SO_2ClF or HSO_3F—SbF_5—SO_2ClF		> 1 [s]				50)

Table 1. (continued)

	Rearranging compound	Solvent	t °C	H	Cl	C_2H_5	C_6H_5	Ref.
22	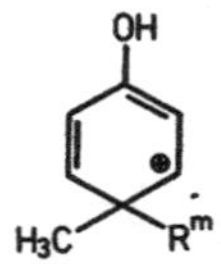	CF_3SO_3H[t]	35		≫1			[66]
		CF_3COOH	38.5			55		[15]
		aq. H_2SO_4	25			49		[20]
		CF_3COOH	38.5				260	[15]
23	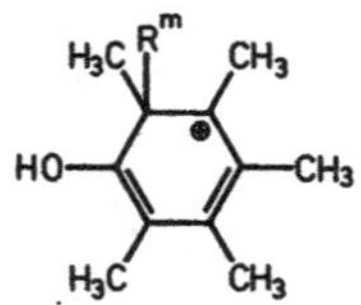	HSO_3F[u]	10		$6.8 \cdot 10^2$			[68]
		$CF_3COOH—H_2SO_4$	25			35		[67]
		$CF_3COOH—H_2SO_4$	25				12	[67]
24	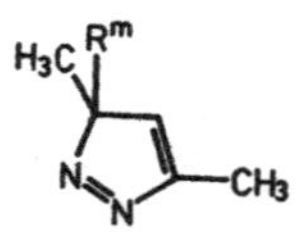	CH_3COOH	60			≈44[v]	≈880[v]	[69]
25	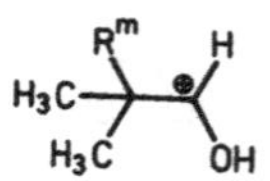	HSO_3F	− 50	4		8		[70]
		$HSO_3F—SbF_5$	− 50	4				[70]
		$HSO_3F—SbF_5$	25	0.1				[70]
		$HF—SbF_5$	− 50	1.9				[70]
26	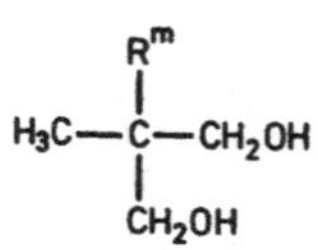	10.N H_2SO_4	108–109			1.40		[71, 72]
							56.7	[72]
27	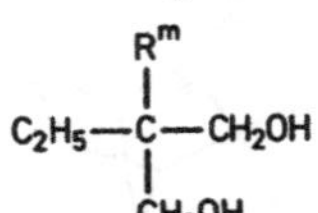	10 N H_2SO_4				1.50	31.6	[71]

28	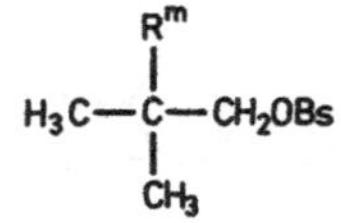	w)	100		1.15		73)
			100			$6.0 \cdot 10^2$	74)
			25			$3.0 \cdot 10^3$	74)
29	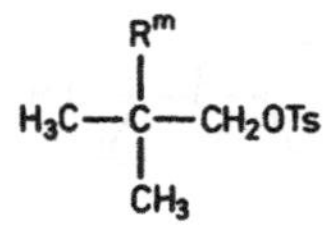	100% C_2H_5OH	45			$5.8 \cdot 10^2$	75)
		90% C_2H_5OH	45			$6.8 \cdot 10^2$	75)
		80% C_2H_5OH	45			$6.8 \cdot 10^2$	75)
		70% C_2H_5OH	45			$6.0 \cdot 10^2$	75)
		CH_3COOH	45			$7.6 \cdot 10^2$	75)
		HCOOH	45			$9.1 \cdot 10^2$	75)
		CF_3COOH	75	<1.5			76)
		96% H_2SO_4	30	<0.45			77)
		HSO_3F	−44	<1.4			78)

a) The MT are either original or calculated from the kinetic data presented in the indicated references, the statistic factors being taken into account. The MT of the CH_3 group is taken as a unit unless otherwise stated. For simplicity here and further on in the structural formulas the n-, π- and σ-electron delocalization is not reflected; for the σ-electron delocalization see, e.g., [25].
b) Data for the 1,2-methyl shift are for a solution without the diluent SO_2ClF [26], see Note [c].
c) Kinetic data for the 1,2-methyl shift are for a solution in HSO_3Cl [28]. The rate of methyl shifts in long-lived carbocations is known (see Sect. 4) to be practically independent of the medium.
d) Estimated from the data of [32,33], cf. [34].
e) The MT of hydrogen is taken as a unit [36], no data for $R^m = CH_3$ are available.
f) The rate constant of the 1,2-methyl shift has been calculated from the kinetic analysis of a three-step degenerate rearrangement of a 1,1,2,3,4-pentamethylnaphthalenonium ion, the values of $k_{\alpha\beta}$ and $k_{\alpha\alpha}$ being determined for solutions in CF_3SO_3H while that of $k_{\beta\alpha}$ — for a solution in $HSO_3F{-}SO_2ClF{-}SO_2F_2$ [38], see Note[c].
g) Kinetic data for the 1,2-methyl shift are for a solution in CF_3COOH [40].
h) Kinetic data for the 1,2-methyl shift are for a solution in $HSO_3F{-}SO_2ClF$ [42], see Note[c].
i) Kinetic data for the 1,2-methyl shift are for a solution in CF_3SO_3H [49], see Note[c].
j) The MT of the ethyl group is likely to be higher than that of the methyl one (cf. [49,50]).
k) For $R^m{=}CH_3$ — an assigned value of the rate constant without medium specification [52].
l) Assuming that for $R^m{=}H$ $\Delta G^{\neq}_{-115\,°C} = \Delta G^{\neq}_{-94\,°C}$.
m) Assuming that for $R^m{=}H$ $\Delta G^{\neq}_{-110\,°C} = \Delta G^{\neq}_{-50\,°C}$ and the kinetic data of [57] refer entirely to the *exo*-methyl shift.

Table 1. (continued)

n) The acid is not specified, it is obviously SbF_5.
o) According to the latest data of M. Saunders and W. Hehre reported in [58], the 2-butyl cation is hydrogen-bridged.
p) More stable in comparison with an open structure is a cyclic isomer corresponding to the intermediate of the 1,2-shift [60].
q) See Note p) [61], cf., however, [62].
r) The rate of an ethyl shift is extremely high even at very low temperatures [63], but for lack of kinetic data the MT cannot be calculated (cf. [64]).
s) Kinetic data for the 1,2-methyl shift are given in [65].
t) Data for the 1,2-methyl shift are for a solution in CF_3COOH [15], see Note[e].
u) Data for the 1,2-methyl shift are for a solution in CF_3COOH—H_2SO_4 [67], see Note[e].
v) Estimated values referring to the 1,2-shift in a carbocation intermediate formed upon protonation of a nitrogen atom.
w) Reaction of acetolysis.

Table 2. Migration Tendencies of Alkyl Groups[a]

	Rearranging compound	Solvent	t °C	C_2H_5	n-C_3H_7	n-C_4H_9	iso-C_3H_7	iso-C_4H_9	sec-C_4H_9	$tert$-C_4H_9	Ref.
1	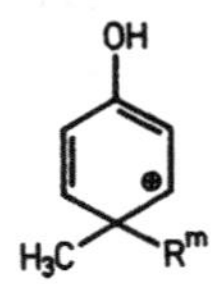	aqueous H_2SO_4	25	49	39						[20]
		aquo-methanol H_2SO_4	25						$\approx 7 \cdot 10^3$		[79]
2	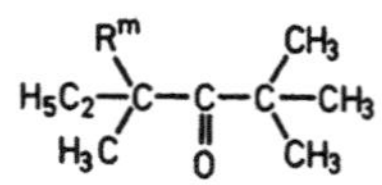	96% H_2SO_4	25	2.3	2.0						[80]
3	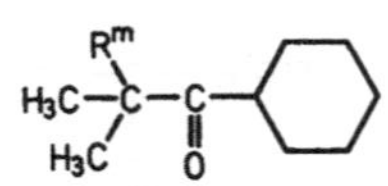	98% H_2SO_4	55	1.2	1.2		0.7				[81]
4	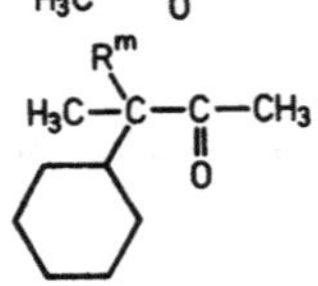	98% H_2SO_4	55		1.0[b]		2.3[b]				[81]

5	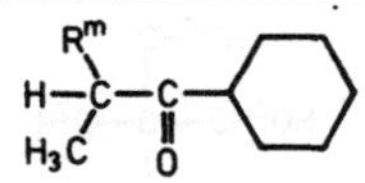	98% H_2SO_4	100 120	$6.2 \cdot 10^{-2}$ c) $7.5 \cdot 10^{-1}$ c)								81) 81)
6	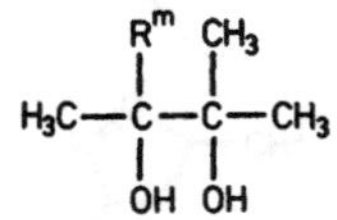	49.22% H_2SO_4	25.0	17							$>4 \cdot 10^3$	19)
7	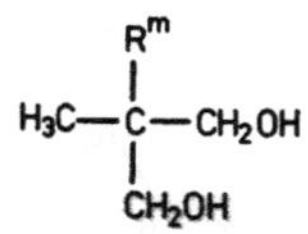	10 N H_2SO_4		1.40	1.80	3.50	0.39					71, 72)
								3.62	0.84	0.72		72)
8	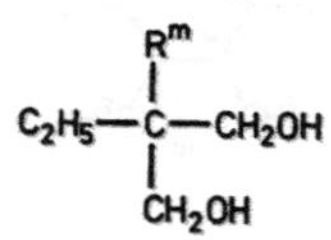	10 N H_2SO_4	108–109	1.50		2.81	0.37					71)
9	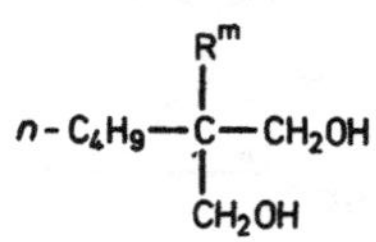	10 N H_2SO_4		1.38		2.14						71)'
10	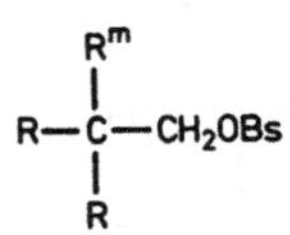	d)	100	1.01–1.18								73)
	R= CH_3, C_2H_5	e)	73.50, 83.50	1.15–1.40								82)

a) See Note a), Table 1.
b) The MT of the ethyl group is taken as a unit.
c) These data seem to be questionable [48].
d) Reaction of acetolysis.
e) Reaction of formolysis.

Table 3. Migration Tendencies of Benzyl and X-Substituted Benzyl Groups[a)]

	Rearranging compound	Solvent	t °C	X: H	p-CH_3	p-CH_3O	p-Cl	p-CF_3	m-F	Ref.
1[b)]	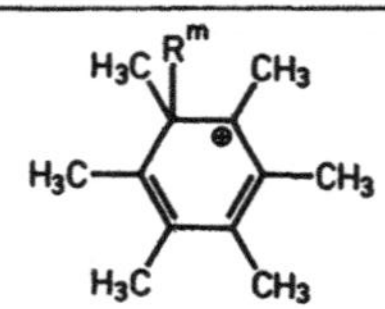	$HSO_3F—SO_2ClF$	—50	$4.8 \cdot 10^6$			$1.9 \cdot 10^6$	$1.1 \cdot 10^4$	$1.3 \cdot 10^5$	89,90)
		$HSO_3F—SO_2ClF—SO_2F_2$	—50		$1.4 \cdot 10^8$					89)
2	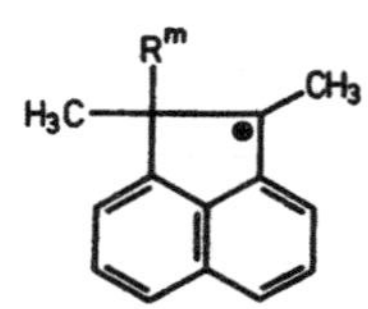	HSO_3F[c)]	0				$8.2 \cdot 10^2$			91)
3	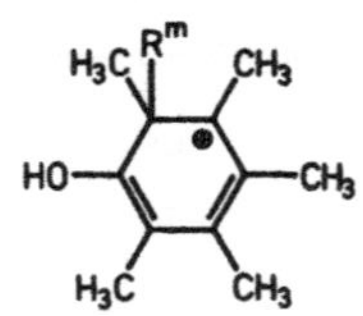	$HSO_3F—SO_2ClF$	—50	$\approx 10^8$[d)]						90)
4	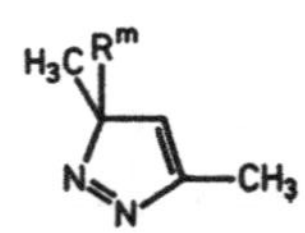	CH_3COOH	60	≈9000[e)]						69)
5	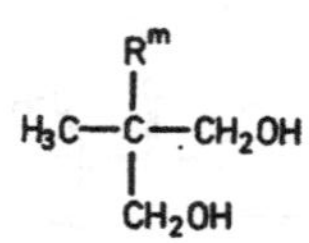	10 N H_2SO_4	108–109	1.07						71)

6	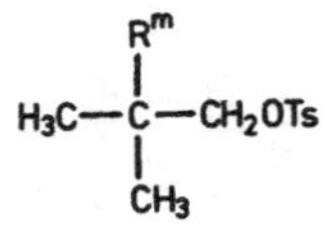	CH_3COOH	109.58	0.40	0.50	0.71	0.14	0.10	94)
			99.85	0.37	0.49	0.67	0.13		94)
		HCOOH	90.07	0.26	0.42		0.09	0.04	94)

a) See Note a), Table 1.
b) Kinetic data on the 1,2-methyl shift are for solution in HSO_3Cl [28], see Note c), Table 1.
c) Kinetic data on the 1,2-methyl shift are for solution in CF_3SO_3H [49], see Note c), Table 1.
d) By analogy with $R^m = CH_3$ [92] the rearrangement is assumed to proceed via consecutive 1,2-shifts of the benzyl group, the limiting step being the first one. The kinetic data on the CH_3 group are for solution in H_2SO_4—CF_3COOH [93]
e) See Note v), Table 1.

i.e. the MT of this group seems to be far lower than that of the methyl one (cf. [59,100]) while according to [15] (cf. [101]) in the dienone-phenol rearrangement of 4-methyl-4-R^m-cyclohexadienones the MT of $COOC_2H_5$ is higher than that of the methyl one (cf. Ref. [69]). It is of interest to note that the MT of the acyl group in the dienone-phenol rearrangement is estimated by the authors of Ref. [102] to be far higher than that of the ethoxycarbonyl group which they attribute to a complete or partial loss of the mesomeric stabilization of the $COOC_2H_5$ group on the way to the transition state.

In general the MT is not qualitatively constant for the *degenerate rearrangements* of long-lived carbocations either, i.e. in case it reflects in the greatest degree the "true" migratory aptitude. The most sound example is the inversion of the MT of H and Cl atoms in passing from 1-R^m-1,2,3,4,5,6-hexamethylbenzenonium ions (Table 1, item 2) to 9-R^m-9,10-dimethylphenanthrenonium ones (Table 1, item 7), see similar data for $R^m = NO_2$ [103], cf. also Ref. [104]. The observed increase in the MT of chlorine atom can be accounted for by the rising back donation effect [105]. This assumption is in accord with the concept of π-complexes developed by Dewar [105-107] according to which the efficiency of such an interaction in the case of an aromatic MCR (1,2-shifts in benzenonium ions) is to be much lower than in the case of an aliphatic MCR (1,2-shifts in aliphatic carbonium ions). The π-bond order (C^9–C^{10}) in the phenanthrene molecule is known [108] to be higher than that in the benzene one and it serves as the basis for the assumption advanced above. In more complicated cases the inconstancy of MT is naturally a rule rather than an exception.

The higher MT of phenyl relative to methyl is probably due to the involvement of π-electrons in the delocalization of the positive charge in the transition state (cf. Ref. [109]). This assumption is in agreement with a distinct tendency to a decrease in the value of *activation entropy* in passing from a methyl migrant to a phenyl one which is seemingly due to a more ordered structure of the activated complex in the latter case [46]. Despite a decrease in the efficiency of such an interaction in passing from aliphatic skeletons to aromatic ones (1,2-shifts in arenonium ions) [107] the inversion of the MT of the phenyl and methyl groups assumed in Ref. [107] failed to be observed in any of the arenonium ions investigated so far.

In addition there is some supplementary evidence on the 1,2-shift rates of such migrants as H, CH_3, C_2H_5, C_6H_5 (data on degenerate rearrangements is presented in our recent review [7], see also Ref. [26,110]). The scattered character of these data, however, prevents their discussion.

In discussing migration tendencies it should be born in mind that many of the MT values cited in the Tables have been calculated on the basis of *assumptions* whose correctness is not always obvious. A considerable part of them belongs to rearrangements which accompany the reactions of dehydration or solvolysis of neopentyl-type systems. These values are only valid if the kinetic data are an adequate reflection of the migrating ability, e.g., if the 1,2-shift of migrant occurs at the limiting stage of rearrangement either via a concerted mechanism or following the detachment of the leaving group, the latter case implying, additionally, that the equilibrium constant of the stage preceding the 1,2-shift (in a more general case — the product of constants) should be independent of the migrant nature. In the case of dehydration reactions the nature of the migrant must not affect the basicity constant of the parent compounds. This condition appears to be largely fulfilled since, e.g., the basicity of carbinols of a similar structure is known to be practically the same [111], cf. Ref. [112,113].

2.2 Mechanism of Rearrangements

Usually the authors dealing with solvolysis believe the mechanism of rearrangements to be concerted, and this assumption is corroborated by other investigations (see Ref. [114–116], cf. Ref. [117,118]). However there is evidence available in the literature testifying against this assumption or at least questioning it. Thus, Schubert and Henson [119] compared the rate constants of solvolysis of neopentyl-2,4-dinitrobenzenesulphonate and its trideuteriomethyl analogues (in trifluoroacetic acid) and concluded that the mechanism of this reaction is not concerted (cf., however, Ref. [120]). The same conclusion was earlier drawn [121] acetolyzing the brosylates of type (*1*), cf. Ref. [122].

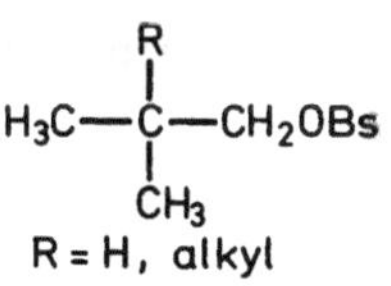

1

Worthy of note is also the point of view of Traylor et al. [123] on the mechanism of solvolysis of neopentyl-type systems. In their opinion the higher rate of reaction in the case of 2,2,3,3-tetramethylbutyl tosylate in comparison with neopentyl tosylate can be accounted for (at least partially) by the vertical stabilization of the developing carbonium centre by the electrons of the C^{β}—C^{γ} bond:

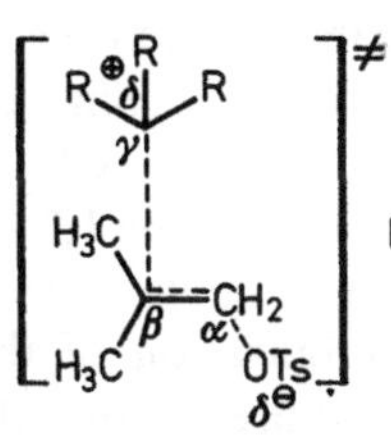

R = CH_3 or H

Such a stabilization, contrary to the participation effect in the case of a concerted mechanism does not imply any movement of the *tert*-butyl group towards the carbonium centre (cf. Ref. [124]).

For the rearrangements accompanying the dehydration reactions there are data, too, testifying against a concerted mechanism. Worthy of note are reports (see Ref. [125–127], cf. Ref. [80,128]) which favour the assumption that the stage of ionization in the pinacol rearrangement is not accompanied by the participation of the migrating alkyl group. On the other hand, the β-disubstituted 1,3-propandiols [71,72,129] seem to be rearranged by the concerted mechanism of dehydration — 1,2-alkyl shift since the obtained results are interpreted by taking into account the structural effects of the migrant and the MCR. In the framework of this approach the low MT of branched alkyl groups, e.g., an isopropyl one, may be due to a lower participation effect as compared with *n*-alkyl groups. Unfortunately, the comparatively low accuracy of the kinetic method restricts the possibilities of interpreting the results. The MT values obtained are listed in Tables 1–3.

As distinct from the above rearrangements, in the rearrangement of ketones the above complicating factors seem to manifest themselves little if at all. Several examples [80,128] have established the independence of the reaction rate constant of the acidity function H_0 in sulphuric acid of more than 95% concentration. The basicity of dialkylketones of similar structure is practically identical [130] (for the discussion of ketone basicity see Ref. [131]).

2.3 Evaluation of NMR Data

The values of MT obtained from the kinetic studies of 1,2-shifts in long-lived carbonium ions by the NMR method are more reliable since the described complicating factors are excluded.

Data [132–135] obtained from the intramolecular rearrangements of heterocyclic compounds support the probability of 1,2-shifts in respective carbocations (*2*):

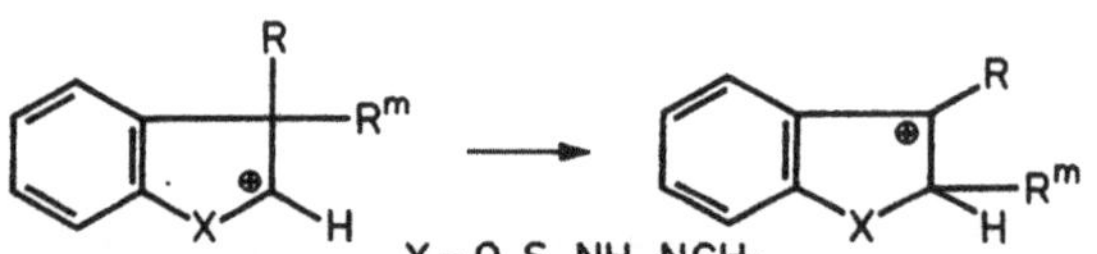

2

Note the higher MT of the substituents containing electron-releasing groups (n-C_3H_7 > C_2H_5 > CH_3; *tert*-C_4H_9 > CH_3; $(CH_3)_2C = CH - CH_2$, $CH_3 - CH = CH - CH_2$ > $CH_2 = CH - CH_2$), cf. [87], as well as the highest MT of the substituents having, at their C^m atoms, groups with readily polarizable π-electrons (p-$CH_3OC_6H_4CH_2$ > $C_6H_5CH_2$ > $CH_2 = CH - CH_2$ > alkyl) (cf. [136]). Similar series (p-$CH_3C_6H_4CH_2$ > $C_6H_5CH_2$ > $CH_2 = CH - CH_2$; $C_6H_5CH_2$ > CH_3; p-ClC_6H_4-CH_2 > CH_3; $CH_2 = CH - CH_2$ > CH_3) are characteristic of the 1,2-shifts of such migrants in other π-delocalized carbonium ions as well (arenonium, acenaphthylenonium ones) [42,89–91,137]. This suggests a resonance stabilization of the transition state with a considerable contribution of the *D*-type structures so that the transition state seems to be similar to the π-complex (cf. Ref. [107,138]).

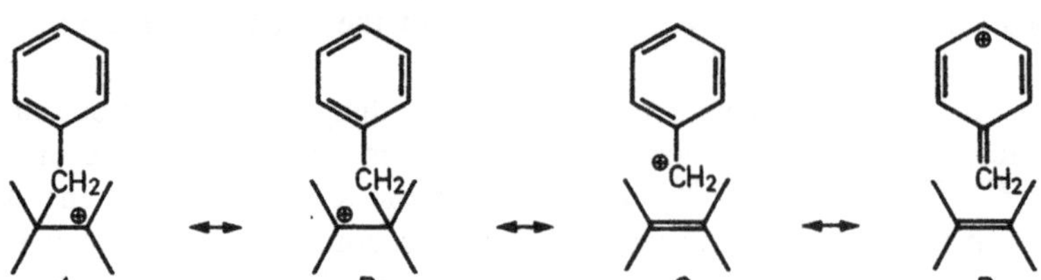

Jackson et al. [139] have advanced the assumption that π-complexes can play a great role in indole reactions. According to Kost et al. [133] the migrants are arranged in a series coinciding with that of the relative stability of respective cations (cf. Ref. [140]). Indeed, according to the data reported in [141], e.g., the stabilization energies of carbonium ions with respect to the CH_3^+ cation (in the gas phase) have the following values: $CH_3CH_2^+$ — 35 kcal/mole and $C_6H_5CH_2^+$ — 55 kcal/mole (cf. Ref. [142]).

The high susceptibility of the rate of 1,2-benzyl shifts in the degenerate rearrangement of 1-R^m-1,2,3,4,5,6-hexamethylbenzenonium ions ($R^m = XC_6H_4CH_2$) to X-substituent variation in the phenyl fragment ($\varrho^+_{-50\,°C} = -4.45$ [89]) agrees with the assumption of the effective participation of the phenyl ring π-electrons in the positive charge delocalization in the 1,2-benzyl shift transition state [83]. In this case the hybridization of the migrant's central carbon atom in all probability fails to change radically which is evidenced by the retention of configuration at this atom upon the 1,2-shift (see below, Sect. 5).

Data on the MT of other migrants not discussed here are rather scarce. The order characteristic of the halogen atoms appears to be as follows: I > Br > Cl[3] [24,143]. As for the fluorine atom, it does not manifest any trend to 1,2-shifts in carbocations at all [144], cf. Ref. [145]. There seems to be only one case of determining the migration ability of the alkoxyl group upon 1,2-shift in a long-lived carbocation [146]: judging by the data of Ref. [31,49,146], the MT of the OCH_3 in the degenerate rearrangement of 1-R^m-1,2-dimethylacenaphthylenonium ions is close to that of the chlorine atom and is far higher than that of the methyl group. Apparently, a similar order ($OCH_3 > CH_3$) is characteristic of solvolytic reactions as well [147].

The largest body of data on the MT of various migrants have been obtained at present for the six-fold degenerate rearrangement of 1-R^m-1,2,3,4,5,6-hexamethylbenzenonium ions (*3*) which proceeds by consecutive 1,2-shifts of R^m (1,2- or 1,6-sigmatropic shifts by the Woodward-Hoffmann classification [148]):

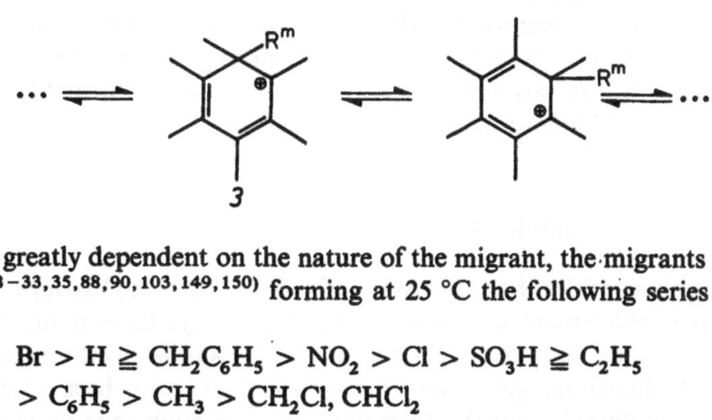

3

The MT is greatly dependent on the nature of the migrant, the migrants investigated in Ref. [4,28–33,35,88,90,103,149,150] forming at 25 °C the following series [7]:

$$Br > H \geqq CH_2C_6H_5 > NO_2 > Cl > SO_3H \geqq C_2H_5$$
$$> C_6H_5 > CH_3 > CH_2Cl, CHCl_2$$

The allyl group is likely to be located on the right of the nitro group [137], the CH_3O and CH_2CHO groups on the right of the benzyl one [4,33]. The rather high MT of some electronegative migrants (Br, Cl, NO_2, SO_3H) (cf. C_6H_5, and also the data on the MT of the $COOCH_3$ and $COOC_2H_5$ groups [15,69,101]) may be due to the involvement of the proper filled orbitals of these migrants into the interaction with antibonding MO of the MCR in the 1,2-shift transition state [7] ("back-dona-

3 Evidence on the MT of the iodine atom in long-lived carbocations seem to be lacking.

tion" effect [105]). The assumption of such an interaction is supported by quantum chemical calculations (see Ref. [151]). On the contrary, the electronegative groups CH_2Cl, $CHCl_2$ incapable of displaying such an effect possess a very low MT, cf. Ref. [31]. The high MT of the hydrogen atom agrees with a well-known ability of this atom to form bridge structures [7], cf. Ref. [152,153].

2.4 Solvolytic Rearrangement of sec-Alkyl Tosylates

Worthy of note are some generalizations as to the migration ability of different migrants made by British investigators [16] in studying the solvolytic rearrangements of secondary alkyl tosylates:

1. Among electronegative groups, the more electronegative the group, the slower the migration.
2. The stronger the C^o—R^m bond, the less the 1,2-shift is favoured in comparison with other competing processes.
3. Perhaps due to the greater length of the C^o—R^m bond and to the fact that small C—R^m—C bond angles are more favourable (along with the smaller strength of the C^o—R^m bond), the migrants with the central atom belonging to the second row of elements (P, S) migrate more easily.

Since MT, as noted above, varies with varying structure of the rearranging ion up to inversion, it would be wrong to generalize all these conclusions. For example, item 1 entails the assumption that the nitro group fails to migrate owing to high electronegativity [16], but in the rearrangements of arenonium ions this group migrates at a sufficiently high rate, seemingly due to the "back donation" effect [4].

The second conclusion, however, is corroborated by the data obtained in studying the rearrangements of long-lived carbocations: on an example of arenonium ion rearrangements we have found a linear relationship between the free activation energy of the 1,2-shift of various migrants and two parameters, one of these being the energy of the C^o—R^m bond rupture (see Sect. 3).

2.5 Effects of Substituted Phenyl Groups

To be separately considered are the MT's of substituted phenyl groups, particularly meta- and para-substituted ones for which, supposedly, there is no complicating steric effect. As model compounds for quantitatively establishing the migration ability of substituted phenyl groups, *symmetrical pinacols* (*4*) have been studied [154–156]. This model is preferred since the formation of carbonium ions upon dehydration of pinacols of this type seems to arouse no doubts (see Ref. [157,158], cf., however, Ref. [159]). Besides, in this case the detachment of a water molecule from any carbon atom results in the same carbonium ion (*5*).

The ratio of rearrangement products — ketones *6* and *7* — was used as a measure of the migration ability of substituted phenyl groups (competitive migratory aptitude, CMA). Belgian investigators [160] (cf. Ref. [161,162]) have studied the rearranging ability of tetraaryl-substituted glycols, each of them containing four identical aryl groups: phenyl, p-tolyl, p-anisyl or p-chlorophenyl. The p-chlorophenyl deri-

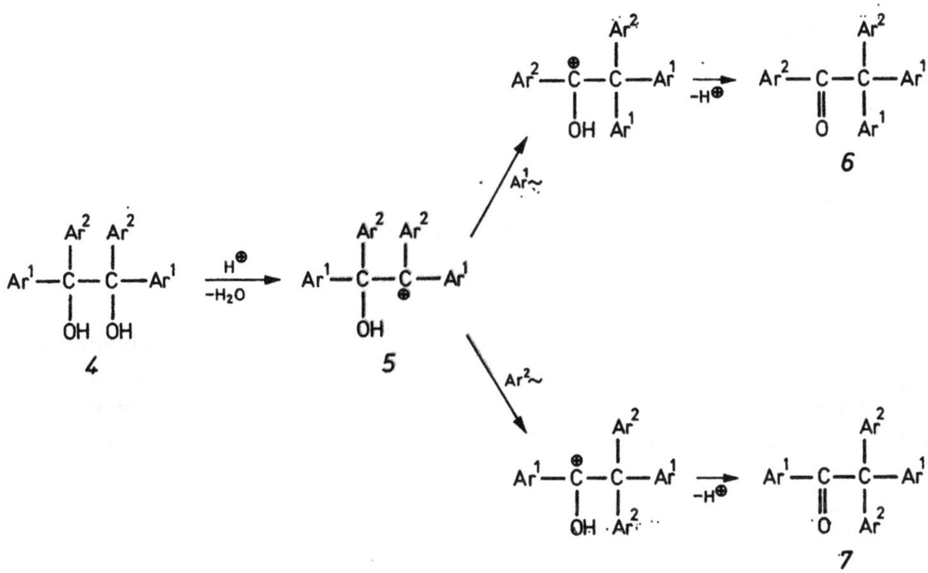

vative was found to undergo no rearrangement.[4] The relative rearrangement rates of the other glycols provide a satisfactory qualitative agreement of the results obtained with the data on the CMA of the substituted phenyl groups[154,155]: p-CH_3O > p-CH_3 > H > p-Cl. It should be noted, however, that none of the series obtained is a series of MT because in neither case is the condition of the structural constancy

The pinacol rearrangement seems to be unsuitable for determining the MT of substituted phenyl groups for the following reasons: In asymmetrical pinacols there are two potential carbonium centres differing from each other, i.e. a mixture is formed. Of course, one can use pinacols with electron-releasing groups at one carbon atom and with an electron-withdrawing one at another; this should lead the dehydration to the formation of ions with a carbonium centre at the carbon carrying the electron-releasing groups. However, one should not except that after a 1,2-shift to this carbonium centre there will occur a 1,2-shift of one of the electron-releasing groups in the opposite direction. Also, one should note that the separation of data relating specifically to the migration step from overall kinetic data is not a simple problem; it may be even more complicated in a number of cases due to the formation of α-oxides or 1,2-shifts of a hydroxy group (see Ref. [161-163]).

Therefore, in order to determine the MT's of substituted phenyl groups it seems expedient to measure directly the rate of the 1,2-shift in *β-aryl-substituted carbonium ions* (*8*). True, this approach is not free of complications either. Of these the principal ones are the very high rate of 1,2-shift which fails to be measured by ordinary methods and the isomerization of the parent carbonium ion with an aryl group at the β-carbon atom into an α-aryl-substituted carbonium ion (*9*) or into a pheno-

4. Since there is not too much difference between the MT of the p-chlorophenyl group and that of the phenyl one (in the rearrangement of arenomium ions, see Table 4), this is likely to be due not so much to the low MT of the former as to other reasons, e.g., to the smaller equilibrium concentration of the corresponding carbocations.

nium ion (*10*):

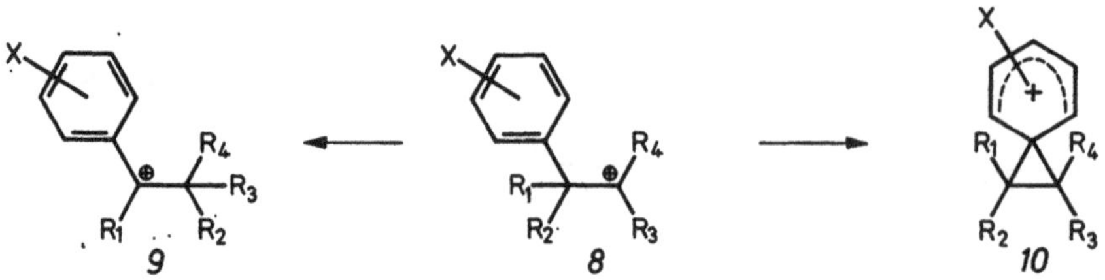

The former complication can be excluded by choosing as a model a carbonium ion capable of a degenerate rearrangement, i.e. containing the same substituents at C^{o} and C^{t} atoms of the skeleton ($R^1 = R^3$, $R^2 = R^4$), and by using the dynamic NMR method. The latter complication is more difficult to exclude since under the "long-life" conditions the equilibrium with the predominance of isomeric ions — (*9*) and/or (*10*) — can be reached very rapidly. This is corroborated, e.g., by the data reported by Olah et al. for the ions with $R^1 = R^3 = H$, $R^2 = R^4 = CH_3$ or $R^1 = R^2 = R^3 = R^4 = CH_3$ [61,62] (cf. Ref. [164,165]).

β-Phenethyl ions (*8*) can be expected to predominate in the equilibrium when the summary effect of any of the two combinations: $R^1 + R^m$ or $R^2 + R^m$ which stabilizes the carbonium centre is less than the summary effect of R^3 and R^4.

This was found [166,167] (cf. Ref. [168]) to be fulfilled for 9-aryl-9,10-dimethyl-phenanthrenonium ions (*11*) (where the aryls are substituted phenyl groups) with the substituents varied within a wide range: from five fluorine atoms and a p-CF_3 group to a p-CH_3O group in the migrant[5] and from a CH_3 group to CF_3 at the 3- and 6-positions of the MCR ($R^m = C_6H_5$). In the case of ion (*11*) (Ar = *p*-$CH_3OC_6H_4$, X = H), by decreasing the medium acidity, it is possible to shift the equilibrium towards the isomeric ion, in all probability, the 9,9-dimethyl-10-*p*-anisylphenanthrenonium one (*12*, Ar = *p*-$CH_3OC_6H_4$, X = H) [167] which reflects the participation of the p-anisyl group in interaction with the medium.

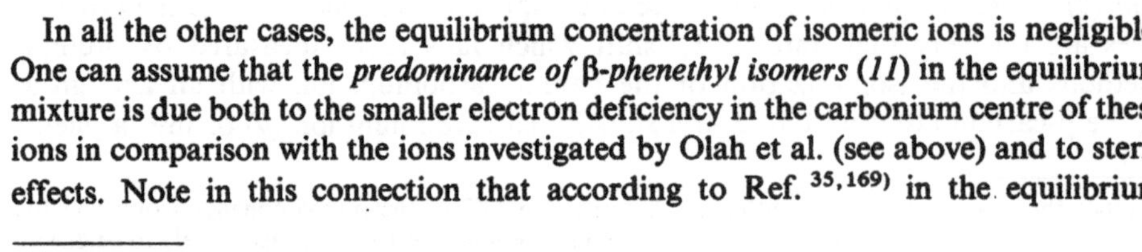

In all the other cases, the equilibrium concentration of isomeric ions is negligible. One can assume that the *predominance of β-phenethyl isomers* (*11*) in the equilibrium mixture is due both to the smaller electron deficiency in the carbonium centre of these ions in comparison with the ions investigated by Olah et al. (see above) and to steric effects. Note in this connection that according to Ref. [35,169] in the equilibrium

5 The electron-releasing effect of this group seems to be somewhat less owing to the formation of hydrogen bonds.

mixture of phenylhexamethylbenzenonium ions, the 1- and 3-phenyl isomers — (*13*) and (*14*) — are predominant i.e. the ions with a phenyl group located at the carbons with the smallest electron deficiency, cf. Ref. [67,170]:

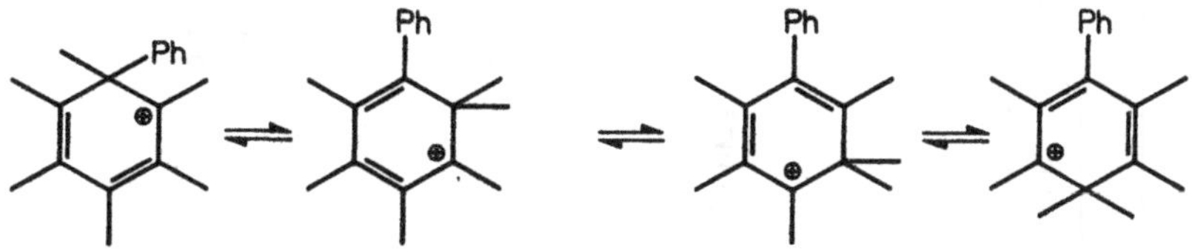

13 14

On the other hand, as found by Borodkin and later confirmed by Rudnev, Bushmelev and Shakirov, the equilibrium mixture of phenyldimethylacenaphthylenonium ions is characterized by the predominance of the 1,1-dimethyl-2-phenyl isomer (*15*):

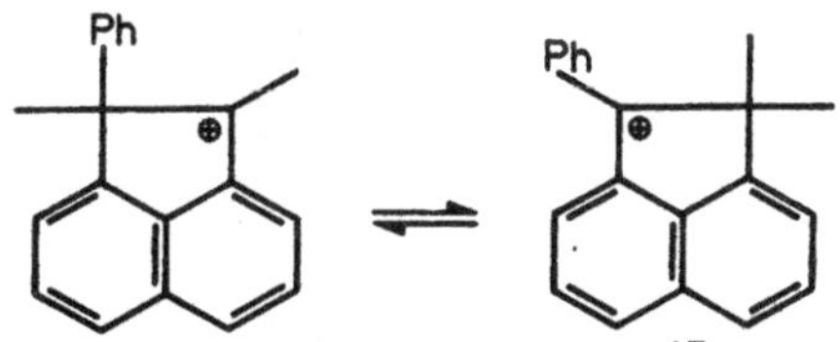

15

This is likely to show that the equilibrium mixture of phenyl-substituted cyclic carbonium ions is determined by a delicate balance of electronic and steric effects.

Despite the complications connected with the above isomerizations it proved possible to obtain some quantitative data on the migration tendency of *meta- and para-substituted phenyl groups* in degenerate rearrangements of 1-X-phenyl-1,2,3,4,5,6-hexamethylbenzenonium ions (*16*) and 1-X-phenyl-1,2-dimethylacenaphthylenonium ions (*17*).

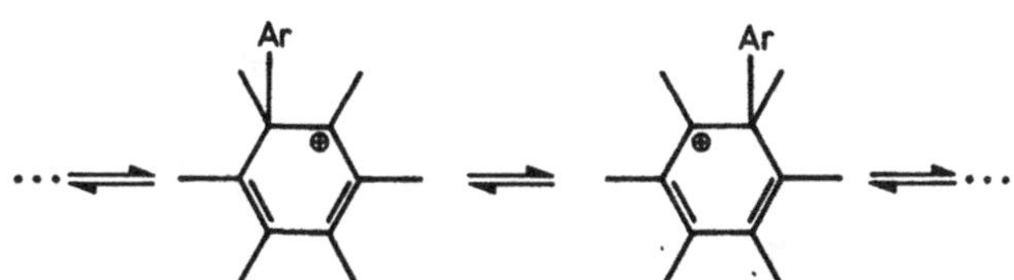

16

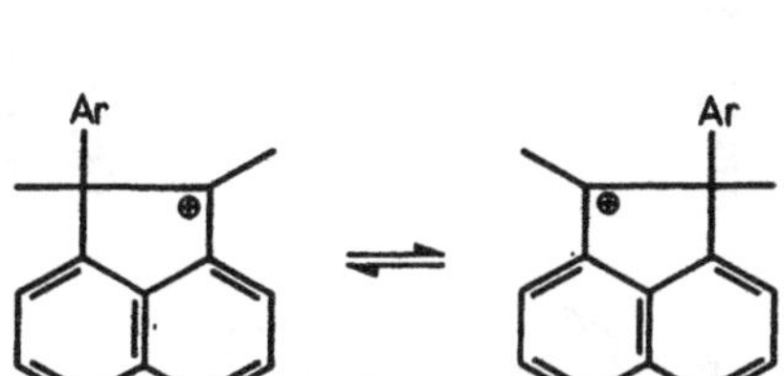

17

The data on the MT of substituted phenyl groups in these rearrangements along with those on the MT for the degenerate rearrangement of ions (*11*), X = H are listed in Table 4.

Table 4. Migration Tendencies of Phenyl and X-Substituted Phenyl Groups[a)]

	Rearranging compound	Solvent	t °C	X									Ref.
				p-CH_3O	p-CH_3	p-Cl	p-F	p-CF_3	m-CH_3	m-Cl	m-F	m-CF_3	
1	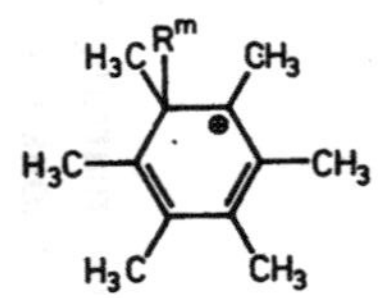	CF_3SO_3H—SO_2ClF	—50		7.6	$7.1 \cdot 10^{-1}$						$1.4 \cdot 10^{-3}$	171)
2	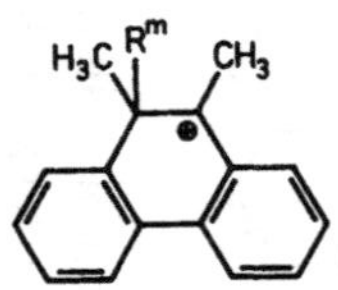	HSO_3F—SO_2ClF	—50	$4 \cdot 10^3$ b)	16	$2.5 \cdot 10^{-1}$	1.0	$1.3 \cdot 10^{-3}$			$8 \cdot 10^{-3}$		45)
			75	71 b)	2.3	$5.0 \cdot 10^{-1}$							45)
3	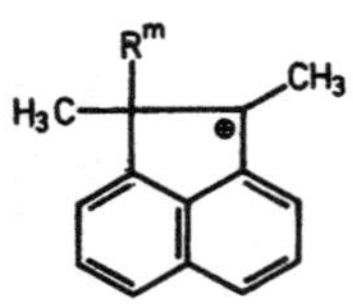	CF_3SO_3H	—50		53	$1.9 \cdot 10^{-1}$		$6.0 \cdot 10^{-5}$		$2.1 \cdot 10^{-3}$		$1.8 \cdot 10^{-4}$	173c)
4	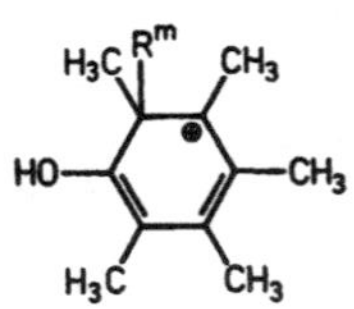	CF_3COOH—H_2SO_4	—50		12	$2.3 \cdot 10^{-1}$		$8.7 \cdot 10^{-3}$				$8.9 \cdot 10^{-3}$	170)
			75		3.5	$3.3 \cdot 10^{-1}$		$4.4 \cdot 10^{-2}$				$5.2 \cdot 10^{-2}$	170)
		CF_3COOH—CF_3SO_3H	—50	24									170)
			75	13									170)
5	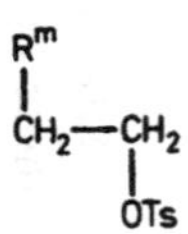	C_2H_5OH	75	$1.40 \cdot 10^2$	14.8								174)
		50% aq. C_2H_5OH	75	53.5	5.0				$9.7 \cdot 10^{-1}$				174)
		CH_3COOH	75	$1.11 \cdot 10^2$	8.32				1.4				174)
		HCOOH	75	45.9	7.48	$1.4 \cdot 10^{-1}$			1.94				174)
		CF_3CH_2OH	75	77.9	12.0								174)
		CF_3COOH	75	6.69	9.5	$1.5 \cdot 10^{-1}$			2.65				174)

6	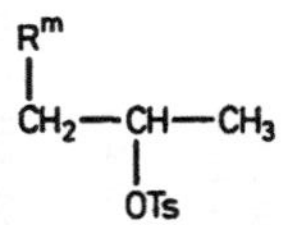	C_2H_5OH	50	70.4					174)
		80% C_2H_5OH	50	34.2	4.30				174)
		CH_3COOH	50	37.0	4.14				174)
		HCOOH	50	32.4	5.84	$1.7 \cdot 10^{-1}$			174)
7	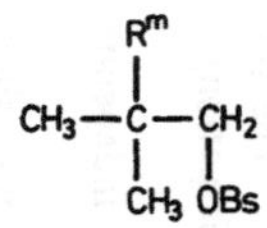	CH_3COOH	75	88.5	7.27			1.93	175)
8	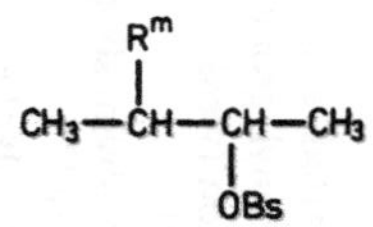 (threo-isomer)	CH_3COOH	75	87.8	5.94	$1.4 \cdot 10^{-1}$		1.72	176, 177)
9	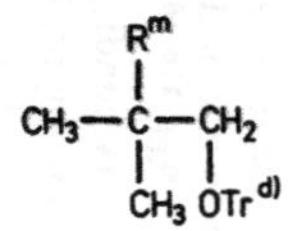	70% CF_3CH_2OH	25				$5.7 \cdot 10^{-3}$		178)
10	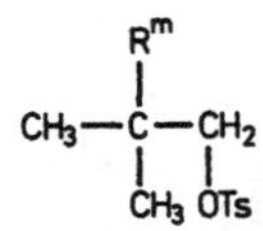	CH_3COOH	75	92.2	7.35	$1.9 \cdot 10^{-1}$			179)

a) See Note a), Table 1; in the case of items 1–4 the MT of the phenyl group is taken as a unit, in the case of items 5-10 the ratio k_Δ^X/k_Δ^H is taken as a measure for MT.
b) The value is calculated from the correlation for $\sigma^+_{OCH_3} = -0.778$ [172].
c) For $R^m = C_6H_5$ see [51].
d) OTr is 2,2,2-trifluorethanesulphonate.

There are numerous data on the *solvolysis rates of β-arylalkyl systems* (see Ref. [165,174,178,180–183]). From these data it is possible to obtain evidence on the rate constants of solvolysis for primary and secondary systems accompanied by the participation of the neighbouring aryl groups[6] (k_Δ or Fk_Δ where F is a fraction of a tight ion pair which does not return). The relative values of these constants (k_Δ^X/k_Δ^H) can be used as a measure of the MT's of aryl groups.[7] Some typical examples are presented in Table 4.

The analysis of Table 4 leads to the conclusion that in a very wide range of rearranging structures and rearrangement conditions and, as a consequence, of changes of rates, the qualitative constancy of the MT's of meta- and para-substituted phenyl groups is retained (cf. similar data for the rearrangements of vinyl cations [192]).[8] The decreasing MT of the p-anisyl group observed in passing from neutral to acid media is obviously due to a specific solvation of the methoxy group (cf. Ref. [174,193]). The reasons for the considerable decrease in the MT of electron-withdrawing groups (X = p-CF_3, m-CF_3) in passing from series 1,2,4 to 3 are not quite clear. Probably the migrants capable of a rather strong back donation effect are less susceptible to steric requirements of a transition state with increased strain in the case of degenerate rearrangement of 1-R^m-1,2-dimethylacenaphthylenonium ions (Series 3), cf. Ref. [173]. This assumption is in agreement with the evidence of a far higher MT of the phenyl group and the chlorine atom in ions of this type as compared with those of the benzene series (Table 1).

The relative rates of 1,2-X-phenyl shifts usually reveal linear relationships with such substituent constants as σ^+ (see, e.g., [45,165,170,171,173,175,194], cf. Ref. [174]) which emphasizes the close relation of these reactions ("ipso-substitution" at the aromatic carbon atom) to the conventional electrophilic aromatic substitution reactions. This is in agreement with a very low MT of a pentafluorophenyl group [45], cf. Ref. [195].

The similarity in the MT's of substituted phenyl groups in terms of linear free energy relationships is expressed in the close values of ϱ^+ [7]. It is of interest to point out that in passing from Series 2 to 1 (Table 4) the decrease in the rate of degenerate 1,2-aryl shifts by about seven orders of magnitude yields practically the same values of ϱ^+ (for −50 °C they are −4.5 [45] and −4.57 ± 0.50 [171], respectively). In passing to Series 4 the rate decreases further by about six orders of magnitude [170], the value of the ϱ^+, however, changes rather little ($\varrho^+_{-50\,°C}$ = −3.37 ± 0.19 [170]). These data indicate rather a low susceptibility of the MT of meta- and para-substituted phenyl groups to changes in the π-electron deficiency in the carbonium centre [170,171]. In the case of Series 3 ϱ^+ is markedly higher (in absolute value): $\varrho^+_{-50\,°C}$ = −6.53 which may be due to the greater strain of the transition state [173].

When dealing with the rearrangements of 1,2,2-triarylethylamines in the deamination reaction the authors [196], by making several assumptions, came to the tentative

6 The solvolysis mechanism of tertiary systems is not concerted, at least in the absence of powerful electron-releasing substituents in the migrant — a β-phenyl group [165]. The transition state in these cases is assumed [184] to be similar to a π-bridged ion (weak participation) rather than to a phenonium ion (strong participation), cf. Ref. [174,185–191].

7 When using the values of Fk_Δ the necessary condition is obviously the independence of F on X. This condition seems to be fulfilled, at least approximately [174].

8 The MT of substituted phenyl groups in β-arylalkyl solvolysis is only slightly sensitive to solvent variation [174].

conclusion that the increasing *nucleophility of an aryl migrant* is accompanied by a decreasing susceptibility to charge variations in the carbonium centre. However, the above data referring directly to 1,2-aryl shifts in long-lived carbocations fail to agree with this conclusion.

The high values of the coefficient ϱ^+ (in absolute value) indicate a high susceptibility of the rearrangement rate to the electron effects of substituents X and agree with the assumption that the transition state is similar to the phenonium ion (*A*) (cf. [10]). It is quite possible that *phenonium ions* are formed as intermediates in the 1,2-shifts of aryl groups in arenonium ions, but their relative stability is very low: they fail to manifest themselves in the NMR spectra [35,45,46,67,104,156,166,167,170,197,198]. This is also in agreement with the fact that for ions (*11*) Ar = p-$CH_3C_6H_4$ and p-$CH_3OC_6H_4$, X = H [45], Ar = C_6H_5, X = CF_3 (3.6), Br (2.7) [46], as well as for 9-methyl-9,10-diphenylphenanthrenonium ion [104] the rate of the aryl ring rotation around the C^9—C^{Ar} bond is far lower than that of the 1,2-aryl shift (cf. Ref. [61,199]). This shows that a nonclassical phenonium ion (*B*) is not involved in the 1,2-aryl shift process.

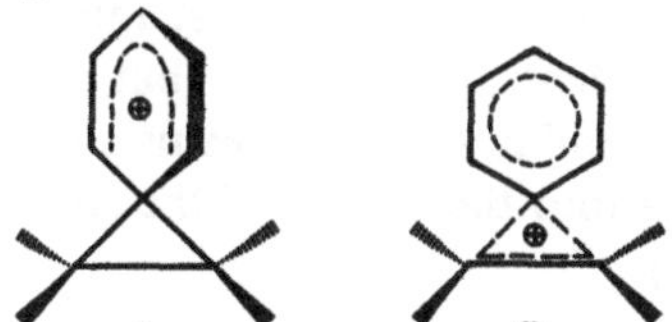

A B

The data on the absence of substantial differences in the MT's of substituted phenyl groups in the deamination of β-arylethyl systems [200] could be regarded as an indication that the "hot" primary carbonium ions generated in this reaction display a far lower selectivity, cf. Ref. [161]. There are data, however, testifying against the assumption of intermediate formation of these ions in the deamination of primary systems (see Ref. [165,201] and the references cited therein, cf. Ref. [24]).

Strictly speaking one cannot preclude the possibility that in passing to ions with β-phenyl groups containing powerful electron-withdrawing substituents the mechanism of rearrangement is transformed toward a "σ-aryl" shift with the formation, as an intermediates or an activated complex, of the nonclassical ion (*B*) rather than the classical ion (*A*) (see Ref. [61]). It can be assumed, however, that for the ions studied (Table 4) this possibility is not yet realized.

In conclusion it should be noted that the MT's of substituted phenyl groups under discussion are not identical with an abstract *intrinsic migratory aptitude* which implies a complete absence of any modification of the MCR of the carbonium ion upon variation of the migrant. This was first pointed out by Stiles and Mayer [19] introducing the concept of migration tendency in application to alkyl migrants.

In the case of β-aryl-substituted carbonium ions one can expect an even greater modification of the MCR due to a more effective donor-acceptor interaction between the aryl group and the carbonium centre. According to the available data [104,166,198,202], in the β-aryl-substituted arenonium ions such an interaction does occur, but the degree of the carbonium centre modification seems to be insignificant [166,197], cf. Ref. [62].

Thus, in general the MT is not a universal characteristic of the migrant even in the qualitative sense. The intrinsic migratory aptitude is no more than an abstraction (cf. Ref. [19,80]). The concept of MT — considered out of touch with the MCR of the rearranging carbocation — cannot serve as the basis of the approach to the quantitative description of carbocation rearrangements, cf. Ref. [80,203]. Hence it is necessary to find out how MT varies with varying MCR of the carbocation.

3 Structural Effects of the Migrant-Carrying Remnant

Section 2 has partly dealt with the dependence of the MT, i.e. the relative 1,2-shift rate constants of different migrants on the structure of the rearranging ion. The present section will consider in greater detail the effect of the migrant-carrying remnant (MCR) of carbonium ions on the rates of their rearrangements. MCR (migrant-carrying remnant) means a system of atoms and bonds of the ion from which the migrant is segregated. This question is separated from the interrelation between the structure of carbonium ions and the rates of their rearrangements so far as the structural effects of the migrant and the MCR are independent. Since the migrant is usually connected with the MCR by a σ-bond, in a first approximation their electron effects can be considered independent and additive, but the steric effects may cause deviations from additivity. A similar situation seems to arise when the structural effects of the MCR are divided into the effects of the substituents at C^o and those at C^t atoms (see Ref. [80,128,204]).

3.1 Structural Effects at the C^t Atom

As 1,2-shifts can be regarded as intramolecular electrophilic substitution reactions, it is quite natural to assume that in the absence of complicating factors the rate of 1,2-shift increases with increasing carbonium centre electrophility which can, as a first approximation, be characterized by the value of the positive charge on the C^t atom. Such suggestions have been repeatedly made in the literature; at present experimental data have been accumulated which support their validity.

3.1.1 Influence of Hydroxy Group and Chlorine Atom

Table 5 lists some data to illustrate it. For example, the rates of 1,2-shifts decrease sharply on substitution of the methyl group at the α-position of aliphatic carbocations by a hydroxy group (items 1,2). A similar effect is also observed on introduction of a hydroxy group into a resonance position (to the C^6 atom) of benzenonium ions with different R^m (item 5)[9] although it is not so large. The chlorine atom at the carbocation centre also displays a retardation effect in the case of a 1,2-methyl shift (items 3,6); apparently the same is characteristic for the ethyl migrant [206]. However, on introduction of a chlorine atom to more distant resonance positions this effect is not observed (items 4,5). The decisive factor here can be the polar effect of the chlorine atom leading to a decrease in the π-electron deficiency in the carbocation centre and to a destabilization of the transition state of the 1,2-methyl shift in the former case and to an increase of the deficiency in the latter (cf. Ref. [37,103,206,207]).

---

9 The same effect is also characteristic of the ions with meta- and para-substituted phenyl groups as migrants (cf. Ref. [170,171]).

Table 5. Effects of Substituents R Located at Resonance Positions Relative to the Carbocation Centre

	Rearranging compound	Migrant	Solvent	t °C	Free activation energy of 1,2-shift for varied R[a)] CH$_3$	OH	Cl	Ref.
1	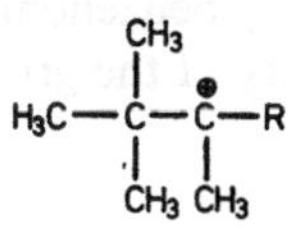	CH$_3$	SO$_2$ClF—SO$_2$F$_2$[b)]	−136	3.8			59)
			96% H$_2$SO$_4$	100		29.0[c)]		205)
2	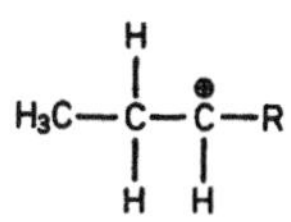	H	SO$_2$ClF—SO$_2$F$_2$[b)]	−140	$\leq$1.9			59)
			HSO$_3$F—SbF$_5$	25		$\approx$25		70)
3	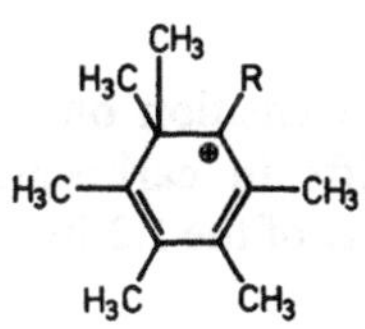	CH$_3$	HSO$_3$Cl	25	18.1			28)
			HSO$_3$F	40			$\geq$21	37)
4	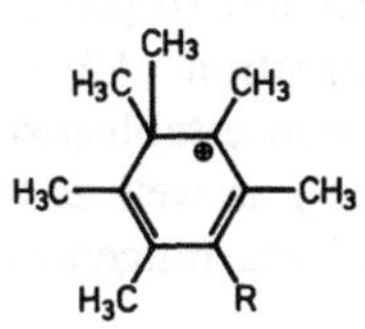	CH$_3$	HSO$_3$Cl	25	18.1			28)
			HSO$_3$F	40			~17.7	37)
5	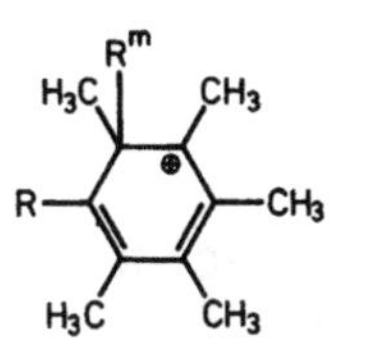	CH$_3$	HSO$_3$Cl	25	18.1			28)
			CF$_3$COOH—H$_2$SO$_4$	25		25.1		67)
			HSO$_3$F	40			~18.4	37)
		C$_2$H$_5$	CF$_3$COOH	25	16.0			4)
			CF$_3$COOH—H$_2$SO$_4$	25		22.9		67)
		C$_6$H$_5$	CF$_3$SO$_3$H—SO$_2$ClF	25	17.2			35)
			CF$_3$COOH—H$_2$SO$_4$	25		23.6		67)
		Cl	HSO$_3$F	10	14.7			31)
						21.5		68)
6	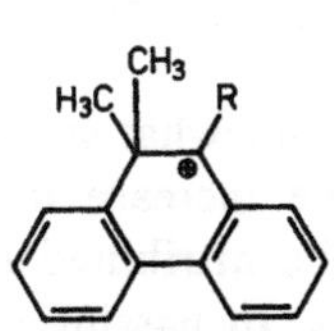	CH$_3$	HSO$_3$F—SO$_2$ClF	− 10	11.5			42)
			HSO$_3$F	− 10			$\geq$15	37,206)

a) Values of $\Delta G^{\neq}$ in kcal/mole; if necessary, they were calculated from the formula: $\Delta G^{\neq} = 2.3\,RT\left(\lg\frac{kT}{h} - \lg k\right)$, k, s^{-1} (taking into account statistic factors).

b) The acid is not specified; it is obviously SbF$_5$.

c) The precursor of the ion is 3,3-dimethylbutane-2-on-[1-^{14}C], cf. [70].

3.1.2 Influence of Alkyl Groups

Substitution of the α-hydrogen atom by an alkyl (ethyl) group results in strong retardation of the 1,2-methyl shift; thus, e.g., for ion (*18*) at 25° (solution in HSO_3F) the effect ($\Delta\Delta G^{\neq}$) amounts to about 6 kcal/mole [70], cf. similar data for hydroxybenzenonium ions [208], as well as those on the retarding effect of the methyl groups substituting hydrogen in the resonance positions of methylbenzenonium ions [26]. It is natural to assign these features to the stronger sensitivity of the ground

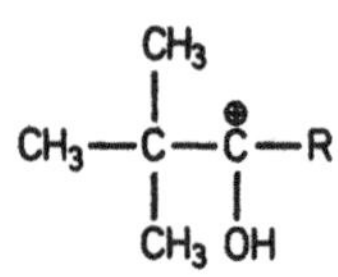

18

state to the electron effect of the alkyl groups relative to the transition one. The data obtained by the MNDO method for 1,2-hydrogen shifts in carbocations $RCH_2—\overset{+}{C}HR'$ [209] agree with the above experimental data (the rate of the 1,2-hydride shift falls in the series: $R' = H > CH_3 > Cl > OCH_3$).

As expected, within the limits of the alkyl series the effects are not so great, the ethyl group displays an accelerating effect on the methyl migration ($\Delta\Delta G^{\neq}_{-78\,°C} = 1.2$ kcal/mole)[10] in comparison with the methyl one as has been established for ions (*19*) [49,65] (cf. similar data for the 1,2-ethyl shift in ions (*20*) as well as those on the migration of the cyclohexyl group in the acid-catalyzed rearrangement of ketones (*21a*) [81]).

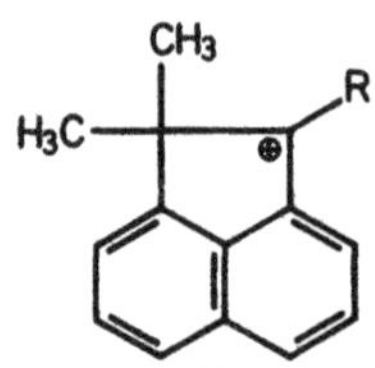

19

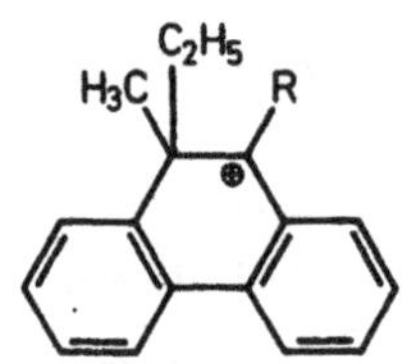

20

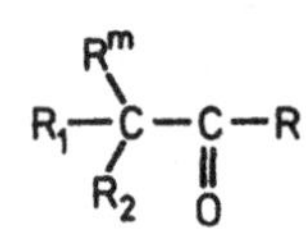

21

a R = CH_3
b R = C_2H_5

a R^m = cyclohexyl, R, R_1, R_2 = CH_3 or C_2H_5
b R^m = cyclohexyl, R_1 = H, R, R_2 = CH_3 or C_2H_5

Judging by the data of Ref. [81] the isopropyl group displays a somewhat stronger accelerating effect than the ethyl one in ions of this type. The increase in the accelerating effect in the series $CH_3 < C_2H_5 <$ *iso*-C_3H_7 may be attributed to a decrease in the effect stabilizing the rearranging ion in this series. In passing from ketones (*21a*) to (*21b*) there occurs a leveling and even an inversion of the effects of the methyl and the ethyl groups [81][11]. These data indicate that, just as MT, the

10 A similar effect is displayed by the *p*-chlorobenzyl group ($\Delta\Delta G^{\neq}_{-28\,°C} = 0.9$ kcal/mole) [49,91].

11 Unfortunately, the data for ketones (*21a*) and (*21b*) belong to different temperatures so they are difficult to compare.

"pulling ability" of substituents, i.e. their effect from the α-position with respect to the carbocation centre on the 1,2-shift rate in the general case is not constant even qualitatively.

3.1.3 Influence of Double Bonds and Aromatic Systems

The double bond in the above position exerts a retardation effect. Thus, the rate of the 1,2-methyl shift in ion (*22*) falls very sharply in passing from R = CH_3 [59] to R = $(CH_3)_2C{=}CH$ [210], cf. Ref. [211]. A similar effect is observed for the 1,2-hydride

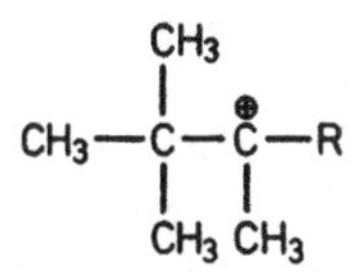

22

shift in passing from the cyclopenthyl cation ($\Delta G^{\neq}_{-130\,°C} < 5.5$ kcal/mole [64,212], cf. Ref. [213]) to a cyclopentenyl one ($\Delta G^{\neq}_{100\,°C} = 20.2$ kcal/mole [214], cf. Ref. [215–217]).

It is well known that the positive charge is very effectively delocalized by the aromatic rings located at the carbonium centre. Accordingly the rearrangement of the 9-*tert*-butylfluorenyl cation (*23a*) which is likely to proceed via consecutive 1,2-shifts of the methyl group, with the first shift being the limiting step [218–220] is far slower than that of 1,2-methyl shifts in the *tert*-butyldimethylcarbonium ion (*24a*) [59], $\Delta\Delta G^{\neq} \approx 9$ kcal/mole.

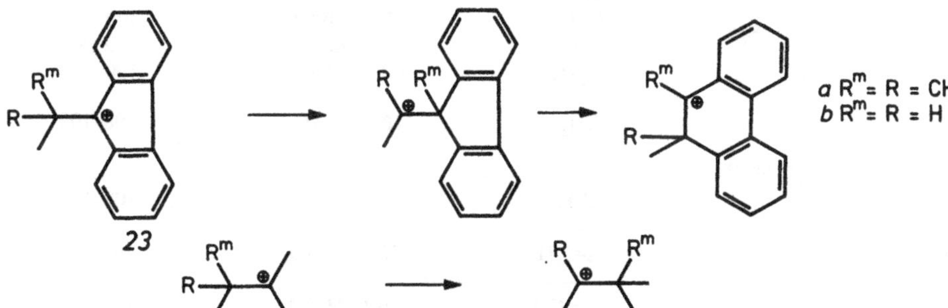

This conclusion seems to be also true for the 1,2-hydrogen shift: judging by the PMR spectra, the 9-ethylfluorenyl cation (*23b*) undergoes no rearrangement at −10 °C [218], cf. [221] when the rate constant of the 1,2-hydride shift in the ethyl-

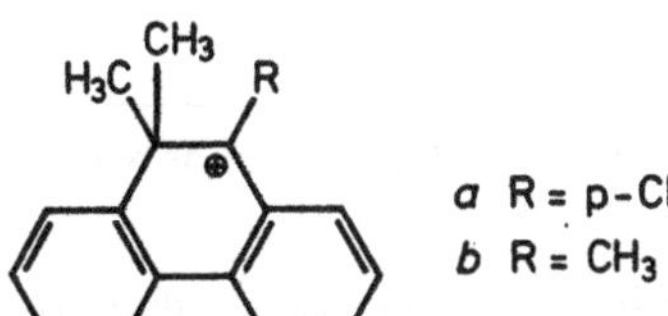

25

dimethylcarbonium ion (*24b*) amounts to about $3 \cdot 10^3\ s^{-1}$ [64], cf. Ref. [222]. The assumption of the great effect of electron deficiency in the carbocation centre on the rate of 1,2-shifts is also corroborated by the fact that the rate constant of the 1,2-methyl shift in ion (*25a*) is by about two orders of magnitude lower than that in ion (*25b*) [42,45], cf. similar data for 1,2-aryl shifts [104].

3.2 Structural Effects at the C° Atom

In order to characterize the structural effects of the substituents at the C° atom it is necessary to compare the rearrangement rates of carbonium ions having a *constant environment* at C^t [12]. It is natural to expect the rearrangement rate to rise with increasing electron-donation of substituents at the C° atom, in accordance with the expected decrease in the electron density on this atom in the transition state. The accelerating effect can also be displayed by bulky groups since in the course of the 1,2-shift the hybridization of the C° atom changes from sp^3 to sp^2.

3.2.1 Solvolysis and Dehydration

A great part of the data on the structural effects of the substituents at the C° atom is concerned with carbonium ion rearrangements in the reactions of *solvolysis* and

Table 6. Effects of Substituents Located at the C° Atoms in Reactions of Solvolysis and Dehydration

	Rearranging compound	Migrant R^m	Solvent	t °C
1	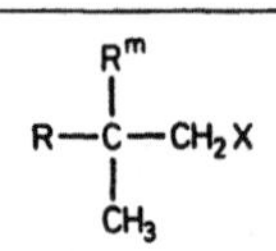	H[b)] CH_3[c)] CH_3[b)] C_2H_5[c)]	98% HCOOH CH_3COOH CH_3COOH CH_3COOH	65 100 99.58 100
2	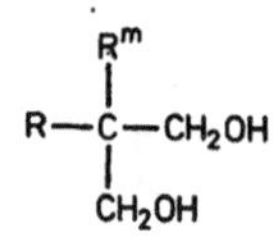	CH_3	10 N H_2SO_4	108–109
		C_2H_5	10 N H_2SO_4	108–109
		n-C_4H_9	10 N H_2SO_4	108–109
		iso-C_3H_7	10 N H_2SO_4	108–109
		C_6H_5	10 N H_2SO_4	108–109
3	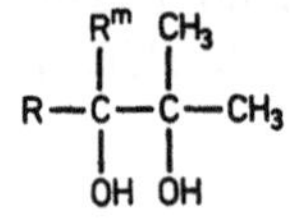	CH_3	49.22% H_2SO_4	25

a) The rate of the 1,2-migrant shift in the case of R = CH_3 is taken as a unit; b) X = OTs;

12 Strictly speaking, the variation of substituents at the C° atom means at the same time that of the environment at the C^t atom as well since these atoms are directly bonded with each other.

dehydration. In these cases the validity of any conclusions relative to the substituent effects at C° is first of all dependent on how correct it is to consider the observed effects on the rate of such reactions as the effects on the rate of 1,2-shifts.

Tables 6 and 7 present some examples of structural effects of substituents at the C° atom. Analysis of these data shows that the substitution of the methyl group at C° by an ethyl one usually results in a somewhat increased rate of 1,2-shifts in the solvolysis of primary alkyl sulphonates.[13] A similar effect is observed for the 1,2-shifts of different groups in the dehydration of β-disubstituted 1,3-propandiols. As the chain of *n*-alkyl groups at the C° atom grows longer the rate of 1,2-shifts tends to rise; this agrees with the increase in the electron-donating ability of these groups.

The data on the effect of substituted benzyl groups at the C° atom on the rate of solvolysis of tosylates (*26*) accompanied by a 1,2-methyl shift (Table 7) agrees with the assumption of the inductive nature of the accelerating effect [94]; a linear relationship is found between the logarithms of the relative rate constants of the 1,2-shift of a CH_3 group and the σ°-constants of p-substituted phenyl groups [226]. The

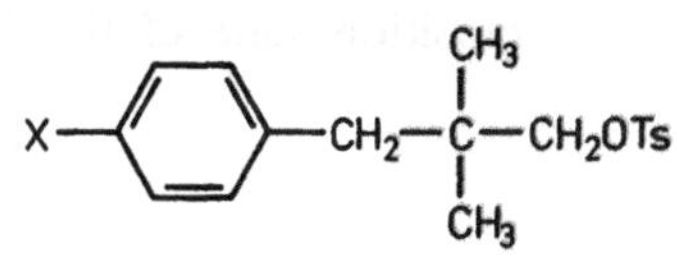

26

Relative rate of 1,2-shift for varied R[a]									Ref.
C_2H_5	*n*-C_3H_7	*n*-C_4H_9	*iso*-C_3H_7	*iso*-C_4H_9	*sec*-C_4H_9	*tert*-C_4H_9	$C_6H_5CH_2$	C_6H_5	
2.3									[122]
2.2									[73]
								0.7	[223]
2.3									[73]
2.19	2.17	3.50	7.41						[71, 72]
				4.25	9.66	34.39		0.3	[72]
							3.2	0.57	[71]
2.35		3.45	10.8					0.50	[71]
1.76		2.1							[71]
2.05									[71]
1.21									[71]
4.7						54			[19]

c) X = OBs.

13 A more comprehensive table of experimental data was given in [21].

Table 7. Relative Rates of the 1,2-Methyl Shift [94) a)]

Rearranging compound	Solvent	t °C	R			
			CH_3	CH_3	Cl	CF_3
p—R—$C_6H_4CH_2$—C(CH_3)$_2$—CH_2OTs	CH_3COOH	109.58	1.2	1.4	0.65	0.43
	CH_3COOH	99.58	1.2	1.4	0.65	
	HCOOH	90.07	1.2		0.58	0.38

a) The rate of the 1,2-shift of the CH_3 group for R=H is taken as a unit.

effect seems to be rather small ($\varrho \approx -0.7$) because of the low value of electron deficiency on the C° atom. Much more noticeable is the accelerating effect of branched alkyl groups most likely due to the relief of the steric strain ("back strain" [227)]) cf. Ref. [71,123)]. It is interesting to note that the phenyl group, as distinct from methyl, displays a retardation effect (cf. Ref. [228)]). Apparently this reflects a low degree of electron deficiency on the C° atom in the transition state of the 1,2-shift (cf. Ref. [223)]) concerted with ionization.

3.2.2 Long-Lived Carbonium Ion Rearrangements

In considering the effects of substituents at the C° atom on the rate of 1,2-shifts in *long-lived carbonium ions* one cannot fail to see a striking increase in the rate on substitution of a hydrogen by a methyl group recorded for different migrants and in carbonium ions of various structure. For instance, the substitution of the hydrogen at C° by a methyl group results in a sharp increase in the rate of the 1,2-shift of the benzyl group in ions (*27a, b*) [132,229)].

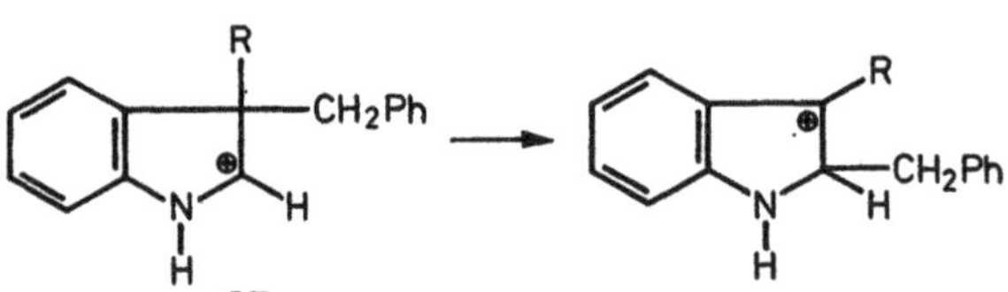

27

a R = H *b* R = CH_3

Introduction of a methyl group to the cyclopentyl ring accelerates the rearrangement of cyclopentylphenyl ketone (*28a*) [230)] whose limiting stage appears to be that of a 1,2-shift in an intermediate α-hydroxycarbonium ion, cf. Ref. [231)]:

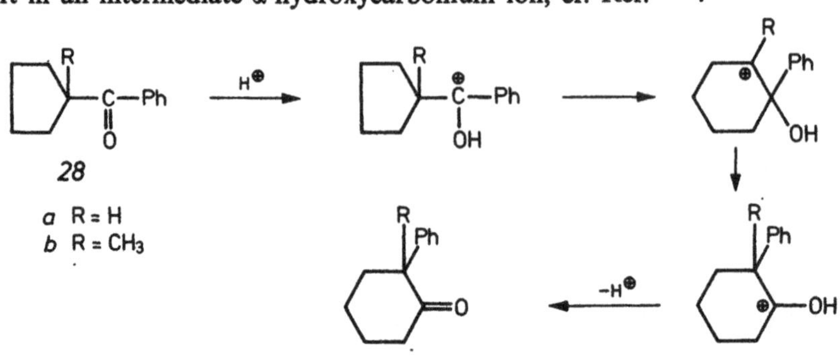

28

a R = H
b R = CH_3

A similar effect is observed for the rearrangement of 2-benzonorbornenyl cations (*29a, b*) [232]:

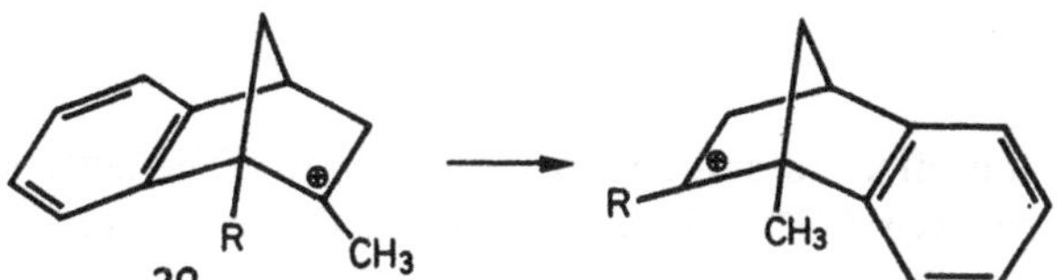

29 *a* R = H *b* R = CH_3

1,2-Methyl shifts in methylbenzenonium [26] and methylhydroxybenzenonium [233] ions are accelerated by the introduction of methyl groups into a meta-position relative to the C^{o} atom, so that in the rearranged ion they turn out to be located in resonance (ortho- or para-)positions.

Olah et al. [50] indicate that the relative "pushing ability" of a methyl group is higher than that of a hydrogen in the rearrangement of ions (*30a, b*) (1,2-hydride shifts). The same conclusion can be drawn from a comparison of the data [49,50] on the rates of rearrangement of ions (*31a, b*) by means of 1,2-methyl shifts.

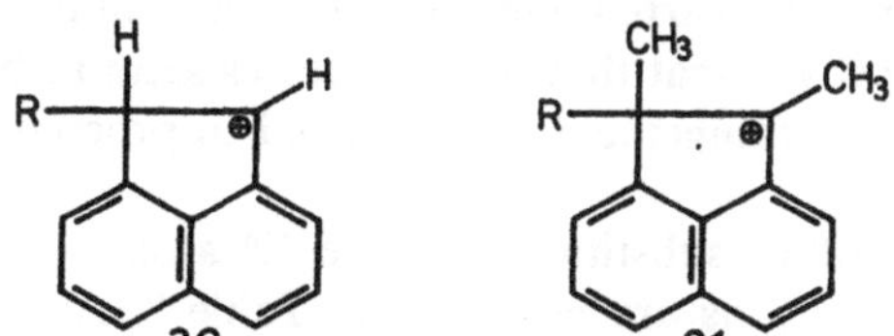

30 31

a R = H *b* R = CH_3

In accordance with the above mentioned data are those reported in Ref. [234,235] indicating that the 9-isopropyl-9-fluorenyl cation (*32b*) does not reveal any tendency to rearrangement even at —20 °C while the rearrangement of a 9-*tert*-butylfluorenyl analogue (*32c*) into a 9,9,10-trimethylphenanthrenonium ion (*33c*) proceeds very quickly even at lower temperatures [218,220], $t_{1/2}$ (—70 °C) ~ 2 min [218].

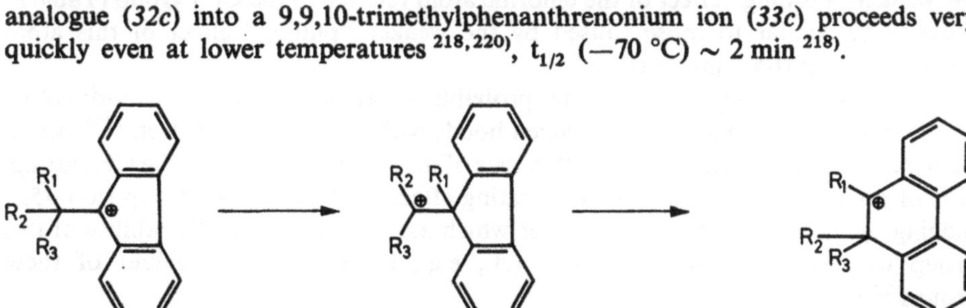

32 33

a $R_1 = CH_3$, $R_2 = R_3 = H$
b $R_1 = R_2 = CH_3$, $R_3 = H$
c $R_1 = R_2 = R_3 = CH_3$

Brouwer and Van Doorn [70] have quantitatively estimated the accelerating effect of a methyl group for 1,2-shifts of a hydrogen (deuterium) and a methyl group to the α-hydroxycarbonium centre in ions (*34a–d*):

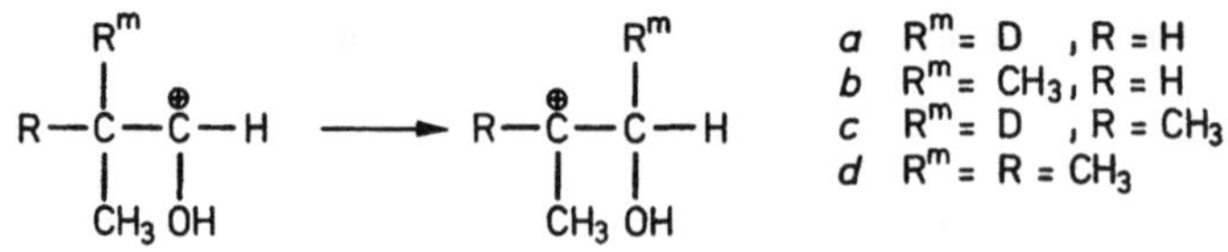

34

They have found this effect to amount to about 10^5–10^6. The authors have noted that the difference between $\Delta G^{\neq}$ values of 1,2-shifts for ions (*34a, b*), on the one hand, and (*34c, d*), on the other (7 kcal/mole), is nearly as great as that in the stability of secondary and tertiary carbonium ions (10 kcal/mole). Berson and Foley [236] considered the relationship between the free activation energy of a reaction involving the development of a carbonium centre and the accelerating effect of a CH_3 group; they concluded rhe latter to gradually decrease with decreasing free activation energy. The data of Brouwer and Van Doorn agree with this conclusion.

The relative stability of primary ions is well known to be minimal. Accordingly, the 1,2-shifts which were expected to result in a primary carbonium centre turn out to be retarded. For instance, the 9-ethylfluorenyl cation (*32a*) is stable at —10 °C while a 9-*tert*-butyl analogue (*32c*) undergoes a rapid rearrangement even at —70 °C [218,220] (a rearrangement scheme has been given above). Although formal schemes of 1,2-shifts may result in primary carbonium ions, the instability of the latter reflected in the transition state is so great that such processes seem to be practically unrealizable. In such cases the rearrangement is considered to proceed via the protonated cyclopropane [84].

Some data on the effect of other substituents at the C° atom on the rates of 1,2-shifts in long-lived carbonium ions are presented in Table 8. A rather modest "pushing" effect of the methoxy group in hydroxybenzenonium ions is due [85] to a relatively weak *conjugative interaction* between this group and the positive charge in the transition state. The importance of such an interaction is evidenced by the fact that, contrary to the 1-hydroxycyclohexadienyl cation the 1-acetoxy analogue fails to aromatize via a 1,2-shift (the so-called "NIH shift") [237]. The reasons for the weaker "pushing" effect of the chlorine atom relative to the CH_3 group (Table 8) seem to be similar to those caused by the weaker "pulling" effect of this atom located at the carbonium centre [206].

The hydroxy group, despite the probable decrease in the electron-donating ability due to the formation of hydrogen bonds with the medium (cf. Ref. [225]) seems to increase the rate of 1,2-methyl shift in ions of the type (*35*) relative to the CH_3 group. It is of interest to note that the accelerating effect of the methoxy group (ion *35c*), judging by the data of [225] is far weaker which agrees with the smaller ability of this group to stabilize cationoid centres (cf., e.g., the σ_p^+-constant values of these groups [239,240]).

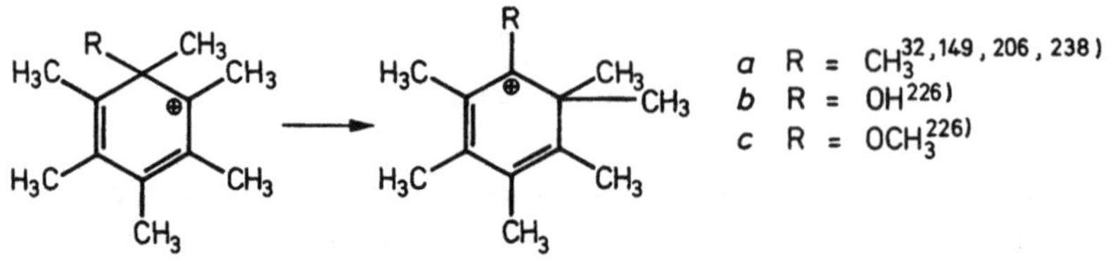

35

The relative effects of substituents at the C° atom on the rate of rearrangements (CH_3 > H, CH_3O > CH_3, CH_3 > Cl) are in agreement with the results of *MO calculations* for the 1,2-hydrogen shifts in the ions $RCH_2-\overset{+}{C}HR'$ [209].

The phenyl group, just as in the case of solvolysis and dehydration, shows a somewhat lower "pushing" effect on the 1,2-methyl shifts relative to the CH_3 group (see also the data on the rearrangements of pentamethylhydroxybenzenonium ions [67]). A similar effect is also characteristic of substituted phenyl groups, as follows from a comparison of the relative rates of 1,2-methyl shifts in 9,9,10-trisubstituted ions of the phenanthrene series (*36*, *37*, *38*, *33c*) [42,45,104].

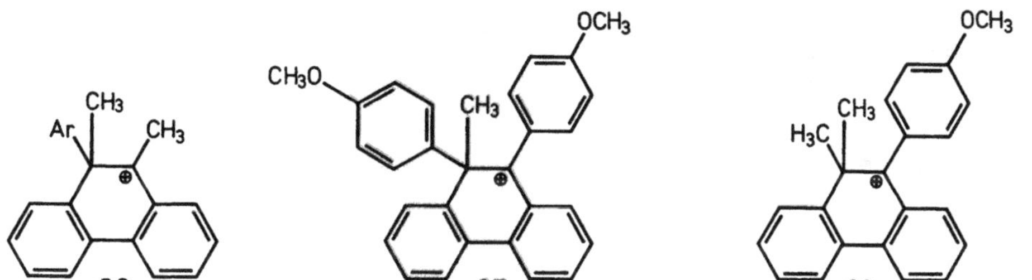

36 *37* *38*

a Ar = m-FC_6H_4 *b* Ar = p-$CF_3C_6H_4$ *c* Ar = C_6F_5

A particularly *strong retardation effect* is expressed in the case of the pentafluorophenyl group: the rate of the 1,2-shift of the CH_3 group in the 9,9,10-trimethylphenanthrenonium ion (*33c*) exceeds that of the 1,2-shift in the 9-pentafluorophenyl-9,10-dimethylphenanthrenonium ion (*36c*) by about two orders of magnitude (at 40 °C).

In considering the effects of substituted phenyl groups at the C^9 atom of 9-aryl-9,10-dimethylphenanthrenonium ions one should bear in mind that these groups seem to be incapable of displaying the resonance effect to a full extent due to unfavourable orientation caused by steric hindrance. This is shown by the predominance — in the equilibrium mixture — of isomeric ions with aryl groups at the C^9, rather than at the C^{10} atom [167]. It is interesting to note that, contrary to the accelerating effect of the methoxy group mentioned above in the case of benzenonium ions, the p-anisyl group (in 9-R-9,10-dimethylphenanthrenonium ions) seems to display a retardation effect on the 1,2-methyl shift (cf. the data of Ref. [42,45]). This may be due both to the unfavourable orientation and to the involvement of the anisyl group in the formation of *hydrogen bonds* with the acid medium decreasing its donating ability.

The alternative explanation assuming the transition state of the 1,2-shift to be similar to the initial one (cf. Ref. [223]) is hardly acceptable for the case of rearrangements of long-lived carbonium ions. The 1,2-methyl-shift-retardation effect of the phenyl group and of the substituted phenyl groups in arenonium ions may be, to a certain degree, due to the predominance of conformation with the aryl group in the quasi-axial position; this is favourable for the intramolecular donor-acceptor interaction with the carbocation centre (cf. Ref. [104,170,166,198,202]).

Table 8. Effects of Substituents at the C° Atom of Rearranging Carbocations

	Rearranging compound	Migrant R^m	Solvent	t °C	Relative rates of 1,2-shift for varied R[a)] C_2H_5	n-C_3H_7	C_6H_5	Cl	CH_3O	Ref.
1	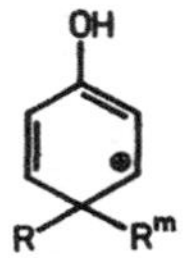	CH_3	aq. H_2SO_4	25	≈1	≈0.8				20)
			aq. $HClO_4$	25					66	85, 224)
2		CH_3	CF_3COOH	38.5	1.1		<0.3			15)
		$COOC_2H_5$	CF_3COOH	38.5	0.4		$1.4 \cdot 10^2$			15)
3	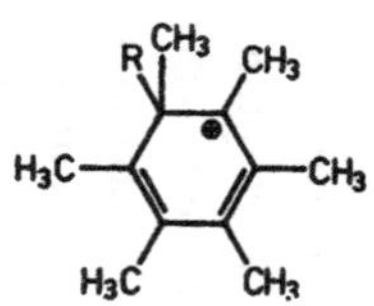	CH_3	HSO_3F	− 10				~0.1[b)]		206)
		CH_3	CF_3SO_3H	80			<0.1[b)]			35)
		CH_3	HSO_3F	− 25					>1[b)]	225)
4	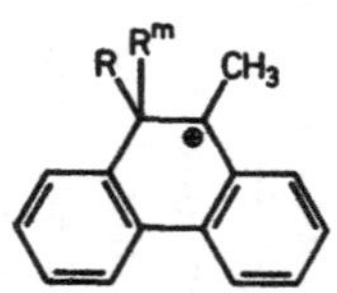	CH_3	HSO_3F	− 10				$\leqq 5.6 \cdot 10^{-4}$ [c)]		37, 206)
		C_2H_5	$HCl—Al_2Cl_6—CH_2Cl_2$	25	<1					48)
		C_6H_5	$HSO_3F—SO_2ClF$	−110			≫1			45, 104)
5	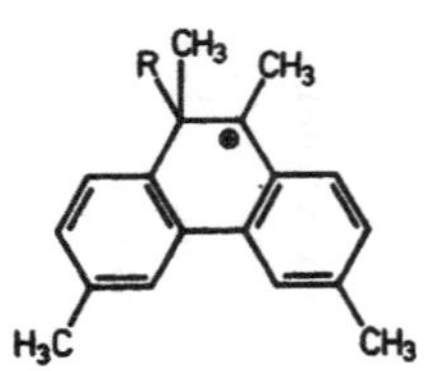	CH_3	$HSO_3F—SO_2ClF$	− 50			<1			42, 46)
6	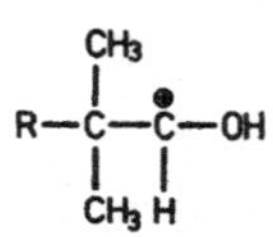	CH_3	HSO_3F	25	~1.5					70)

7	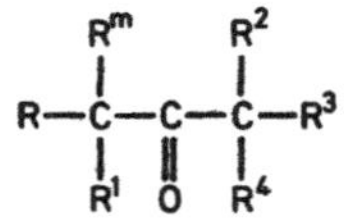	CH_3	96% H_2SO_4	25	0.4–1.1[d]		80)
			96% H_2SO_4	25	1.0[e]	1.1[e]	80)
		C_2H_5		25	0.5–2.3[d]		80)
				25	1.5[f]	2.3[f]	80)
8	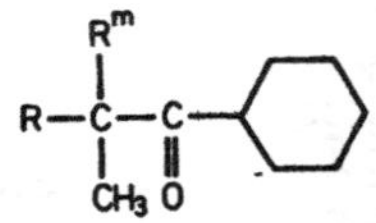	CH_3	98% H_2SO_4	55	0.4	0.8	81)
		C_2H_5	98% H_2SO_4	55	0.9		81)
9	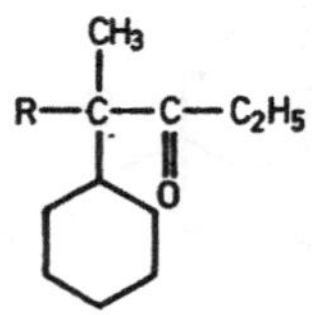	CH_3	98% H_2SO_4	55	0.02		81)

a) The rate of the 1,2-migrant shift in the case of R = CH_3 is taken as a unit.
b) Kinetic data for R = CH_3 are for solution in HSO_3Cl [28].
c) Kinetic data for R = CH_3 are for solution in $HSO_3F—SO_2ClF$ [42].
d) R^1–R^4 are various combinations of the groups CH_3 and C_2H_5.
e) $R^1 = R^2 = R^3 = R^4 = CH_3$; f) $R^1 = C_2H_5$, $R^2 = R^3 = R^4 = CH_3$.

Above, the phenyl and related groups at the C° atom were assumed to display an accelerating effect relative to the methyl one in the case of favourable orientation and electron deficiency on this atom (cf. Ref. [228,241,242]). However, the opposite view is also possible. For example, according to the data of [243] the α-CH_3 group stabilizes some types of carbonium ions more effectively than the phenyl one does. One can expect this order to be retained in the transition state of the 1,2-methyl shift as well (cf. Ref. [106], p. 285).

It is interesting to note that in the case of 1,2-shifts of π-migrants ($COOC_2H_5$, C_6H_5) the phenyl group displays an accelerating effect relative to the methyl one. In Ref. [104] a suggestion is made that the accelerating effect of the C_6H_5 group in the series of phenanthrenonium ions should result from the growth of strain at C° of the rearranging cation.

The effects of the ethyl and the *n*-propyl groups at the C° atom on the rates of carbocation rearrangements under the long-life conditions (Table 8) are usually close to that of the CH_3 group; although, as distinct from the available data on the reactions of solvolysis and dehydration, the qualitative constancy of the effects ($C_2H_5 > CH_3$) is not observed here. A considerable decrease in the "pushing" effect of the ethyl group in the case of ketone in item 9 of Table 8 is probably due to the intervention of conformational factors [81].

Judging by the estimates made in Ref. [48] the "pushing" effect of the ethyl group on the *1,2-ethyl shift* in 9,9,10-trialkylphenanthrenonium ions (the alkyls are CH_3 and C_2H_5) is also far lower than that of the methyl one. A similar order of influence seems to be also characteristic of the 1,2-methyl shift in ions of this type (cf. the data of Ref. [42,44] and [43,244]).

In this connection it should be noted that the concentration of 9,9-dimethyl-10-ethylphenanthrenonium ion mixed in equilibrium with 9-ethyl-9,10-dimethylphenanthrenonium ion is so low that signals of the former are absent from the PMR spectrum of the solution [43]. It is hardly possible to attribute such a difference in the relative stability of these isomers exclusively to the hyperconjugation effect.

Just as in the case of the rearrangements accompanying dehydration reactions the bulky groups at the C° atom display a significant accelerating effect relative to the methyl group (see, e.g., the data on the accelerating effect of the isopropyl and cyclohexyl groups in the acid-catalyzed rearrangement of ketones [81]).

Thus, the substituents at the C° atom exert a marked effect on the rate of 1,2-shifts which corroborates Stiles' and Mayer's viewpoint [19] as to the incorrectness of the approach to the estimation of migration ability based on the comparison of competitive migrations in the same compound-precursor of the carbonium ion (cf. Ref. [245]).

In the absence of complicating factors the competitive migratory aptitude (CMA) equals the migration etendency (MT) multiplied by the coefficient Q reflecting the differences in the influence of the substituents at C° atom (cf. [71,94]):

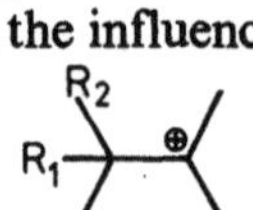

A

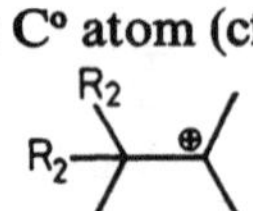

B

$$CMA = \frac{k_A^{R_1}}{k_A^{R_2}} = Q \cdot MT \qquad MT = \frac{k_A^{R_1}}{k_B^{R_2}} \; ; \quad Q = \frac{k_B^{R_2}}{k_A^{R_2}}$$

where $k_N^{R^i}$ is the rate constant of the 1,2-shift of the migrant R^i in the carbonium ion N.

It is obvious that the assumptions concerning the rearrangement direction which are made on the basis of the values either of the term Q or MT are in general incorrect although under certain correlations between these terms the conclusions may be in accordance with the experimental data. In Prof. Temnikova's paper contributed to the conference on the chemistry of carbocations [8] the CMA is treated in terms of the extended Hammond postulate and some example of such an accordance are cited.

By making a number of rather rough assumptions (in terms of the additive approach) the authors of Ref. [246] have been able to give a qualitative interpretation of the CMA of the chlorine atom, as well as of the methyl and phenyl groups in the dienone-phenol rearrangement of 4,4-disubstituted cyclohexadiene-2,5-ones (under the conditions of "complete" protonation of the carbonyl group).

3.3 Quantitative Description of Rearrangements

Above (Tables 2, 8) some data have already been cited from the works by French investigators Dubois and Bauer [80,128,204]. These works are aimed at developing an approach to the quantitative description of the structural factors of the carbonium ion migrant-carrying remnant. Below they will be considered in somewhat greater detail.

3.3.1 Dubois' and Bauer's Approach

The authors used as structural models bis-*tert*-alkyl ketones rearranging in an acid medium according to the following scheme, the limiting step being that of the 1,2-shift of an alkyl group (CH_3 or C_2H_5) in α-hydroxycarbonium ions of type (*39*):

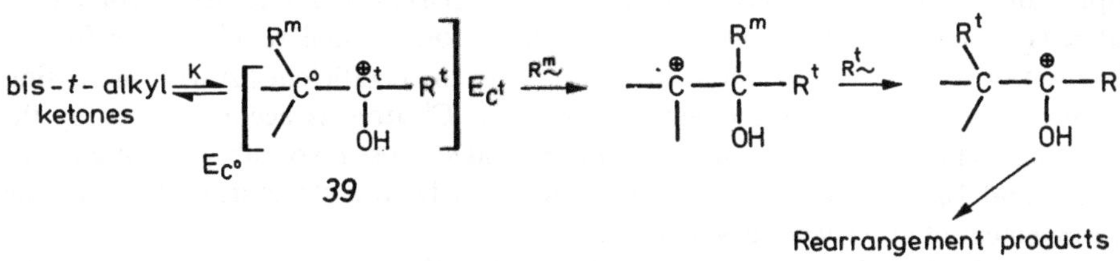

39

where E_{C^o} and E_{C^t} are the environments of the C^o and C^t atoms, respectively.

As a result of studying the kinetics and composition of the rearrangement products Dubois and Bauer found the partial rate constants for the migration of the CH_3 and C_2H_5 groups, i.e. the values of the MT of these groups. The data obtained by the authors on the MT against a background of identical MCRs have been discussed above in Section 2. Here the quantitative data on the MT of alkyl groups against a background of different MCR will be discussed, i.e. the structural effects of the MCR on the rate of the 1,2-alkyl shifts.

Dubois and Bauer have suggested that the overall structural effect of the MCR can be divided into two effects one of which is due to the substituents ("environment") at the C^o atom, the other — to those at C^t, the interactions between these sets of substituents being, for a first approximation, negligible. The variation of one of the environments while the other remained constant made it possible to establish linear relationships between the logarithms of the partial rate constants for migration and the number of carbon atoms in the varied environment:

$$\lg k_p^{R^m} = \lg k_p'^{R^m} + an_o$$
$$\lg k_p^{R^m} = \lg k_p'^{R^m} + bn_t$$

where $k_p^{R^m}$ is the partial rate constant for migration of the group R^m; $k_p'^{R^m}$ is that of the group R^m in a compound chosen as standard — in a structure where $E_{C^o} = CH_3$, CH_3 ($n_o = 0$) and $E_{C^t} = tert\text{-}C_4H_9$ ($n_t = 0$), $n_{o(t)}$ is the difference between the number of carbons in the set of substituents at the $C^{o(t)}$ atom for the given compound and for the standard one; a and b are the susceptibility coefficients; the equation

$$\lg k_p^{R^m} = \lg k'^{R^m}_p + an_o + bn_t + i$$

with the interaction term $i \approx 0$ is valid in a number of cases.

The increasing number of alkyl fragments, i.e. of carbons in the environment of the C^o atom accelerated the migration of both the ethyl and the methyl groups, this effect displaying itself far more strongly in the case of ethyl migrant (for $R^m = CH_3$ $a = 0.017$, for $R^m = C_2H_5$ $a \doteq 0.26$). The increasing number of carbons in the environment of the C^t atom leads, on the contrary, to a retardation of the rate; the susceptibility coefficient (b) for the methyl migrant is —0.63. This value is very close to that for the ethyl one (—0.66).

The authors have found that the variations reactivity cannot be interpreted as simple correlations with Taft's steric or polar parameters and have concluded that it is necessary to take into account the conformational requirements for the migration of R^m. In their opinion, the decrease in the coefficient (a) with increasing degree of ramification in the environment of the C^t atom is accounted for by the destabilization of conformers with a migrant located most favourably for migration. Dubois' and Bauer's approach was used to describe the structural effects in the rearrangement of alkylcyclohexyl ketones [81].

In the author's opinion, this approach is not without shortcomings such as the necessity of obtaining the values of the partial 1,2-shift rate constants from the overall values of the rate constants of rearrangement and the difficulty in identifying the structural factors determining the rate of rearrangement.

3.3.2 Fundamental Role of Degenerate Rearrangements

More promising appears an approach based on the direct measurement of "*elementary*" *rate constants* of 1,2-shifts in carbocations generated under the "long-life" conditions. The approach embodying the notions of a fundamental role of kinetic characteristics of *degenerate rearrangements* finds its substantiation in Marcus' theory [95,247], cf. Ref. [96]. This theory was initially worked out in application to electron and proton transfer processes and subsequently used to describe the transfer of heavy particles (see, Ref. [93,97], cf. Ref. [248]).

The fundamental role of the kinetic characteristics of degenerate processes is demonstrated by the results of using Marcus Eq. (1) in the research on isomerization of aromatic compounds [249–256].

$$\Delta G^{\neq}_{ij} = \Lambda + \frac{\Delta G_{ij}}{2} + \frac{\Delta G^{2}_{ij}}{16\,\Lambda} \tag{1}$$

The quantum chemical calculations of the enthalpies of activation and reaction for the rearrangements of carbocations of the type $RCH_2 - \overset{+}{C}HR'$ ($R^m = H$) with widely varied substituents R and R' [209] corroborate the applicability of Eq. (2) analogous with that of Marcus to carbocation rearrangements. In so doing they emphasize the great significance of the quadratic term in those cases when $\Lambda \lessapprox \Delta G_{ij}$.

$$\Delta H^{\neq}_{ij} = \Lambda' + \frac{\Delta H_{ij}}{2} + \frac{\Delta H^{2}_{ij}}{16\,\Lambda'} \tag{2}$$

As a rule, the opposite ($\Lambda > \Delta G_{ij}$) is valid; this makes it possible to neglect the quadratic term. If the following Equation $\Lambda = (\Delta G^{\neq}_{ij} + \Delta G^{\neq}_{jj})/2$ is additionally fulfilled then the Marcus equation becomes equivalent to the expression for the additivity principle. The possibility of using the additive approach at least to a rough estimation of the rates of carbocation rearrangements can be illustrated by the following data [38]: the experimental value of the free activation energy for the nondegenerate 1,2-shift of the methyl group in the 1,2,2,3,4-pentamethylnaphthalenonium ion ($\Delta G^{\neq}_{ij}$) is 11.2 kcal/mole, and the one calculated from Eq. (3) — 12.9 kcal/mole.

$$\Delta G^{\neq}_{ij} = (\Delta G^{\neq}_{ii} + \Delta G^{\neq}_{jj} + \Delta G_{ij})/2 \tag{3}$$

where $\Delta G^{\neq}_{ii}$ and $\Delta G^{\neq}_{jj}$ are the free activation energies of the 1,2-shifts of the methyl group in degenerate rearrangements of the heptamethylbenzenonium and the 9,9,10-trimethylphenanthrenonium ions, respectively; ΔG_{ij} is the free energy of the rearrangement:

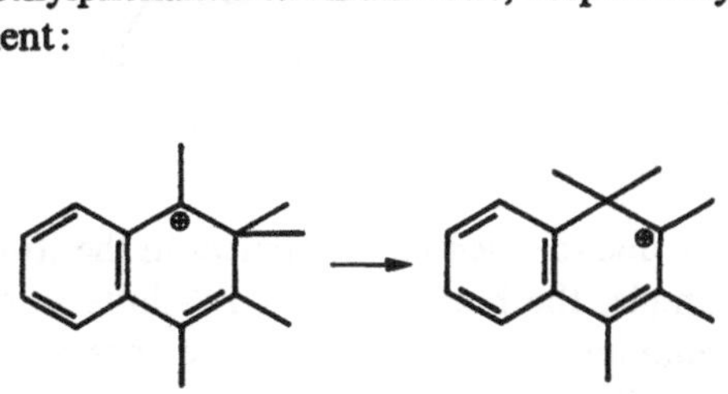

The possibility of using the simple additive approach to the approximate estimation of the rates of nondegenerate 1,2-methyl shifts is also demonstrated by the data on the rates of 3,2-methyl shifts in ions of the norbornyl type (*40a*) and

(*41a*) [52]. According to the additive approach[14]

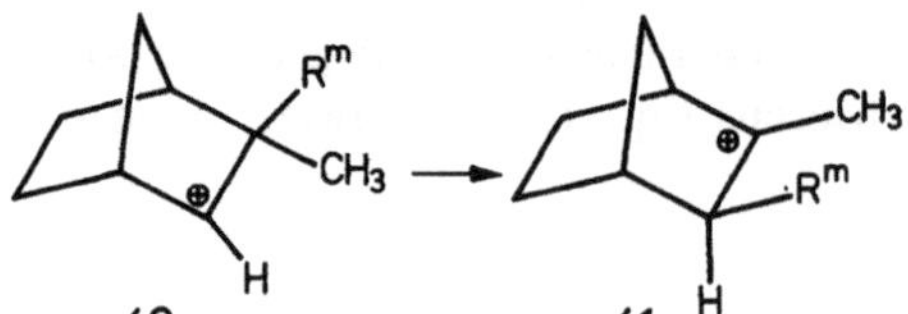

40 *41*

a $R^m = CH_3$ *b* $R^m = H$

$$\Delta G^{\neq}_{MTS} = (\Delta G^{\neq}_{MSS} + \Delta G^{\neq}_{MTT} + \Delta G_{MTS})/2$$

Substituting into this Equation the data on $\Delta G^{\neq}_{MSS}$, $\Delta G^{\neq}_{MTT}$ and ΔG_{MTS} [52] we obtain: $\Delta G^{\neq}_{MTS}$ = 14.45 kcal/mole — a value which is rather close to the experimental one (13.1 kcal/mole [52]). Note, however, that in the case of the 3,2-hydrogen shift in ions (*40b*) and (*41b*) the discrepancy between the calculated value and the experimental one is far larger, it amounts to 3 kcal/mole.

3.3.3 Structure-Reactivity Relationships

Comparison of the data on the rates of the 1,2-shift of the same migrant, say, the methyl group in degenerate rearrangements of different carbonium ions shows them to be greatly dependent on the structure of the rearranging carbocation:

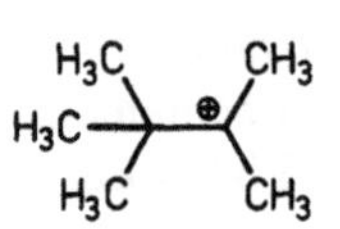

$k^{-136°} = 2.8 \cdot 10^6 s^{-1}$ [59]

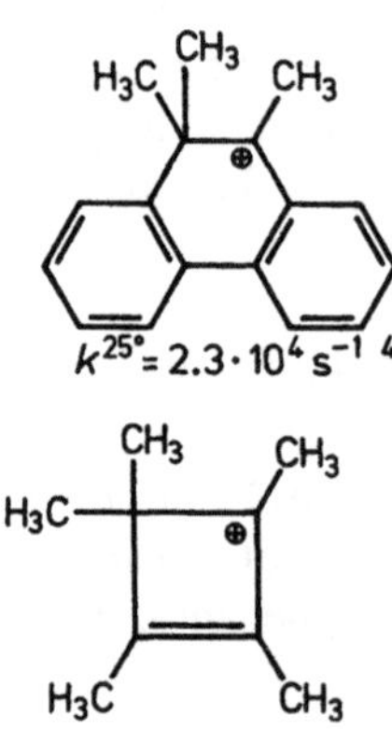

$k^{25°} = 2.3 \cdot 10^4 s^{-1}$ [42]

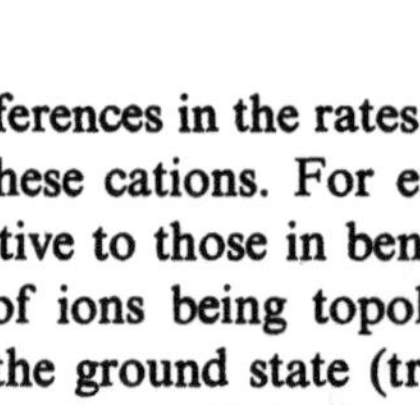

$k^{25°} = 3.6 \cdot 10^{-1} s^{-1}$ [28]

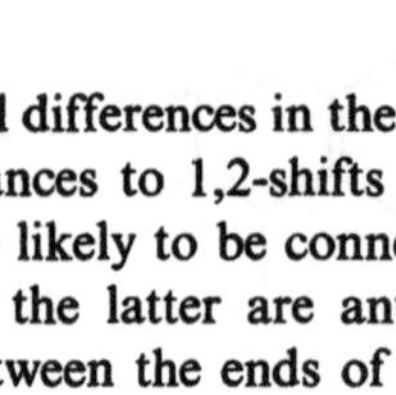

$k^{70°} < 2 \cdot 10^{-5} s^{-1}$ [257, 258]

These differences in the rates are due to essential differences in the structure of the MCRs of these cations. For example, the hindrances to 1,2-shifts in cyclobutenyl cations relative to those in benzenonium ones are likely to be connected with these two types of ions being topologically opposite: the latter are antihomoaromatic systems in the ground state (true, the overlap between the ends of the pentadienyl system and, hence, the extent of antihomoaromaticity seems to be very small,

14 Use is made of Prof. Sorensen's designations [52].

cf. Ref. [259]), and in the transition state — aromatic, while in the case of cyclobutenyl cations, on the contrary, they are homoaromatic systems in the ground state [260,261] and antiaromatic in the transition state. According to the theory of pericyclic reactions [148] the 1,2-shifts in cyclobutenyl cations, i.e. sigmatropic 1,4-shifts are forbidden.

Comparison of data on the rates of degenerate 1,2-hydrogen shifts in long-lived carbocations [30,49,50,59,262,263,54,264] (summarized in Ref. [7]) shows that the substitution of the methyl groups by hydrogens at C° and C', i.e. in the nearest environment results in the retardation of rearrangement in the case of cyclic carbocations (with the effect tending to grow in passing from six-member cycles to five- and four-member ones) and in its acceleration in the case of open structures (from the 2,3-dimethyl-2-butyl cation to the 2-butyl one[15]). The reasons for this regularity are not clear.

For a consistent quantitative description of rearrangements it seems reasonable to divide the whole set of degenerate carbocations into series of structurally related ones within which the changes in rearrangement rates can be treated as caused mainly by one structural effect. There is every reason to believe (cf. Ref. [30]) that in many cases the extent of electron deficiency in the carbonium centre may serve as a valuable parameter for such a purpose. For the arenonium ions with an aromatic transition state the decisive factor may prove to be the value of π-electron deficiency which for a series of related structures may be characterized by the chemical shift of the carbonium carbon (δC^+) in the ^{13}C NMR spectrum [266–268], cf. Ref. [269,270].

An example of degenerate 1,2-methyl shifts are arenonium ions of the type *A*; it has been found that the electronic influence of the MCR of the carbocation on the rate of rearrangement is characterized not only by δC^+ but also by such structural parameters as the σ^+-constants of substituents [42] and the infrared frequencies of $C = 0$, ν_{CO} of ketones related to the rearranging carbocations [271].

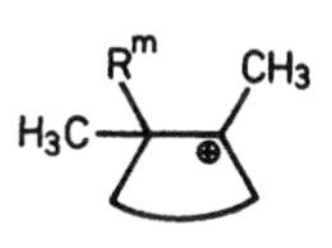

π-electron system
A

The most appropriate parameter seems to be the chemical shift of the carbon atom — the carbocation centre (δC^+). Its main advantage is its close relationship with the value of π-electron deficiency in the carbocation centre which can be estimated by the MO method. For the series of structurally related arenonium ions [42–45] a linear correlation is found between the free activation energy of the

15 The 2-butyl cation which undergoes an extremely rapid degenerate rearrangement [59] and displays a relative isotope splitting which is intermediate between the values characteristic of classical, and those of nonclassical carbocations [265] was regarded as a partially bridged one [25]. This assumption agreed with the ^{13}C NMR data [25], cf. Ref. [63]. According to M. Saunders and W. Hehre's data cited in Ref. [58], this cation is hydrogen-bridged.

degenerate 1,2-methyl shift[16] and δC^+ [38]:

$$\Delta G^{\neq}(-110\ °C) = 53.7 - 0.183\ \delta C^+ \qquad (4)$$
$$r\ -0.997 \qquad s\ 0.39$$

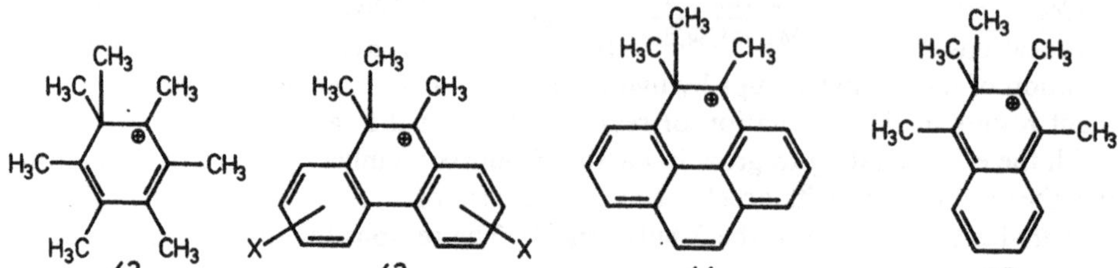

42 43 44 45

a X=H
b X=Br(2.7)
c X=Br(3.6)
d X=CH_3(3.6)
e X=CF_3(3.6)

The experimental value of the free activation energy for the 1,2-shift of the methyl group in the 1,2,2,3,4-pentamethylbiphenylenonium ion (*46*) deviates from this correlation towards greater values ($\Delta\Delta G^{\neq} \approx 4.5$ kcal/mole [40])[17] which may be due, at least partially, to the antiaromatic character of the transition state of the 1,2-shift (cf. Ref. [41,274]).

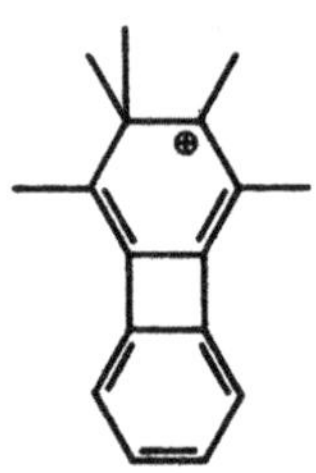

46

Similar correlations were earlier established for the same series of ions, with the exception of ion (*45*) [42]:

$$\Delta G^{\neq}(-110\ °C) = 55.8 - 0.192\ \delta C^+ \qquad (5)$$
$$r\ -0.993 \qquad s\ 0.39$$
$$\Delta G^{\neq}(25\ °C) = 53.7 - 0.181\ \delta C^+ \qquad (6)$$
$$r\ -0.995 \qquad s\ 0.29$$

16 In typical examples for $R^m = CH_3$ and for other migrants the mechanism of 1,2-shift (allowed sigmatropic 1.2, 1.6, 1.10 or 1.14 shifts according to the Woodward-Hoffmann classification [148]) was specially established by such methods as that of spin saturation transfer [32,150,272,273], the method of labelled atoms [35] and others.

17 It is taken that $\Delta G^{\neq}$ (25 °C) = $\Delta G^{\neq}$ (−110 °C).

In this work the relationships were interpreted in terms of the PMO theory. From the interpretation it follows that the linear relationship must hold between the free activation energy of the rearrangement and the length of the double carbon-carbon ($C^o = C^t$) bond of the compounds formed as the result of the methyl cation departure from the respective carbonium ions. Indeed, such a relationship has been found [7,38,42]. Correlations according to Eq. (4)–(6) were successfully used to quantitatively describe the isomerization of methylbenzenes [249–253] and methylnaphthalenes [254–256] in terms of the Marcus equation.

The models of "π-delocalized" ions, contrary to the aliphatic carbocations made it possible to measure directly the rates of 1,2-shifts of such migrants as, e.g., a chlorine atom [31,37,68] and aryl groups [35,46,171,173], cf. Ref. [275–276], and to establish the structure-reactivity relationships.

The correlations according to Eq. (4) have also been established for such migrants as H [29,39], C_6H_5 [46], Cl [37], NO_2 [103], the MT displaying no qualitative constancy even in this, most "refined" case of degenerate rearrangements of long-lived carbonium ions though their free-energy barrier seems to be "purely kinetic, intrinsic" (cf. Sect. 2). It is not surprising that for nondegenerate rearrangements whose free energy barrier, according to the Marcus theory, is determined in addition to this thermodynamic term, the MT inconstancy is characteristic as well. One of the few exceptions is the MT of such groups as CH_3, C_2H_5 and CH_2Cl; it reveals a qualitative constancy of MT in the degenerate rearrangements of arenonium ions [48,83]. The qualitative constancy of MT is characteristic of the groups CH_3 and C_2H_5 in the case of nondegenerate rearrangements as well. [19]

In terms of the PMO theory, the differences in the angular coefficients of the plots of the free activation energy vs. the charge in the carbonium centre are interpreted for the 1,2-shifts of different migrants [217].

Along with the above partial parameters (δC^+, ν_{CO} etc.) characterizing first of all the electronic effects of the MCR, the carbocation rearrangements can be quantitatively described by using a generalized parameter — the free activation energy (or $\lg k$ of the rate) of the 1,2-shift of a migrant chosen as a standard [29], cf. Ref. [37]. As an example one can mention the correlation covering, along with the arenonium ions (*42, 43a, c–e, 45, 46*) and their structural analogues with R^m = H [28, 29, 38–42], carbocations of other types (*47–50*):

$$\Delta G^{\ddagger}_{H}\ (25\ °C) = 2.18 \pm 0.73 + (0.527 \pm 0.051)\, \Delta G^{\ddagger}_{CH_3}\ (25\ °C)^{20} \qquad (7)$$

r 0.960 s 0.92 n 11

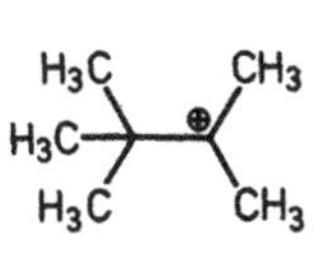

47 [59]

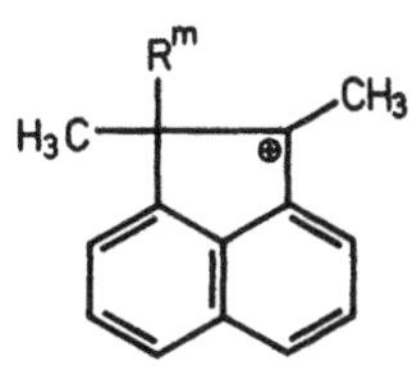

48 [49]

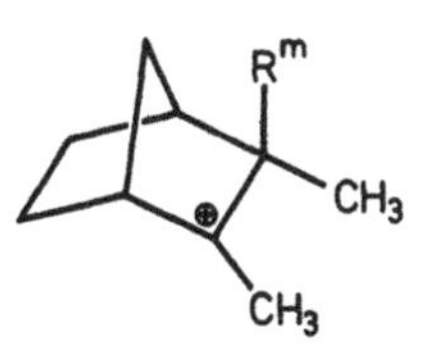

49 [54]

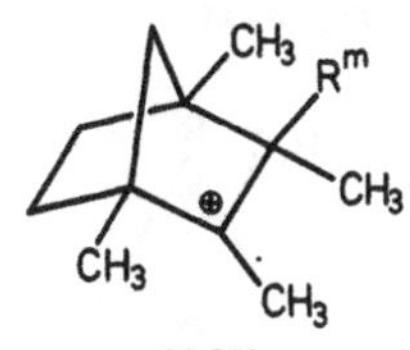

50 [56,57]

19 No data appear to be available on the MT of the group CH_2Cl for nondegenerate rearrangements.
20 In a number of cases $\Delta G^{\ddagger}$ is assumed to be independent of temperature.

It is interesting to note that this correlation allows one to estimate at least roughly the rates of 1,2-shifts of hydrogens in neutral compounds — methylcyclopentadienes (*51b*) and (*52b*)[21], cf. Ref. [29] — calculated values of $\Delta G^{\neq}_{H}$ amount to 24.2 and 24.4 kcal/mole, experimental ones — to 22.5 and 23.7 kcal/mole, respectively [279,282].

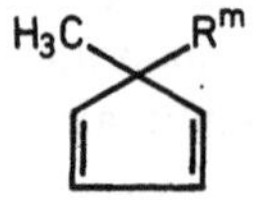

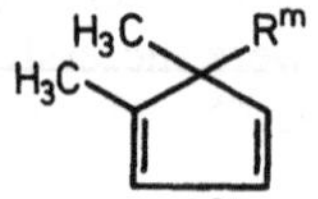

51
a $R^m = CH_3$ [278]
b $R^m = H$ [279]

52
a $R^m = CH_3$ [280,281]
b $R^m = H$ [282]

It should be noted that at the great variation in the structure of the MCR of the carbocations covered by Eq. (7) the nearest environment of C^o and C^t remained partially constant: these atoms are in all cases directly bonded with methyl groups.

Of definite interest is the problem of the validity of such relationships in the case of more significant changes in the nearest environment. According to Sorensen's estimation [52] the free activation energy for the 1,2-shift of the methyl group from C^3 to C^2 in the *exo*-3-methyl-2-norbornyl cation (*53*) whose $C^o(C^3)$ and $C^t(C^2)$ atoms are directly bonded with the hydrogens rather than with the methyl groups of the MCR amounts to 14.5 kcal/mole. The substitution of this value into Eq. (7) yields the value $\Delta G^{\neq}_{H} = 9.9$ kcal/mole rather close to the experimental one ($\Delta G^{\neq}_{H} = 11.5$ kcal/mole [53]).

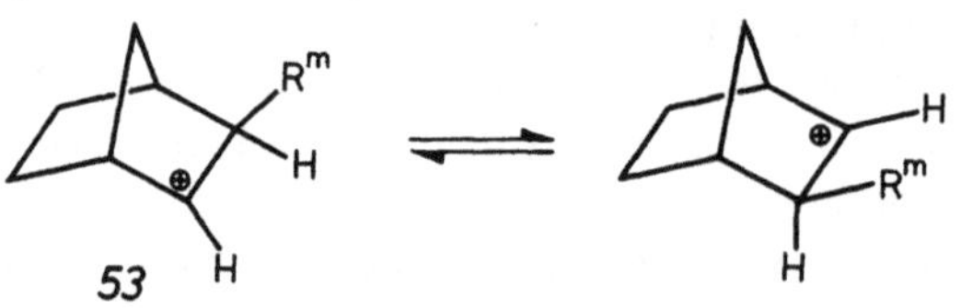

53

a $R^m = CH_3$ *b* $R^m = H$

Similar relationships are fulfilled in choosing the hydrogen atom as a standard migrant [37]. It should be noted that on extrapolation of these relationships to the value $\Delta G^{\neq}_{H} = 3.1$ kcal/mole characteristic of the 1,2-hydrogen shift in the isopropyldimethylcarbonium ion [59] the value $\Delta G^{\neq}$ for $R^m = CH_3$ drops to zero while the values $\Delta G^{\neq}$ for $R^m = Cl, C_6H_5, NO_2$ get into the region of negative numbers, the respective values coming in the following order: $|Cl| > |C_6H_5|, |NO_2|$. On the basis of these data it appears possible to assume [37], cf. Ref. [29] that the tendency of migrants to form cyclic structure of the type *A* decreases in the order $Cl > C_6H_5, NO_2 > CH_3 > H$.

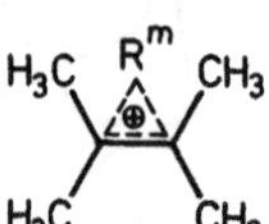

A

21 1.5-Sigmatropic shifts according to the Woodward-Hoffmann classification [148]. The nondegenerate character of these rearrangements does not seem to be a complicating factor since the free energy of rearrangement is very low relative to the free activation energy (cf. Ref. [282]).

The experimental data for $R^m = Cl$ [60], compared to those for $R^m = CH_3$, H [59] agree with this assumption (cf. also the data of quantum chemical calculations of similar structures for $R^m = NO_2$ [283], H [284–286], CH_3 [285–287], C_6H_5 [288], Cl [151, 286, 287, 289, 290], see also Ref. [292]). On the contrary, the experimental data for $R^m = CH_3$ ($\Delta G^{\neq}_{110\,°C} = 3.8$ kcal/mole [59] taking into account the statistical factor taken to be equal to 3) compared to those for R^m = H, do not.

The value [31, 292] (cf. Ref. [37]) of the free activation energy for the 1,2-chlorine shift in the 1-chloro-1,2-dimethylacenaphthylenonium ion is found to be, just as in the case of 1,2-shifts of the hydrogen and the methyl group [49], larger than that predicted on the basis of the corresponding data for arenonium ions.[22] The possible reasons for the deviations observed are a higher steric strain in the transition state (cf. Ref. [50]) and "electronic retardation" (cf. Ref. [7]) — lower energy of interaction of the three-membered cycle formed by the migrant and the C^o and C^t atoms with the π-electron system of the remaining part of the cation (Structure *A*) in the case of acenaphthylenonium ions in comparison with arenonium ones [31].

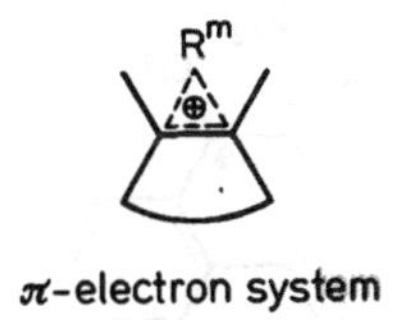

A

Note that the deviation in the case of R^m = Cl is less than that for $R^m = CH_3$ which may be due to lower strain of the transition state [31] or (in other terms) to lower susceptibility of the migrant possessing p-AO accessible for interaction with p-AO of cationic centre to the effect of orbital orientation (cf. Ref. [293, 294]).

In terms of the discussed approach to the quantitative description of carbocation rearrangements it proved possible to cover by general correlations the MT of different migrants and its dependence on the electronic structure of the MCR of the rearranging carbocation [295]. The basis of these correlations is a simple model and the corresponding terms: E_1 which describes the loosening of the C^o—R^m bond, and E_2 which describes the affinity of the migrant to the carbocation centre.

For degenerate rearrangements the following linear correlation was assumed to be fulfilled:

$$\Delta G^{\neq} = a + bE_1 + cE_2 \qquad (8)$$

The used parameters characterizing the terms E_1 and E_2 were the C—R^m bond rupture energy (E_b) in the compounds CH_3—R^m and the proton affinity of these compounds (PA). For the arenonium ions of the benzene, naphthalene, phenanthrene and pyrene series with $R^m = CH_3$, C_2H_5, X-C_6H_4, Cl, Br the following correlation was obtained [295]:

$$\Delta G^{\neq}_{25°} = 0.9 + 0.329\,E_b - 0.00129\,\text{PA}\,(\delta C^+ - 130) \qquad (9)$$

r 0.987 s 0.7 n 29

22 Judging by the data available [46, 51] a similar deviation is also typical of $R^m = C_6H_5$.

A parameter characterizing the term E_1 is the deformation vibration force constant of the fragment C^t—C^o—R^m (k_{def}). If this parameter is used instead of the bond rupture energy the resulting correlation is of a somewhat lower quality, but it covers the arenonium ions with $R^m = H$ as well:

$$\Delta G^{\neq}_{25^\circ} = 17.1 + 14.5\, k_{def} - 0.00127\, PA\, (\delta C^+ - 130) \qquad (10)$$
$$r\ 0.963 \qquad s\ 1.1 \qquad n\ 33$$

These correlations make it possible to calculate the PA of the central atoms of the migrant in the CH_3—R^m compounds and to use them for the prediction of the rearrangement rates in other carbocations. Thus, the PA value of the nitrogen atom in nitromethane (104.7 kcal/mole) calculated on the basis of the value of $\Delta G^{\neq}$ for the rearrangement of ion (*54*) was used to calculate the $\Delta G^{\neq}$ values of the 1,2-nitro group shifts in ions (*55a, b*) (7.5 and 8.6 kcal/mole, respectively). These values are rather close to the experimental ones (7.0 and 9.9 kcal/mole, respectively[103]).

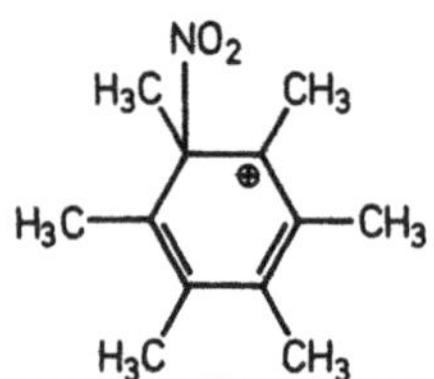

54

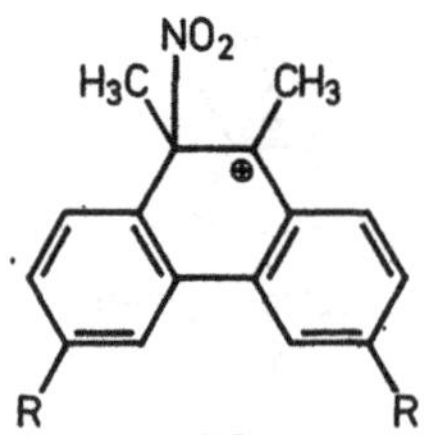

55

a R = H *b* R = CH_3

According to the preliminary data[296], the correlations obtained make it possible not only to perform similar estimations for the rearrangements of long-lived carbocations, but also to predict the rates of anchimerically assisted solvolytic reactions.

3.4 Steric Effects

To conclude this section consider briefly some aspects of the steric effects. These effects seem to play an important role in carbocation rearrangements. As an example illustrating, perhaps, the decisive role of steric factors, one can mention the rearrangement of the tertiary carbocation (*56*) to the secondary one[297]:

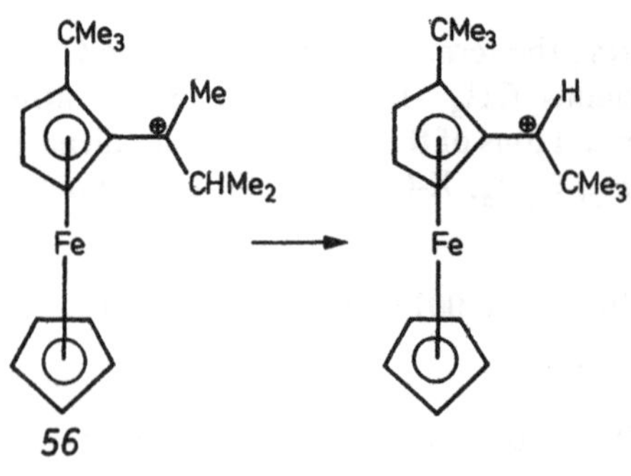

56

Such an unusual direction of rearrangement is accounted for by the steric compression between the *tert*-butyl group and the fragment $C^+(CH_3)\ CH(CH_3)_2$.

In the framework of the assumption that the free-energy barrier is dependent on the two terms: "intrinsic barrier" and the free energy of rearrangement (see above)[23] one should consider the effect of steric factors on each component. Here only the "intrinsic barrier" steric effects will be discussed.

The value of the intrinsic barrier is naturally assumed to grow with increasing steric strain in the transition state of rearrangement. It is probably this factor which accounts for the retardation of the degenerate 1,2-hydrogen shift in the acenaphthylenonium ion (*30a*):

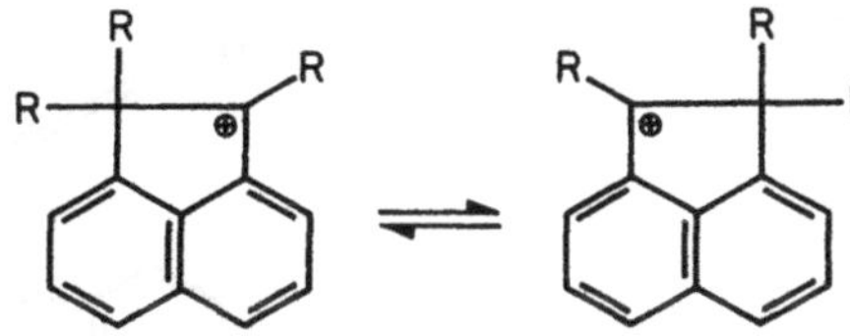

30a R=H *31b* R=CH₃

Note, however, that contrary to the data of [50] on an abnormally large retardation of the 1,2-methyl shift in ion (*31b*) it occurs at a rather high rate ($k^{25\,°C} \approx 0.2\ s^{-1}$ [49]) although far slower than in the 9,9,10-trimethylphenanthrenonium ion. The growth of steric strain in the transition state is assumed to account for the lower rate of the 1,2-methyl shift in the 4,5,9,9,10-pentamethylphenanthrenonium ion (*57*) relative to the 9,9,10-trimethylphenanthrenonium one [300].

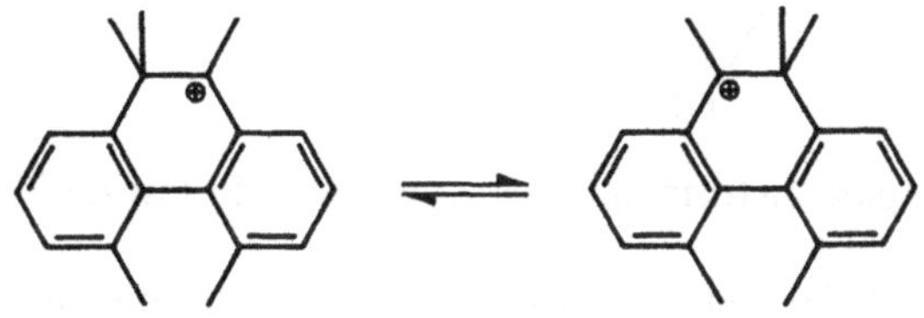

57

It is of interest to note that due to the structural features of ion (*57*) the rates of 1,2-shifts of nonequivalent — quasi-axial and quasi-equatorial methyl groups turn out to be equal [300,301]; this shows the limited character of the "classical" treatment of the bond alignment effect which virtually assumed that the more favourable orientation of the interacting orbitals of the migrant and the carbocation centre in the ground state necessarily correspond with the maximum overlap in the transition state [302,303] (cf. Ref. [304]) and, hence, with the higher rate of 1,2-shift.

The steric strain in the transition state seems to be responsible for the retardation of the degenerate 1,2-shifts in ions (*58*) and (*59*) relative to (*60*) and (*61*), respectively [17,41], cf. Ref. [305,306].

23 The methods of describing the steric effects in "strongly bonded" systems based on molecular mechanics (see Ref. [298,299]) are not discussed here.

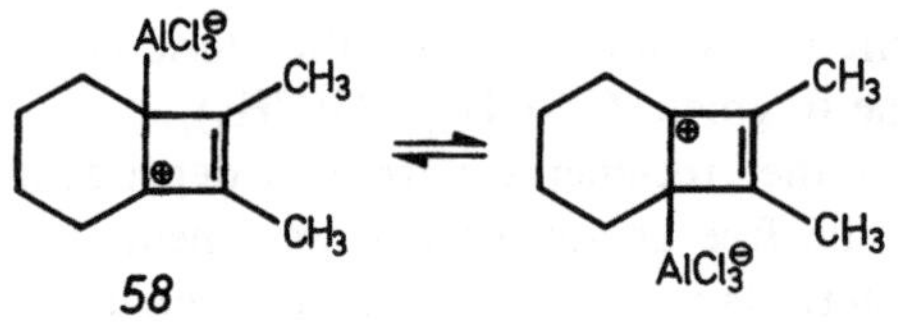

58

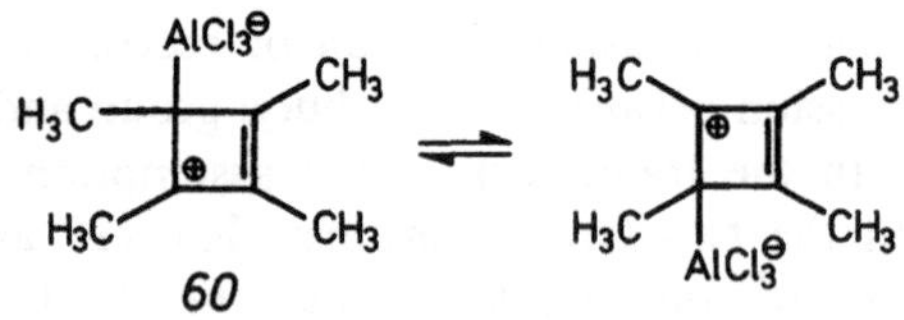

60

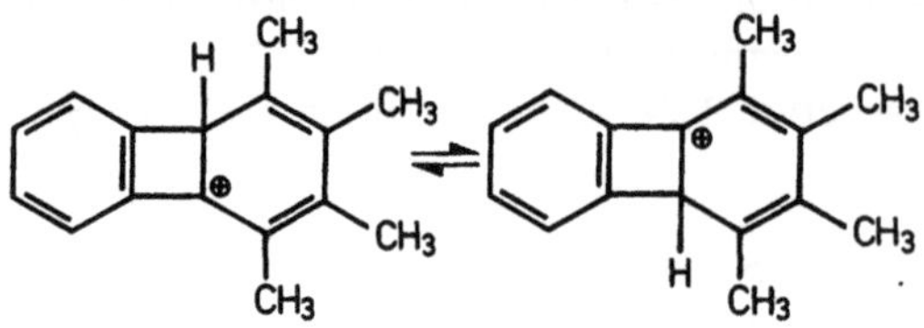

59

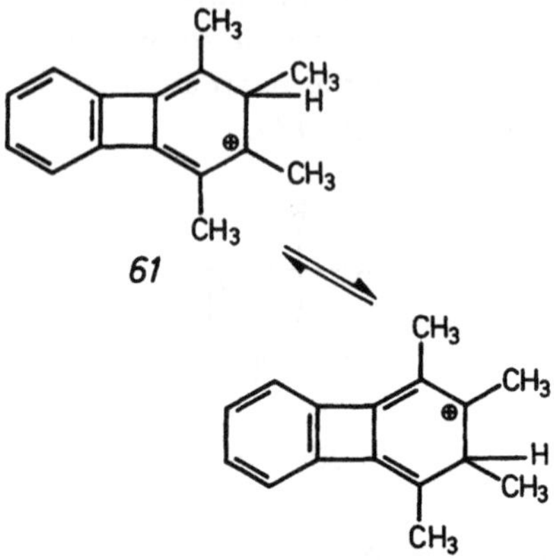

61

A similar reason may account for a certain decrease of the rearrangement rate in passing from ion (*62*) to (*63*) [59]. As follows from the data of Ref. [17], the effect of

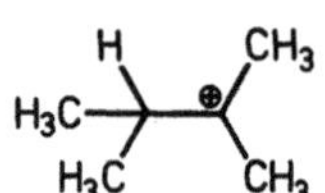

62

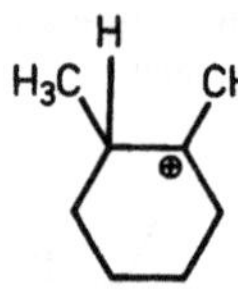

63

retardation disappears in passing from the six-membered to the less sterically demanding seven-membered ring.

The steric effects reveal themselves in considering the data on the relative rates of 1,2-hydride shifts in ions of type (*64*):

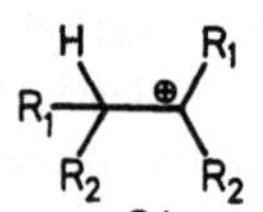

64

a $R_1 = H$, $R_2 = CH_3$ [59]
b $R_1 = R_2 = CH_3$ [59]
c $R_1 = C_6H_5$, $R_2 = CH_3$ [307]
d $R_1 = R_2 = C_6H_5$ [307]
e $R_1 = R_2 = p\text{-}CH_3C_6H_4$ [206]

Substitution of two methyl groups for hydrogens (passage from ion (*64a*) to (*64b*)) results in some retardation of rearrangement (taking account of the statistical factor $\Delta\Delta G^{\neq}_{-138\,°C} = 1.2$ kcal/mole [59]), a far stronger effect is observed in passing to ion (*64c*): according to the data of Ref. [307], this ion is static (PMR method) even at -80 °C ($\Delta G^{\neq}_{-80\,°C} > 10$ kcal/mole). A similar effect is observed in passing to ions (*64d, e*) (for (*64d*) $\Delta G^{\neq}_{37\,°C} > 17$ kcal/mole, for (*64e*) $\Delta G^{\neq}_{0\,°C} > 14.7$ kcal/mole, estimates on the data of Ref. [206,307], respectively).

The effects observed might be associated with the decrease in the electron deficiency of the carbonium centre in the above series of carbocations, but a comparison of these data with those on the rates of degenerate 1,2-hydride shifts in some other carbocations suggests that other factors, too, seemingly steric, act here. Indeed, as distinct from ion (*64c*), the 9-H-9,10-dimethylphenanthrenonium ion (*33b*) undergoes a very rapid rearrangement by 1,2-hydride shift ($k^{-80\,°C} \approx 10^3\ s^{-1}$, $\Delta G^{\neq}_{-80\,°C} \approx 8.5$ kcal/mole [29]) despite the fact that the electron deficiency in the carbonium centre of this ion seems to be even somewhat lower than in the case of ion (*64c*) (cf. Ref. [276]).

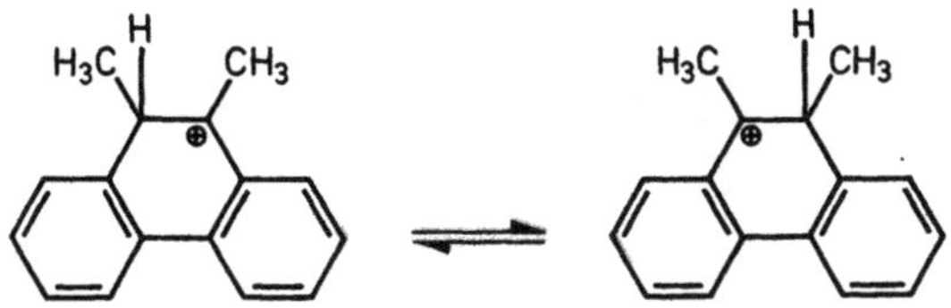

33 b

One can assume that in the transition state of the 1,2-hydride shift in ion (*64c*) there occurs an essential increase in the strain brought about by the unfavourable steric interaction of hydrocarbon fragments. This assumption is evidenced by the fact that the 1,2-shift of a far bulkier migrant — a methyl group in ion (*64c*) does occur [307] despite its obvious endothermicity (cf. Ref. [308]):

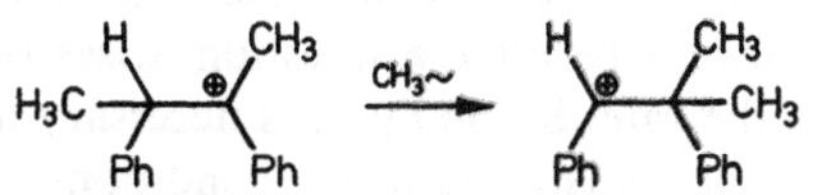

64 c

An additional argument may be the fact that the rearrangement according to the scheme *i* proceeds at a higher rate than that according to the scheme *ii* [275,276,309] (although some interference of electronic effects is not excepted here):

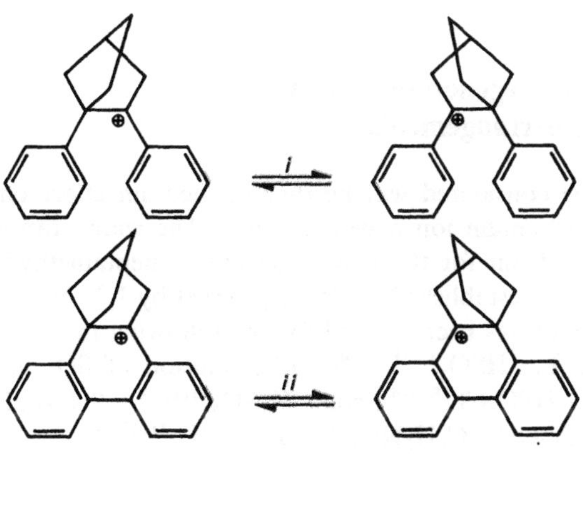

The rate of 1,2-shifts is assumed to decrease in passing from open carbocations to cyclic ones and with decreasing dimensions of the cycle [306]. In Ref. [306] the reasons for this regularity are discussed. But the latter does not seem to be of a general character, in the case of migrants with small steric requirements, such as a hydrogen and of MCR with relatively bulky groups (alkyls, aryls and the like) it is believed to be violated. It is indicative, in this respect, that the rates of 1,2-hydride shifts in ions (*64a*) and (*65*) are equal while in ion (*66*) the rate is markedly lower [59]; this is tentatively attributed by the authors [59] to the increase in the barrier of rotation around the central bond of ion (*66*) relative to (*64a*).

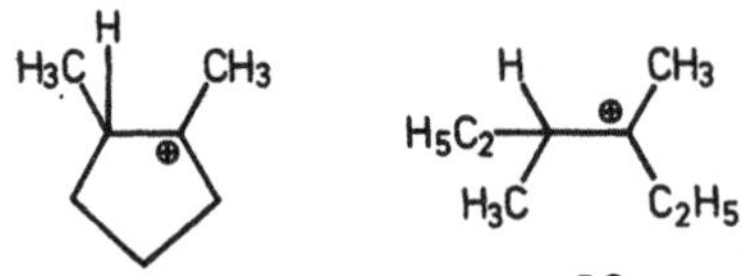

65 66

4 Effects of Media on the Rates of Carbocation Rearrangements

In the framework of the present review the problem of the medium effects on the rates of rearrangements of carbonium ions has two main aspects. Firstly, in order to establish the relationships between the structure of long-lived carbonium ions and the rates of 1,2-shifts it is necessary to standardize the conditions. The best way would be to study the kinetics of rearrangements under the conditions of a constant medium. Since this is not always possible a question arises of estimating the effect of the medium changes on the rate. Secondly, it is necessary to determine to what extent these relationships can be transferred to the conditions of the "short life" of carbocations. The problem of taking into account the specific medium effect (hydrogen bonds and the like, cf. Ref. [208,310–312]) will not be specially discussed here. As this type of effect implies significant changes in the ion structure it can conventionally be placed among the structure factors. It is but natural that no distinct boundary-line can be drawn between the specific and the non-specific medium effects.

4.1 Medium Effect on Degenerate Long-Lived Carbonium Ion Rearrangements

The first question to be considered will be that of medium effect on the rate of degenerate long-lived carbonium ion rearrangements. The widest range of medium variation has been carried out for the rearrangement of heptamethylbenzenonium ion which has been reliably established [272,32] to proceed by 1,2-shifts of the methyl group. This rearrangement has been studied for the following media:
a) acids — HSO_3F—SbF_5, HSO_3F [313,314], 20% solution of SO_3 in HSO_3F [315], HSO_3F—$C_6H_5NO_2$ [313], HSO_3Cl [28,314], conc. H_2SO_4 [314,315], 9.4 M H_2SO_4 [149,314], CF_3COOH [32,238], CF_3COOH—$C_6H_5NO_2$ [313], HCOOH [313,314];

b) neutral solvents — solutions of heptamethylbenzenonium heptachlorodialuminate in hexafluorobenzene [238,316], in nitrobenzene [313,314], CH_2Cl_2 [314], a solution of tetraphenylborate in nitrobenzene [313].

For the solutions of the respective salts of heptamethylbenzenenonium in HSO_3Cl, 9.4 M H_2SO_4 and CF_3COOH the kinetics of the degenerate rearrangement were studied by dynamic NMR. In all the other cases comparisons seem to have been made of the spectra recorded at some definite temperatures. The rate of rearrangement proved to be practically insensitive to the medium changes.

The same conclusion can be drawn from the comparison of the rates of degenerate rearrangement of the 9,9,10-trimethylphenanthrenonium ion by 1,2-methyl shifts in the following acid media:
$HCl—Al_2Cl_6—CH_2Cl_2$ [44], $SbF_5—SO_2ClF$ [218], HSO_3F [220,313] and $HSO_3F—SO_2$-ClF [42,313] as well as of the 1,1,2-trimethylacenaphthylenonium ion in CF_3SO_3H and HSO_3F [49].

The rates of 1,2-aryl shifts in degenerate rearrangements are also practically insensitive to acid medium changes as established for the 1-phenyl-1,2,3,4,5,6-hexamethylbenzenonium (solutions in CF_3SO_3H, $CF_3SO_3H—SO_2ClF$, CF_3COOH) [35], the 9-phenyl-9,10-dimethylphenanthrenonium (solutions in $HCl—Al_2Cl_6—CH_2Cl_2$ [45], $HSO_3F—SO_2ClF$ [46][24]) and the 9,9,10-tri-p-X-phenylphenanthrenonium ions ($X=H$, CH_3, OCH_3, CF_3) [198] (e.g., for $X = CH_3$ solutions in $HCl—AlCl_3—CH_2Cl_2$ or in $CF_3SO_3H—SO_2ClF$). Since it is not excepted that 1,2-aryl shifts involve the formation of unstable intermediates of the phenonium type, i.e. the elementary stage of rearrangement is a nondegenerate process, one can conclude that the data on the practical insensitivity of 1,2-aryl shift rates to acid medium changes can be related precisely to such processes.

According to the available data [317], the medium change insensibility of the rate is also characteristic of the degenerate 1,2-hydrogen shifts. For instance, the rearrangement rate of the hexamethylbenzenonium ion and some other benzenonium ions is the same for the acid systems $HF—BF_3$ and $HF—SbF_5$ [30]. The rearrangement rate of the hexamethylbenzenonium ion also remains practically unchanged in passing to the acid system $HSO_3F—SbF_5$ [317] (or to $HSO_3F—SbF_5—SO_2ClF$ [29]) and on variation of the component ratio of this system from 17:1 to 1:1 [313].

A similar conclusion follows from the data on the rates of degenerate 1,2-hydrogen shifts in the following ions:
9-H-3,6-9,10-tetramethylphenanthrenonium (solutions in $HSO_3F—SbF_5—SO_2ClF—$ $—SO_2F_2$ and in $HSO_3F—SO_2ClF—SO_2F_2$) [29],
2-H-1,2,3,4-tetramethylbiphenylenonium (solutions in $CF_3SO_3H—SO_2ClF$ and in HSO_3F) [41],
1-H-1,2-dimethylacenaphthylenonium (solutions in $HSO_3F—SbF_5—SO_2ClF$ and in $HSO_3F—SO_2ClF$) [49,50],
2,3-dimethyl-2-norbornyl (*exo*-3,2-hydride shift, solutions in $HSO_3F—SO_2ClF—$ $—SO_2F_2$, $HSO_3F—SO_2F_2$ or in $SbF_5—SO_2ClF—SO_2F_2$ [54]).

The data available for some carbocations, e.g., for 1-H-1,2-dimethyl-3,4,5,6-tetrahalogenbenzenonium (halogen—fluorine or chlorine) [36], 1-H-1,2-3,4-tetramethylnaphthalenonium [39] as to the insensitivity of the degenerate rearrangement

24 The same conclusion seems to be valid for solutions in HSO_3F and conc. H_2SO_4 as well [304].

rate constants to the dilution of acid solutions also points to the insensitivity of 1,2-hydrogen shift rates to medium changes.

The same is typical of other migrants as well. Thus, the rate of the 1,2-chlorine shift in the degenerate rearrangement of the 1-chloro-1,2,3,4,5,6-hexamethylbenzenonium ion has been found in Ref. [31] to change very little even in the case of great changes in the solvent polarity: passage from a weakly polar solvent — CH_2Cl_2 (anion $AlCl_4^-$) — to a strongly polar one — HSO_3F — results in a very slight retardation of the 1,2-shift. Although the effect does correspond to the expected one (in the qualitative sense, cf. Ref. [106], p. 222) it is very small, if any ($\Delta \lg k^{25\,°C} = 0.1$). In Ref. [17] the rate of 1,2-shifts of the $AlCl_3^-$ fragment in ion (*60*) is found to be the same for the solutions in CH_2Cl_2 and $CHCl_2CHCl_2$.

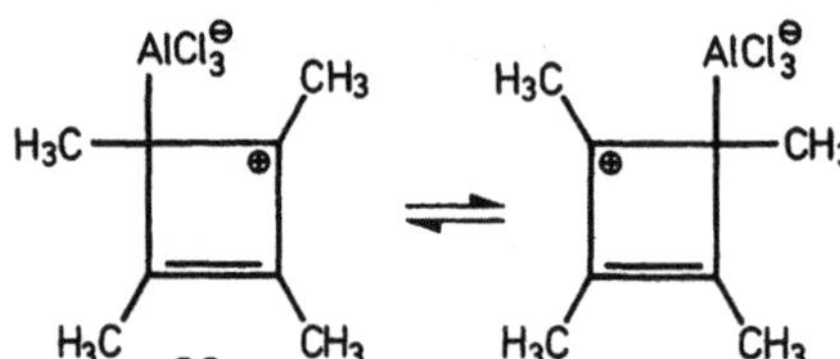

60

The reason for the observed insensitivity of degenerate rearrangement rates to medium changes seems to consist in that the 1,2-shift is a chemical reaction in the course of which not only the total charge of the reacting molecule remains constant, but also its redistribution is relatively small.

The rates of degenerate rearrangements proceeding through the 1,2-shift of migrants capable of specific interaction with the medium are, just as expected, sensitive to the latter's variation. For instance, the authors of Ref. [103] have found that an addition of small amounts of antimony pentafluoride to a solution of 1-nitro-1,2,3,4,5,6-hexamethylbenzenonium ion in HSO_3F—SO_2ClF does not practically affect the rate of the 1,2-nitrogen shifts, but an increase in the concentration of SbF_5 results in a retardation of rearrangement (cf. Ref. [318]). A similar retardation is also characteristic for such groups as methoxyl [146] and alkoxycarbonyl [99].

In considering the effect of the medium on the rate of degenerate rearrangements one should bear in mind that the free-energy barrier of these processes depends on the two terms: the intrinsic barrier and the term reflecting the influence of the thermodynamical factor, i.e. the difference in the relative stability of the initial and the final states. The medium independence of the free-energy barriers of degenerate rearrangements indicates the medium independence of the intrinsic barriers of nondegenerate rearrangements. As for the second term, since in the case of nondegenerate rearrangements, contrary to degenerate ones, the energetic levels of the initial, and the final, state are different, then in general one can expect some effect of the medium on the relative position of these levels and, hence, on the rate of rearrangement. The relative stability of carbocations, however, is known to be usually weakly sensitive to medium changes (see Ref. [35,169,242,319,320]).

Accordingly the rates of nondegenerate carbocation rearrangements, as well as of multi-step degenerate ones representing a sequence of nondegenerate steps are weakly sensitive to medium changes. The data illustrating this are listed in Tables 9 and 10. In addition to these data one can cite those of Ref. [38] showing the

Table 9. Data on Rearrangement Rates of Aliphatic Carbocations

	Rearranging carbocation	Rearranged carbocation	Solvent	t °C	k, s^{-1}	$\Delta G^{\neq}$, kcal/mole	Ref.
1	$CH_3CH_2-\overset{+}{C}(CH_3)-CH_3$	Degenerate rearrangement	$HF-SbF_5$[a]	0	20	14.3 ± 0.1	64,321,322)
			SbF_5-SO_2ClF[b]	0.2	8.5	14.8 ± 0.1	64,222)
2	$CH_3CH_2CH_2-\overset{+}{C}(CH_3)-CH_3$	$CH_3CH_2-\overset{+}{C}(CH_3)-CH_2CH_3$	$HF-SbF_5$[c]	0	~20	~14.3	64,321)
			$HSO_3F-SbF_5-SO_2ClF$	−77.5	$7.8 \cdot 10^{-4}$	14.05 ± 0.1	326)

a) The rearrangement rate is independent of the solution acidity varied by changing the amount of SbF_5 [321].
b) Rates and activation parameters are the same in $HSO_3F-SbF_5-SO_2ClF$ [64,222].
c) Visual comparison of PMR spectra shows the interconversion rate of these ions to remain practically unchanged when changing the component ratio in the indicated acid system and when passing to HSO_3F-SbF_5, $HF-SbF_5-SO_2ClF$, SbF_5-SO_2ClF, SbF_5 [323], cf. Ref. [324,325].

Table 10. Data on Rearrangement Rates of Cyclic Allylic Ions[a]

	Rearranging carbocation	Rearranged carbocation	Solvent		t °C	k, s^{-1}	Note	Ref.
1	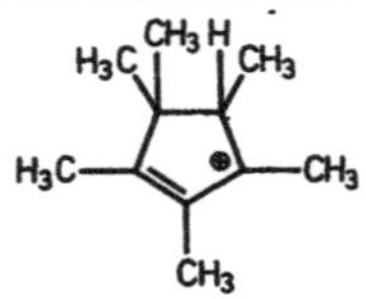	Degenerate rearrangement	HSO_3F—SbF_5	(5:1)	150	5.5		[215]
			HSO_3F—SbF_5	(1:1)	150'	6.2		[215]
2	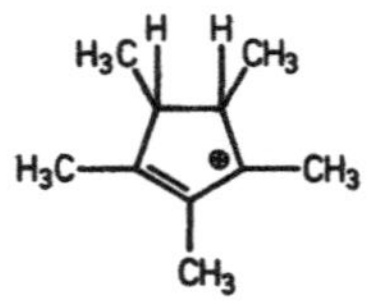	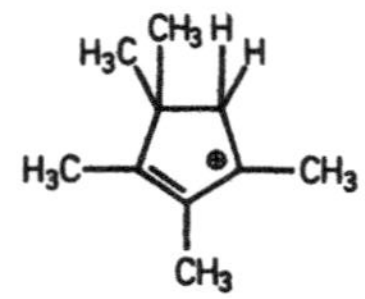	98% H_2SO_4		106	$2.2 \cdot 10^{-4}$	b)	[215]
			HSO_3F—SbF_5	(5:1)	106	$7.9 \cdot 10^{-5}$		[215]
3	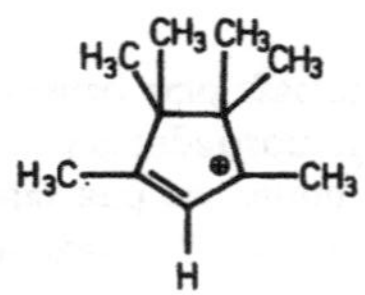	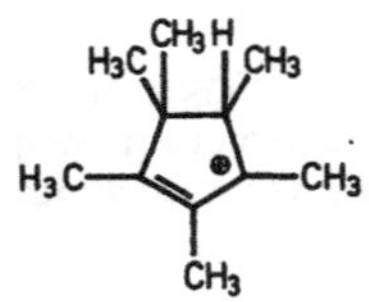	HSO_3F		34	$2.2 \cdot 10^{-3}$	c)	[215]
			HSO_3F—SbF_5	(5:1)	34	$2.15 \cdot 10^{-3}$		[215]
			HSO_3F—SbF_5	(1:1)	34	$1.75 \cdot 10^{-3}$		[215]
			HF—SbF_5	(9:1)	34	$1.9 \cdot 10^{-3}$		[215]
			HSO_3F		40	$4.6 \cdot 10^{-3}$		[215]
			HSO_3F—SbF_5	(1:1)	40	$4.05 \cdot 10^{-3}$		[215]
4	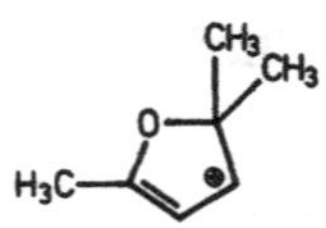	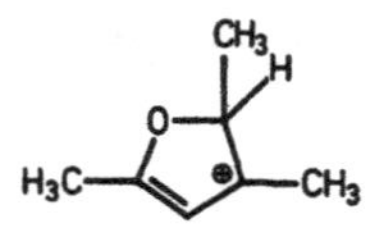	98% H_2SO_4		100	$6(\pm 2) \cdot 10^{-4}$		[328]
			20% oleum		100	$4.3 \cdot 10^{-4}$		[328]
			HSO_3Cl		100	$5.0(\pm 0.5) \cdot 10^{-4}$		[328]
			HSO_3F or HSO_3F—SbF_5	(5:1)	100	$(2 \div 10) \cdot 10^{-4}$		[328]

a) The mechanism of rearrangements consists in consecutive 1,2-H- and 1,2-CH_3-shifts.
b) A mixture of *cis*- and *trans*-isomers.
c) Insensitivity of rearrangement rates to medium changes (from 96% to 70% H_2SO_4) has also been noted by Deno et al. [327].

insensitivity of the rate of the 1,2-methyl shift in the 1,1,2,3,4-pentamethylnaphthalenonium ion (*67*) to acid medium changes (solutions in HSO_3F, $HSO_3F—SO_2ClF$, CF_3SO_3H, conc. H_2SO_4, CF_3COOH).

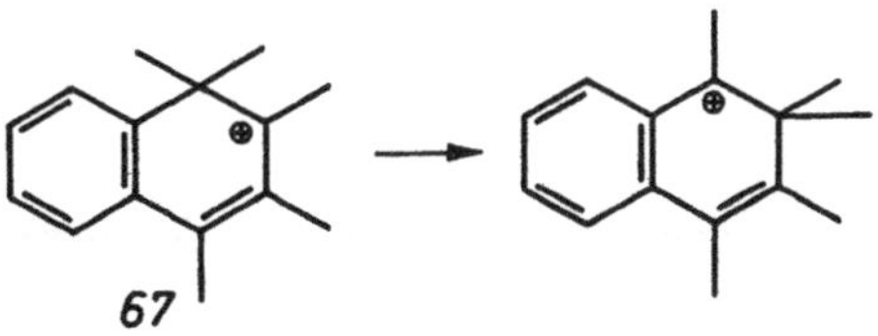
67

Even in the case of rearrangement of the hydroxycarbonium ions for which one can expect complications due to the effects of specific solvation the rearrangement rates do not, as a rule, change much with changing medium. As an example one can cite the rearrangement of the di-*tert*-butylhydroxycarbonium ion (*68*) proceeded by 1,2-alkyl shifts [70,329]:

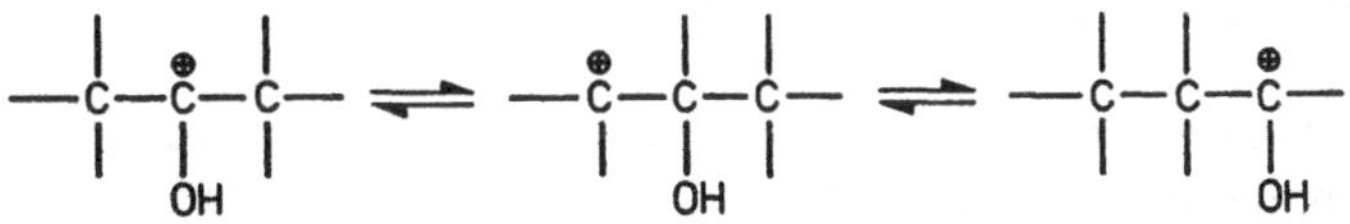
68

The kinetics of this rearrangement were investigated for 1.4 M solutions of hexamethylacetone in the following acid media: $HF—SbF_5$ (9:1), $HSO_3F—SbF_5$ (5:1), HSO_3F, H_2SO_4 and HF [70]. The rate constant values obtained do not differ much from one another: $6 \cdot 10^{-4}$, $1.0 \cdot 10^{-3}$, $6.3 \cdot 10^{-4}$, $3.5 \cdot 10^{+4}$ and $2.3 \cdot 10^{-4}$ s^{-1}, respectively (35 °C).

A similar result has been obtained for the rearrangement of the 1,1,3,4,5,6-hexamethyl-2-hydroxybenzenonium and 1-phenyl-1,3,4,5,6-pentamethyl-2-hydroxybenzenonium ions to the corresponding 4-hydroxy isomers proceeding by means of consecutive 1,2-shifts of the methyl and the phenyl group, respectively [67], see also Ref. [315], cf., however, the kinetic data for the dienone-phenol rearrangement of the 4,4-dimethylcyclohexadienone [15,208,311].

Of great interest is the problem of the effect on the rate of 1,2-shifts in the long-lived carbocations of the phase state of the medium. The evidence on this problem is at present very scarce. According to the preliminary data [65] the rearrangement rate of ion (*69*) in passing from a liquid solution to a frozen one in CF_3SO_3H does not practically change (cf. Ref. [330]) while the degenerate rearrangement rate of ion (*13*) (the anion $AlCl_4^-$) in passing from a solution in CH_2Cl_2 to the crystalline solid

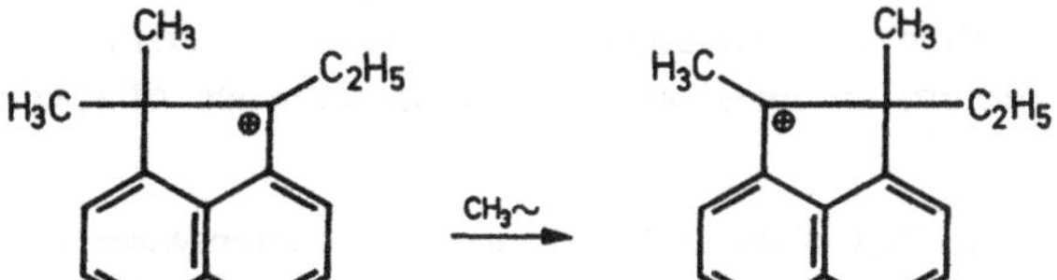

69

state sharply decreases (the constant rate ratio is about 10^4 at 22 °C [331]).

13

A similar effect seems to be also characteristic of the nondegenerate rearrangement of ion (*13*) to the 3-phenyl isomer [331]:

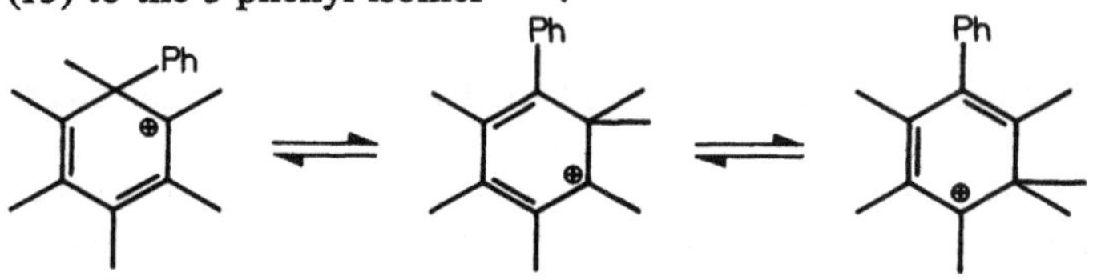

13

It is obvious that the effects of retardation are due to the action of the crystalline forces [331].

Worthy of note is also the comparison of the carbocation rearrangement rates in a solution and in a gas phase. The relative heats of carbocation formation in the gas phase are known [242] (cf. Ref. [332–336]) to be remarkably close to those under the long-life conditions (solutions in SO_2ClF). If the entropy factor is assumed to be insignificant with the degenerate rearrangement rates remaining unchanged in passing from the solution to the gas phase then the rates of nondegenerate rearrangements are expected to be also close to each other under the long-life conditions and in the gas phase.

According to the authors of Ref. [337], in passing from the gas phase to the solution the rates of degenerate 1,2-hydride shifts in ethyl and cyclohexadienyl ions should decrease due to more effective solvation of the carbocation in the ground state, see also Ref. [338], cf., however, Ref. [339]. This conclusion agrees with the results of quantum chemical calculations. The experimental data available, however, are either insufficient for verifying this conclusion or contradict it. For example, for the 1,2-hydride shifts in the ethyl cation only the upper limits of the activation barrier are known: $E_a \leqq 1.9$ kcal/mole for the solution in SbF_5—SO_2 [340] (cf. Ref. [63]) and $E_a \leqq 5$ kcal/mole for the gas phase [341], cf. Ref. [342,343] and for the 1,2-hydride shift in the cyclohexadienyl cation, according to the data of Ref. [344], the activation barrier in the gas phase is no less than 35 kcal/mole while for the solution in HF—SbF_5—SO_2ClF—SO_2F_2 it is only 10 kcal/mole [345]. There are not enough data for the other migrants to draw any definite conclusions (see [346–349]).

Thus, the first of the questions raised above — whether the changes in the medium affect the rates of long-lived carbonium ions rearrangement — is answered by the available data in the negative: the rearrangements of such ions are insensitive or rather poorly sensitive to medium changes, at least in the absence of specific solvation effects, cf. the data of [350][25].

25 The data about a striking effect of the acid medium on the rearrangement rates of long-lived carbonium ions, e.g., *tert*-amyl cation [324,325,351] seem to be hardly reliable as follows from Prof. Olah's private communication (cf. Ref. [323]).

4.2 Transfer of Results from Long-Life to Short-Life Conditions

The second of the raised questions — whether it is possible to "transfer" the regularities of the medium effect on long-lived carbonium ion rearrangements to the "short-life" conditions cannot, it appears, so far be given an unequivocal answer. Some of the data available can be regarded as testifying to the possibility of such a transfer. For example, in Ref. [352] it was pointed out that the energetic differences between carbocations in superacids and solvolytic media are practically equal, cf. Ref. [332,353]. This allows one to assume the free-energy term in the Marcus Equation (see p. 309) to be independent of the medium so that the rearrangement rates of carbocations under the conditions of long life and those of short life should be equal if the intrinsic barrier of rearrangement does not depend on the medium either. Although no reliable data on the rates of degenerate 1,2-shifts under short-life conditions seem to be available this assumption can be considered true.

The available data on the change in the rate ratio of the 1,2-shifts of the *exo*- and *endo*-methyl groups in ion (*50b*) as the conditions of long life change for those of short life [57] (see also the data on the rates of the *exo*-methyl and the WM-6.2-WM shifts in the camphenhydro cation [52], cf. Ref. [354,355]) are interpreted as a consequence of the changes in the fine structure of the carbocation with changing medium; so they can be referred to the effect of specific solvation.

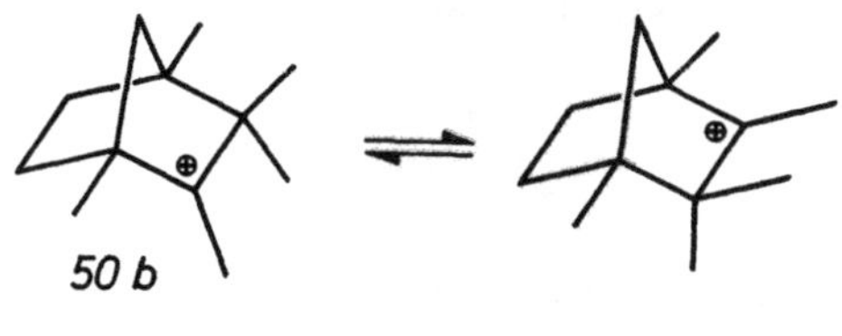

50 b

In the case of the rearrangements accompanying the solvolytic and related reactions which involve considerable changes of the reacting molecule on the way to the cationoid transition state the rates of the processes are sensitive to the medium effects, but even in these cases the MT appears to be but weakly sensitive to the medium changes (Tables 1–4).

A special place is held by carbocations generated in a solution in a "free", "non-solvated" form. The rearrangement rates of such ions are extremely high. For example, the rate of rearrangement of the "free" ions of the composition $C_4H_9^+$ into a *tert*-butyl cation is by 7–10 orders of magnitude higher than that of rearrangement under the long-life conditions [356], cf. Ref. [357]. One can assume the reason for so sharp an increase in the rate to be due to the transformation of the carbocation from the non-equilibrium solvated ("non-solvated") ground state into the equilibrium solvated (or a state close to that) final state ("thermodynamic acceleration effect"). As to the intrinsic barrier, its value seems to change little in passing from the conditions of "long life" to those of "free" carbocations: for instance, the degenerate rearrangement rates of the 2-butyl cation[26] under the long-life conditions [59] and in the "free" state [356] seem to be close to each other.

---

26 See the footnote, p. 311.

5 Some Aspects of the Mechanism of Carbocation Rearrangements

Since the problem of the carbocation rearrangement mechanism has been paid much attention to in the literature including the reviews, this section will briefly deal only with some aspects of the matter.

5.1 Retention or Inversion of Configuration

It has been already pointed out that the 1,2-shifts of migrants to the carbonium centre can be regarded as reactions of intramolecular electrophilic substitution. In the case of alkyl migrants we deal with these reactions at the saturated carbon atom. It is quite natural to raise the problem as to the stereochemistry of these reactions: is the alkyl migrant (R^m) attacked "along" the C^o—C^m bond, i.e. with the retention of configuration or from the backside so that there occurs an inversion? The retention of the migrating group configuration was established for 1,2-shifts to electron deficient nitrogen and oxygen [358], as well as to the neutral C atom in carbenes [359], but no data on the stereochemistry of the migrants in 1,2-shifts to the carbonium centre have been available before the publication of Ref. [360]. The authors of this work have investigated the pinacol rearrangement of (4S)-2,3,4-trimethyl-2,3-hexanediol (*70*) possessing an optically active *sec*-butyl group and found the configuration of the latter completely retained.

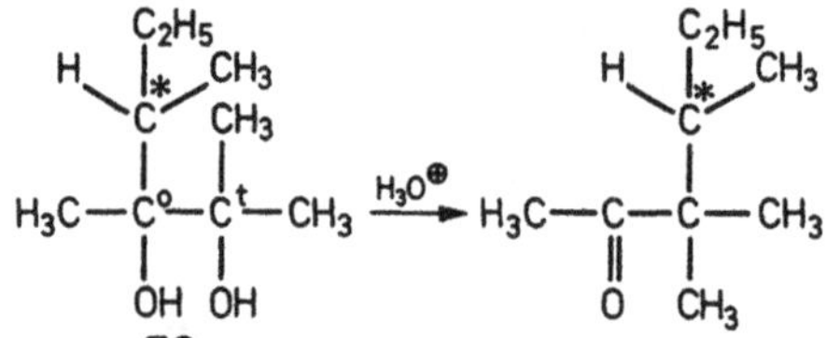

70

Since under the conditions of rearrangement (boiling of diol with 40% sulphuric acid for 15 minutes) the *sec*-butyl group seems to make several migrations from C^o to C^t and back [359], the authors have concluded the migrant configuration to be retained in the retropinacol rearrangement as well. This conclusion appears to hold for all the rearrangements whose key stage is the 1,2-shift of the substituent with the $C^m_{sp^3}$ atom to the carbonium centre[27]; this is in accordance with the theory of sigmatropic reactions developed by Woodward and Hoffmann and with quantum chemical calculations [5]. It was confirmed by the research on the stereochemistry of the tetragonal migrants CH_2R in the carbocationic 1,2-shift under the long-life conditions: the retention of the configuration at the tetragonal carbon atom was recorded in the case of both the σ-donor and the π-donor fragment R (migrants C_2H_5 [83] and p-$ClC_6H_4CH_2$ [91], respectively).

Note also, that the 1,2-ethyl shift does not involve the states that could result in the "randomization" of the hydrogen atoms of the methyl and methylene fragments. This follows from the retention of multiplicity of the signals of the ethyl

27 A similar conclusion is made in Ref. [361,362], cf. Ref. [363,364].

groups migrating from the achiral centre in the PMR spectra (see, e.g., the corresponding data for the heptaethylbenzenonium ion [365,366]).

In Ref. [83,91] the stereochemistry of the tetragonal migrant was studied implying suprafacial 1,2-shift of the latter; note, however, that the result on whose basis the configuration was concluded to be retained, i.e.: the averaging of the diastereotopic methylene fragment hydrogens is also in formal agreement with the case of antarafacial migration with inversion. The study of long-lived carbocation rearrangements made it possible to obtain experimental evidence for the suprafacial route of the 1,2-shift of the tetragonal alkyl migrant, in accordance with the theory of pericyclic reactions [148,367,368]: in studying the degenerate rearrangement of the 4,5,9,9,10-pentamethylphenanthrenonium ion the 1,2-shift of the methyl group was found [300] to proceed just by the suprafacial route.

5.2 Rearrangement by Simultaneous 1,2-Shifts of Two Groups

The idea that in the course of carbonium ion rearrangements there occurs a direct 1,2-shift of only one of the substituents located in the β-position relative to the carbocation centre is generally accepted. Of great interest in this connection are the assumptions made by some authors (see Ref. [24,81,99,205,369,370], cf. Ref. [371]) that rearrangement can proceed by means of two simultaneous 1,2-shifts of two groups. Thus, the authors of [205] have suggested the following concerted scheme of the degenerate rearrangement of the protonated 3,3-dimethyl-2-butanone (*71*) with the transition state (*A*), although they believe it to be less probable than the alter-

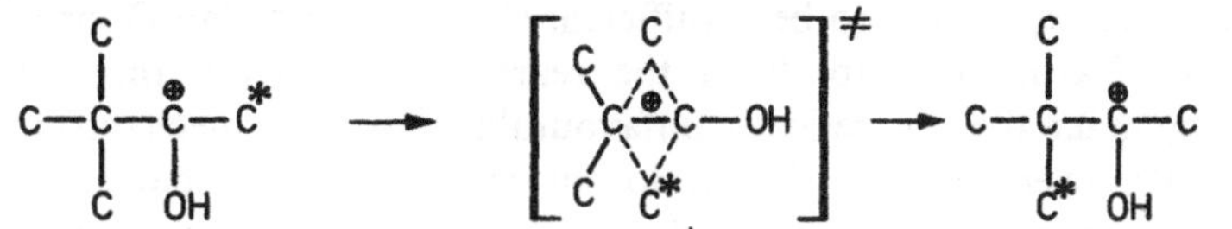

71 *A*

native two-step scheme:

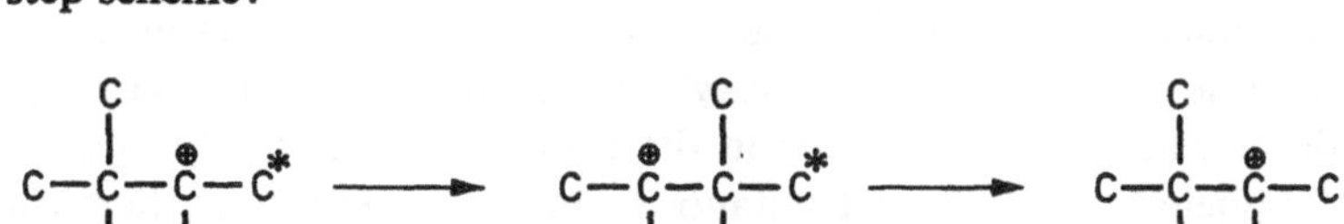

71 *72*

A special term ("dyotropic reaction" [372], see also Ref. [373,374]) was proposed to denote the noncatalyzed processes in the course of which two σ-bonds migrate simultaneously and intramolecularly, and a general theory of such reactions was developed [375]. In accordance with the definitions given in Ref. [372], the rearrangement of ion (*71*) by the concerted route should be considered as a dyotropic reaction of type I.

This scheme is not so easy to disprove as it may seem at first sight. The point is that the life time of the intermediate β-hydroxycarbonium ion (*72*) in the alternative two-step scheme is far shorter than that of the parent α-hydroxycarbonium ion, so such methods as the kinetic isotopic one seem to be inapplicable here. Some data obtained from the study of long-lived carbonium ion rearrangements can be regarded as testifying against the scheme of dyotropic shifts. For instance, the "confusion" of the CH_3 groups of the 9-ethyl-9,10-dimethylphenanthrenonium ion (*73*), is fulfilled by means of a 1,2-ethyl shift rather than according to the dyotropic scheme, since it is accompanied by a simultaneous averaging of the protons of the aromatic ring A with the respective protons of the ring B [43,244], cf. Ref. [198,376].

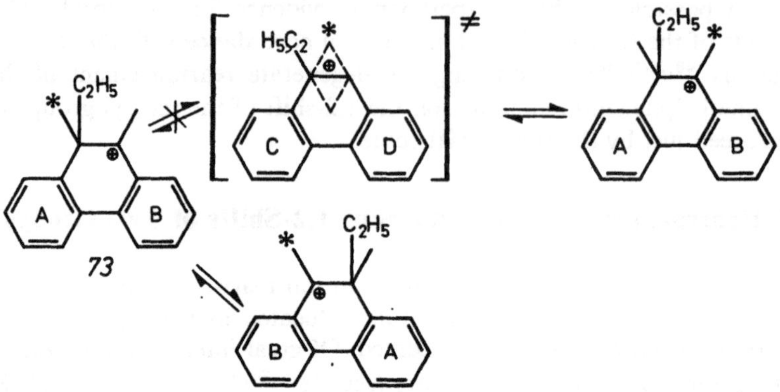

It is evident that if the concerted mechanism were realized the states of the protons in the aromatic rings would undergo no changes. It should be emphasized that these data by no means preclude the feasibility of a concerted mechanism in the rearrangement of the noncyclic ion (*71*). The differences in the stability of ions (*71*) and (*72*), however, appear to be insufficient (Brouwer and Van Doorn [70] estimated $\Delta\Delta G^{25°*} \approx 12$ kcal/mole) to "turn the rearrangement from the classical route" and thereby make the energetically unfavourable dyotropic mechanism [372] possible. Note that Brouwer and Van Doorn do not consider this mechanism at all holding the two-step scheme to be proved. This scheme has been established in Ref. [128] for the rearrangement of bis-*tert*-alkylketones in sulphuric acid.

In connection with the subject under discussion note should be made of the work by the French investigators [377]. They have found that the rearrangement of the optically active S(+)-2-phenyl-2-cyclohexyl-3-pentanone under the action of aluminium chloride or sulphuric acid results in the formation of partially racemized R(—)-2-methyl-2-phenyl-1-cyclohexyl-1-butanone. On this ground the authors concluded that the dyotropic mechanism which would be accompanied by inversion rather than by racemization is at least not the only one.

Rather a curious kind of dyotropy is virtually suggested in [378], it implies the movement of two migrants towards the same carbocation centre:

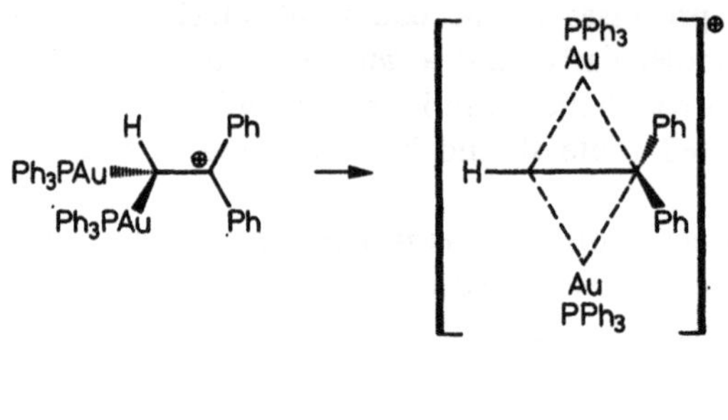

The authors of Ref. [379] claim to have discovered the first case of dyotropic Wagner-Meerwein rearrangement occurring when the ring of 2-oxetanones expands to 2-oxolanones under the action of magnesium bromide in ether.

A number of publications suggest an important role of ion radicals and/or radicals in the general scheme of cationoid rearrangements (see Ref. [380], cf. Ref. [381,382]). A few years ago Perrin proposed an ion-radical mechanism for the degenerate rearrangement of the 1-nitro-1,2,3,4,5,6-hexamethylbezenonium ion [383].

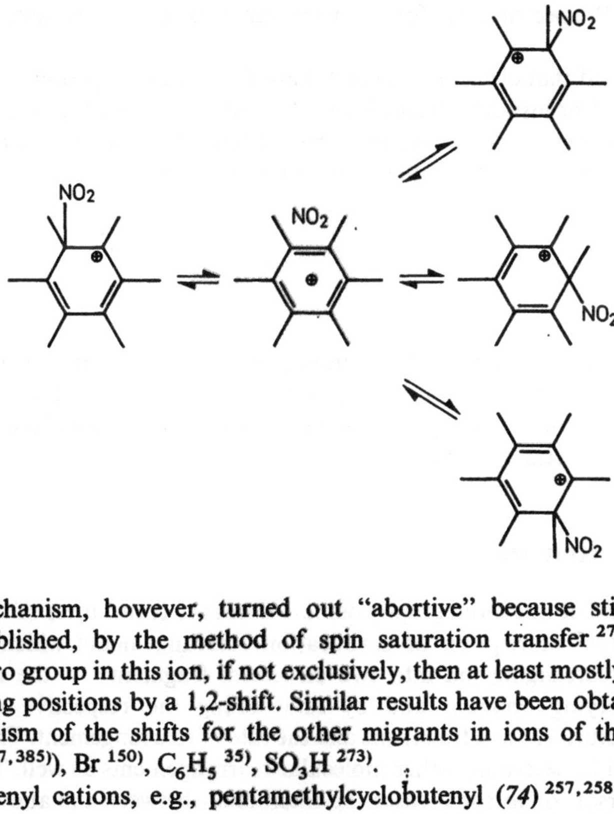

This mechanism, however, turned out "abortive" because still earlier it was firmly established, by the method of spin saturation transfer [273] (cf. Ref. [384]), that the nitro group in this ion, if not exclusively, then at least mostly, migrates to the neighbouring positions by a 1,2-shift. Similar results have been obtained in studying the mechanism of the shifts for the other migrants in ions of this type: CH_3 [32] (cf. [28,239,317,385]), Br [150], C_6H_5 [35], SO_3H [273].

Cyclobutenyl cations, e.g., pentamethylcyclobutenyl (*74*) [257,258,260], are known to display no tendency to rearrangement by a 1,2-methyl shift (forbidden sigmatropic shift of order [1.4] according to the Woodward and Hoffmann classification [148]).

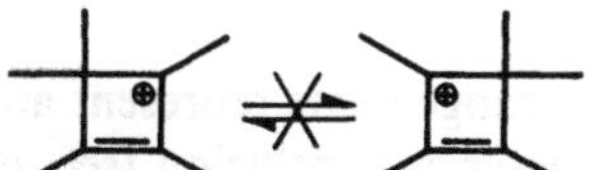

74

Also "static" are other carbocations of this type (for more details see Ref. [21]). However such an unusual migrant as the $AlCl_3^-$ group is capable of performing 1,2-shifts in ions of this type [17]. The possible reason for these differences is that a pseudopericyclic mechanism is realizable (cf. Ref. [293,294]) in the case of $R^m = AlCl_3^-$ [17].

5.3 Rearrangement by One-Step Process or via Intermediates

Of considerable interest is the problem of the intimate mechanism of 1,2-shifts of various migrants in carbocations. Are these shifts one-step processes or do intermediates form in their course? The experimental data available are insufficient to obtain an unequivocal answer to this question. One can only point out that even if intermediates do form, then they, at least in the case of the carbocations with a π-delocalized MCR, are unstable (cf. Ref. [386]): they do not show themselves in any way in the NMR spectra even for such migrants as a chlorine atom [31] or aryl groups [166,198].

It is not excluded that cyclic intermediates are formed in the process of 1,2-hydrogen shifts as well. For instance, according to the authors of Ref. [63], from the analysis of the ^{13}C NMR spectra it follows that the 1,2-hydride shift in *sec*-butyl cation[28] occurs with the formation of an intermediate bridged ion:

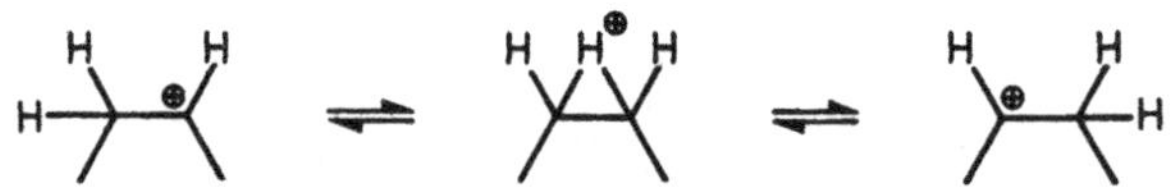

In the case of the dimethylisopropyl carbonium ion the relative stability of the bridged intermediate (if it is formed at all) is far lower. The authors hold that in the case of 1,2-methyl shifts from a tertiary to tertiary carbon bridged nonclassical ions can form as high lying metastable states.

6 Concluding Remarks

The subject matter of this monography demonstrates the fruitfulness of the approach to the quantitative description of carbocation rearrangements which consists in "dividing" the rearranging ion into two structural fragments — the migrant and the migrant carrying remnant and by establishing the respective relationships between the structure of these fragments and the rate of rearrangement. This approach may prove useful in describing other molecular rearrangements as well. True, it is by no means universal for in some cases the interaction between the above fragments even in the ground state may happen to be so strong that the very notions "migrant" and "MCR" will be "blurred". The necessity for the quantitative description of the rearrangement of such "strongly bonded" systems stimulates the search for other approaches.

Carbocation rearrangements represent an active, developing area, more and more new data appear deserving inclusion (see, e.g. Ref. [387–389]). The author's desire to make the review the most up-to-date comes in conflict with the necessity to finish somewhere after all. That is why he is compelled to confine himself here to mere enumeration of some of the latest works on the subject. The author regretfully assumes that there are certain works worthy of attention that have escaped his notice.

28 See, however, the footnote at p. 311.

7 Acknowledgements

The author is indebted to the many colleagues and associates with whom he had discussed problems dealt with in this review. The author would like to mention the name of Prof. Koptyug the scientific discussions with whom have been particularly valuable and stimulating.

8 References

1. B. V. Stolyarov, in: Contemporary Problems of Organic Chemistry (I. A. D'yakonov, B. V. Ioffe, Eds.) Leningrad State University Press, 1969, p. 5 (in Russian)
2. W. Kirmse: Top. Curr. Chemistry, Berlin, Heidelberg, New York, Springer, 1979, vol. 80, p. 125
3. H. Hogeveen, E. M. G. A. van Kruchten, in: Top. Curr. Chemistry, Berlin, Heidelberg, New York, Springer, 1979, vol. 80, p. 89
4. V. G. Shubin, V. A. Koptyug: Izv. Sib. Otd. Akad. Nauk SSSR. Ser. Khim. Nauk, No. 4, issue 2, p. 131 (1976)
5. T. S. Sorensen, A. Rauk, in: Pericyclic Reactions (A. P. Marchand and R. E. Lehr, Eds.) New York, Academic Press, 1977, vol. 2, p. 1
6. D. Bethell, in: Reactive Intermediates: A Serial Publication, (M. Jones, R. A. Moss, Eds.) New York, Wiley, 1978, vol. 1, p. 117
7. V. A. Koptyug, V. G. Shubin: Zh. Org. Khim. *16*, 1977 (1980)
8. T. I. Temnikova: Izv. Sib. Otd. Akad. Nauk SSSR. Ser. Khim. Nauk, No 7, issue 3, p. 5 (1980)
9. V. G. Shubin: Izv. Sib. Otd. Akad. Nauk SSSR. Ser. Khim. Nauk, No 7, issue 3, p. 18 (1980)
10. D. Bethell, V. Gold: Carbonium Ions, An Introduction, London, Academic Press, 1967
11. Carbonium Ions (G. A. Olah, P. v. R. Schleyer, Eds.) New York, Wiley-Interscience, 1968–1976, vol. 1–5
12. Mechanisms of Molecular Migrations, (B. S. Thyagarajan, Ed.) 1978–1971, vol. 1–4
13. Contemporary Problems of the Chemistry of Carbonium Ions, (V. A. Koptyug, Ed.) Novosibirsk, Nauka, 1975 (in Russian).
14. D. J. Dagli, R. A. Gorski, J. Wemple: J. Org. Chem. *40*, 1741 (1975)
15. J. N. Marx, J. C. Argyle, L. P. Norman: J. Am. Chem. Soc. *96*, 2121 (1974)
16. P. Brownbridge, P. K. G. Hodgson, R. Shepherd, S. Warren: J.C.S., Perkin I, *1976*, 2024
17. P. B. J. Driessen, H. Hogeveen: J. Am. Chem. Soc. *100*, 1193 (1978)
18. M. Liler: Reactions Mechanisms in Sulphuric Acid and Other Strong Acid Solutions, New York, Academic Press, 1971, p. 243
19. M. Stiles, R. P. Mayer: J. Am. Chem. Soc. *81*, 1497 (1959)
20. J. W. Pilkington, A. J. Waring: J.C.S., Perkin II, *1976*, 1349
21. V. G. Shubin, in: Contemporary Problems of the Chemistry of Carbonium Ions (V. A. Koptyug, Ed.) Novosibirsk, Nauka, 1975, p. 179 (in Russian)
22. A. A. Schegolev, M. I. Kanischev: Uspekhi Khim. *50*, 1046 (1981)
23. P. J. Stang, Z. Rappoport, M. Hanack, L. R. Subramanian: Vinyl Cations, New York, Academic Press, 1979
24. O. A. Reutov, A. S. Gudkova: Rev. roum. chim. *20*, 1443 (1975)
25. P. v. R. Schleyer, D. Lenior, P. Mison, G. Liang, G. K. S. Prakash, G. A. Olah: J. Am. Chem. Soc. *102*, 683 (1980)
26. D. M. Brouwer: Rec. Trav. Chim. *87*, 611 (1968)
27. G. A. Olah, Y. K. Mo: J. Am. Chem. Soc. *94*, 9241 (1972)
28. V. A. Koptyug, V. G. Shubin, A. I. Rezvukhin, D. V. Korchagina, V. P. Tret'yakov, E. S. Rudakov; Dokl. Akad. Nauk SSSR *171*, 1109 (1966)
29. G. I. Borodkin, M. M. Shakirov, V. G. Shubin, V. A. Koptyug: Zh. Org. Khim. *14*, 989 (197
30. D. M. Brouwer, C. MacLean, E. L. Mackor: Disc. Faraday Soc. *1965*, 121
31. V. P. Loktev, D. V. Korchagina, M. M. Shakirov, V. G. Shubin: Izv. Sib. Otd. Akad. Nauk SSSR. Ser. Khim. Nauk, No 12, issue 5, p. 146 (1978)

32. B. G. Derendyaev, V. I. Mamatyuk, V. A. Koptyug: Izv. Akad. Nauk SSSR, Ser. Khim. *1971*, 972
33. A. A. Tabatskaya: Dissertation, Novosibirsk (1972)
34. S. Hünig, P. Schilling: Chem. Ber. *108*, 3355 (1975)
35. G. I. Borodkin, M. M. Shakirov, V. G. Shubin: Zh. Org. Khim. *13*, 2152 (1977)
36. V. P. Loktev, V. G. Shubin: Izv. Sib. Otd. Akad. Nauk SSSR. Ser. Khim. Nauk, No 2, issue 1, p. 108 (1981)
37. V. P. Loktev, D. V. Korchagina, V. G. Shubin: Izv. Sib. Otd. Akad. Nauk SSSR, Ser. Khim. Nauk, No 2, issue 1, p. 112 (1981)
38. S. V. Morozov, M. M. Shakirov, V. G. Shubin, V. A. Koptyug: Zh. Org. Khim. *15*, 770 (1979)
39. S. V. Morozov, M. M. Shakirov, V. G. Shubin: Zh. Org. Khim. *16*, 2103 (1980)
40. S. V. Morozov, M. M. Shakirov, V. G. Shubin, V. A. Koptyug, in: Carbocation Chemistry, Novosibirsk, 1979, p. 83
41. S. V. Morozov, M. M. Shakirov, V. G. Shubin: Zh. Org. Khim. *17*, 154 (1981)
42. G. I. Borodkin, M. M. Shakirov, V. G. Shubin, V. A. Koptyug: Zh. Org. Khim. *12*, 1297 (1976)
43. V. G. Shubin, D. V. Korchagina, B. G. Derendyaev, V. I. Matatyuk, V. A. Koptyug: Zh. Org. Khim. *6*, 2066 (1970)
44. V. G. Shubin, D. V. Korchagin, A. I. Rezvukhin, V. A. Koptyug: Dokl. Akad. Nauk SSSR *179*, 119 (1978)
45. V. G. Shubin, D. V. Korchagina, B. G. Derendyaev, G. I. Borodkin, V. A. Koptyug: Zh. Org. Khim. *9*, 1041 (1973)
46. G. I. Borodkin, M. M. Shakirov, V. G. Shubin, V. A. Koptyug: Zh. Org. Khim. *14*, 321 (1978)
47. D. V. Korchagina, B. G. Derendyaev, V. G.: Izv. Akad. Nauk, SSSR, Ser. Khim. *1971*, 441
48. D. V. Korchagina, B. G. Derendyaev, V. G. Shubin, V. A. Koptyug: Zh. Org. Khim. *12*, 384 (1976)
49. V. A. Bushmelev, M. M. Shakirov, V. A. Koptyug: Zh. Org. Khim. *13*, 2161 (1977)
50. G. A. Olah, G. Liang, P. Westerman: J. Am. Chem. Soc. *95*, 3698 (1973)
51. The data on the 1,2-phenyl shift have been taken from M. I. Rudnev's diploma thesis (Novosibirsk, 1980) and specified by V. A. Bushmelev
52. T. S. Sorensen: Acc. Chem. Res. *9*, 257 (1976)
53. M. Saunders, P. v. R. Schleyer, G. A. Olah: J. Am. Chem. Soc. *86*, 5680 (1964)
54. A. J. Jones, E. Huang, R. Haseltine, T. S. Sorensen: J. Am. Chem. Soc. *97*, 1133 (1975)
55. R. Haseltine, E. Huang, K. Ranganayakulu, T. S. Sorensen: Can. J. Chem. *53*, 1056 (1975)
56. L. Huang, K. Ranganayakulu, T. S. Sorensen: J. Am. Chem. Soc. *95*, 1936 (1973)
57. R. P. Haseltine, T. S. Sorensen: Can. J. Chem. *53*, 1067 (1975)
58. P. von R. Schleyer, J. Chandrasekhar: J. Org. Chem. *46*, 225 (1981)
59. M. Saunders, M. R. Kates: J. Am. Chem. Soc. *100*, 7082 (1978)
60. G. A. Olah: Halonium Ions, New York, Willy-Interscience, 1975
61. G. A. Olah, R. D. Porter: J. Am. Chem. Soc. *93*, 6877 (1971)
62. G. A. Olah, R. J. Spear, D. A. Forsyth: ibid. *99*, 2615 (1977)
63. G. A. Olah, D. J. Donovan: ibid. *99*, 5026 (1977)
64. D. M. Brouwer, H. Hogeveen, in: Progr. Phys. Org. Chem. (A. Streitwieser, Jr., R. W. Taft, Eds.) New York, Wiley, 1972, vol. 9, p. 179
65. G. I. Borodkin, Ye. B. Panova, V. G. Shubin, in: Abstr. Papers All-Union Conf. Low Temperature Chemistry, Moscow, Moscow State University Press, 1979, p. 68
66. A. Fisher, G. N. Henderson: Chem. Commun. *1979*, 279
67. R. N. Berezina, D. V. Korchagina, V. G. Shubin: Zh. Org. Khim. *16*, 371 (1980)
68. R. N. Berezina, S. N. Akimova, D. V. Korchagina, V. G. Shubin: Izv. Sib. Otd. Akad. Nauk SSSR. Ser. Khim. Nauk, No 7, issue 3, p. 144 (1981)
69. L. Radom, D. Poppinger, R. C. Haddon, in: Carbonium Ions, (G. A. Olah, P. von R. Schleyer, Eds.) New York, Wiley — Interscience, 1976, vol. 5, p. 2303
70. D. M. Brouwer, J. A. van Doorn: Rec. Trav. Chim. *90*, 1010 (1971)
71. M. Mazet: Thése, Limoges, 1969
72. Th. Yvernault, M. Mazet: Bull. Soc. Chim. France *1971*, 2652
73. R. L. Heidke, W. H. Saunders, Jr.: J. Am. Chem. Soc. *88*, 5816 (1966)
74. Cited from R. S. Bly, R. T. Swindell, J. Org. Chem. *30*, 10 (1965)

75. D. D. Roberts, R. S. Snyder, Jr.: J. Org. Chem. *45*, 4052 (1980)
76. I. L. Reich, A. Diaz, S. Winstein: J. Am. Chem. Soc. *91*, 5635 (1969)
77. P. C. Myhre, K. S. Brown: ibid. *91*, 5639 (1969)
78. A. Diaz, I. L. Reich, S. Winstein: ibid. *91*, 5637 (1969)
79. J. D. Palmer, A. J. Waring: J.C.S., Perkin II, *1979*, 1089
80. P. Bauer, J. E. Dubois: J. Am. Chem. Soc. *98*, 6999 (1976)
81. R. Magne, E. Torreilles, L. Giral: Bull. Soc. Chim. France *1973*, 2813
82. D. N. Miller: Diss. Abstr. *29B*, 112 (1968)
83. G. I. Borodkin, Ye. B. Panova, M. M. Shakirov, V. G. Shubin, in: Development of Organic Synthesis on the Basis of Studies on General Regularities and Mechanisms of Reactions in Organic Chemistry, Abstr. of Papers, Leningrad, Nauka, 1980, p. 41
84. H. Langhals, G. Range, E. Wistuba, C. Rüchardt: Chem. Ber. *114*, 3813 (1981)
85. V. P. Vitullo, E. A. Logue: J. Am. Chem. Soc. *98*, 5906 (1976)
86. M. Attinà, F. Cacace, P. Giacomello: ibid. *102*, 4768 (1980)
87. Z. Dillenberger, H. Schmid, H.-J. Hansen: Helv. Chim. Acta *61*, 1856 (1978)
88. V. G. Shubin, A. A. Tabatskaya, B. G. Derendyaev, V. A. Koptyug: Izv. Akad. Nauk SSSR, Ser. Khim. *1968*, 2417
89. G. I. Borodkin, Ye. B. Panova, M. M. Shakirov, V. G. Shubin, in: Carbocation Chemistry, Novosibirsk, 1979, p. 85
90. L. I. Sazonova, M. M. Shakirov, V. G. Shubin: Zh. Org. Khim. *13*, 2456 (1977)
91. G. I. Borodkin, Ye. B. Panova, M. M. Shakirov, V. G. Shubin: Chem. Commun. *1979*, 354
92. V. G. Shubin, V. P. Chzhu, A. I. Rezvukhin, A. A. Tabatskaya, V. A. Koptyug: Izv. Akad. Nauk SSSR, Ser. Khim. *1967*, 2365
93. W. J. Albery, M. M. Kreevoy, in: Adv. Phys. Org. Chem. (V. Gold, D. Bethell, Eds.) London, Academic Press, 1978, vol. 16, p. 87
94. J. R. Owen, W. H. Saunders, Jr.: J. Am. Chem. Soc. *88*, 5809 (1966)
95. R. A. Marcus: J. Phys. Chem. *72*, 891 (1968)
96. J. R. Murdoch: J. Am. Chem. Soc. *94*, 4410 (1972)
97. E. S. Lewis, S. Kukes, C. D. Slater: ibid. *102*, 1619 (1980)
98. B. Miller: ibid. *92*, 6252 (1970)
99. D. Berner, H. Dahn, P. Vogel: Helv. Chim. Acta *63*, 2538 (1980)
100. H. O. House, E. J. Grubbs, W. F. Gannon: J. Am. Chem. Soc. *82*, 4099 (1960)
101. T. Suehiro, S. Yamazaki: Bull. Chem. Soc. Jpn. *48*, 3655 (1975)
102. J. H. Zaidi, A. J. Waring: Chem. Commun. *1980*, 618
103. V. P. Loktev, D. V. Korchagina, V. G. Shubin: Izv. Sib. Otd. Akad. Nauk SSSR. Ser. Khim. Nauk, No 14, issue 6, p. 86 (1980)
104. V. P. Loktev, D. V. Korchagina, V. G. Shubin: Izv. Sib. Otd. Akad. Nauk SSSR. Ser. Khim. Nauk, No 4, issue 2, p. 118 (1977)
105. M. J. S. Dewar: The Molecular Orbital Theory of Organic Chemistry, New York, McGraw-Hill, 1969
106. M. J. S. Dewar, R. C. Dougherty: The PMO Theory of Organic Chemistry, New York, Plenum Press, 1975
107. M. J. S. Dewar, in: Molecular Rearrangements (P. de Mayo, Ed.) New York, Wiley-Interscience, 1963, vol. 1, p. 295
108. C. A. Coulson, A. Streitwieser, Jr.: Dictionary of π-Electron Calculations, Oxford, Pergamon, 1965, pp. 35, 41
109. H. E. Zimmerman: Acc. Chem. Res., *5*, 393 (1972)
110. L. A. Paquette, K. Ohkata, R. V. C. Carr: J. Am. Chem. Soc. *102*, 3303 (1980)
111. A. Collumeau: Bull. Soc. Chim. France *1968*, 5087
112. E. M. Arnett, G. Scorrano, in: Adv. Phys. Org. Chem., (V. Gold Ed.) London, Academic Press, 1976, vol. 13, p. 83
113. M. I. Kabachnik: Uspekhi Khim. *48*, 1523 (1979)
114. T. Ando, S.-G. Kim, K. Matsuda, H. Yamataka, Y. Yukawa, A. Fry, D. E. Lewis, L. B. Sims, J. C. Wilson: J. Am. Chem. Soc. *103*, 3505 (1981)
115. T. Ando, H. Yamataka, H. Morisaki, J. Yamakawi, J. Kuramochi, Y. Yukawa: J. Am. Chem. Soc. *103*, 430 (1981)
116. V. J. Shiner, Jr., J. J. Tai: ibid. *103*, 436 (1981)

117. D. Howells, S. Warren: J. C. S., Perkin II, *1974*, 994
118. H. S. Mosher: Tetrahedron *30*, 1733 (1974)
119. W. M. Schubert, W. L. Henson: J. Am. Chem. Soc. *93*, 6299 (1971)
120. D. J. Raber, J. M. Harris, P. v. R. Schleyer, in: Ions and Ion Pairs in Organic Reactions, (M. Szwarc, Ed.) New York, Wiley, 1974, vol. 2, p. 247
121. R. M. Fritz: Diss. Abstr. *23*, 3119 (1963)
122. Yu. G. Bundel, I. Yu. Levina, N. Yu. Smorodina, O. A. Reutov: Zh. Org. Khim. *6*, 409 (1970)
123. T. G. Traylor, W. Hanstein, H. J. Berwin, N. A. Clinton, R. S. Brown: J. Am. Chem. Soc. *93*, 5715 (1971)
124. J. Laureillard, A. Laurent, E. Laurent: Bull. Soc. Chim. France *1973*, 249
125. P. D. Bartlett, T. T. Tidwell: J. Am. Chem. Soc. *90*, 4421 (1968)
126. S. Wold: Acta Chem. Scand. *23*, 2978 (1969)
127. M. J. Tompkins: Diss. Abstr. Internat. *40*, No 12, Pt. 1, p. 5683-B (1980)
128. J. E. Dubois, P. Bauer: J. Am. Chem. Soc. *98*, 6993 (1976)
129. M. Mazet: Bull. Soc. Chem. France *1969*, 4309
130. E. M. Arnett, in: Progr. Phys. Org. Chem. (S. G. Cohen, A. Streitwieser, R. W. Taft, Eds.) New York, Interscience, 1963, vol. 1, p. 223
131. V. A. Palm, Ü. L. Haldna, A. J. Talvik, in: The Chemistry of the Carbonyl Group (S. Patai, Ed.) London, Interscience, 1966, p. 421
132. K. M. Biswas, A. H. Jackson: Tetrahedron *25*, 227 (1969)
133. V. A. Budylin, A. N. Kost, E. D. Matveeva: Khim. Geterotsikl. Soedin. *1972*, 55
134. A. N. Kost, V. A. Budylin, E. D. Matveeva, D. O. Sterligov: Zh. Org. Khim. *6*, 1503 (1970)
135. A. N. Kost, V. A. Budylin, O. A. Mochalova: Vestn. Mosk. Univ., Khim. *16*, 467 (1975)
136. Kwo-Hrong Lai, B. Miller: J. Org. Chem. *37*, 2505 (1972)
137. G. I. Borodkin, Ye. B. Panova, M. M. Shakirov, V. G. Shubin, in: Carbocation Chemistry, Novosibirsk, 1979, p. 84
138. M. J. S. Dewar, A. P. Marchand, in: Ann. Rev. Phys. Chem. (H. Eyring, Ed.) Palo Alto, Annual Reviews, Inc., 1965, vol. 16, p. 321
139. A. H. Jackson, B. Naidoo, P. Smith: Tetrahedron *24*, 6119 (1968)
140. W. R. Dolbier, Jr., K. E. Anapolle, L. McCullagh, K. Matsui, J. M. Riemann, D. Rolison: J. Org. Chem. *44*, 2845 (1979)
141. R. W. Taft, R. H. Martin, F. W. Lampe: J. Am. Chem. Soc. *87*, 2490 (1965)
142. J. L. Franklin in: Carbonium Ions (G. A. Olah, P. von R. Schleyer, Eds.) New York, Wiley-Interscience, 1968, vol. 1, p. 77
143. A. Streitwieser, Jr.: Chem. Rev. *56*, 571 (1956)
144. J. Burdon, I. W. Parsons: Tetrahedron *36*, 1423 (1980)
145. P. N. Dobronravov, V. D. Steingarts in: Carbocation Chemistry, Novosibirsk, 1979, p. 126
146. V. A. Bushmelev, M. M. Shakirov, A. V. Chikinev: Izv. Akad. Nauk SSSR, Ser. Khim. *1979*, 1667
147. S. Winstein, C. R. Lindegren, L. L. Ingraham: J. Am. Chem. Soc. *75*, 155 (1953)
148. R. B. Woodward, R. Hoffmann: The Conservation of Orbital Symmetry, New York, Academic Press, 1970
149. M. Saunders, in: Magnetic Resonance in Biological Systems (A. Ehrenberg, B. G. Malmström, T. Vänngård, Eds.) Oxford, Pergamon, 1967, p. 90
150. A. N. Detzina, V. I. Mamatyuk, B. G. Derendyaev, V. A. Koptyug: Zh. Org. Khim. *12*, 610 (1976)
151. W. J. Hehre, P. C. Hiberty: J. Am. Chem. Soc. *96*, 2665 (1974)
152. R. P. Kirchen, K. Ranganayakulu, A. Rauk, B. P. Singh, T. S. Sorensen: ibid. *103*, 588 (1981)
153. R. P. Kirchen, N. Okazawa, K. Ranganayakulu, A. Rauk, T. S. Sorensen: ibid. *103*, 597 (1981)
154. W. E. Bachmann, J. W. Ferguson: ibid. *56*, 2081 (1934)
155. W. E. Bachmann, F. H. Moser: ibid. *54*, 1124 (1932)
156. A. C. Sapia: Diss. Abstr. Internat. *36B*, 2816 (1975)
157. K. Matsumoto: Bull. Chem. Soc. Jpn. *41*, 1356 (1968)
158. W. M. Schubert, P. H. LeFevre: J. Am. Chem. Soc. *94*, 1639 (1972)
159. K. Tamura, T. Moriyoshi: Bull. Chem. Soc. Jpn. *47*, 2942 (1974)
160. P. Depovere, R. Devis: Bull. Soc. Chim. France *1969*, 479
161. Y. Pocker, B. P. Ronald: J. Am. Chem. Soc. *92*, 3385 (1970)

162. Y. Pocker, B. P. Ronald: J. Org. Chem. *35*, 3362 (1970)
163. A. Fry, M. Oka: J. Am. Chem. Soc. *101*, 6353 (1979)
164. G. A. Olah, R. J. Spear, D. A. Forsyth: ibid. *98*, 6284 (1976)
165. C. J. Lancelot, D. J. Cram, P. v. R. Schleyer, in: Carbonium Ions (G. A. Olah, P. v. R. Schleyer, Eds.) New York, Wiley-Interscience, 1972, vol. 3, p. 1347
166. G. I. Borodkin, M. M. Shakirov, V. G. Shubin: Zh. Org. Khim., *14*, 374 (1978)
167. V. G. Shubin, D. V. Korchagina, G. I. Borodkin, B. G. Derendyaev, V. A. Koptyug: Zh. Org. Khim. *9*, 1031 (1973)
168. R. L. Shriner, L. Geipel: J. Am. Chem. Soc. *79*, 227 (1957)
169. V. A. Koptyug, L. M. Mozulenko: Zh. Org. Khim. *6*, 102 (1970)
170. R. N. Berezina, L. E. Ugryumova, D. V. Korchagina, V. G. Shubin: Zh. Org. Khim. *18*, 592 (1982)
171. G. I. Borodkin, L. M. Pletneva, M. M. Shakirov, V. G. Shubin: Zh. Org. Khim. *15*, 652 (1979)
172. L. M. Stock, H. C. Brown, in: Adv. Phys. Org. Chem. (V. Gold, Ed.) London, Academic Press, 1963, vol. 1, p. 35
173. G. I. Borodkin, S. M. Nagy, M. M. Shakirov, V. G. Shubin: Zh. Org. Khim. *17*, 202 (1981)
174. F. L. Schadt III, C. J. Lancelot, P. v. R. Schleyer: J. Am. Chem. Soc. *100*, 228 (1978)
175. R. Heck, S. Winstein: J. Am. Chem. Soc. *79*, 3432 (1957)
176. H. C. Brown, C. J. Kim, C. J. Lancelot, P. von R. Schleyer: ibid. *92*, 5244 (1970)
177. J. A. Thompson, D. J. Gram: ibid. *91*, 1778 (1969)
178. V. J. Shiner, Jr., R. C. Seib: ibid. *98*, 862 (1976)
179. M. G. Jones, J. L. Coke: ibid. *91*, 4284 (1969)
180. R. S. Bly, E. K. Ni, A. K. K. Tse, E. Wallace: J. Org. Chem. *45*, 1362 (1980)
181. A. M. Avedikian, A. Kergomard, J. C. Tardivat, J. P. Vuillerme: Bull. Soc. Chim. France, *1974*, 2653
182. J. B. Lambert, H. W. Mark, E. S. Magyar: J. Am. Chem. Soc. *99*, 3059 (1977)
183. T. Ando, N. Shimizu, S.-G. Kim, Y. Tsuno, Y. Yukawa: Tetrahedron Lett. *1973*, 117
184. H. C. Brown, C. J. Kim: J. Am. Chem. Soc. *90*, 2082 (1968)
185. V. R. Kartashov, V. P. Pushkarev, I. V. Bodrikov: Zh. Org. Khim *8*, 24 (1972)
186. Ye. V. Skorobogatova, Dissertation, Gorky (1973)
187. D. S. Noyce, R. L. Castenson: J. Am. Chem. Soc. *95*, 1247 (1973)
188. V. R. Kartashov: Dissertation, Moscow (1974)
189. C. A. Kingsbury, D. C. Best: Bull. Chem. Soc. Jpn. *45*, 3440 (1972)
190. B. G. Ramsay, N. K. Das: J. Am. Chem. Soc. *94*, 4233 (1972)
191. D. F. Eaton, T. G. Traylor: ibid. *96*, 7109 (1974)
192. C. C. Lee, E. C. F. Ko, Z. Rappoport: Can. J. Chem. *58*, 884 (1980)
193. G. L. Nelson, G. C. Levy, J. D. Cargiolo: J. Am. Chem. Soc. *94*, 3089 (1972)
194. J. M. Harris, F. L. Schadt, P. von R. Schleyer, C. J. Lancelot: ibid. *91*, 7508 (1969)
195. R. Filler, Y. S. Rao: J. Org. Chem. *39*, 3421 (1974)
196. M. J. McCall, J. M. Townsend, W. A. Bonner: J. Am. Chem. Soc. *97*, 2743 (1975)
197. A. I. Rezvukhin, D. V. Korchagina, V. G. Shubin: Izv. Akad. Nauk SSSR, Ser. Khim *1976*, 1253
198. V. P. Loktev, D. V. Korchagina, V. G. Shubin: Izv. Sib. Otd. Akad. Nauk SSSR. Ser. Khim. Nauk, No 2, issue 1, p. 83 (1977)
199. M. Shanshal: J. C. S., Perkin II *1972*, 335
200. J. D. Roberts, C. M. Regan: J. Am. Chem. Soc. *75*, 2069 (1953)
201. N. C. Deno, in: Isotopes in Organic Chemistry (E. Buncel, C. C. Lee, Eds.) Amsterdam, Elsevier, 1975, vol. 1, p. 1
202. G. I. Borodkin, D. V. Korchagina, B. G. Derendjaev, V. G. Shubin: Tetrahedron Lett. *1973*, 539
203. J. Kagan, D. A. Agdeppa, Jr., S. P. Singh, D. A. Mayers, C. Boyajian, C. Poorker, B. E. Firth: J. Am. Chem. Soc. *98*, 4581 (1976)
204. J. E. Dubois, P. Bauer: J. Am. Chem. Soc. *90*, 4510, 4511 (1968)
205. K. Bhatia, A. Fry: J. Org. Chem. *34*, 806 (1969)
206. V. P. Loktev, D. V. Korchagina, V. G. Shubin: Izv. Sib. Otd. Akad. Nauk SSSR. Ser. Khim. Nauk, No 12, issue 5, p. 135 (1978)

207. G. A. Olah, H. C. Lin, D. A. Forsyth: J. Am. Chem. Soc. *96*, 6908 (1974)
208. K. L. Cook, A. J. Waring: J.C.S., Perkin II *1973*, 88
209. G. Frenking, H. Schwarz: Z. Naturforsch. *36b*, 797 (1981)
210. T. S. Sorensen, K. Ranganayakulu: Tetrahedron Lett. *1970*, 659
211. C. D. Poulter, S. Winstein: J. Am. Chem. Soc. *91*, 3649 (1969); *94*, 2297 (1972)
212. G. A. Olah, J. Lukas: ibid. *90*, 933 (1968)
213. G. A. Olah, A. M. White, J. R. DeMember, A. Commeyras, C. Y. Lui: ibid. *92*, 4627 (1970)
214. M. Saunders, R. Berger: ibid. *94*, 4049 (1972)
215. D. M. Brouwer, J. A. van Doorn: Rec. Trav. Chim. *89*, 333 (1970)
216. G. A. Olah, G. Liang, Y. K. Mo: J. Am. Chem. Soc. *94*, 3544 (1972)
217. R. Bladek, T. S. Sorensen: Can. J. Chem. *50*, 2806 (1972)
218. D. V. Korchagina, B. G. Derendyaev, V. G. Shubin, V. A. Koptyug: Zh. Org. Khim. *7*, 2582 (1971)
219. V. G. Shubin, D. V. Korchagina, V. A. Koptyug: Izv. Akad. Nauk SSSR, Ser. Khim. *1969*, 1201
220. H. Volz, G. Zimmermann, B. Schelberger: Tetrahedron Lett. *1970*, 2429
221. G. A. Olah, G. K. S. Prakash, G. Liang, P. W. Westerman, K. Kunde, J. Chandrasekhar, P. v. R. Schleyer: J. Am. Chem. Soc. *102*, 4485 (1980)
222. M. Saunders, E. L. Hagen: ibid. *90*, 2436 (1968)
223. W. H. Saunders, Jr., R. H. Paine: ibid. *83*, 882 (1961)
224. V. P. Vitullo, E. A. Logue: J. Org. Chem. *37*, 3339 (1972)
225. L. M. Mozulenko, V. A. Koptyug: Zh. Org. Khim. *8*, 2152 (1972)
226. R. W. Taft, Jr.: J. Phys. Chem. *64*, 1805 (1960)
227. H. C. Brown, H. Bartholomay, Jr., M. D. Taylor: J. Am. Chem. Soc. *66*, 435 (1944)
228. H. E. Zimmermann, A. Zweig: ibid. *83*, 1196 (1961)
229. A. H. Jackson, P. Smith: Tetrahedron *24*, 2227 (1968)
230. T. E. Zalesskaya, O. A. Netsetskaya: Zh. Org. Khim. *5*, 1605 (1969)
231. O. A. Netsetskaya: Dissertation, Leningrad (1973)
232. G. A. Olah, G. Liang: J. Am. Chem. Soc. *98*, 6304 (1976)
233. A. J. Waring, J. H. Zaidi, J. W. Pilkington: J.C.S., Perkin II *1981*, 935
234. V. P. Loktev, V. G. Shubin: Izv. Sib. Otd. Akad. Nauk SSSR. Ser. Khim. Nauk, No 9, issue 4, p. 114 (1981)
235. V. P. Loktev, V. G. Shubin, in: Theoretical and Applied Aspects of the Chemistry of Aromatic Compounds. III All-Union Symposium on Organic Synthesis. Abstr., Moscow, Nauka, 1981, p. 94
236. J. A. Berson, J. W. Foley: J. Am. Chem. Soc. *93*, 1297 (1971)
237. K. S. Feldman, A. McDermott, P. C. Myhre: ibid. *101*, 505 (1979)
238. V. A. Koptyug, V. G. Shubin, A. I. Rezvukhin: Izv. Akad. Nauk SSSR, Ser. Khim. *1965*, 201
239. C. G. Swain, E. S. Lupton, Jr.: J. Am. Chem. Soc. *90*, 4328 (1968)
240. O. Exner, in: Correlation Analysis in Chemistry. Recent Advances (N. B. Chapman, J. Shorter, Eds.) New York, Plenum, 1978, p. 439
241. R. Jost, J. Sommer, C. Engdahl, P. Ahlberg: J. Am. Chem. Soc. *102*, 7663 (1980)
242. E. M. Arnett, C. Petro: ibid. *100*, 5408 (1978)
243. J. W. Larsen, P. A. Bouis, C. A. Riddle: J. Org. Chem., *45*, 4969 (1980)
244. B. G. Derendyaev: Org. Magn. Res. *4*, 27 (1972)
245. D. Howells, S. Warren: J.C.S., Perkin II *1973*, 1645
246. V. A. Koptyug, V. I. Buraev: Zh. Org. Khim. *16*, 1882 (1980)
247. R. A. Marcus: J. Am. Chem. Soc. *91*, 7224 (1969)
248. E. D. German, R. R. Dogonadze: Int. J. Chem. Kinetics *6*, 457, 467 (1974)
249. V. A. Koptyug, V. I. Buraev: Zh. Org. Khim. *14*, 18 (1978)
250. V. A. Koptyug, V. I. Buraev, I. S. Isaev, O. N. Perevyazkina: Zh. Org. Khim. *14*, 328 (1978)
251. V. A. Koptyug, V. I. Buraev, I. S. Isaev: ibid. *14*, 1922 (1978)
252. V. I. Buraev, I. S. Isaev, V. A. Koptyug: ibid. *15*, 782 (1979)
253. V. I. Buraev: Dissertation, Novosibirsk (1978)
254. V. I. Rodionov, V. A. Koptyug: Zh. Org. Khim. *15*, 2555 (1979)
255. V. I. Rodionov, M. M. Shakirov, I. S. Isaev, V. A. Koptyug: ibid. *16*, 1515 (1980)
256. V. I. Rodionov: Dissertation, Novosibirsk (1981)

257. V. A. Koptyug, I. A. Shleider, I. S. Isaev, L. V. Vasil'eva, A. I. Rezvukhin: Zh. Org. Khim. *7*, 1089 (1971)
258. I. A. Schleider: Disseration, Novosibirsk (1972)
259. G. A. Olah, J. S. Staral, G. Asencio, G. Liang, D. A. Forsyth, G. D. Mateescu: J. Am. Chem. Soc. *100*, 6299 (1978)
260. T. J. Katz, E. H. Gold: ibid. *86*, 1600 (1964)
261. G. A. Olah, J. S. Staral, R. J. Spear, G. Liang: ibid. *97*, 5489 (1975)
262. G. A. Olah, D. J. Donovan, G. K. S. Prakash: Tetrahedron Lett. *1978*, 4779
263. J. S. Staral, J. D. Roberts; J. Am. Chem. Soc. *100*, 8018 (1980)
264. M. Saunders, P. v. R. Schleyer, G. A. Olah: ibid. *86*, 5680 (1964)
265. M. Saunders, M. R. Kates, K. B. Wiberg, W. Pratt: ibid. *99*, 8072 (1977)
266. V. Koptyug, A. Rezvukhin, E. Lippmaa, T. Pehk: Tetrahedron Lett. *1968*, 4009
267. V. A. Koptyug, E. T. Lippmaa, A. I. Rezvukhin, T. U. Pehk: Izv. Akad. Nauk SSSR, Ser. Khim. *1969*, 285
268. V. I. Mamatyuk, A. N. Detzina, V. A. Koptyug: Zh. Org. Khim *12*, 739 (1976)
269. V. I. Mamatyuk, V. A. Koptyug: Zh. Org. Khim. *13*, 818 (1977)
270. D. G. Farnum, in: Adv. Phys. Org. Chem. (V. Gold, Ed.) London, Academic Press, 1975, vol. 11, p. 123
271. G. I. Borodkin, I. K. Korobeinicheva, V. G. Shubin: Zh. Org. Khim. *12*, 2381 (1976)
272. B. G. Derendyaev, V. I. Mamatyuk, V. A. Koptyug: Tetrahedron Lett. *1969*, 5
273. V. I. Mamatyuk, B. G. Derendyaev, A. N. Detzina, V. A. Koptyug: Zh. Org. Khim. *12*, 2487 (1974)
274. N. V. Bodoev, V. I. Mamatyuk, A. P. Krysin, V. A. Koptyug: Izv. Akad. Nauk SSSR, Ser. Khim. *1978*, 1199
275. R. M. Cooper, M. C. Grossel, M. J. Perkins: J.C.S., Perkin II *1972*, 594
276. M. C. Grossel, M. J. Perkins: J.C.S., Perkin II *1976*, 1334
277. V. D. Sutula, V. A. Koptyug: Dokl. Akad. Nauk SSSR *247*, 1424 (1979)
278. S. McLean, D. M. Findlay: Can. J. Chem. *48*, 3107 (1970)
279. S. McLean, C. J. Webster, R. J. D. Rutherford: ibid. *47*, 1555 (1969)
280. J. W. de Haan, H. Kloosterziel: Rec. Trav. Chim. *87*, 298 (1968)
281. W. C. Herndon, J. M. Manion: J. Org. Chem. *33*, 4504 (1968)
282. S. McLean. P.: Haynes: Tetrahedron *21*, 2329 (1965)
283. F. Bernardi, W. J. Hehre: J. Am. Chem. Soc. *95*, 3078 (1973)
284. H.-J. Köhler, H. Lischka: ibid. *101*, 3479 (1979)
285. K. Raghavachari, R. A. Whiteside, J. A. Pople, P. v. R. Schleyer: J. Am. Chem. Soc. *103*, 5649 (1981)
286. M. J. S. Dewar, G. P. Ford: ibid. *101*, 783 (1979)
287. H. Lischka, H.-J. Köhler: ibid. *100*, 5297 (1978)
288. W. J. Hehre: ibid. *94*, 5919 (1972)
289. A. C. Hopkinson, M. H. Lien, K. Yates, I. G. Csizmadia: Theoret. Chim. Acta. *44*, 385 (1977)
290. A. C. Hopkinson, M. H. Lien, I. G. Csizmadia, K. Yates: ibid. *55*, 1 (1980)
291. L. Radom, D. Poppinger, R. C. Huddon, in: Carbonium Ions (G. A. Olah, P. von R. Schleyer, Eds.) New York, Wiley-Interscience, 1976, vol. 5, p. 2303
292. V. P. Loktev: Dissertation, Novosibirsk (1981)
293. J. A. Ross, R. P. Seiders, D. M. Lemal: J. Am. Chem. Soc. *98*, 4325 (1976)
294. J. P. Snyder, T. A. Halgren: ibid. *102*, 2861 (1980)
295. G. I. Borodkin, V. A. Koptyug, V. G. Shubin: Dokl. Akad. Nauk SSSR *255*, 587 (1980)
296. G. I. Borodkin, V. G. Shubin: Zh. Org. Khim. *18*, 899 (1982)
297. T. S. Abram, W. E. Watts: J.C.S., Perkin I *1975*, 116
298. T. M. Gund, P. v. R. Schleyer, P. H. Gund, W. T. Wipke: J. Am. Chem. Soc. *97*, 743 (1975)
299. E. Ōsawa, K. Aigami, Y. Inamoto: J.C.S., Perkin II *1979*, 181
300. V. A. Bushmelev, M. M. Shakirov, B. G. Derendyaev, V. A. Koptyug: Zh. Org. Khim. *15*, 1934 (1979)
301. V. A. Bushmelev, M. M. Shakirov, D. M. Brouwer, V. A. Koptyug: ibid. *15*, 1347 (1979)
302. P. v. R. Schleyer: Angew. Chem., Internat. Edit. *8*, 529 (1969)
303. D. M. Brouwer, H. Hogeveen: Rec. Trav. Chim. *89*, 211 (1970)
304. D. V. Korchagina: Dissertation, Novosibirsk (1974)

305. R. P. Kirchen, T. S. Sorensen: J. Am. Chem. Soc. *100*, 1487 (1978)
306. J. Chandrasekhar, P. von R. Schleyer: Tetrahedron Lett. *1979*, 4057
307. G. A. Olah, P. W. Westerman, J. Nishimura: J. Am. Chem. Soc. *96*, 3548 (1974)
308. P. van Pelt: Thesis, Eindhoven (1975)
309. G. A. Olah, G. Liang: J. Am. Chem. Soc. *96*, 195 (1974)
310. K. L. Cook, A. J. Waring: J.C.S., Perkin II *1973*, 84
311. V. P. Vitullo, N. Grossman: J. Am. Chem. Soc. *94*, 3844 (1972)
312. V. A. Bushmelev, A. V. Chikinev, M. M. Shakirov: Zh. Org. Khim. *18*, 1691 (1982)
313. M. M. Shakirov: Diploma Thesis, Novosibirsk (1972)
314. M. M. Shakirov, V. G. Shubin, A. I. Rezvukhin: Izv. Sib. Otd. Akad. Nauk SSSR. Ser. Khim. Nauk, No 9, issue 4, p. 155 (1974)
315. R. F. Childs: Chem. Commun., *1969*, 946
316. V. G. Shubin: Dissertation, Novosibirsk (1967)
317. D. M. Brouwer, E. L. Mackor, C. MacLean, in: Carbonium Ions (G. A. Olah, P. von R. Schleyer, Eds.) New York, Wiley-Interscience, 1970, vol. 2, p. 837
318. T. Birchall, R. J. Gillespie: Can. J. Chem. *43*, 1045 (1965)
319. D. Mirda, D. Rapp, G. M. Kramer: J. Org. Chem. *44*, 2619 (1979)
320. L. M. Mozulenko: Dissertation, Novosibirsk (1972)
321. D. M. Brouwer: Rec. Trav. Chim. *87*, 210 (1968)
322. D. M. Brouwer, E. L. Mackor: Proc. Chem. Soc. *1964*, 147
323. D. M. Brouwer: Rec. Trav. Chim. *88*, 9 (1969)
324. G. A. Olah, J. Lukas: J. Am. Chem. Soc. *89*, 2227 (1967)
325. G. A. Olah, J. Lukas: ibid. *89*, 4739 (1967)
326. D. M. Brouwer, J. A. van Doorn: unpublished, cited from Ref. [64]
327. N. C. Deno, N. Friedman, J. D. Hodge, J. J. Houser: J. Am. Chem. Soc. *85*, 2995 (1963)
328. D. M. Brouwer, J. A. van Doorn, A. A. Kiffen: Rec. Trav. Chim. *91*, 1359 (1972)
329. S. Barton, C. R. Porter: J.C.S., *1956*, 2483
330. P. C. Myhre, C. S. Yannoni: J. Am. Chem. Soc. *103*, 230 (1981)
331. G. I. Borodkin, S. M. Nagy, V. G. Shubin: Izv. Akad. Nauk SSSR, Ser. Khim., *1982*, 2639
332. W. L. Jorgensen: J. Am. Chem. Soc. *100*, 1049 (1978)
333. G. M. Kramer: Int. J. Mass Spectr. Ion Phys. *19*, 139 (1976)
334. H.-J. Rauscher, D. Heidrich, H.-J. Köhler, D. Michel: Theoret. Chim. Acta *57*, 255 (1980)
335. J. F. Wolf, R. G. Härch, R. W. Taft: J. Am. Chem. Soc. *97*, 2904 (1975)
336. V. Gold: J.C.S., Faraday I *1978*, 1611
337. D. Heidrich, U. Göring, W. Förster, C. Weiss: Tetrahedron *35*, 651 (1979)
338. C. Wesdemiotis, R. Wolfschütz, H. Schwarz: ibid. *36*, 275 (1980)
339. I. A. Abronin, G. M. Zhidomirov: Izv. Akad. Nauk SSSR, Ser. Khim. *1981*, 1752
340. J. H. Vorachek, G. G. Meisels, R. A. Geanangel, R. H. Emmel: J. Am. Chem. Soc. *95*, 4078 (1973)
341. P. Ausloos, R. E. Rebbert, L. W. Sieck, T. O. Tierman: J. Am. Chem. Soc. *94*, 8939 (1972)
342. T. H. Morton: J. Am. Chem. Soc. *102*, 1596 (1980)
343. D. G. Hall, T. H. Morton: J. Am. Chem. Soc. *102*, 5686 (1980)
344. D. Kuck, H.-Fr. Grützmacher: Org. Mass Spectr. *13*, 90 (1978)
345. G. A. Olah, R. H. Schlosberg, R. D. Porter, Y. K. Mo, D. P. Kelly, G. D. Mateescu: J. Am. Chem. Soc. *94*, 2034 (1972)
346. F. Cacace, P. Giacomello: J.C.S., Perkin II *1978*, 652
347. F. Cacace, M. Speranza: J. Am. Chem. Soc. *98*, 7305 (1976)
348. V. D. Nefedov, E. N. Sinotova, G. P. Akulov, M. V. Korsakov: Zh. Org. Khim. *6*, 1214 (1970)
349. J. T. Gleghorn, F. W. McConkey, and (in part) K. Lundy: J. Chem. Res. (M) *1978*, 4731
350. L. A. Telkowski, M. Saunders, in: Dynamic Nuclear Magnetic Resonance Spectroscopy (L. M. Jackson, F. A. Cotton, Eds.) New York, Academic Press, 1975, p. 523
351. G. A. Olah, E. B. Baker, J. C. Evans, W. S. Tolgyesi, J. S. McIntyre, I. J. Bastien: J. Am. Chem. Soc. *86*, 1360 (1964)
352. P. von R. Schleyer, in: Reaction Transition States (J.-E. Dubois, Ed.) New York, Gordon and Breach, 1975, p. 197
353. E. M. Arnett, C. Petro, P. von R. Schleyer: J. Am. Chem. Soc. *101*, 522 (1979)

354. M. P. Polovinka, S. A. Osadchii, D. V. Korchagina, Zh. V. Dubovenko, V. A. Barkhash: Zh. Org. Khim. *17*, 1629 (1981)
355. G. Fraenkel, P. D. Ralph, J. P. Kim: Can. J. Chem. *43*, 674 (1965)
356. J. T. Keating, P. S. Skell, in: Carbonium Ions (G. A. Olah, P. v. R. Schleyer, Eds.) New York, Wiley-Interscience, 1970, vol. 2, p. 573
357. M. Saunders, E. L. Hagen, J. Rosenfeld: J. Am. Chem. Soc. *90*, 6882 (1968)
358. D. Cram, in: Steric Effects in Organic Chemistry (M. S. Newman, Ed.) New York, Wiley, 1956, p. 249
359. J. F. Lane, E. S. Wallis: J. Am. Chem. Soc. *63*, 1674 (1941)
360. J. J. Beggs, M. B. Meyers: J.C.S. *1970B*, 930
361. S. M. Funtova: Dissertation, Moscow (1972)
362. Yu. G. Bundel: Dissertation, Moscow (1973)
363. T. Shono, K. Fujita, S. Kumai: Tetrahedron Lett. *1973*, 3123
364. W. Kirmse, W. Gruber: Chem. Ber. *106*, 1365 (1973)
365. D. V. Korchagina, B. G. Derendyaev. V. G. Shubin: Zh. Org. Khim. *10*, 1457 (1974)
366. G. A. Olah, R. J. Spear, G. Messina, P. W. Westerman: J. Am. Chem. Soc. *97*, 4051 (1975)
367. T. L. Gilchrist, R. C. Storr: Organic Reactions and Orbital Symmetry, Cambridge University Press, 1972, p. 215
368. R. E. Lehr, A. P. Marchand: Orbital Symmetry. A Problem-Solving Approach, New York, Academic Press, 1972, p. 13
369. H. D. Zook, S. C. Paviak: J. Am. Chem. Soc. *77*, 2501 (1955)
370. Z. N. Parnes, S. V. Vitt, D. N. Kursanov: Zh. Obsch. Khim. *28*, 410 (1958)
371. T. D. Turbitt, W. E. Watts: J. Organomet. Chem. *57*, C78 (1973)
372. M. T. Reetz: Angew. Chem., Internat. Edit. *11*, 129, 130 (1972)
373. M. T. Reetz, in: Adv. Organomet. Chem. (F. G. A. Stone, R. West, Eds.) New York, Academic Press, 1977, vol. 16, p. 33
374. M. T. Reetz, M. Kliment, M. Plachky: Angew. Chem., Internat. Edit. *13*, 813 (1974)
375. M. T. Reetz: Tetrahedron *29*, 2189 (1973)
376. G. I. Borodkin, M. M. Shakirov, V. G. Shubin, V. A. Koptyug: Zh. Org. Khim. *12*, 1303 (1976)
377. B. Calas, L. Giral: Bull. Soc. Chim. France *1972*, 2895
378. V. I. Sokolov, K. I. Granberg, V. V. Bashilov, D. A. Lemenovskii, P. V. Petrovskii, O. A. Reutov: Dokl. Akad. Nauk SSSR *214*, 393 (1974)
379. J. Mulzer, G. Brüntrup: Angew. Chem., Internat. Edit. *18*, 793 (1979)
380. V. G. Lipovich: Izv. Sib. Otd. Akad. Nauk SSSR. Ser. Khim. Nauk, No 7, issue 3, p. 12 (1980)
381. Yu. N. Porshnev, V. I. Erikhov, M. I. Cherkashin: Dokl. Akad. Nauk SSSR *248*, 118 (1979)
382. E. P. Ivakhnenko, V. B. Panov, O. Yu. Okhlobystin: ibid. *243*, 659 (1978)
383. C. L. Perrin: J. Am. Chem. Soc. *99*, 5516 (1977)
384. C. E. Barnes, P. C. Myhre: ibid. *100*, 975 (1978)
385. B. H. Meier, R. R. Ernst: ibid. *101*, 6441 (1969)
386. M. J. Perkins, P. Ward, in: Mechanisms of Molecular Migrations (B. S. Thyagarajan, Ed.) New York, Wiley-Interscience, 1971, vol. 4, p. 55
387. J. J. Dannenberg, B. J. Goldberg, J. K. Barton, K. Dill, D. H. Weinwurzel, M. O. Longas: J. Am. Chem. Soc. *103*, 7764 (1981)
388. E. Wistuba, C. Rüchardt: Tetrahedron Lett. *1981*, 4069
389. D. D. Roberts: J. Org. Chem. *77*, 561 (1982)

Author Index Volumes 101–117

Contents of Vols. 50–100 see Vol. 100
Author and Subject Index Vols. 26–50 see Vol. 50

The volume numbers are printed in italics

Bei Fragen zur Produktsicherheit wenden Sie sich bitte an:
If you have any questions regarding product safety,
please contact:

Walter de Gruyter GmbH
Genthiner Straße 13
10785 Berlin
productsafety@degruyterbrill.com